Symbiosis of Plants and Microbes

For Katja und Gesche

Symbiosis of Plants and Microbes

D. Werner
Philipps-Universität Marburg

Germany

CHAPMAN & HALL
London · Glasgow · New York · Tokyo · Melbourne · Madras

Published by Chapman & Hall, 2–6 Boundary Row, London SE1 8HN

Chapman & Hall, 2–6 Boundary Row, London SE1 8HN, UK

Blackie Academic & Professional, Wester Cleddens Road, Bishopbriggs, Glasgow G64 2NZ, UK

Chapman & Hall, 29 West 35th Street, New York NY10001, USA

Chapman & Hall Japan, Thomson Publishing Japan, Hirakawacho Nemoto Building, 6F, 1-7-11 Hirakawa-cho, Chiyoda-ku, Tokyo 102, Japan

Chapman & Hall Australia, Thomas Nelson Australia, 102 Dodds Street, South Melbourne, Victoria 3205, Australia

Chapman & Hall India, R. Seshadri, 32 Second Main Road, CIT East, Madras 600 035, India

---

First edition 1992

Original German language edition – *Pflanzliche und mikrobielle Symbiosen* – 1987, Georg Thieme Verlag.

Typeset in 10/12 Plantin by Excel Typesetters Company, Hong Kong
Printed in Great Britain at the University Press, Cambridge

ISBN 0 412 36230 9

A catalogue record for this book is available from the British Library

Library of Congress Cataloging-in-Publication data available

Contents

Preface

Fifty years ago (1943) the first edition of Reinhold Schaede's book *The Symbiosis of Plants* was published, and more than 25 years ago the third edition by F.H. Meyer. Since then a large number of monographs on certain symbioses, such as the mycorrhizas, and several more general volumes on symbioses have become available. The volume by D.C. Smith and A.E. Douglas (1987) is the most recent one.

Research and teaching on symbiosis is attractive, since it includes methods and results from various disciplines: cell biology, biochemistry and molecular genetics as well as ecology, morphology and taxonomy. Botany and microbiology are, of course, basic subjects.

A better understanding of symbioses of plants and microbes contributes to the praxis of disciplines, such as plant nutrition, plant production, phytopathology, forestry and soil microbiology. Also, for several environmental and global problems symbiosis has an impact. The different forms of mycorrhizas are essential components for the stress tolerance of trees and forests against acid rain and mineral deficiencies in the leaves. N_2-fixing symbioses are part of the nitrogen budget in the environment since nitrate affects in increasing areas of the world the quality of drinking water. Lichens and other symbioses such as corals or legume nodules are sensitive indicators for pollutants in the air, the water of tropical oceans or agricultural soils.

The present book covers, in eleven chapters, the most important symbioses of plants and microbes. Chloroplasts and mitochondria can be considered as evolutionary very old endosymbionts, and are included in Chapter one. To compare the functions of endosymbiotic bacteria in plants with those in animals, also a number of microbe-animal symbioses are covered in Chapter six. Since the molecular understanding of a symbiosis is most advanced in the *Rhizobium*/*Bradyrhizobium*-legume system, Chapter three is more comprehensive than the others.

For supply of original figures and personal communications, I am very grateful to the following colleagues: Professor D. Bauer, Missouri, USA; Dr F.J. Bergersen, Canberra, Australia; Professor J. Beringer, Bristol. UK; Professor Bonfante-Fasolo, Turin, Italy; Professor J.A. Breznak, Michigan,

USA; Dr H.E. Calvert, Yellow Springs, USA; Professor T. Cavalier-Smith, London, UK; Professor J. Denarié, Toulouse, France; Dr Haymann, Rothamstead, UK; Professor A. Henssen, Marburg, Germany; Professor B. Hock, Munich, Germany; Professor R. Honegger, Zürich, Switzerland; Professor B.D.W. Jarvis, Palmerston North, New Zealand; Professor Lalonde, Quebec, Canada; Professor B.J.J. Lugtenberg, Leiden, The Netherlands; Professor P. Martin, Hohenheim, Germany; Dr T.W. Mew, IRRI Institute, Philippines; Dr I.M. Miller, Glasgow, UK; Professor K.H. Nealson, Milwaukee, USA; Professor F. Oberwinkler, Tübingen, Germany; Professor J.D. Palmer, Bloomington, USA; Professor W. Reisser, Göttingen, Germany; Dr W. Rühle, Professor A. Wild, Mainz, Germany; Dr G. Schaller, Munich, Germany; Professor F. Schönbeck, Hannover, Germany; Professor M. Siegel, Lexington, USA; Professor W.D.P. Stewart, Dundee, UK; Professor H. Stolp, Bayreuth, Germany; Professor D.G. Strullu, Angers, France; Professor C.K. Stumm, Nijmegen, The Netherlands; Professor S. Tzean, Taipea, Taiwan; Professor D. Webb, Kingston, Canada; Dr J.F. Witty, Aberystwyth, UK.

I am very grateful to Lucette Karner and Dr Sigrid Schenk, from my laboratory, for their dedicated help in preparing the manuscript, and to Dr Robert Mellor who was involved in parts of the translation. My thanks also go to the publisher Chapman & Hall, London.

Dietrich Werner
Marburg

1 An overview: the meaning of plant and microbe symbioses and the endosymbiont theory of evolution of chloroplasts and mitochondria

1.1 COMMENSALISM, PARASITISM AND SYMBIOSIS (MUTUALISM)

The word symbiosis was introduced by the botanist De Bary 1879 to describe the most intimate living relationship between different species. This rather broad definition included parasitic as well as symbiotic states. In the past 100 years the definition, although becoming more concrete, is still used differently depending on the standpoint of the author (e.g. ecologists and physiologists) (Gäumann, 1951; Schaede 1943; Matthes, 1978; Lewin, 1982; Smith and Douglas, 1987). The situation may be further complicated as one relationship may pass through several phases, e.g. starting as commensalism and progressing through parasitism until a symbiotic state is reached. One example of this is the *(Brady)rhizobium*–Fabales symbiosis (Chapter 3). In this book definitions are based on the following (see Figure 1.1):

1. *Commensalism*: one partner living on the other with no obvious effect on the second;
2. *Parasitism (antagonism)*: one partner living on the other with detrimental effect on the second;
3. *Symbiosis (mutualism)*: advantages for both partners.

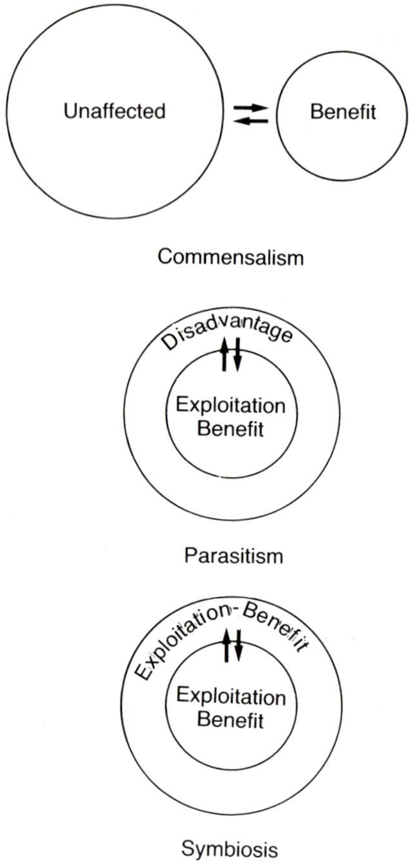

Fig. 1.1. Definition of commensalism, parasitism and symbiosis.

In order to remain precise, it must be stressed that in all cases (especially the third) we are talking about species in extremely close contact with each other. Certainly the idea of mutual aid, inter- and intraspecies help at longer distance (Kropotkin, 1901), is very valuable and this concept, loosely spoken of as symbiosis, is found in many other disciplines, for example Schäffle (1906), v. Sydow (1923), and Meinecke (1924). These different uses of the word **symbiosis** have undergone their own development until today when the original biological definition is, in many cases, completely lost.

The biology of symbiosis uses information from several disciplines, such as cell biology, biochemistry, botany, zoology, microbiology, genetics, ecology and theoretical biology (Schimper, 1883; Gäumann, 1951; Buchner, 1953; Schaede and Meyer, 1962; Kaplan, 1978; Matthes, 1978; Smith, 1979; Whatley *et al.*, 1979; Follmann, 1981; Margulis, 1981; Schweiger,

Table 1.1 Symbioses and availability of biological macroelements

Element	*Symbiosis*
C	Endosymbiont theory of chloroplast evolution
	Similar living prokaryote: *Synechococcus* sp.
H	Endosymbiont theory of chloroplast evolution
	Exploitation of photolytic H_2O-cleavage by eukaryotes
O	Endosymbiont theory of mitochondria evolution
	Similar living prokaryote: *Rhodopseudomonas globiformis*
N	*Rhizobium*–legume symbiosis
	Frankia (actinomycetes)-symbiosis with species from eight plant families
P	Ectotrophic mycorrhiza, VA-mycorrhiza (*Glomus* spp.)
	Symbioses with most spermatophytes

1981; Law and Lewis, 1983; Schenk and Schwemmler, 1983; Schlösser, 1983; Herrmann, 1984; von Wettstein and Oliver, 1984; Mohr and Schopfer, 1985; Gilbert, 1986; Lüttge *et al.*, 1988; Remmert, 1989; Kleinig and Sitte, 1992).

1.2 THE IMPORTANCE OF PLANT SYMBIOSES

In principle, symbiosis is important for the use of the five major biological **macroelements** (Table 1.1). According to the endosymbiont theory, symbioses with cyanobacteria (blue–green algae) resulted in the availability of **carbon** (as CO_2) and **hydrogen** (from H_2O) for eukaryotic algae, Archegoniatae and spermatophytes. The same theory states that the first mitochondria, derived from endosymbiotic prokaryotes, made **oxygen** for energy and respiration available to eukaryotes. The symbioses concerned with **nitrogen** and **phosphorus** are more amenable to experimental approaches. The assimilation of atmospheric N_2 and its use by eukaryotes through symbiosis is presented in Chapters 3–5. The availability of the phosphorus is markedly increased in almost all spermatophytes through associations with various forms of mycorrhiza (Chapters 10 and 11).

1.3 THE EVOLUTION OF CHLOROPLASTS AND MITOCHONDRIA: THE ENDOSYMBIONT THEORY

Schimper (1883) was the first scientist to note that chloroplasts reminded one of symbiotic algae. This passing comment, published as a footnote in a manuscript on the development of plastids, is now recognized as the origin of the endosymbiotic theory of chloroplast evolution. A general theory, which also included mitochondria, was formulated by Altmann in 1890.

Table 1.2 Biochemical and cytological characteristics of chloroplasts, mitochondria, prokaryotes and eukaryotes

Characteristic	*Chloroplasts*	*Mitochondria*	*Prokaryote*	*Eukaryote (cytoplasm, nucleus, membranes)*
DNA	circular (150 kbp)	circular* (1500–2000 kbp)	circular	chromosomal
Introns	rare	rare	rare	+
Cap-structures at 5′ position of mRNA (7-methylguanine)	–	–	–	+
70 S ribosomes (30 S/50 S)	+	+	+	–
80 S ribosomes (40 S/60 S)	–	–	–	+
Inhibition of protein biosynthesis by chloramphenicol	+	+	+	–
Inhibition of protein biosynthesis by cycloheximide	–	–	–	+
Initiation factors	prokaryotic	prokaryotic	prokaryotic	eukaryotic
rRNA sequences	similar to cyanobacteria	similar to Rhodospirillaceae		
Cardiolipin	–	+ (inner membrane)	+ (plasma membrane)	–
Sterol lipids		– (inner membrane)	– (plasma membrane)	+
Fatty acid synthase	7 components		7 components	2 multifunctional polypeptides

* linear in some protozoa.

Many decades passed before experiments could be formulated to test the theory. However, since current knowledge of a large number of biochemical, cytological and genetic characteristics of chloroplasts and mitochondria shows that there is homology with prokaryotes and little or none with eukaryotes, the endosymbiont theory is now regarded as very plausible. In molecular biology, **chloroplasts** and **mitochondria** show similarities to prokaryotes in the important characteristics of DNA circularity, the rare occurrence of introns, protein biosynthesis initiation sequences, rRNA sequence and size, as well as sensitivity to chloramphenicol and not cycloheximide (Table 1.2).

Biochemical characteristics include membrane analysis. Prokaryotic plasmamembrane and the mitochondrial inner membrane contain cardiolipin instead of sterollipids. Although cardiolipin is not found in chloroplasts, fatty acid synthase components are nearly identical in these and prokaryotes.

The endosymbiont theory further demands that the outer membrane of the symbiosis-derived organelle is more similar to eukaryotic membranes (especially plasma membrane) than the inner, prokaryote-derived membrane. This is in fact so, and can be proved by direct experimentation; the development can be followed in more recent symbioses (see Chapters 3, 5 and 7). Thus, although the thickness of spinach outer and inner chloroplast membranes is the same (5 nm, compared to 7 nm for the plasma membrane) as the lipid:protein ratio 2.4:1 for the outer membrane but 0.8:1 for the inner membrane, the difference is a concentration of phosphatidylcholine five times higher in the outer membrane. Of significance also are the carotenoid content and proportions of the two membranes. The outer membrane has a larger neoxanthin proportion compared to violaxanthin, whereas antheraxanthin is only found in the inner membrane (Table 1.3). The specific activity of the Mg^{2+}-dependent ATPase in the inner membrane is double that of the outer, the UPD-Gal-diacylglycerol galactosyltransferase and the acyl-CoA thioesterase are seven to eight times higher. On the other hand acyl-CoA synthetase is much more active in the outer chloroplast membrane (Table 1.3).

The outer and inner membranes of mitochondria are similarly characterized by different lipid:protein ratios. Marker enzyme activities for the outer membrane are: monoamineoxidase, acyl-CoA-synthetase, NADH–cytochrome b_5 reductase, kynurenine 3-mono-oxygenase and phospholipase A_2. The inner membrane is characterized with respect to oxidative phosphorylation, with, for example, up to 12 units of ATP-synthase complex, and 12 or 13 polypeptides of cytochrome *c* oxidase. Whereas all polypeptides of the outer membrane are coded for on nuclear genes, around 10% of the inner membrane proteins are products of the mitochondrial genome. Within this 10% are the three large subunits of cytochrome *c* oxidase, the cytochrome *b* complex and two or three subunits of ATP-

Table 1.3 Characteristics and components of the outer and inner membranes of chloroplasts (based on Block *et al.*, 1983)

Characteristics/Components	*Chloroplasts (spinach)*	
	Outer membrane	*Inner membrane*
Lipid : protein ratio	2.5–3 : 1	0.8–1 : 1
Thickness	5.5 nm	5.5 nm
	(Plasmalemma: 7 nm)	
Particle density	660	4200
Digalactosyldiacylglycerol (% of total polar lipids)	29	30
Monogalactosyldiacylglycerol (%)	17	49
Phosphatidylcholine (%)	32	6
Sulpholipid (%)	6	5
Phosphatidylglycerol (%)	10	8
Carotenoids (μg/mg protein)	2.9	7.2
Violaxanthin (μg/mg protein)	1.4	3.4
Neoxanthin (μg/mg protein)	0.75	0.94
Antheraxanthin (μg/mg protein)	not detectable	0.36
Mg^{2+}-dependent ATPase (nmol P_i released $\cdot h^{-1} \cdot mg^{-1}$ protein)	1350	3300
UDP-Gal:diacylglycerol galactosyltransferase (nmol [^{14}C]galactose incorporated h^{-1} mg^{-1} protein)	120	880
Acyl-CoA thioesterase (nmol [^{14}C]palmitic acid relased h^{-1} mg^{-1} protein)	50	350
Acyl-CoA Synthetase (nmol [^{14}C]oleoyl synthesized h^{-1} mg^{-1} protein)	3900	900

synthase. These data, summarized in Tables 1.2, 1.3 and 1.4, lead to the theory of the development of eukaryotic cells from a procyte through invagination of the plasma membrane or production of new membrane to form specialized organelles. Because of the advantages of superior C, H, O and N nutrition, and the metabolic efficiencies of compartmentation, the prokaryote gradually loses its free-living ability and many functions of its own DNA are taken over by nuclear genes or plasmids.

Precise comparisons between chloroplasts and mitochondria and prokaryotes can be made by means of sequence analyses of rRNA and protein. A phylogenetic tree of 5S rRNA sequence made from counting point mutations per 100 nucleotides shows the close ancestry between chloroplasts and cyanobacteria on one side and mitochondria and Rhodospirillaceae on the other. In comparison, the branch representing eukaryotic host cells comes from the region of the Archaebacteria (Fig. 1.2).

Table 1.4 Characteristics and components of the outer and inner membranes of mitochondria

Characteristics/components	*Mitochondria*	
	Outer membrane	*Inner membrane*
Lipid : protein ratio	1.5 : 1	0.25 : 1
Marker enzymes	Monoamine oxidase	Glycerol-3-P dehydrogenase
	Acyl-CoA synthetase	ATP-synthase complex (EC 3.6.1.3)
	NADH–cytochrome b_5 reductase	Cytochromes a,a_3,b,c,c_1
	Kynurenine 3-mono-oxygenase	Cytochrome c oxidase (EC 1.9.3.1) with 12 or 13 polypeptides in a ratio of 1 : 1
Coding by nuclear genes	Phospholipase A_2	90% of proteins
	All proteins	

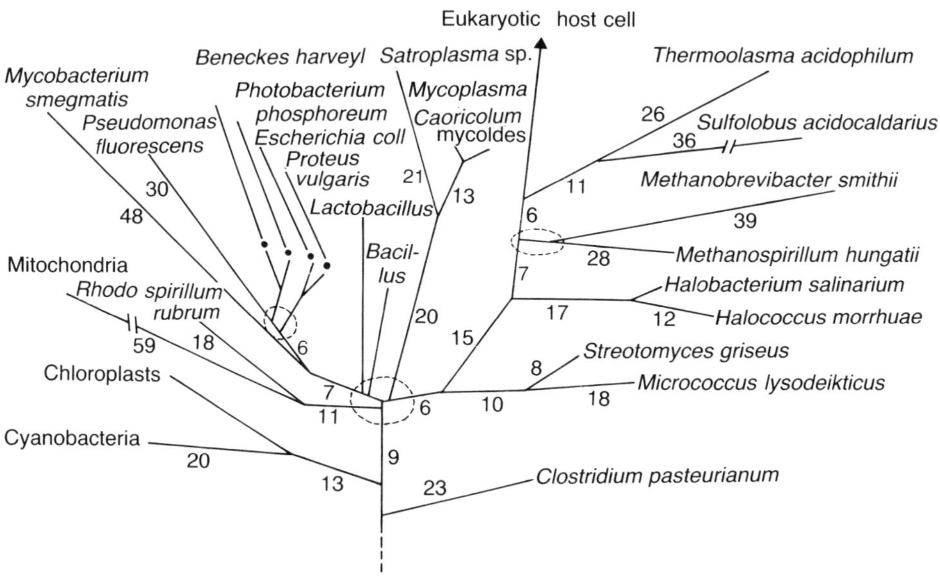

Fig. 1.2. Phylogenetic tree for prokaryotes, chloroplasts and mitochondria as deduced from sequence analysis of ribosomal 5S rRNA. Figures indicate the number of point mutations per 100 nucleotides. The branching at positions marked by dashed circles is uncertain. (Based on George *et al.*, 1983.)

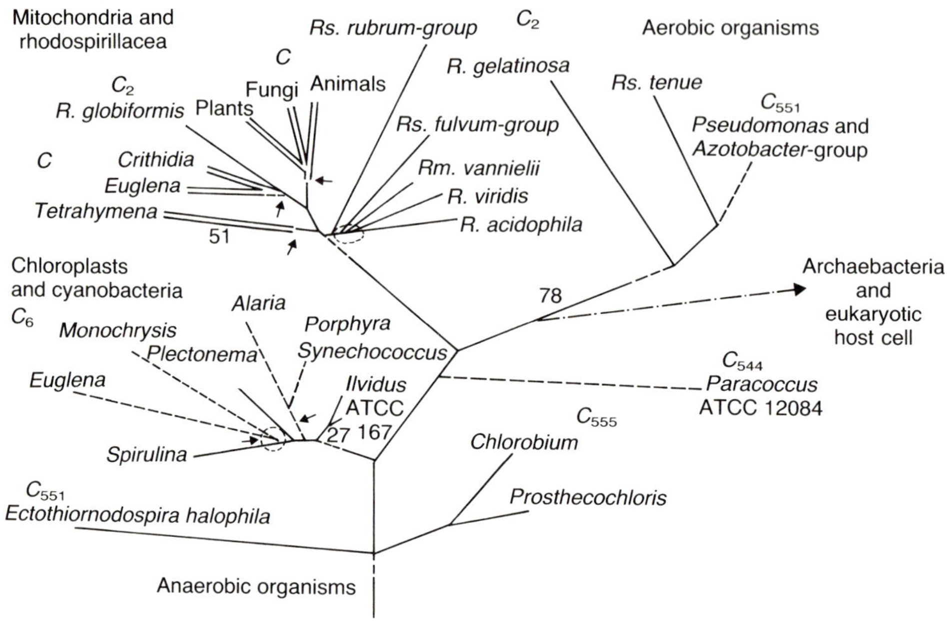

Fig. 1.3. Phylogenetic tree for chloroplasts and cyanobacteria, mitochondria and Rhodospirillaceae and for other prokaryotes, as deduced from sequence analysis of *c*-type cytochromes. (Based on Dayhoff, 1979.)

Similar conclusions were reached from phylogenetic studies based on sequences of cytochrome *c* (Fig. 1.3). From a study of several hundred 5S rRNA sequences, mitochondria and chloroplast evolution was discovered to separate in the phylogenetic tree around 800 million years ago (Hori and Osawa, 1987). Chloroplasts derived from *Euglena*, *Monochrysis*, *Alaria* and *Porphyra* show homology with the cyanobacteria *Synechococcus* and *Spirulina*. Mitochondria from plants, animals, fungi, as well as from *Euglena* and *Tetrahymena*, show a particularly close similarity to *Rhodopseudomonas globiformis*.

Earlier opinions regarding a particularly close relationship between mitochondria and *Paracoccus denitrificans* have not been confirmed. This is also the case for suggestions that *Prochloron* is particularly close to chloroplasts. The latter theory was based on correlations in the pigment content of the cyanobacterium *Prochloron*, inasmuch that it contains not only chlorophyll *a* but also chlorophyll *b*. Figure 1.4, however, shows that on the basis of sequence homology of 16S RNA, *Prochloron* lies phylogenetically far away from chloroplasts not only of *Nicotiana* and *Zea mays*, but also of *Chlamydomonas* and *Euglena*. Also the relationship between chloroplasts and *Prochlorothrix* was determined by 16S RNA to be not very close (Turner

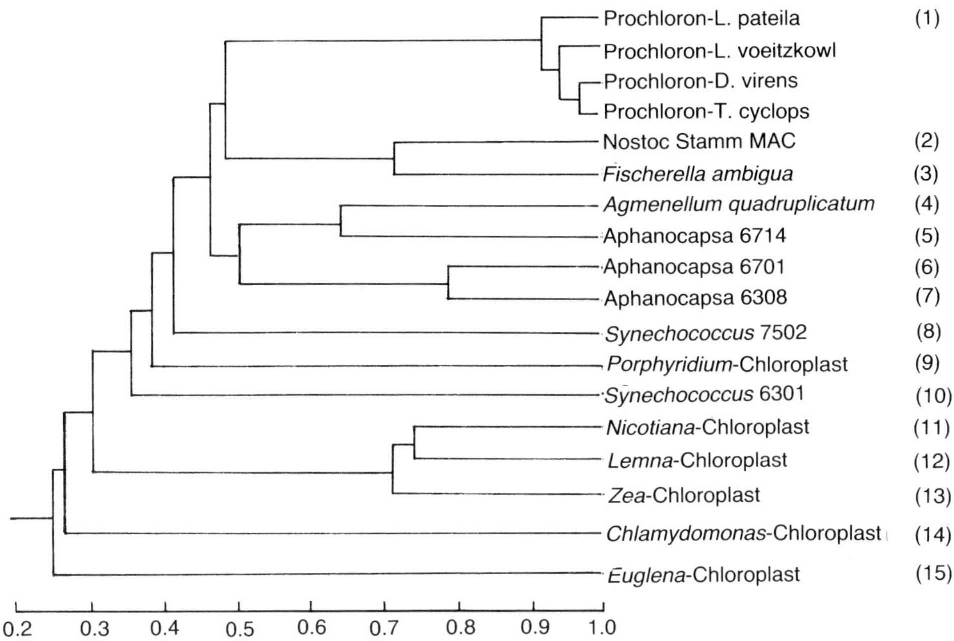

Fig. 1.4. Phylogenetic relationship of cyanobacteria (including *Prochloron* species) and chloroplasts from spermatophytes and eukaryotic phototrophic protists, based on sequence analysis of 16S RNA. (Based on Seewalt and Stackebrandt, 1982.)

et al., 1989). However, in the *psbA*-gene coding for the D1 protein of membranes of photosystem II, other authors found in *Prochlorothrix* the seven amino acid deletions, characteristic for higher plants but absent in cyanobacteria (Morden and Golden, 1989). Since specific deletions have a higher significance for phylogenetic relationships than single nucleotide sequences (Meyer *et al.*, 1986), the conclusion so far is in favour of a relationship of Prochlorophyta to chloroplasts. The similarity in nucleotide sequence of 12 to 16S RNA from chloroplasts and mitochondria is much higher when compared to 16S RNA from *Escherichia coli* as with 18S RNA from *Xenopus laevis* (Fig. 1.6). Homology is somewhat higher between plant mitochondria and *E. coli* than between mitochondria from fungi and animals.

Introns found in the genes for $tRNA^{Ala}$ and $tRNA^{Ileu}$ of chloroplasts of all higher plants studied were also looked for in several algal groups. All algae and eubacteria have uninterrupted genes, with the exception of *Nitella* and *Coleochaete*, members of the Charaphyceae, indicating that this algal group is part of the lineage giving rise to the evolution of land plants (Lüttge *et al.*, 1988).

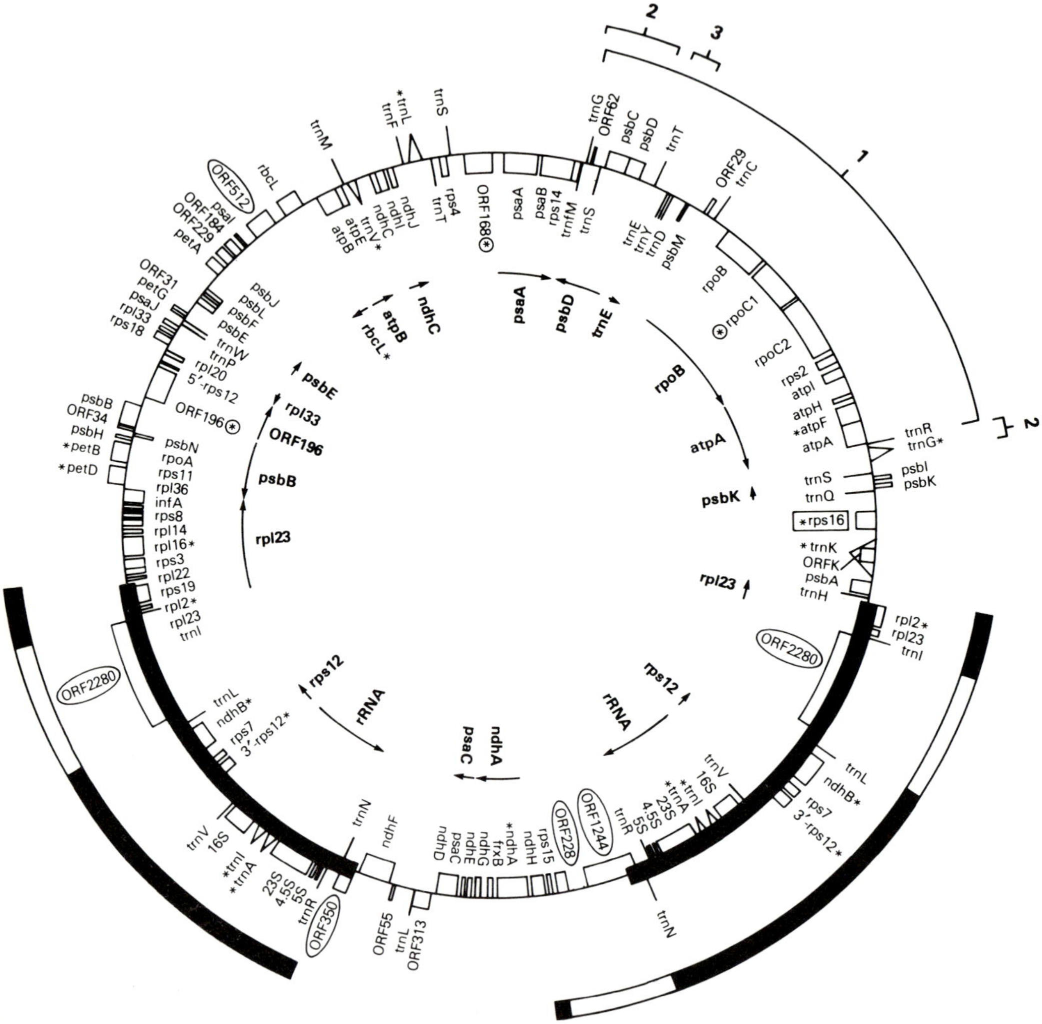

Fig. 1.5. Gene map of the 156 kb tobacco chloroplast genome, showing selected rearrangements found in rice and *Marchantia*. Genes on the outside of the circle are transcribed counterclockwise; those on the inside are transcribed clockwise. Arrows on the inside of the circle indicate sets of genes thought to constitute operons. In general, the operons are named according to their first gene. The asterisked operon (*rbc*L) consists of a single gene in land plants and chlorophytes, but two genes (*rbc*L and *rbc*S) in all other algae. Circled gene names indicate genes present in tobacco and *Marchantia* but absent from rice. The solid boxed name indicates a gene present in tobacco and rice but absent from *Marchantia*. Asterisks denote genes that have the same intron(s) in all three sequenced genomes; circled asterisks denote genes that are split in only two of the three genomes. The thickened parts of the circle represent the 25.3 kb inverted repeat of tobacco. The thick lines outside the tobacco inverted repeat represent the extent of the inverted repeat in rice (20.8 kb), with regions deleted from rice indicated by diagonal hatching. The numbered brackets

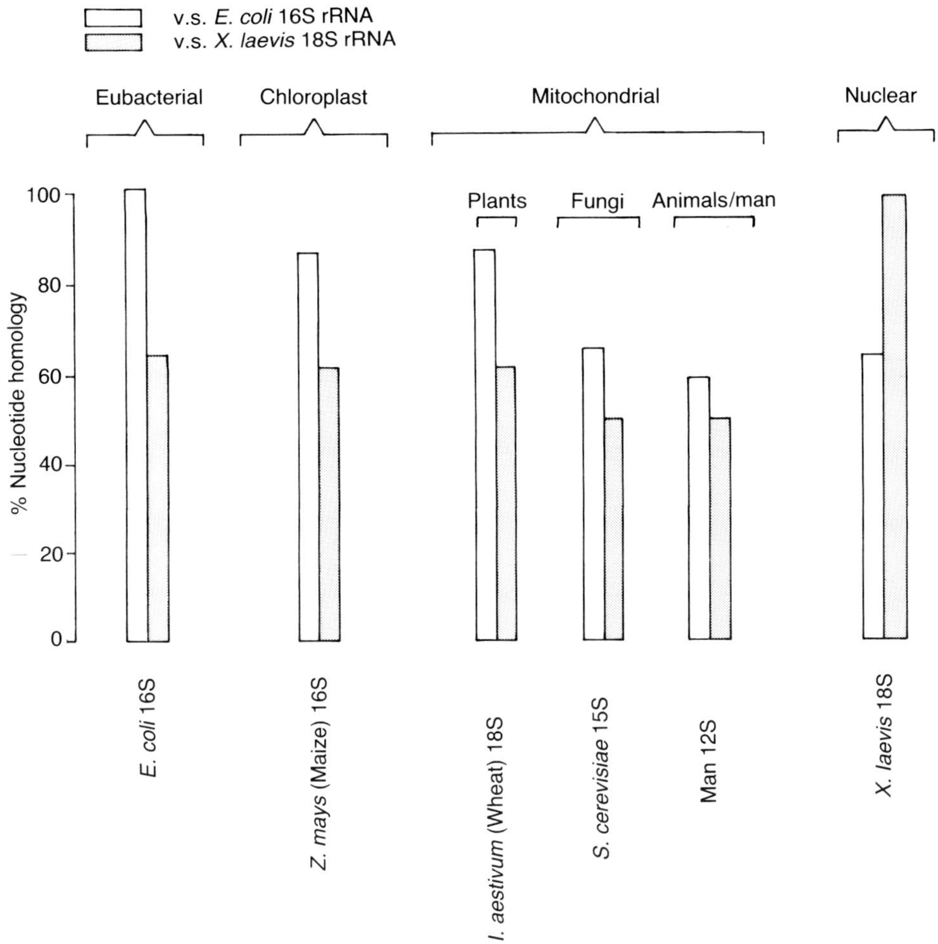

Fig. 1.6. Sequence homologies of plastid and mitochondrial RNA with RNA from *E. coli* and *Xenopus laevis* (nuclear RNA). Data are given as percentage nucleotide homology in core regions of 12–18S RNA. (Based on Gray, 1983.)

---

outside the circle indicate regions that have been moved by three overlapping inversions in rice relative to tobacco. Gene nomenclature: rRNA genes are indicated by 16S, 23S, 4.5S and 5S; tRNA genes by *trn*, followed by the one-letter amino acid code; genes for RNA polymerase subunits by *rpo*, followed by a subunit-specific letter, genes for the 50S and 30S ribosomal proteins by *rpl* and *rps*, respectively, followed by the number of the corresponding *E. coli* protein; genes for components of the thylakoid membrance complexes ATP synthase, photosystem I, photosystem II, and cytochrome b_6/f complex by *atp*, *psa*, *psb*, and *pet*, respectively, followed by a subunit-specific letter; genes for NADH dehydrogenase subunits by *ndh*, followed by the letter of the corresponding mitochondrial subunit; gene for the large subunit of ribulose-1,5-bisphosphate carboxylase by *rbc*L; open reading frames that are conserved between at least two of the three sequenced genomes are indicated by ORF, followed by their lengths in tobacco in codons. (From Palmer, 1991.)

Relationships between chloroplasts and mitochondria also exist insofar as plant mitochondria contain non-expressed chloroplast DNA sequences. This includes chloroplast genes for 16S rRNA (Stern and Lonsdale, 1982), several tRNAs (Joyce *et al.*, 1988) and the large subunit of ribulose-bisphosphate carboxylase (Rubisco) (Lonsdale *et al.*, 1983; Manhart and Palmer, 1990). The functional genes in the mitochondrial DNA are dispersed in the genome and account for not more than 15% of the total size (Lonsdale, 1987). It is not known whether the other 85% are completely without function (sequence independent functions). In addition to the master chromosome and smaller subgenomic circles, plant mitochondria contain small circular and linear plasmids, ranging in size (in maize) between 1.4 and 7.2 kb (Pring and Lonsdale, 1985). The complete mitochondrial DNA in plants varies between 70 kb in yeast to 2500 kbp in some Curcurbitaceae (Douce and Neuberger, 1989).

Chloroplast DNA, which has been fully sequenced, is, with a length of 120–160 kbp (kilobase pairs), much more uniform than that of plant mitochondria. The chloroplast DNA from *Nicotiana tabacum* contains 155 844 bp with 146 genes (Shinozaki *et al.*, 1986). A region of 25 kb is present as an inverted repeat, coding in the first place for ribosomal RNA with an arrangement similar to prokaryots. Some legumes such as *Pisum sativum* and *Vicia faba* have only one transcriptional unit for the chloroplast rRNA.

Chloroplasts are polyploid with up to 100 copies of the pt DNA and also polyenergide with many nucleoids per chloroplast.

The 156 kb chloroplast gene map from tobacco with selected rearrangements found in rice and the moss *Marchantia* is shown in Fig. 1.5. As indicated by the thickened parts in the inner and the outer circle, the inverted repeats in tobacco (25.3 kb) and in rice (20.8 kb) are different in size (Palmer, 1991).

The striking similarity of the plastid genome in several classes of green plants contrasts to the loss of many photosynthesis genes in **plastids from parasitic plants**. *Conopholis* has a **plastid genome of only 51 kb** and lacks 29 photosynthesis related genes, together with 34 other genes (Wimpee *et al.*, 1991).

DNA GC content is a systematic criterion often used in bacterial taxonomy. The G + C proportion of total plastid DNA from monocots, dicots, members of the pteridophytes, chlorophytes and rhodophytes, lie together in the relative narrow region between 30 and 40 mol% GC content (Fig. 1.7). Some of the cyanobacteria have similar values, but values up to 70% are also found. Clearly different GC contents have been found for chloroplasts from Euglenophyta and Cryptophyceae. The results obtained from RNA and protein sequencing, indicate that GC content is not a strongly conserved property in chloroplast evolution. This suggests, therefore, that the homology of GC content between nuclear and plastid DNA in monocots

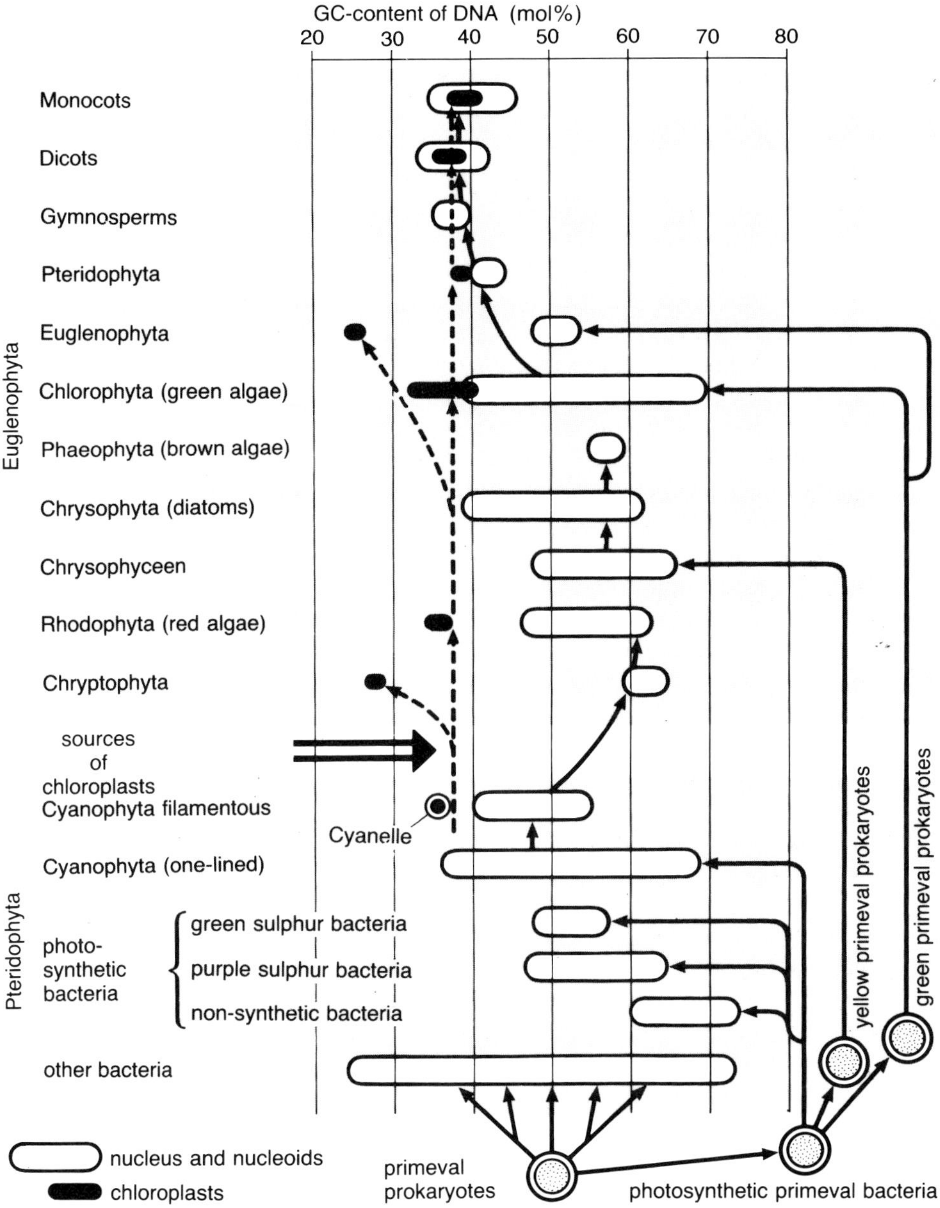

Fig. 1.7. GC-content of chloroplast and nuclear DNA of plants and prokaryotes. (Based on Ishida, 1978.)

and dicots is possibly a coincidence. The large differences found between the Rhodophyceae and the Cryptophyceae are thus explainable.

Only around 100 chloroplast polypeptides are coded for by plastid DNA. This is only half the number of those coded for by nuclear DNA. Nuclear-encoded proteins are synthesized on free cytoplasmic (poly)ribosomes and must be subsequently transported into the chloroplast (Fig. 1.8). This separation of genetic information between nucleus and plastid demands, with regard to the endosymbiotic theory, a fragmentation of the original prokaryotic organelle genome and recombination with host cell DNA. This appears to be the case with the small subunit of ribulose-1,5-bisphosphate carboxylase (RUBISCO), enzymes of chlorophyll biosynthesis and plastid RNA polymerase, the genetic information for which has been integrated into the nucleus. Exactly how the gene transfers have been achieved and why these genes in particular have been transferred, is not yet known. In addition, the haploid nuclear genome of spinach has integrated the plastome genome five times, although not active due to methylation (Scott and Timmis, 1984). A conserved self-splicing group I intron has been found in *Anabaena*, indicating the bacterial origin of chloroplast introns (Xu *et al.*, 1990).

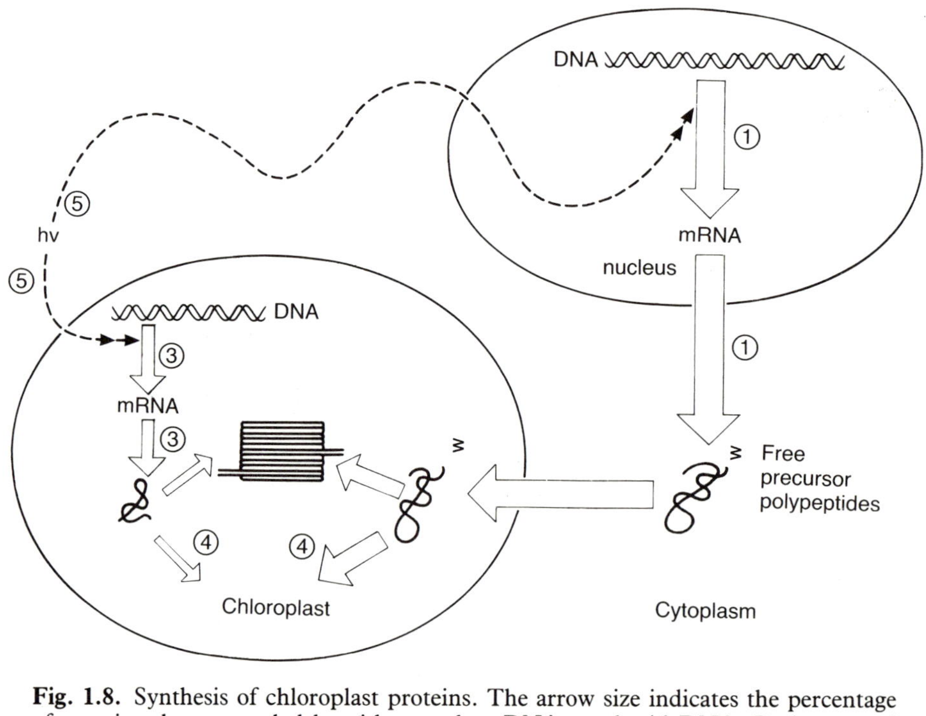

Fig. 1.8. Synthesis of chloroplast proteins. The arrow size indicates the percentage of proteins that are coded by either nuclear DNA or plastid DNA. Both types of protein biosynthesis can be regulated by light (hv). (Based on Ellis, 1983.)

The early separation of red algae in the phylogenetic tree is supported by the finding that both subunits of ribulose-1,5-bisphosphate carboxylase are located on the plastome in one operon in *Porphyridium aerugineum* (Valentin and Zetsche, 1989). Studies of a recent symbiosis, such as *Geosiphon*, in comparison to chloroplasts and mitochondria, were the basis for the development of Schnepf's (1984) theory for the organization of the eukaryotic cell.

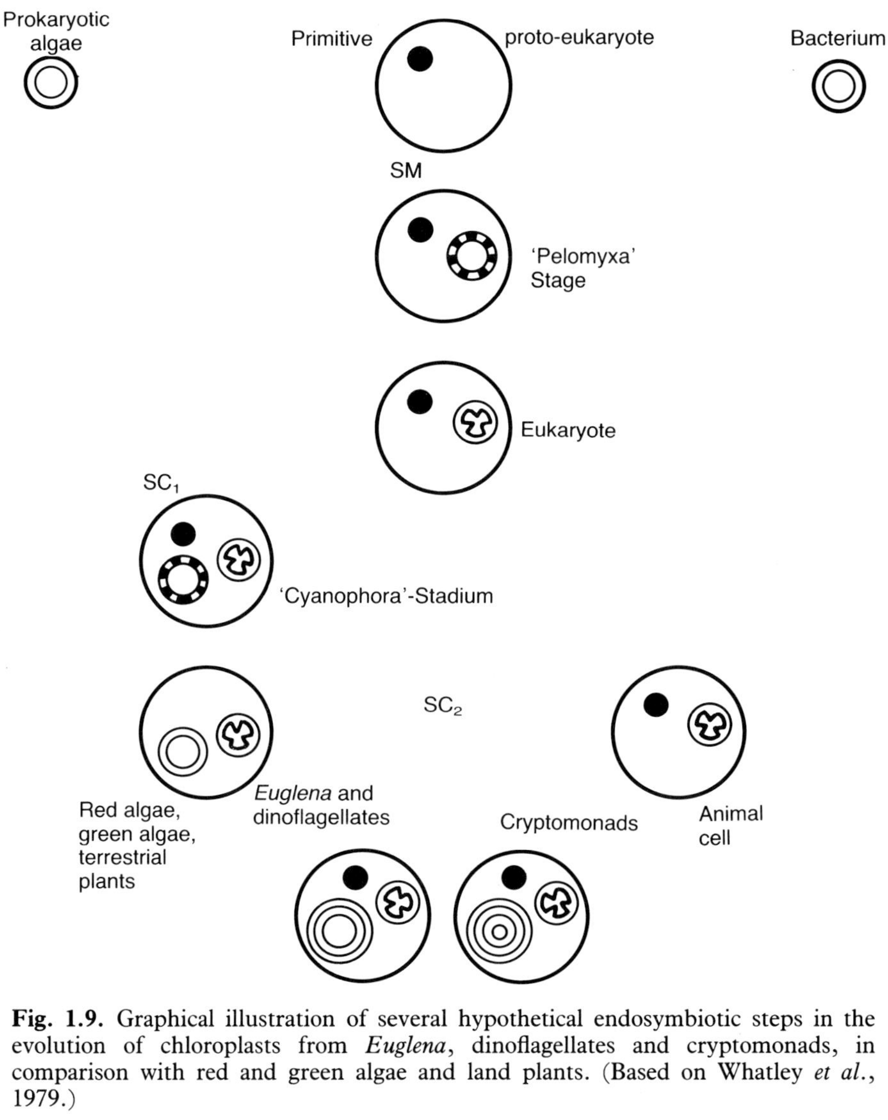

Fig. 1.9. Graphical illustration of several hypothetical endosymbiotic steps in the evolution of chloroplasts from *Euglena*, dinoflagellates and cryptomonads, in comparison with red and green algae and land plants. (Based on Whatley *et al.*, 1979.)

Table 1.5 The eight kingdoms of cellular organisms. (Based on Cavalier-Smith, 1988)

Empire 1.	Bacteria
	(syn. Prokaryota; DNA not separated from ribosomes, an envelope; endomembrane system and cytoskeleton absent; DNA attached to 'envelope skeleton')
	Kingdom 1. Eubacteria (acylglycerol membrane lipids: murein)
	Subkingdom 1. Negibacteria (plasma membrane + outer membrane)
	Subkingdom 2. Posibacteria (outer membrane absent)
	Kingdom 2. Archaebacteria (isoprenoid ether lipids; no murein; outer membrane absent)
Empire 2.	Eukaryota
	(nucleated cells; with endomembrane systems, cytosis and internal cytoskeleton; acyl glycerol membrane lipids; no peptidoglycan walls)
	Superkingdom 1. Archezoa (70S ribosomes; Golgi dictyosomes, mitochondria, chloroplasts, and peroxisomes absent)
	Kingdom 3. Archezoa
	Superkingdom 2. Metakaryota (80S ribosomes; Golgi dictyosomes, mitochondria, peroxisomes, and often chloroplasts, typically present)
	Kingdom 4. Protozoa (Chloroplast envelope of three membranes; includes Mycetozoa, predominantly phagotrophic)
	Kingdom 5. Chromista (tubular ciliary mastigonemes and/or chloroplast endoplasmic reticulum present)
	Kingdom 6. Plantae (plastids invariably present; chloroplast envelope of two membranes; plastids not in ER)
	Kingdom 7. Fungi (typically with chitinous walls and no chloroplasts or phagocytosis; excludes Mycetozoa and pseudofungi)
	Kingdom 8. Animalia (triploblastic multicellular phagotrophs)

Expression of genes in the nucleus, coding for proteins of the plastids, is regulated mainly at the transcriptional level. Two especially well-studied proteins, the light-harvesting chlorophyll *a*/*b* protein and the small subunit of RUBISCO are regulated by phytochrome and a second blue light receptor (Taylor, 1989). Gene expression in the plastids is often under translational control, also regulated by light (Klein *et al.*, 1987).

The endosymbiotic theory then, as described, also gives a plausible explanation for the existence of two very different chloroplast membranes in green algae and all higher plants. At first sight the theory fails to explain why euglenoids and dinoflagellates have three chloroplast membranes. Indeed, cryptophytes have four (Chapter 8). The solution is multiple endosymbiosis (Sitte, 1990). Thus chloroplasts of *Euglena* spp. and dinoflagellates may have come about through the progressive reduction of endosymbiotic eukaryotic algae (Gibbs, 1978), Fig. 1.9. Thus the third membrane has its origins in a second endosymbiotic step. However, it is unclear if this third membrane is equivalent to the plasmamembrane of the hypothetical host cell, or is a modified vacuole membrane. The existence of four membranes around the chloroplasts of cryptomonads is thus the result of an animal cell taking up a complete eukaryotic algal cell. Thus these plastids are comparable to the *Paramecium bursaria–Chlorella* symbiosis (Chapter 8). Besides chloroplast and mitochondria, several other characters of cell organelles have been used to establish the eight kingdoms of cellular organisms according to Cavalier-Smith (1988) (Table 1.5).

The occurrence of DNA in centrosomes (Hall *et al.*, 1989), the organization centres of microtubuli, could mean that basic cellular processes like chromosome orientation in mitosis could also be related to an early endosymbiotic process. In recent symbiosis as treated in the following chapters of this book, some regulation of symbiotic developments may have evolved by transposable genetic elements (Dyer, 1989).

1.4 REFERENCES

Altmann, R. (1890) *Die Elementarorganismen und ihre Beziehung zu den Zellen.* Veit und Comp., Leipzig.

Block, M.A., Dorne, A.-J., Joyard, J. and Douce, R. (1983) Preparation and characterization of membrane fractions enriched in outer and inner envelope membranes from spinach chloroplasts. *J. Biol. Chem.*, **258**, 13281–6.

Buchner, P. (1953) *Endosymbiose der Tiere mit pflanzlichen Mikroorganismen.* Birkhäuser Verlag, Basel.

Cavalier-Smith, T. (1988) Eukaryote cell evolution, in *Proceedings of the XIV International Botanical Congress* (eds W. Greuter and B. Zimmer) Koeltz, Königstein/Taunus, pp. 203–23.

Dayhoff, M.O. (1979) *Atlas of Protein Sequences.* NBRC Foundation, Washington.

Douce, R. and Neuburger, M. (1989) The uniqueness of plant mitochondria. *Ann. Rev. Plant Physiol. Plant Mol. Biol.*, **40**, 371–414.

Dyer, B.D. (1989) Symbiosis and organismal boundaries. *Am. Zool.*, **29**, 1085–93.

Ellis, R.J. (1983) Chloroplast protein synthesis: principles and problems. *Subcell. Biochem.*, **9**, 237–61.

Follmann, H. (1981) *Chemie und Biochemie der Evolution: Wie und wo entstand das Leben?* Quelle und Meyer, Heidelberg.

Gäumann, E. (1951) *Pflanzliche Infektionslehre*, 2. Aufl. Birkhäuser, Basel.

George, D.G., Hunt, L.T. and Dayhoff, M.O. (1983) Sequence evidence for the symbiotic origins of chloroplasts and mitochondria, in *Endocytobiology* II. *Intracellular Space as Oligogenetic Ecosystem* (eds H.E.A. Schenk and W. Schwemmler) W. de Gruyter, Berlin, New York, pp. 845–61.

Gibbs, S.P. (1978) The chloroplasts of *Euglena* may have evolved from symbiotic green algae. *Can. J. Bot.*, **56**, 2883–9.

Gilbert, W. (1986) Origin of life: the RNA world. *Nature*, **319**, 618.

Gray, M.W. (1983) The bacterial ancestry of plastids and mitochondria. *BioScience*, **33**, 693–9.

Hall, J.L., Ramanis, Z. and Luck, D.J.L. (1989) Basal body/centriolar DNA: molecular genetic studies in *Chlamydomonas. Cell*, **59**, 121–32.

Herrmann, R.G. (1984) Zur Organisation des Plastoms (Eine Übersicht). *Ber. Dtsch. Bot. Ges.*, **97**, 335–50.

Hori, H. and Osawa, S. (1987) Origin and evolution of organisms as deduced from 5S ribosomal RNA sequences. *Mol. Biol. Evol.*, **4**, 445–72.

Ishida, M.R. (1978) *Origin of Life*, Japan Scientific Society Press, Tokyo.

Joyce, P.B.M., Spencer, D.F., Bonen, L. and Gray, M.W. (1988) Genes for $tRNA^{Asp}$, $tRNA^{Pro}$, $tRNA^{Tyr}$ and two $tRNA^{Ser}$ in wheat mitochondrial DNA. *Plant Mol. Biol.*, **10**, 251–62.

Kaplan, R.W. (1978) *Der Ursprung des Lebens*, 2. Aufl. Thieme Verlag, Stuttgart.

Klein, R.R., Mason, H. and Mullet J.E. (1987) Light-regulated translation of chloroplast proteins. Transcripts of psaA-psaB, psbA and rbcL are associated with polysomes in dark-grown and illuminated barley seedlings. *J. Cell Biol.*, **106**, 289–302.

Kleinig, H. and Sitte, P. (1992) *Zellbiologie*, 3. Aufl., G. Fischer Verlag, Stuttgart.

Kropotkin, P. (1987) *Mutual Aid, a Factor of Evolution.* 1st edn 1901, Freedom Press, London.

Law, R. and Lewis, D.H. (1983) Biotic environments and the maintenance of sex – some evidence from mutualistic symbioses. *Biol. J. Linn. Soc.*, **20**, 249–76.

Lewin, R.A. (1982) Symbiosis and parasitism – definitions and evaluations. *BioScience*, **32**, 254–9.

Lonsdale, D.M. (1987) The molecular biology and genetic manipulation of the cytoplasm of higher plants. *Genet. Eng.*, **6**, 50–104.

Lonsdale, D.M., Hodge, T.P., Howe, C.J. and Stern, D.B. (1983) Maize mitochondrial DNA containing sequences homologous to the ribulose-1,5-bisphosphate carboxylase large subunit gene of chloroplast DNA. *Cell*, **34**, 1007–14.

Lüttge, U., Kluge, M. and Bauer, G. (1988) *Botanik. Ein grundlegendes Lehrbuch.* Verlag Chemie, Weinheim.

Manhart, J.R. and Palmer, J.D. (1990) The gain of two chloroplast tRNA introns marks the green algal ancestors of land plants. *Nature*, **345**, 268–70.

Margulis, L. (1981) *Symbiosis in Cell Evolution*, Freeman, San Francisco.

Matthes, D. (1978) *Tiersymbiosen*, G. Fischer Verlag, Stuttgart.

Meinecke, F. (1924) *Die Idee der Staatsräson in der neueren Geschichte.* Oldenburg Verlag München, Berlin.

Meyer, T.E., Cusanovich, M.A. and Kamen, M.D. (1986) Evidence against the use of bacterial amino acid sequence data for construction of all-inclusive phylogenic trees. *Proc. Natl. Acad. Sci. USA*, **83**, 217–20.

Mohr, H. and Schopfer, P. (1985) *Pflanzenphysiologie*, 3. Aufl. Springer-Verlag, Berlin.

Morden, C.W. and Golden, S.S. (1989) *psb*A genes indicate common ancestry of prochlorophytes and chloroplasts. *Nature*, **337**, 382–5.

Palmer, J.D. (1991) Comparison of chloroplast and mitochondrial genome evolution in plants, in *Plant Gene Research*, Vol. 6: *Organelles* (ed. R.G. Herrmann), Springer, Berlin, Heidelberg, New York.

Pring, D.R. and Lonsdale, D.M. (1985) Molecular biology of higher plant mitochondrial DNA. *Int. Rev. Cytol.*, **97**, 1–46.

Remmert, H. (1989) *Ökologie. Ein Lehrbuch.* 4. Aufl., Springer-Verlag, Berlin.

Schaede, R. (1943) *Die pflanzliche Symbiose.* G. Fischer Verlag, Jena.

Schaede, R. (1962) *Die pflanzliche Symbiose.* G. Fischer Verlag, Stuttgart.

Schaede, R. and Meyer, F.H. (1962) *Die pflanzlichen Symbiosen.* G. Fischer Verlag,

Stuttgart.

Schäffle, A.E.F. (1906) *Abriß der Soziologie*. H. Laupp, Tübingen.

Schenk, H.E.A. and Schwemmler, W. (1983) *Endocytobiology* II. W. de Gruyter, Berlin.

Schimper, A.F.W. (1883) Über die Entwicklung der Chlorophyllkörner und Farbkörper. *Bot. Ztg.*, **41**, 105–11, 121–31, 137–46.

Schlösser, E. (1983) *Allgemeine Phytopathologie*. Thieme Verlag, Stuttgart.

Schnepf, E. (1984) Zur Feinstruktur von *Geosiphon pyriforme*. Ein Versuch zur Deutung cytoplasmatischer Membranen und Kompartimente. *Arch. Microbiol.*, **49**, 112–32.

Schweiger, H.D. (1981) *International Cell Biology 1980–1981*. Springer-Verlag, Berlin.

Scott, N.S. and Timmis J.N. (1984) Homologies between nuclear and plastid DNA in spinach. *Theor. Appl. Genet.*, **67**, 279–88.

Seewalt, E. and Stackebrandt, E. (1982) Partial sequence of 16S ribosomal RNA and the phylogeny of *Prochloron. Nature*, **295**, 618–20.

Shinozaki, K., Ohme, M., Tanaka, M. *et al.* (1986) The complete nucleotide sequence of the chloroplast genome: its gene organization and expression. *EMBO J.*, **9**, 2043–9.

Sitte, P. (1990) Phylogenetische Aspekte der Zellevolution. *Biol. Rundsch.*, **28**, 1–18.

Smith, D.C. (1979) From extracellular to intracellular: the establishment of a symbiosis. *Proc. R. Soc. Lond.* [Biol.], **204**, 115–30.

Smith, D.C. and Douglas, A.E. (1987) *The biology of symbiosis*. E. Arnold Publ., London.

Stern, D.B. and Lonsdale, D.M. (1982) Mitochondrial and chloroplast genomes of maize have a 12-kilobase DNA sequence in common. *Nature*, **299**, 698–702.

Sydow, E. von (1928) *Die Kunst der Naturvölker und der Vorzeit*. Propylaen-Verlag, Berlin.

Taylor, W.C. (1989) Regulatory interactions between nuclear and plastid genomes. *Ann. Rev. Plant Physiol. Plant Mol. Biol.*, **40**, 211–33.

Turner, S., Burger-Wiersma, T., Giovannoni, S.J., Mur, L.R. and Pace, N.R. (1989) The relationship of a prochlorophyte *Prochlorothrix hollandica* to green chloroplasts. *Nature*, **337**, 380–2.

Valentin, K. and Zetsche, K. (1989) The genes of both subunits of ribulose-1,5-bisphosphate carboxylase constitute an operon on the plastome of a red alga. *Curr. Genet.*, **16**, 203–9.

von Wettstein, D. and Oliver R.P. (1984) Zur Molekularbiologie der Photosynthese. *Ber. Dtsch. Bot. Ges.*, **98**, 335–50.

Whatley, J.M., John, P. and Whatley, F.R. (1979) From extracellular to intracellular: the establishment of mitochondria and chloroplasts. *Proc. R. Soc. Lond.* [*Biol*], **204**, 165–87.

Wimpee, C., Garvin, D., Wrobel, R. *et al.* (1992) An old symbiosis and a new one: The impact of the abandonment of photosynthesis on the plastid genome of parasitic plants. *Symbiosis*, **12**, Lectures from the International Symbiosis Congress, Jerusalem, November 1991.

Xu, M.-Q., Kathe, S.D., Goodrich-Blair, H. *et al.* (1990) Bacterial origin of a chloroplast intron: conserved self-splicing group I introns in cyanobacteria. *Science*, **250**, 1566–70.

2
Specific associations between micro-organisms and plants

2.1 RHIZOSPHERE

The rhizosphere is the contact zone of roots and soil, which is directly influenced by root metabolism. The rhizosphere can be further separated into the **endorhizosphere**, the cell layers of the root itself, the **rhizoplane**, the surface of the root system and the **ectorhizosphere**, the area around the root (Lynch, 1990). Important physiological processes are the uptake of mineral nutrients by the root and the production of root exudates (Barber and Loughman, 1967; Marschner, 1985, 1986). The latter brings about an enrichment of micro-organisms in this area. Competition for limiting concentrations of mineral nutrients such as Fe^{III}/Fe^{II} between plant roots and micro-organisms may have contributed to the evolution of specific phytosiderophores or other strategies for iron uptake in plants (Marschner *et al.*, 1989). Processes related to rhizosphere organisms have been summarized by Brown (1981), Klingmüller (1985) and Lynch (1990), as either beneficial, harmful or neutral to variable. New and further developed methods are necessary to study and detect all micro-organisms that are involved in these processes.

1. By **DNA probe techniques** as little as 10 cells of *Pseudomonas putida* per g of soil have been detected (Fredrickson *et al.*, 1988);
2. With a **luminescence technique**, the same low concentration of bacteria has been detected with a plastid-born lux casette in genetically modified strains (Rattray *et al.*, 1990);
3. A combination of classical **ELISA tests** with monoclonal antibodies increases the specificity for identification (Dewey, 1988).

2.1.1 Root exudates

The activity and composition of the microbial population in the rhizosphere

Table 2.1 Root exudates from barley (*Hordeum vulgare*) in solution culture or cultured on glass beads (ballotini). (Based on Barber and Gunn, 1974)

	Solution culture	*Culture on glass beads**
Shoot dry weight (mg dry matter per plant)	57	52
Root dry weight (mg dry matter per plant)	32	36
Amino acids exuded (mg per plant)	0.14	0.23
Carbohydrates exuded (mg per plant)	1.45	3.03
Total exudates (% of root weight)	5.0	9.0
Total exudates (% of plant weight)	1.8	3.7

* Culture medium identical to solution culture.

Table 2.2 Exudation of organic compounds from roots grown under sterile conditions. (Based on Newman and Watson, 1977)

Plant species	*Root age (days)*	*Compounds analysed*	*Exudation per root increment (mg/g)**
Barley	0–6	Sugars	15
Barley	6–27	Carbohydrates and amino acids	42–82
Maize	6–27	Carbohydrates and amino acids	6–60
Alfalfa	0–28	Neutral carbohydrates	247
Locust	8–18	Carbohydrates including organic acids and amino acids	38
Pine	8–18	Carbohydrates including organic acids and amino acids	67–244

* A value of 250 mg corresponds with an exudation rate of about 25 μg cm^{-2} root surface day^{-1}.

is limited and influenced by the type and amount of root exudates. Between 5 and 10% of root mass (Table 2.1) and up to 25% of root growth increment (Table 2.2) can be used by chemo-organotrophic micro-organisms. Exudation may, however, be even higher in soil under field conditions. This is suggested in experiments using glass beads instead of hydroculture where a doubling of the exudation rate is found (Table 2.1), and is probably due to increased exudation when roots come into contact with solid surfaces. This contact is probably quantitatively and qualitatively more intense in soil than in laboratory media and is thus an area for further research.

In order to determine organic carbon turnover in soil, plants were either grown in a $^{14}CO_2$ atmosphere in normal soil, or grown in a $^{12}CO_2$ atmosphere in soil containing organic ^{14}C. Roots were surrounded by concentric layers of perlon net through which they could not grow but through which organic compounds could diffuse or could be transported in both directions

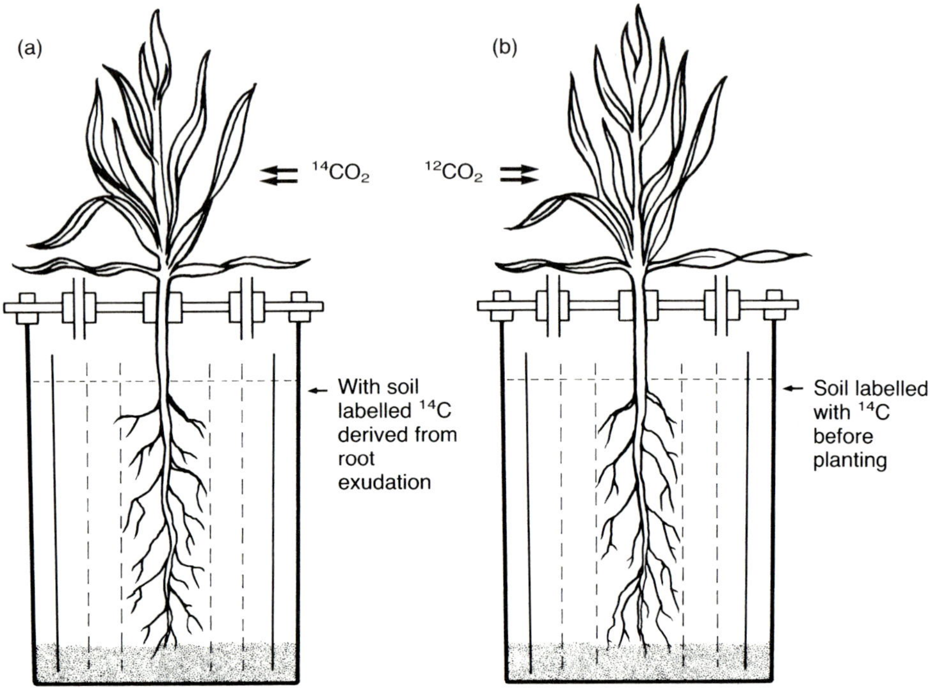

Fig. 2.1. Effect of plant roots on soil carbon metabolism at various distances from the root (sandwich method: separation of soil layers by plant impenetrable plastic nets). (a) Labelled root exudates after incorporation of $^{14}CO_2$ by the shoot. (b) Labelling of soil carbon by ^{14}C before planting and determination of the labelled carbon taken up by the plant. (Based on Helal and Sauerbeck, 1984.)

with water movements (Fig. 2.1). By labelling the whole plant with ^{14}C Helal and Sauerbeck (1989) determined that 19% of the photosynthetic product was released as organic material into the rhizosphere. About 75% of this material was utilized by the micro-organisms and 25% remained in the soil. Root effects were noticed even at a distance of more than 20 mm from the root surface. In another study, about 12% of photosynthetic products were found in the root exudate (Fig. 2.2).

2.1.2 pH alterations in the rhizosphere

One important factor in the specificity of plant root/rhizosphere micro-organism associations is the pH value around the root. This is affected not only by root exudates, but also by nutrient uptake by the root system, transport processes and by the release of H^+ associated with root growth. The largest changes are to be found on the root surface. The clearest changes in pH values in acidifying systems are found between pH 4.5 and

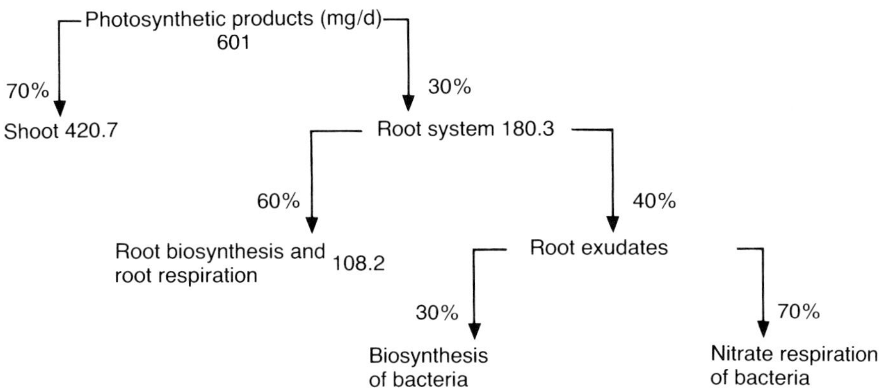

Fig. 2.2. Partitioning of assimilation products in 4-week-old maize plants. (Based on Haller and Stolp, 1985.)

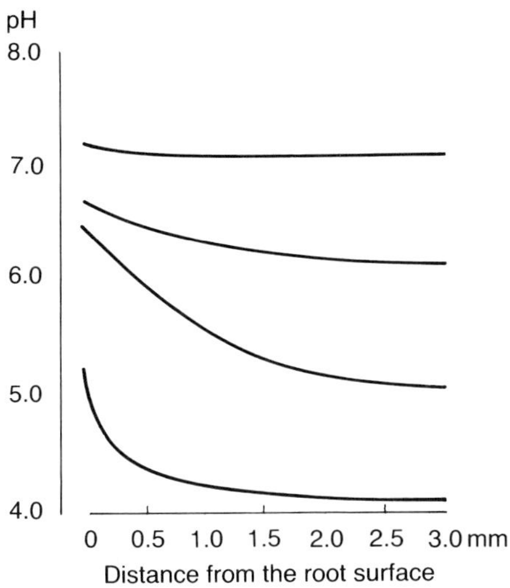

Fig. 2.3. pH-reduction in the rhizosphere due to excretion of H^+ by plant roots, in relationship to the surrounding (initial) soil pH. Computer simulation based on experimentally confirmed assumptions as follows: mean root diameter, 0.4 mm; anion uptake, 10^{-9} mol g^{-1} root fresh weight s^{-1}; flux of H^+ and HCO_3^- at root surface, 3×10^{-12} mol cm^{-2} s^{-1}. (Based on Nye, 1981 and Isermann, 1983.)

6.0 (Fig. 2.3). When HCO_3^- is released by the plant, large changes are found in the pH 4.0–5.0 region (Fig. 2.4). These pH changes are evident up to a distance of 2 mm from the root surface. An elegant method of measuring pH changes in the neighbourhood of the root is with indicator dyes in a rhizotron. A rhizotron is a flat plastic case with replaceable

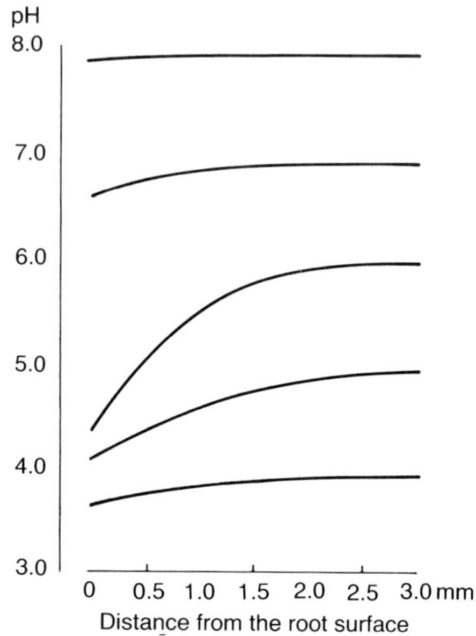

Fig. 2.4. pH increase in the rhizosphere due to excretion of HCO_3^- by plant roots, in relation to the surrounding (initial) soil pH. Computer simulation based on the following, experimentally confirmed assumptions: mean root diameter, 0.4 mm; anion uptake, 10^{-9} mol g^{-1} root fresh weight s^{-1}; flux of H^+ and HCO_3^- at root surface: 3×10^{-12} mol cm^{-2} s^{-1}. (Based on Nye, 1981 and Isermann, 1983.)

windows on the bottom. The cases are set up at an angle of 45° so that the root grows downwards until it comes into contact with the window, whereupon further growth is along the window, and the development of roots and root hairs can be directly observed. For pH-change experiments the rhizotron is carefully turned over and the window removed. A thin layer of indicator-containing agar is then poured onto the exposed surface. The agar must be just over the gelling temperature so as to damage the sensitive plant tissue as little as possible. The window is cleaned and the colour change in the agar layer can be observed visually. Nitrate-fed plants show increasing pH values whereas ammonium-fed plants show a decrease in pH. When the nitrogen balance is equal, e.g. when NH_4^+ and NO_3^- are used in equal proportions, the pH value of the soil remains constant as shown in experiments using meadow grass (*Lolium perenne*) (Fig. 2.5). Here, even after the grass was cut seven times, the pH values around the root remained unchanged at 6.8–7.0. Clover (*Trifolium pratense*) with symbiotic nitrogen fixation on the other hand, showed continual acidification around the roots during the seven mowing treatments. The result is a pH change from around pH 7.0 to pH 4.3, a total of 50 mval protons per kg soil must have

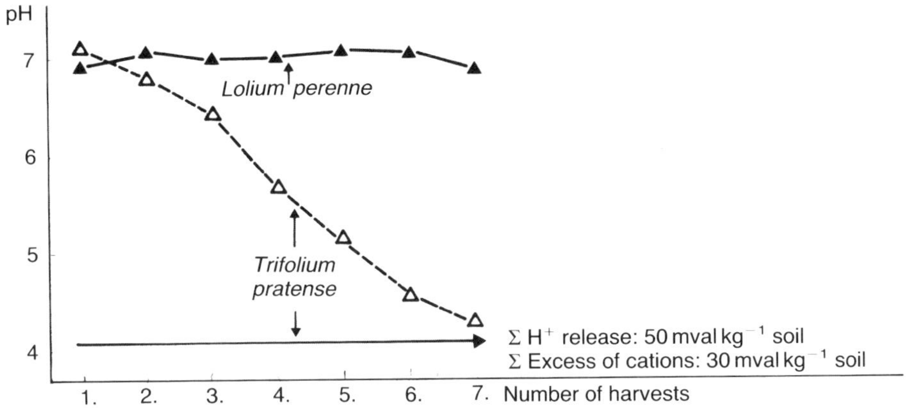

Fig. 2.5. pH values in the rhizosphere of *Lolium perenne* (ammonium nitrate as N-source) and *Trifolium pratense* (symbiotic N_2-fixation as N-source). (Based on Mengel and Steffens, 1982) (mval = milli valent).

been set free. Simultaneously a cation excess equivalent to 60% of the amount of protons was measured in the aboveground parts of the plant.

A more precise measurement of pH differences along roots is possible using antimony pH electrodes in the soil (Fig. 2.6) (Fischer *et al.*, 1989). Each of about 30 electrodes is connected to a register and a data processor.

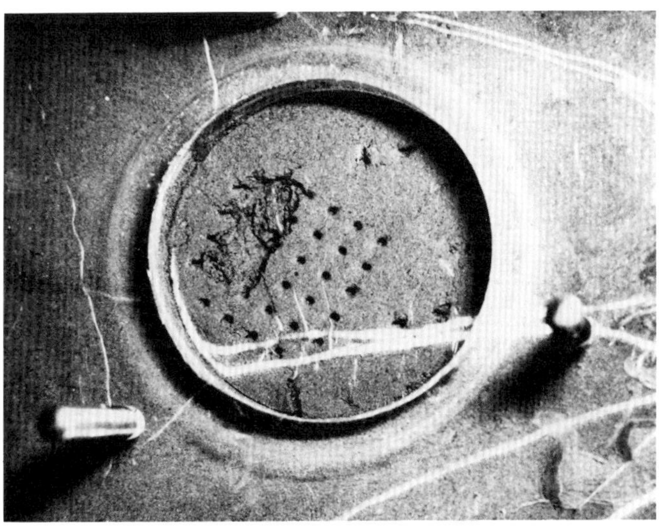

Fig. 2.6. Pt-Sb electrodes with groundnut roots growing towards the electrode contact zone. This experimental set-up allows *in situ* measurements of pH changes in the rhizosphere. (By courtesy of Fischer and Schaller, TU München-Weihenstephan.)

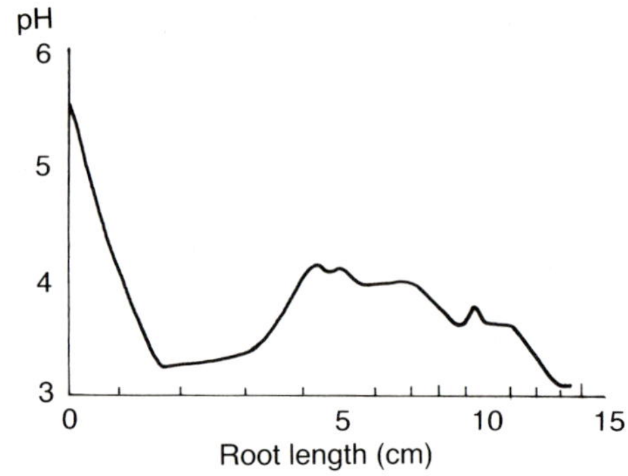

Fig. 2.7. Changes in pH at the surface of a groundnut root, in relationship to the distance from the root tip. (Based on Fischer *et al.*, 1989.)

Thus gradual changes in pH at each electrode can be continually measured. The important feature in this set-up is that the electrodes are not brought to the root, which would cause a change in the system, but that the roots grow to and over the electrodes. As soon as contact is made between the root surface and the electrode, a change in the pH can be measured. Due to the close rows of electrodes there is a very high probability that many contacts will be formed along one root (even though some electrodes never make contact). Figure 2.7 shows the pH change along a root measured with the help of the described set-up. After a sharp decline from pH 5.5 to pH 3.2 in the first 1.5 cm a zone between 4 and 10 cm of higher pH follows. Between 10–15 cm distance another drop in pH is observed. By using this multi-electrode system the effect of day/night cycles can be followed as well as effects of soil composition, temperature and cultivars. With this apparatus one of the most important ecophysiological criteria is met, namely that the biological system is not, or only slightly, influenced by the measurement. *In situ* pH measurements and microbial composition and associations in the rhizosphere can thus be correlated.

A number of other processes in the soil affect the proton balance (Ulrich, 1987):

1. nitrification (production of HNO_3)
2. nitrate uptake (consumption of HNO_3)
3. denitrification (consumption of HNO_3) (Trolldenier, 1989)
4. decrease of cation exchange capacity of soil organic matter due to decomposition/mineralization (consumption of protons)
5. silicate weathering (consumption of protons)

The net rate of proton production ($Q_H{}^+$) is the balance of these processes and is equal to

$$Q_H{}^+ = 1 - (2 + 3 + 4 + 5)$$

2.1.3 Specificity between bacterial populations around roots

Farming management affects the microbial biomass in the soil. With conventional farming, the total biomass of soil organisms is about 700 kg C ha^{-1}; under integrated farming 900 kg. Of this biomass bacteria constituted by far the largest proportion with more than 90%, fungi about 5% and protozoa less than 2% (Brussaard *et al.*, 1990). The ratio of microbial biomass to total organic carbon is also affected by the farming system (Anderson and Domsch, 1989). A change from 'r'-strategists (fast-reacting organisms after introduction of usable biomass) to 'K'-strategists (the autochthonous slow-growing population) has been observed during the readjustment of a disturbed soil (Lynch, 1984; Maire, 1987). Rhizosphere organisms and plant pathogens are usually K-selected (Andrews, 1984; Paul and Clark, 1989).

In many soils a large percentage of the bacteria present belongs only to a few genera (Paul and Clark, 1989): *Arthrobacter* spp. can make up to 40%, *Pseudomonas* spp., *Streptomyces* spp. and *Bacillus* spp. can each contribute between 5 and 20% of the bacterial plate count. However, many very specialized bacteria may not be registered by this simple technique. The maximal bacterial biomass is in the order of 0.4% of the soil volume (Harman and Lumsden, 1990). The fungal biomass can be somewhat larger in special soil types (Anderson and Domsch, 1973).

Thorough taxonomic analyses of the composition of rhizosphere microflora are seldom. Apart from some dominant forms bacteria present in the rhizosphere are mostly divided into groups (Table 2.3). Thus on roots of

Table 2.3 Composition of the rhizosphere microflora of barley and wheat from a sand/clay soil near Bayreuth (Germany). Data are expressed as a percentage of the total number of identified cells (n = 340). (Based on Kleeberger et al., 1983)

	Barley roots		*Wheat roots*	
	Untreated (10^7–10^8 cells*)	*Surface sterilized* (10^6–10^7 cells)	*Untreated* (10^7–10^8 cells)	*Surface sterilized* (10^6–10^7 cells)
Coryneform bacteria	61	13	50	16
Pseudomonas fluorescens	30	47	41	46
Pseudomonas putida	4	6	5	6
Enterobacter agglomerans type I	–	5	1	7
Enterobacter agglomerans type II	–	7	2	11
Citrobacter freundii	–	1	–	1
Serratia sp.	–	3	1	3
Other Gram-negative bacteria	–	18	–	8
Spore-forming bacteria	5	–	–	2

*No. of cells per g root fresh wt.

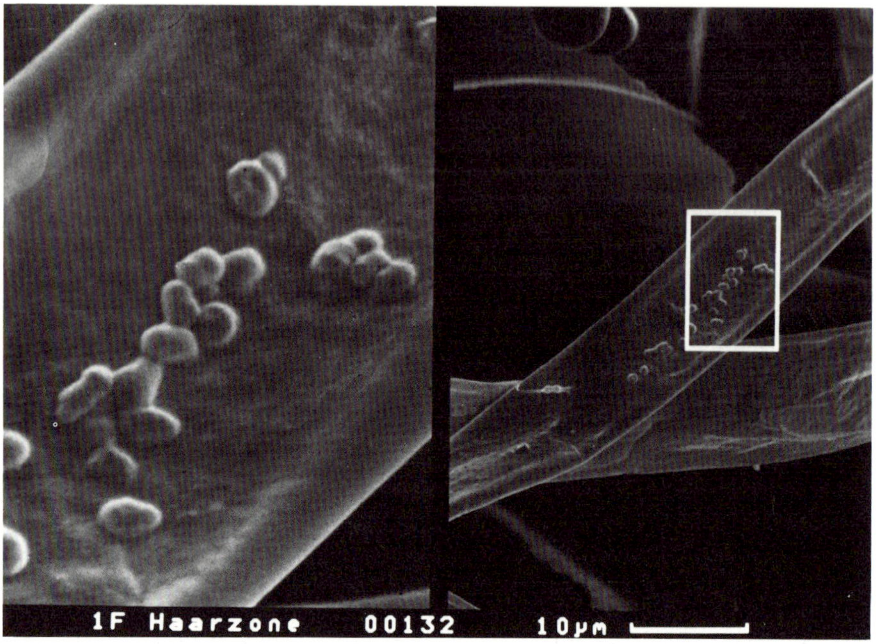

Fig. 2.8. Cells of *Arthrobacter globiformis* on root hairs of *Triticum aestivum*. (Original photograph by K. Fecher and B. Mosler, Marburg.)

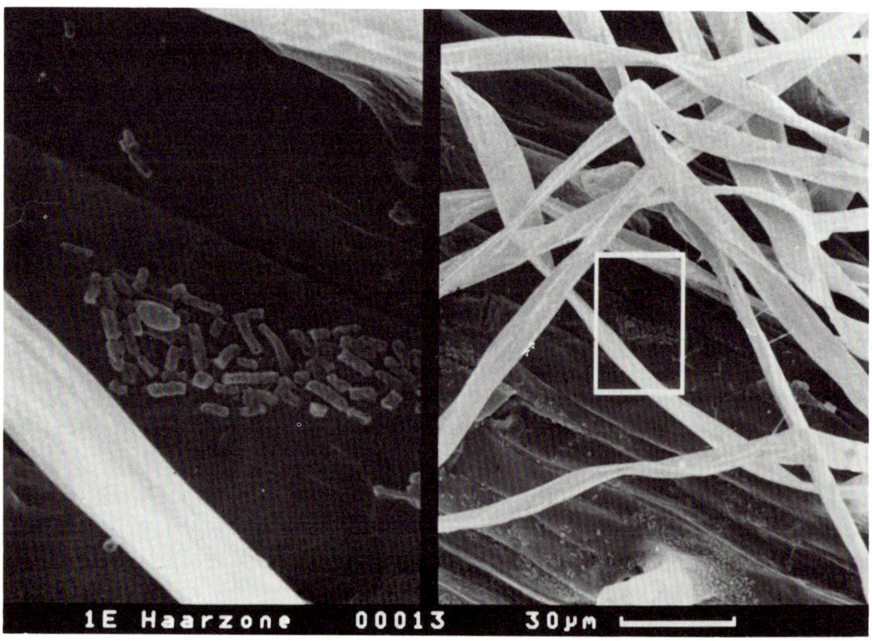

Fig. 2.9. Cells of *Klebsiella oxytoca* Kll on roots in the root hair zone of *Triticum aestivum*. (Original photograph by K. Fecher and B. Mosler, Marburg.)

barley and wheat grown on sandy soils, over 50% of all bacteria were placed in the coryneform group whereas 30–40% were identified as *Pseudomonas fluorescens*. *Enterobacter* spp. formed less than 5% of the population. The total cell count per g root fresh weight was between 10^7 and 10^8. This is only one order of magnitude more than that found when roots were surface sterilized, although in this case the proportion of coryneforms was lower and of enterobacteria significantly higher (Table 2.3). This may have been due to differences in the strength of contact between bacteria and different parts of the root system, e.g. root hairs (Fig. 2.8) or taproot (Fig. 2.9). The adsorption of bacteria to different surfaces also has to be taken into account (Bitton and Marshall, 1980; Brouwer *et al.*, 1981). Bigger differences have, however, been found between cultivars of *Sorghum*. For the two cultivars shown in Fig. 2.10 the total living cell count of chemo-organotrophic bacteria per gram rhizosphere soil is around $3–4 \times 10^8$. However, the number of nitrogen-fixing bacteria is a factor of 3 higher in one cultivar than the other, and in the same cultivar the number of aerobic nitrogen fixers is 10 times higher. This increase was at the expense of the actinomycetes (reduced to 10%) and bacteria of the type *Arthrobacter* (reduced to 25–30%; Fig. 2.10) (Werner *et al.*, 1989). The contribution of associative nitrogen fixing bacteria to the N-budget of agricultural crops is insignificant for temperate climates (Werner *et al.*, 1989), but may be of some importance in tropical soils, Berkum and Bohlool, 1980).

In addition to the pH value and the exudation of organic material into the soil the partial pressure of oxygen is important in the specificity of bacteria/root associations.

Simulation and experimental reproduction of O_2 gradients in soil is possible using the apparatus shown in Fig. 2.11, which spreads an oxygen gradient (which normally extends in the soil for a few millimetres) over 50 cm. At one end is an anaerobic zone which is continually gassed with N_2. Air is added in increasing quantities and a continuous O_2 gradient is created over the agar surface. Samples, e.g. dilutions from rhizosphere soils, are spread over the total agar surface and incubated under the O_2 gradient. Typical results are shown in Fig. 2.12. A root zone specific oxygen tolerance was also found in different diazotrophic bacteria from Kallar grass (Hurek *et al.*, 1987).

Figure 2.12 shows that the total colony count reaches a maximum in the anaerobic area, and a second maximum in the aerobic area. There is little difference between the two cultivars of *Sorghum*. Nitrogen fixing bacteria reach a minimum at 2.5 kPa oxygen and there is a population of aerobic nitrogen fixers only in the rhizosphere of one *Sorghum* cultivar. The oxygen content of rhizosphere soil is influenced not only by root respiration but also by bacterial respiration and the water content of the soil. One of the most sensitive indicators for this is that aerobic nitrogen fixation increases by a factor of 50 when the water content of soil is raised from 8 to 21% (Fig.

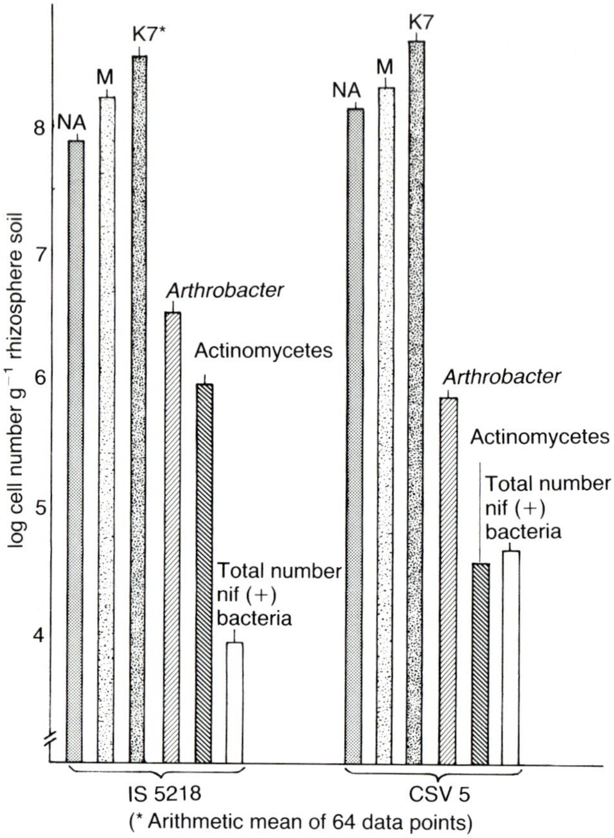

Fig. 2.10. Genotype effect on microbial populations in the rhizosphere of two cultivars of *Sorghum*. NA, M and K7, cell counts on three different complex media as compared to the number of *Arthrobacter*, actinomycetes and N_2-fixing (nif^+) bacteria. (Based on Krotzky *et al.*, 1986.)

2.13). Note that with higher fixing capacity and water-logging the standard deviation of the curves is significantly reduced. It is clear from this example that the physiochemical conditions in the soil influence the specific associations between plants and bacteria. This works both ways however; changes in the specific associations can therefore be a sensitive marker for changes in soil characteristics. So far only from forest soils and sludge of lakes, diazotrophic cellulolytic bacteria have been isolated (Leschine *et al.*, 1988). But due to the fact that larger decaying roots are a rich source of cellulose, also in older rhizospheres these bacteria may be expected.

Denitrifying bacteria in the rhizosphere are also dependent on different factors. Their number may increase by exudation of organic substrates and by higher O_2 consumption by the roots. It may decrease by improved soil

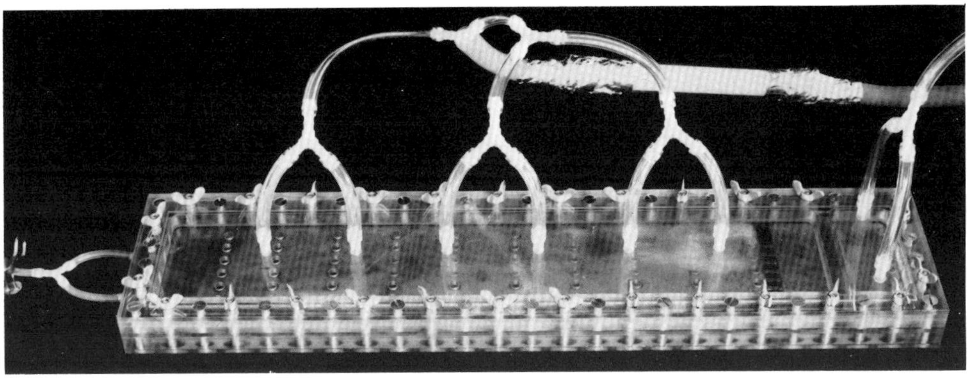

Fig. 2.11. Oxygen gradient system for simultaneous cultivation of rhizosphere bacteria in a continuous gradient of 0.00–0.04 atm pO_2. The gradient is maintained by a constant flow of oxygen-free N_2 (from right) and air (from 6 central inlets). (Original photograph A. Krotzky, Marburg.)

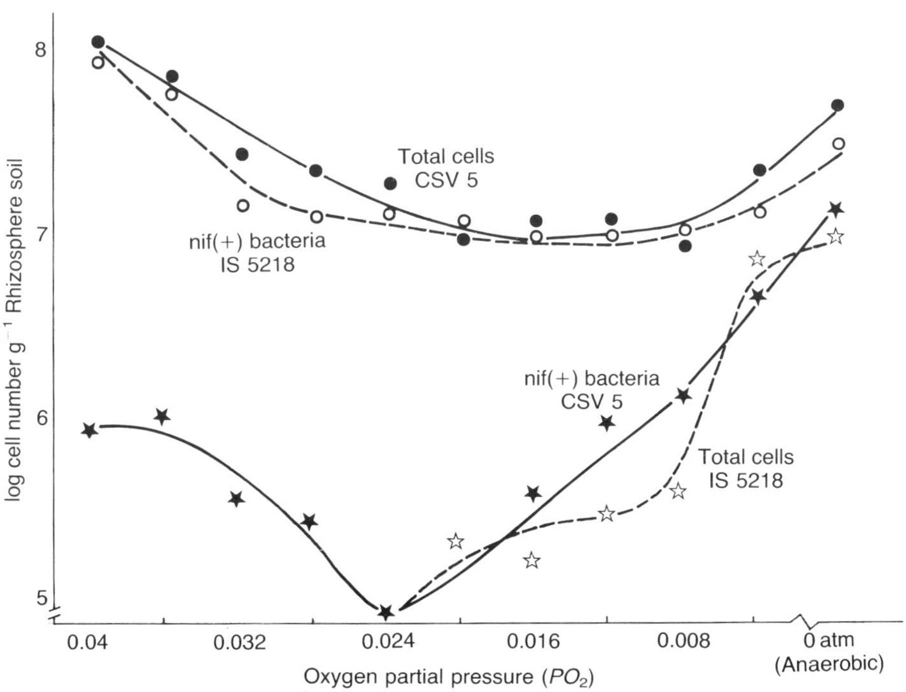

Fig. 2.12. Total cells and number of N_2-fixing bacteria from the rhizosphere of two *Sorghum* cultivars, growing at various oxygen concentrations in an O_2-gradient system. (Based on Krotzky *et al.*, 1986.)

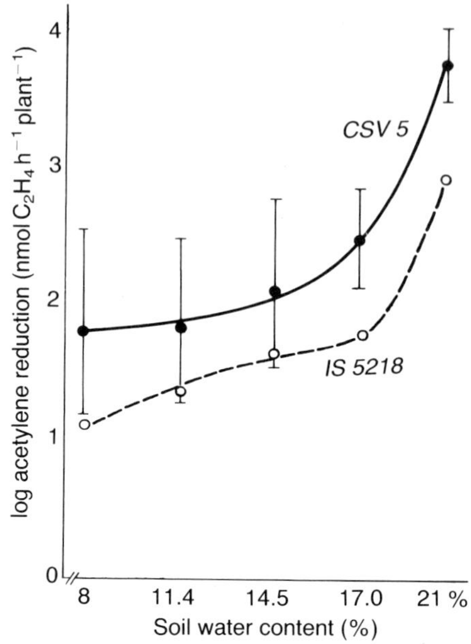

Fig. 2.13. Nitrogenase activity in two cultivars of *Sorghum* (IS 5218 and CSV 5) in relationship to soil water content. (Based on Krotzky *et al.*, 1986.)

aeration due to root growth-related higher water consumption, by nitrate uptake and pH shifts (Ottow *et al.*, 1985; Trolldenier, 1989). Also nitrification in soils can be supported by supply of organic nutrients, leading to NO_2, NO_3 and also to N_2O and NO, as demonstrated for *Alcaligenes faecalis* (Papen *et al.*, 1989).

Total number of bacteria in a certain microhabitat is also affected by other soil organisms. Protozoa, for example, can reduce the number of bacteria by a factor of 8 and increase uptake of bacterial N by the plants by 65% (Kuikman and Van Veen, 1989).

(a) Chemotaxis in the rhizosphere

Motile bacteria and the mobile stages of fungi are attracted by different chemical substances (Eisenbach, 1985; Adler, 1987). In general, there is little correlation between the affinity of an organism for a chemotactic agent and its ability to metabolize it. Amino acids and sugars are among the most widely spread chemotactic agents. In certain cases, however, motile bacteria can be attracted to nucleotides, vitamins and inorganic ions. For associations in the rhizosphere specific attractions are of interest. As a general rule, however, most types of bacteria are attracted by the components of root exudates. *Escherichia coli* is attracted by the L-amino acids alanine, asparagine, aspartic acid, cysteine, glutamic acid, glycine, methionine, serine and threonine, whereas arginine, cystine, glutamine, histidine, isoleucine, leucine, lysine, phenylalanine, tryptophane, tyrosine and valine have no effect. *Pseudomonas lachrymans* is attracted by 16 amino acids, especially arginine, aspartic acid and methionine. In the case of sugars, *E. coli* is attracted by *N*-acetylglucosamine, 6-deoxyglucose, fructose, fucose, galactose, glucose, lactose and maltose, whereas *P. lachrymans* is attracted by ribose, arabinose and glucose. In both species chemotaxis is also found using nucleotides; as well as thymine, adenine, cytosine and guanine and the vitamins biotin and thiamine. Thus it is not surprising that *P. lachrymans* exhibits a tactic response to root exudate. This is, however, true for many micro-organisms. The optimal concentration range for many chemotactic agents is 10^{-3}–10^{-4} mol. Dilutions of 1 order of magnitude strongly decreases the effect. This means that at 2 mm from the root surface the concentrations of sugars and amino acids are too low to be effective. Thus chemotaxis can only be important in associations in the very centre of the rhizosphere.

Somewhat higher specificity appears among fungal zoospores. For example, the zoospores of *Pythium aphanidermatum* are highly attracted by glutamic acid. Active compounds for the zoospores of *Allomyces macrogynus* and *A. arbuscula* are leucine and lysine. These substances have a synergistic effect which is further strengthened by the presence of L-proline. Sugars have no effect on *Allomyces* zoospores. Zoospores from *Phytophthora* species show significant differences in the sensitivity of their zoospores, e.g. those from *P. cactorum* and *P. capsicii* are significantly more attracted by aspartic acid and glutamic acid than those from *P. citrophthora*.

Negative chemotaxis is also exhibited both by bacteria and the motile stages of fungi. This is in most cases induced by higher concentrations of ions and secondary plant products. Thus *Salmonella* has an especially strong negative chemotactic response towards phenol. *Rhodospirillum rubrum* shows negative aerotaxis (reaction against dissolved oxygen) in higher light intensities.

(b) A model for the enrichment of microbes in the rhizosphere

The percentage of the root surface of maize covered by the microflora was 4% in the region where root hairs emerge, 7% in the root hair zone and up to 20% in the oldest parts of the root (Schönwitz and Ziegler, 1989). Newman and Watson (1977) proposed a model to calculate the root-dependent bacterial biomass at various distances from the root surface (Fig. 2.14). The biomass directly on the rhizoplane (0 mm) is by more than three orders of magnitude larger than at a distance of 0.3 mm. Experimental evidence supports this model except that it appears that competition for the root-derived nutrient between bacteria is a more important

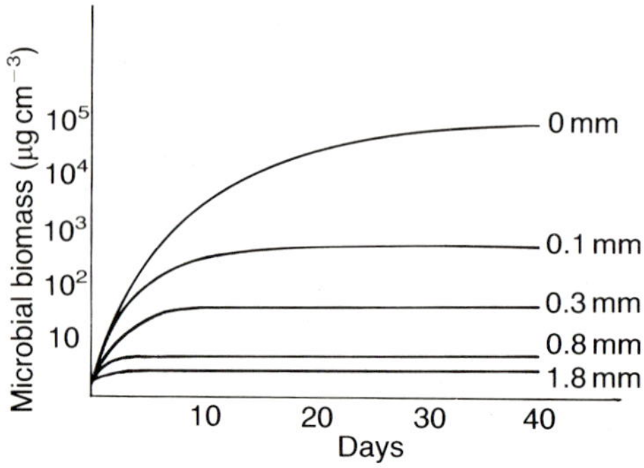

Fig. 2.14. Microbial biomass at various distances from the root surface (0–1.8 mm), during 40 days from onset of exudation 1 μg of microbial biomass is equivalent to approx. 5×10^6 bacteria. (Based on Newman and Watson, 1977)

factor than was previously believed. This was not previously taken into account because the technical difficulties of the experiment were so great that few have been performed. For example the concentrations of nutrients within millimetres of the root is in the nanomolar range. As with mineral nutrients, account has to be taken of the fact that with carbon sources there is a large difference between total concentration and available concentration (i.e. the amount which plants and microorganisms can use). For example, the total concentration of alanine was found to be $800\,\text{nmol}\,\text{l}^{-1}$, whereas the available concentration was about 10 nmol. The microbial rhizosphere population is also involved in the kinetics of degradation and availability of agricultural pesticide residues, leaving only about an average of 20% after 20 months in the soil, almost all of it adsorbed or incorporated into soil components (Führ *et al.*, 1989).

(c) Antagonistic organisms

More than 40 different genera of bacteria and fungi have been used as antagonists against pathogens (Cook and Baker, 1983). However, only a small number of examples and inoculation methods have so far been reliable. One example is the direct application of *Agrobacterium radiobacter* to roots to control crown gall disease formation by *Agrobacterium tumefaciens* (Kerr, 1980). Seeds treated with *Trichoderma* were protected against *Pythium* infection, but only when the seeds were allowed to germinate for 12–24 h to allow the *Trichoderma* to proliferate before the fast-growing *Pythium* started to infect the plant tissue (Chet, 1987).

Of special interest is the antagonistic effect of fluorescent *Pseudomonas* strains against *Gaeumannomyces graminis*, causing the 'take all' or root and foot disease in cereals such as wheat or barley. There are no fungicides available against this disease. The fungus is dispersed in soils and host debris and increases during a 3–6 year continuous cereal cultivation. Application of fluorescent *Pseudomonas* strains, selected *in vitro* against *Gaeumannomyces graminis*, reduced the infestation indices on plants by 42% after 68 days and by 16% after 125 days (Zaspel, 1989). An optimal supply of Mn^{2+} and Cu^{2+} can also reduce the infestation with *Gaeumannomyces* by increasing the resistance (Sieling, 1990). Antagonistic effects have also been observed between fungi and seed pathogenic fungi; however only a small number of the more than 400 seed-borne pathogens have been studied (Agarwal and Sinclair, 1987). A negative correlation has been found in infection of soybean seed by *Cercospora kikuchii* and infection by *Alternaria*, *Fusarium* and *Phomopsis*. A positive correlation was observed between *Phomopsis* and *Fusarium* sp. seed infection (McGee *et al.*, 1980).

2.1.4 The effect of rhizosphere micro-organisms on root morphology and the phosphate content of roots and shoots

The relationship between plants and bacteria is not one-sided. Vitamins are released from roots to quite a significant extent (up to 400 ng per plant), but the growth of vitamin-requiring micro-organisms was not stimulated compared to the soil population (Schönwitz and Ziegler, 1989). Vitamin-producing bacteria are widespread in the rhizosphere, also with many not N_2-fixing species. Under optimal laboratory conditions, bacterial cultures in a concentration equal to that in the rhizosphere produce phytohormones to

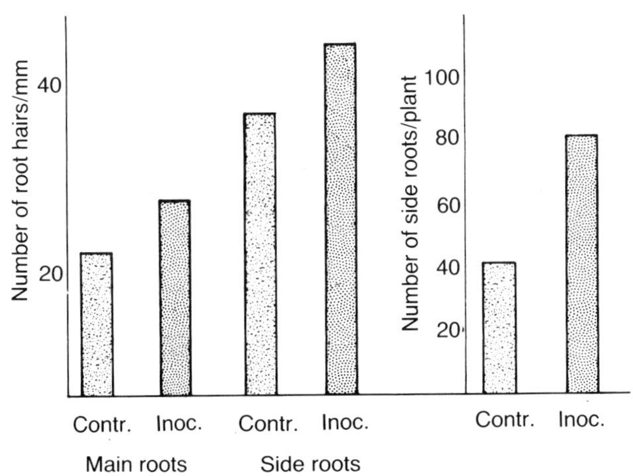

Fig. 2.15. Number of root hairs on main and side roots of silver beet, after inoculation with *Azospirillum brasiliense* (controls without bacterial inoculation). (Based on Martin *et al.*, 1989.)

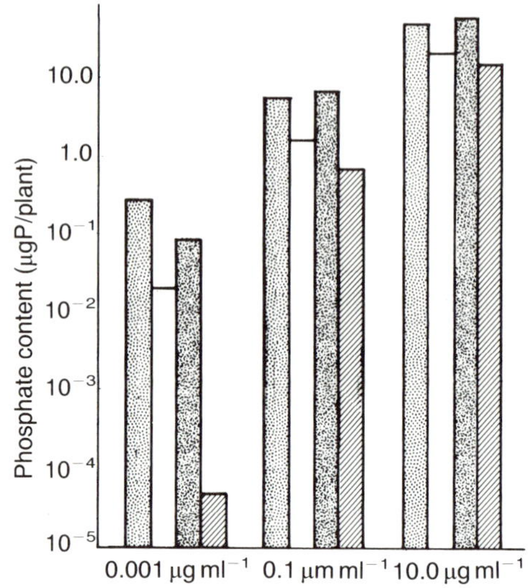

Fig. 2.16. Effect of micro-organisms on the phosphate content in barley, at three phosphate concentrations in the culture solution. Root, sterile (▨), plus micro-organisms (□). Shoot, sterile (▨), plus micro-organisms (▨). (Based on Barber and Loughman, 1967.)

a level comparable to concentrations around field-grown maize roots (Müller *et al.*, 1989). In experiments with young silver beet plants inoculated with *Azospirillum brasilense*, the number of root hairs per millimetre of root on both tap and secondary roots increased by 25%. This effect was even more noticeable on the number of side roots (Fig. 2.15). The most probable explanation for this effect is that *Azospirillum* excretes phytohormones, which then influence root growth. Not only plant morphology but also cytology and metabolism is influenced by rhizosphere micro-organisms. In barley, the phosphate content of the shoot is increased under phosphate-poor conditions (0.001 mg ml^{-1} soil solution) in the presence of rhizosphere micro-organisms (Fig. 2.16). At higher concentrations (100-fold increase) the increase of P in the shoot is still distinct, but effects on the roots are no longer discernible. Thus, rhizosphere microflora can exert a considerable effect on the plant and such effects are more evident in symbioses.

2.2 PHYLLOSPHERE

Our knowledge about the qualitative composition of the microflora on leaves is even more limited than that about the rhizosphere population (Fokkema and van den Heuvel, 1986). The various reports about genera found on leaves in various climatic zones repeatedly give similar results, but it must be remembered, however, that in all cases many cells were not named due to difficulties in identification. The most often occurring bacteria belong to the genera *Pseudomonas*, *Xanthomonas*, *Flavobacterium*, *Achromobacter*, *Aerobacter*, *Bacillus*, *Mycobacterium*, *Beijerinckia*, *Azotobacter* and *Mycoplana*. Yeast genera are *Rhodotorula*, *Cryptococcus* and *Torulopsis*. Among the first colonizers on leaf surfaces are the fungi *Aureobasidium pullulans*, *Cladosporium herbarum* and *Epicoccum nigrum*. Some fungi, such as *Sporobolomyces*, restrict themselves mostly to the upper surface of leaves, whereas others, such as *Cladosporum herbarum* are found on the underside (Pugh and Buckley, 1971). Older leaves have many cyanobacteria, mainly from the genera *Anabaena*, *Calothrix*, *Nostoc*, *Oscillatoria* and *Scytonema*. Interestingly, among chemo-organotrophic bacteria are many yellow-pigmented strains. It is assumed that this pigmentation acts to protect leaf-surface bacteria against u.v. radiation. Some older works imply that strains isolated from the phyllosphere are not able to compete in soil with soil bacteria, but this remains to be confirmed.

Between 10^3 and 10^7 heterotrophic bacteria and yeast cells per gram fresh weight are found on leaves and buds of plants under field conditions. Especially high values have been found on leaves of soybean (Table 2.4). Upper leaves were found to have 10–100 times more bacteria than older lower leaves. This may be an effect of more dust being caught by higher leaves, and also of the higher rate of leaf exudation by the younger tissue. In greenhouses plants had 3–4 orders of magnitude lower cell counts on

Table 2.4 Associations of micro-organisms with the aerial parts of seed plants (phyllosphere). (Based on Leben, 1972)

Plant	*Heterotrophic bacteria g^{-1} fresh weight*	*Yeast cells g^{-1} fresh weight*
Apple buds	3.6×10^2	1.3×10^3
Poplar buds	6.1×10^3	2.0×10^4
Pine buds	2.2×10^4	1.8×10^4
Soybean		
terminal buds	2.0×10^7	
youngest leaf	1.5×10^7	
second leaf	3.0×10^6	
older leaves	$0.03–1.4 \times 10^6$	
flowers	5.0×10^6	
young pods	5.4×10^7	
Soybean (greenhouse)		
terminal buds	$0.2–5.0 \times 10^3$	
young pods	$<1.0 \times 10^2$	

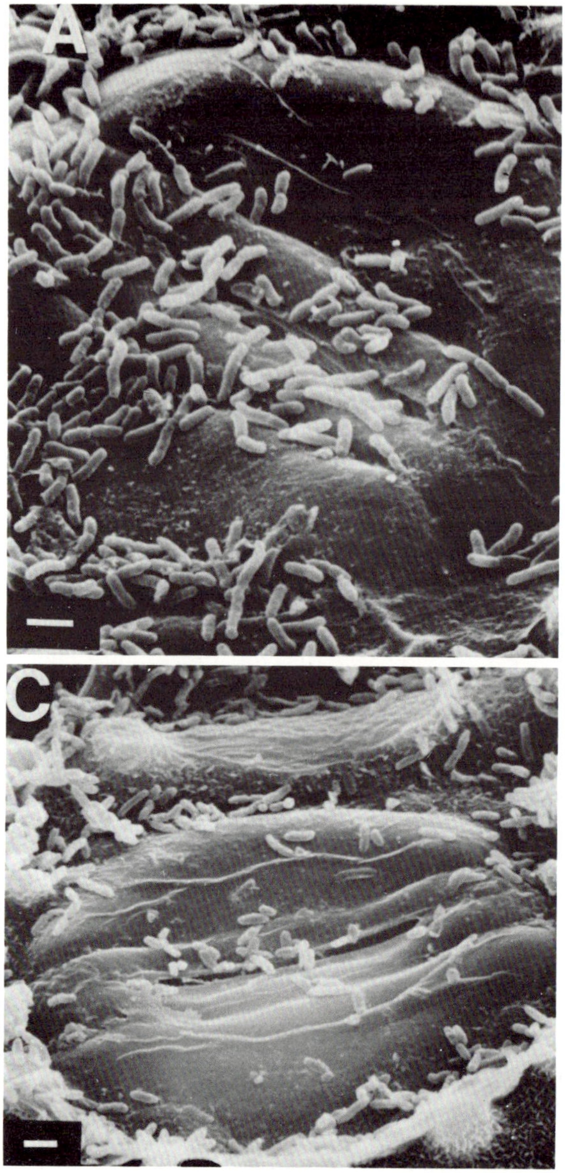

Fig. 2.17. Virulent (A) and avirulent (C) strains of *Xanthomonas campestris* on hydathodes of rice. (From Mew *et al.*, 1984.)

Table 2.5 Cell counts of *Xanthomonas campestris* pv. *oryzae* on hydathodes from rice (cv. CAS 209). This cultivar is resistant to *Xanthomonas* strains PX 086 and PX 0101, but susceptible to strain PX 061. (Based on Mew *et al.*, 1984)

Xanthomonas strain	*Xanthomonas cells per hydathode*		
	1 h	*24 h after infection*	*48 h*
PX 061	5–20	100–200	100–200
PX 086	5–20	10–50	50–80
PX 0101	5–20	5–20	10–60

Table 2.6 Effects of *Beijerinckia* and *Klebsiella* sp. on the growth of rice (cv. IR 26) in N-free sand culture. Leaves were sprayed three times and plants were analysed after 8 weeks. (Based on Nandi and Sen, 1981)

Treatment of 14-day-old rice seedlings	*Fresh weight per plant (mg)*	*Chlorophyll content ($\mu g\ mg^{-1}$ leaf tissue)*	*N-content per plant (mg)*
Control	378	1.22	1.01
+ *Beijerinckia* sp.*	1080	1.54	4.32
+ *Klebsiella oxytoca**	705	1.45	2.82

* Addition of nitrogen by bacterial inoculation: approx. 0.12 mg N per plant

terminal buds and young pods. Colonization of leaf surfaces is also influenced in some cases by virulence and avirulence of different strains (Fig. 2.17). Differences in the multiplication of *Xanthomonas campestris* strains on hydathodes of the same rice cultivar have been found (Table 2.5) 24 h after inoculation with a compatible strain the bacteria had multiplied by a factor of 10, while an incompatible strain on the resistant plant only doubled.

Within the phyllosphere certain areas can become necrotic as leaf spots, caused by phytopathogenic bacteria. Host specificity of these bacteria and **production of phytotoxins** are apparently not correlated. Also from the more than 150 phytotoxins produced by phytopathogenic fungi, only about 10% seem to be host specific. Toxins are often enzyme inhibitors. **Phaseolotoxin**, produced by *Pseudomonas syringae* bv. *phaseolicola*, is an inhibitor of ornithine-carbamyl transferase, rhizobitoxin from *Pseudomonas andropogonis* inhibits the β-cystathionase (Fritsche, 1989). Nitrogen-fixing bacteria can increase rice growth under N-limiting conditions in a way similar to mineral leaf fertilizer (Table 2.6). The fresh weight of rice plants inoculated with *Beijerinckia* or *Klebsiella oxytoca* was twice that of control plants after 14 days. After 8-weeks growth, inoculated plants had 3–4 times the nitrogen content of control plants.

2.2.1 Endophytes in leaves

Endophytic fungi in leaves of grasses belong to the Clavicipitaceae (tribe Balansieae) mainly in the genera *Atkinsonella*, *Balansia*, *Epichloe* and *Myriogenospora* (Siegel *et al.*, 1987; Siegel and Schardl, 1991). The Clavicipitales are closely related to the lichen fungi orders Ostropales and Graphidales (Chapter 9). Several hundred species of Gramineae can be hosts of these fungi (Bacon and De Battista, 1990). *Atkinsonella* sp. and *Epichloe* sp. infect C3 grasses whereas *Balansia* and *Myriogenospora* sp. are symbionts of C4 Gramineae. Asexual stages of *Epichloe* and *Atkinsonella* are included in the genus form '*Acremonium* section *Albolanosa*'.

The life cycle of endophytic fungi, such as *Epichloe typhina*, which can cause some level of disease is summarized in Fig. 2.18 together with other related endophytes. A mycelium can infect meristem cells and multiply there to an active endophytic form. Sporulation can lead to a dispersal of

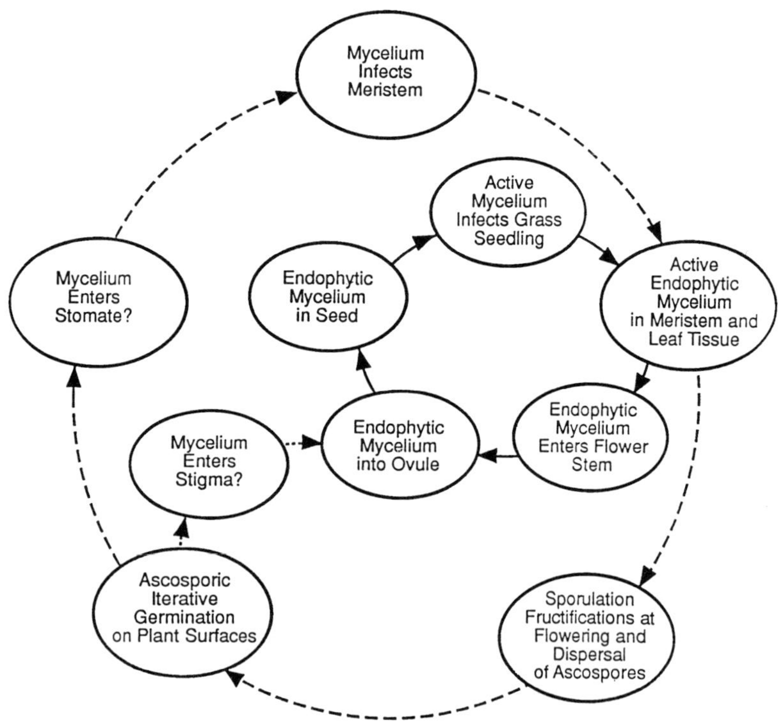

Fig. 2.18. Life cycles of endophytic fungi. (Based on Siegel and Schardl, 1991.)

ascospores, which germinate on other plant surfaces. In another cycle, mycelia enter the flower stem and the ovules, leading to a continuous infection of seeds. It is important to note that roots are not infected. An *Acremonium* type endophyte can be easily detected with an ELISA technique (Dahlman *et al.*, 1990). As with true vesicular–arbuscular-mycorrhizae (Chapter 10), the symtomless endophytes are obligate biotrophs. Many species are potential pathogens, causing, for example, a sterilization of the flower parnicles. However, even this effect can be made up for by improved vegetative growth and protective alkaloids. Endophytic fungi can produce ergot alkaloids and thereby protect their host plants (grasses) against larvae of insects such as the fall armyworm (*Spodoptera frugiperda* – Lepidoptera) (Clay, 1989). The fungi have an interesting general potential as biocontrol agents (Clay and Cheplick, 1989). Peramine, lolitrem, loline and ergovaline are important secondary metabolites produced in the associations. They have a strong antiherbivore activity, some more against insects, some against mammals (Dahlman *et al.*, 1990). In the fertilization of *Epichloe typhina*, the fly *Phorbia phrenione* is an important vector for transfering the conidia of one mating type to plants, harbouring the opposite mating type (Bultman and White, 1988).

The benefits for the macrosymbionts (the host plants) from the endophyte infection can be summarized as follows:

1. reduced damage from grazing by herbivorous mammals (producing unwanted reduced growth or fecundity of the grazing cattle)
2. reduced predation by insects
3. enhanced resistance against nematodes (Pedersen *et al.*, 1988)
4. enhanced resistance against other fungi (Koshino *et al.*, 1989)
5. production of more tillers and biomass, together with fewer inflorescences (Clay *et al.*, 1989)
6. allelopathic effects by *N*-formyl-loline (Petroski *et al.*, 1990)
7. improved recovery from drought stress (Arachevaleta *et al.*, 1989)

Several of these effects can be also observed in the vesicular–arbuscular mycorrhizal symbiosis (Chapter 10).

Endophytes from tree leaves belong to other genera of fungi. *Picea abies* harbours *Lophodermium piceae*, *Abies balsamea* is colonized by *Phyllosticta multicorniculata* and in *Fagus silvatica* leaves *Apiognomonia errabunda* has been identified together with *Diaporthe eres* and *Bisporella* sp. (Petrini, 1991). *Holmiella sabina* and its *Cornicularia*-like anamorph was found in *Juniperus* sp. (Petrini *et al.*, 1979). When needles of conifers age, the community of the endophytic fungi is gradually replaced by saprophytic species, characteristic also for the litter in the soil. This has been demonstrated for needles from the European white fir (Canavesi, 1987) and for *Sequoia sempervirens* (Espinosa-Garcia and Langenheim, 1990).

Table 2.7 Succession of fungi in relationship to the moisture content of cereal caryopses and seeds (Based on Christensen and Kaufmann, 1974)

Fungal species	*Lower moisture limit for fungal growth (moisture content as % of total fresh weight)*		
	Wheat grains	Sorghum *grains*	*Soybean seeds*
Aspergillus restrictus	13.5–14.5	14.0–14.5	12.0–12.5
Aspergillus glaucus	14.0–14.5	14.5–15.0	12.5–13.0
Aspergillus candidus and *A. ochraceus*	15.0–15.5	16.0–16.5	14.5–15.0
Aspergillus flavus	18.0–18.5	19.0–19.5	17.0–17.5
Alternaria sp., *Cladosporium* sp., *Fusarium* sp., *Helmithosporium* sp.	22.0–23.0	23.0–24.0	

2.3 SPERMOSPHERE

Populations of bacteria and fungi associated with seeds of wild plants have hardly been researched. A few types and genera have been reported from cultivated plants and crops (Table 2.7). On cereals, *Aspergillus glaucus* is particularly widespread, being found on over 80% of maize samples and 40% of wheat samples. In this research 8000 samples, grown under a variety of conditions, were studied. *Aspergillus flavus* was found on 35% of the maize samples but never on wheat. In addition, many other fungi were found on 70% of both maize and wheat; these belonged to the genera *Alternaria, Cladosporium, Fusarium* and *Helminthosporium*. Fungi do not grow on cereals when the water content drops below 11–12%. Stored cereals normally contain about 15% water. Thus temperature, which at this water content has a large effect on fungal growth, plays an important part in preserving stored cereals.

2.4 PHYCOSPHERE

The phycosphere is defined as that volume of water around photosynthetically active algae which is physiologically changed by that alga. As in the rhizosphere, in the phycosphere there is a difference by several orders of magnitude between the viable cell count and the number of particles found. At the same time, however, around 30% of the cells found in the phycosphere are metabolically active. It is thus apparent that the culture conditions used to estimate viable cell numbers are in many cases not specific

Table 2.8 Associations of algae and cyanobacteria with heterotrophic bacteria. (References given in Werner, 1987)

Alga/Cyanobacterium	*Heterotrophic bacterium*	*Type of relationship*
Chlamydomonas reinhardii	*Pseudomonas* sp.	Enhanced growth of the alga
Haemotococcus lacustris	*Flavobacterium* sp.	Enhanced growth of the alga
Skeletonema costatum	*Pseudomonas* sp.	Specific uptake up extracellular products in co-culture, $>10^7$ bacteria ml^{-1}
Acrochaetium sp.	*Arthrobacter globiformis*, *A. pascens Corynebacterium fascians*, *Mycoplana bullata*	Association morphologically very close
Anabaena sp.	*Zoogloea* sp.	Bacterial enrichment in *Anabaena* slime by a factor of 10^5
Fischerella sp.	*Legionella* sp.	Specific growth enhancement of *Legionella*

enough to cover the nutritional conditions needed for phycosphere bacteria. It is proposed that part of the carbon and energy sources for chemoorganotrophic bacteria in the phycosphere does not come from exudates but from dying algal cells. Total bacterial cell count follows closely after the maximum of phytoplankton increase (Cole, 1982). Of bacteria isolated from phytoplankton 70% belong to the genera *Aeromonas* and *Vibrio*. Also present are representatives of *Pseudomonas*, *Flavobacterium* and *Achromobacter*. The relationship between the heterotrophic bacteria and the algae or cyanobacteria could be that they help algal growth. This has been proven for *Chlamydomonas reinhardii* and *Pseudomonas* spp and for *Hematococcus lacustris* and flavobacteria (Table 2.8). In the other direction extracellular products from the diatom *Skeletonema costatum* help *Pseudomonas* cells in co-culture to reach levels 10^7 bacterial cells per ml higher than in single culture. In *Anabaena* slime an enrichment of *Zoogloea* has been found, 10^5 times higher than in the surrounding medium. Specific help of algal growth by associated bacteria is often caused by the increased availability of vitamins. Especially important for many phytoplankton algae is the availability of vitamin B_{12}.

2.5 REFERENCES

Adler, J. (1987) How motile bacteria are attracted and repelled by chemicals. An approach to neurobiology. *Biol. Chem. Hoppe-Seyler*, **368**, 163–73.

Agarwal, V.K. and Sinclair, J.B. (1987) *Principles of Seed Pathology*, Vol. I, CRC Press, Boca Raton.

Anderson, J.P.E. and Domsch, K.H. (1973) Quantification of bacterial and fungal contributions to soil respiration. *Arch. Microbiol.*, **93**, 113–27.

Anderson, T.-H. and Domsch, K.H. (1989) Ratios of microbial biomass carbon to total organic carbon in arable soils. *Soil Biol. Biochem.*, **21**, 471–9.

Andrews, J.H. (1984) Relevance of r- and k-theory to the ecology of plant pathogens, in *Current Perspectives in Microbial Ecology* (eds M.J. Klug and C.A. Reddy) American Society of Microbiology, Washington, DC.

Arachevaleta, M., Bacon, W.C., Hoveland, C.S. and Radcliffe, D.E. (1989) Effect of the tall fescue endophyte on plant response to environmental stress. *Agr. J.*, **81**, 83–90.

Bacon, C.W. and De Battista, J.D. (1990) Endophytic fungi of grasses, in *Handbook of Applied Mycology*, Vol. 1. *Soils and Plants* (ed. D.K. Arora) Marcel Dekker, New York, pp. 231–44.

Barber, D.A. and Loughman, B.C. (1967) The effect of micro-organisms on the absorption of inorganic nutrients by intact plants. *J. Exp. Bot.*, **18**, 170–6.

Barber, D.A. and Gunn, K.B. (1974) The effect of mechanical forces on the exudation of organic substances by the roots of cereal plants grown under sterile conditions. *New Phytol.*, **73**, 39–45.

von Berkum, P. and Bohlool, B.B. (1980) Evaluation of nitrogen fixation by bacteria with roots in tropical grasses. *Microbiol. Rev.*, **44**, 491–517.

Bitton, G. and Marshall, K.C. (1980) *Adsorption of Microorganisms to Surfaces*. Wiley, New York.

Brouwer, R., Gasparikova, O., Kolek, J. and Loughman, B.C. (1981) *Structure and Function of Plant Roots*. Nijhoff/Junk, The Hague.

Brown, M. (1981) Rhizosphere microorganisms – opportunists, bandits or benefactors, in *Soil Microbiology* (ed. N. Walker) Butterworths, London.

Brussaard, L., Bouwman, L.A., Geurs, M. *et al.* (1990) Biomass, composition and temporal dynamics of soil organisms of a silt loam soil under conventional and integrated management. *Netherlands J. Agr. Sci.*, **38**, 283–302.

Bultman, T.L. and White, J.F. (1988) 'Pollination' of a fungus by a fly. *Oecologia*, **75**, 317–19.

Canavesi, F. (1987) Beziehungen zwischen endophytischen Pilzen von *Abies alba* Mill. und den Pilzen der Nadelstreue. Dissertation ETH Nr. 8325, Swiss Federal Institute of Technology, Zürich.

Chet, I. (1987) *Trichoderma* application, mode of action, and potential as a biocontrol agent of soilborne plant pathogenic fungi, in *Innovative Approaches to Plant Disease Control* (ed. I. Chet) Wiley, New York.

Chet, I. and Mitchell, R. (1976) Ecological aspects of microbial behavior. *Ann. Rev. Microbiol.*, **30**, 221–39.

Christensen, C.M. and Kaufmann, H.H. (1974) in *Storage of Cereal Grains and their*

Products, 158, American Association of Cereal Chemists, St Paul, MN.

Clay, K. (1989) Clavicipitaceous endophytes of grasses: their potential as biocontrol agents. *Mycol. Res.*, **92**, 1–12.

Clay, K. and Cheplick, G.P. (1989) Effect of ergot alkaloids from fungal endophyte-infected grasses on fall armyworm (*Spodoptera frugiperda*). *J. Chem. Ecol.*, **15**, 169–82.

Clay, K., Cheplick, G.P. and Marks, S. (1989) Impact of the fungus *Balansia henningsiana* on *Panicum agrostoides*: frequency of infection, plant growth and reproduction and resistance to pests. *Oecologia*, **80**, 374–80.

Cole, J.J. (1982) Interactions between bacteria and algae in aquatic ecosystems. *Ann. Rev. Ecol. Syst.*, **13**, 291–314.

Cook, R.J. and Baker, K.F. (1983) *The Nature and Practice of Biological Control of Plant Pathogens*. American Phytopathological Society, St Paul, MN.

Dahlman, D.L., Eichenseer, H. and Siegel, M.R. (1990) Chemical perspectives on endophyte–grass interaction and their implications to insect herbivory, in *Multitrophic Level Interactions Among Microorganisms, Plants and Insects* (eds P. Barbosa, L. Kirschik and E. Jones) Wiley, New York.

Dewey, F.M. (1988) Development of immunological diagnostic assays for fungal plant pathogens. *Brighton Crop Protection Conference – Pests and Diseases*, **2**, 777–86.

Eisenbach, M. (1985) in *Sensing and Response in Microorganisms* (eds M. Eisenbach and M. Balaban) Elsevier, Amsterdam, pp. 287–302.

Espinosa-Garcia, F.J. and Langenheim, J.H. (1990) The leaf fungal endophytic community of a coastal redwood population. Diversity and spatial patterns. *New Phytol.*, **116**, 89–97.

Fischer, W.R., Flessa, H. and Schaller, G. (1989) pH values and redox potentials in microsites of the rhizosphere. *J. Plant Nutr. Soil Sci.*, **152**, 191–5.

Fokkema, N.J. and van den Heuvel, J. (eds) (1986) *Microbiology of the Phyllosphere*. Cambridge University Press, Cambridge, London, New York.

Fredrickson, J.K., Bezdicek, D.F., Brockman, F.J. and Li, S.W. (1988) Enumeration of Tn5 mutant bacteria in soil using a most-probable-number DNA hybridization procedure and antibiotic resistance. *Appl. Environ. Microbiol.*, **54**, 446–53.

Fritsche, W. (1989) Microbial ecology of the phyllosphere: A source of microorganisms producing bioactive compounds, in *Bioactive Metabolites from Microorganisms. Progress in Industrial Microbiology*, **27**, 39–50.

Führ, F., Steffens, W., Mittelstaedt, W. and Brumhard, B. (1989) Pflanzenschutzmittel: Gift in Boden und Grundwasser? Jahresbericht der Kernforschungsanlage Jülich GmbH.

Haller, Th. and Stolp, H. (1985) Quantitative estimation of root exudation of maize plants. *Plant Soil*, **86**, 207–16.

Harman, G.E. and Lumsden, R.D. (1990) Biological disease control, in *The Rhizosphere*, (ed. J.M. Lynch) Wiley, Chichester, New York, pp. 259–80.

Helal, H.M. and Sauerbeck, D.R. (1984) Influence of plant roots on C and P metabolism in soil. *Plant Soil*, **76**, 175–82.

Helal, H.M. and Sauerbeck, D. (1989) Carbon turnover in the rhizosphere. *Z. Pflanzenernähr. Bodenk.*, **152**, 211–16.

Hurek, T., Reinhold, B., Fendrik, I. and Niemann, E.-G. (1987) Root-zone-specific oxygen tolerance of *Azospirillum* spp. and diazotrophic rods closely associated with Kallar grass. *Appl. Environ. Microbiol.*, **53**, 163–9.

Isermann, K. (1983) Bewertung natürlicher und anthropogener Stoffeinträge über die Atmosphäre als Standortfaktoren im Hinblick auf die Versauerung land- und forstwirtschaftlich genutzter Böden. *VDI-Berichte*, **500**, 307–35.

Kerr, A. (1980) Biological control of crown gall through production of agrocin 84. *Plant Dis.*, **64**, 24–30.

Kleeberger, A., Castorph, H. and Klingmüller, W. (1983) The rhizosphere microflora of wheat and barley with special reference to Gram-negative bacteria. *Arch. Microbiol.*, **136**, 306–11.

Klingmüller, W. (1985) Azospirillum III. *Genetics, Physiology, Ecology*, Springer Verlag, Berlin.

Koshino, H., Yoshihara, T., Sakamura, S. *et al.* (1989) A ring B aromatic sterol from stromata of *Epichloe typhina. Phytochemistry*, **28**, 771–2.

Krotzky, A., Berggold, R. and Werner, D. (1986) Analysis of factors limiting associative N_2-fixation (C_2H_2 reduction) with two cultivars of *Sorghum nutans. Soil Biol. Biochem.*, **18**, 201–7.

Kuikman, P.J. and Van Veen, J.A. (1989) The impact of protozoa on the availability of bacterial nitrogen to plants. *Biol. Fertil. Soils*, **8**, 13–18.

Leben, C. (1972) Micro-organisms associated with plant buds. *J. Gen. Microbiol.*, **71**, 327–31.

Leschine, S.B., Holwell, K. and Canale-Parola, E. (1988) Nitrogen fixation by anaerobic cellulolytic bacteria. *Science*, **242**, 1157–8.

Lynch, J.M. (1984) Interactions between biological processes, cultivation and soil structure. *Plant Soil*, **76**, 307–18.

Lynch, J.M. (1990) Introduction: some consequences of microbial rhizosphere competence for plant and soil, in *The Rhizosphere* (ed. J.M. Lynch) Wiley, Chichester, New York, pp. 1–10.

Maire, N. (1987) Evaluation de la vie microbienne dans les sols par un système d'analyses biochimiques standardisé. *Soil Biol. Biochem.*, **19**, 491–500.

Marschner, H. (1985) Nährstoffdynamik in der Rhizosphäre – eine Übersicht. *Ber. Dtsch. Bot. Ges.*, **98**, 291–309.

Marschner, H. (1986) *The Mineral Nutrition of Higher Plants*. Academic Press, New York.

Marschner, H., Treeby, M. and Römheld, V. (1989) Role of root-induced changes in the rhizosphere for iron acquisition in higher plants. *Z. Pflanzenernähr. Bodenk.*, **152**, 197–204.

Martin, P., Glatzle, A., Kolb, W. *et al.* (1989) N_2-fixing bacteria in the rhizosphere: Quantification and hormonal effects on root development. *J. Plant Nutr. Soil Sci.*, **152**, 237–45.

McGee, D.C., Brandt, C.L. and Burris, J.S. (1980) Seed mycoflora of soybeans relative to fungal interactions, seedling emergence, and carryover of pathogens to subsequent crops. *Phytopathology*, **70**, 615.

Mengel, K. and Kirkby, E.A. (1982) *Principles of Plant Nutrition*, 3rd edn, International Potash Institute, Bern.

Mengel, K. and Steffens, D. (1982) Beziehung zwischen Kationen/Anionen-Aufnahme von Rotklee und Protonenabscheidung der Wurzeln. *Z. Pflanzenernähr. Bodenkd.*, **145**, 229–36.

Mew, T.W., I-pin, C. and Huang, J.S. (1984) Scanning electron microscopy of virulent and avirulent strains of *Xanthomonas campestris* pv. *oryzae* on rice leaves. *Phytopathology*, **74**, 635–41.

Müller, M., Deigele, C. and Ziegler, H. (1989) Hormonal interactions in the rhizosphere of maize (*Zea mays* L.) and their effects on plant development. *Z. Pflanzenernähr. Bodenk.*, **152**, 247–54.

Nandi, A.S. and Sen, S.P. (1981) Utility of some nitrogen-fixing microorganisms in the phyllosphere of crop plants. *Plant Soil*, **63**, 465–76.

Newman, E.I. and Watson, A. (1977) Microbial abundance in the rhizosphere: a computer model. *Plant Soil*, **48**, 17–56.

Nye, P.H. (1981) Changes of pH across the rhizosphere induced by roots. *Plant and Soil*, **61**, 7–26.

Ottow, J.C.G., Burth-Gebauer, L. and El-Demerdash, M.E. (1985) Influence of pH and partial oxygen pressure on the N_2O to N_2 ratio of denitrification, in *Denitrification in the Nitrogen Cycle*, (ed. H.L. Golterman) NATO Conference Series, Plenum, New York, London, pp. 101–20.

Papen, H., von Berg, R., Hinkel, I. *et al.* (1989) Heterotrophic nitrification by *Alcaligenes faecalis*: NO_2^-, NO_3^-, N_2O, and NO production in exponentially growing cultures. *Appl. Environ. Microbiol.*, **55**, 2068–72.

Paul, E.A. and Clark, F.E. (1989) *Soil Microbiology and Biochemistry*, Academic Press, London, Sydney, Tokyo, Toronto.

Pedersen, J.F., Rodriguez-Kabana, R. and Shelby, R.A. (1988) Ryegrass cultivars and endophyte in tall fescue affect nematodes in grass and succeeding soybean. *Agr. J.*, **80**, 811–14.

Petrini, O. (1991) Fungal epiphytes and endophytes of tree leaves, in *Microbial Ecology of Leaves* (eds J.H. Andrews and S.S. Hirano) Springer-Verlag, New York.

Petrini, O., Samuels, G.J. and Müller, E. (1979) *Holmiella sabina* (de Not.) comb. nov. (syn. *Eutryblidiella sabina*) and its *Cornicularia*-like anamorph, an endophyte of *Juniperus* species. *Rep. Swiss Bot. Soc.*, **89**, 80–91.

Petroski, R.J., Dornbos, D.L. Jr and Powell, R.G. (1990) Germination and growth inhibition of annual ryegrass (*Lolium multiflorum* L.) and alfalfa (*Medicago sativa* L.) by loline alkaloids and synthetic *N*-acylloline derivatives. *J. Agr. Food Chem.*, **38**, 1716–18.

Pugh, G.J.F. and Buckley, N.G. (1971) The leaf surface as a substrate for colonization by fungi, in *Ecology of Leaf Surface Microorganisms* (eds T.F. Preece and C.H. Dickinson) Academic Press, London, pp. 431–45.

Rattray, E.S.A., J.I. Prosser, K. Killham and L.A. Glover (1990) Luminescence-based nonextractive technique for *in situ* detection of *Escherichia coli* in soil. *Appl. Environ. Microbiol.* **56**, 3368–74.

Schönwitz, R. and Ziegler, H. (1989) Interaction of maize roots and rhizosphere microorganisms. *Z. Pflanzenernähr. Bodenk.*, **152**, 217–22.

Siegel, M.R., Latch, G.C.M. and Johnson, M.C. (1987) Fungal endophytes of grasses. *Ann. Rev. Phytopathol.*, **25**, 293–315.

Siegel, M.R. and Schardl, C.L. (1991) Fungal endophytes of grasses. Detrimental and beneficial associations, in *Microbial Ecology of Leaves* (eds J.H. Andrews and S.S. Hirano), Springer-Verlag, New York.

Sieling, K. (1990) Auftreten und Schadwirkung bodenbürtiger Krankheiten in Abhängigkeit von der Mineraldüngung. *KALI-BRIEFE* **20** (Büntehof), 65–76.

Trolldenier, G. (1989) Plant nutritional and soil factors in relation to microbial activity in the rhizosphere, with particular emphasis on denitrification. *Z. Pflanzenernähr. Bodenk.*, **152**, 223–30.

Ulrich, B. (1987) Stability, elasticity, and resilience of terrestrial ecosystems with respect to matter balance, in *Ecological Studies* Vol. 61, (eds E.D. Schulze and H. Zwölfer) Springer-Verlag, Berlin, Heidelberg, pp. 11–49.

Werner, D. (1987) Pflanzliche und mikrobielle Symbiosen, G. Thieme Verlag, Stuttgart, New York, 241 pp.

Werner, D., Berggold, R., Jaeger, D. *et al.* (1989) Plant microbial and soil factors, determining nitrogen fixation in the rhizosphere. *Z. Pflanzenernähr. Bodenk.*, **152**, 231–6.

Zaspel, I. (1989) Isolierung und Selektion fluoreszierender *Pseudomonas*-Arten als Antagonisten gegen *Gaeumannomyces graminis* (Sacc.) Arx et Olivier. *Arch. Phytopathol. Pflanzenschutz, Berlin*, **25**, 123–30.

3
The *Rhizobium*/ *Bradyrhizobium*–Fabales Symbiosis

3.1 THE MICROSYMBIONT: *RHIZOBIUM*, *BRADYRHIZOBIUM*, *SINORHIZOBIUM* AND *AZORHIZOBIUM*

Soil bacteria belonging to the family Rhizobiaceae are Gram-negative, chemo-organotrophic or chemolithotrophic and obligate aerobes. The rod-shaped cells are not spore-forming and are motile with either one polar or subpolar flagella or between two and six peritrichous flagellae. The G+C content (mol %) of their DNA is between 57 and 65 (T_m). This places them taxonomically near the pseudomonads and azotobacteria.

3.1.1 Taxonomy and systematics

Until 1984 all rhizobia strains were collected in one genus separated according to their infective capacities on legumes. The many well-known differences between the so-called 'fast-growers' and 'slow growers' led them to be classified as two different genera in the 1984 edition of Bergeys Manual of Systematic Bacteriology. Thus there are currently four genera in the family Rhizobiaceae.

1. Genus *Rhizobium* G+C DNA: 59–64 mol %
2. Genus *Bradyrhizobium* G+C DNA: 61–65 mol %
3. Genus *Agrobacterium* G+C DNA: 57–63 mol %
4. Genus *Phyllobacterium* G+C DNA: 60–61 mol %

In addition three new genera have been proposed:

5. Genus *Azorhizobium* (Dreyfus *et al.*, 1988)
6. Genus *Sinorhizobium* (Chen *et al.*, 1988)
7. Genus *Photorhizobium* (Eaglesham *et al.*, 1990). This genus needs further characterization.

Table 3.1 Host range for nodule induction by members of the genera *Rhizobium* and *Bradyrhizobium*. (Based on Jordan (1984) Evans *et al*. (1987) and Newton)

Host plant	*Rhizobium*					*Sinorhizobium fredii*	*Rhizobium* sp. (Galega)	*Bradyrhizobium*		
	leguminosarum			*meliloti*	*loti*			japonicum	*sp.* (Vigna)	*sp.* (Lupinus)
	biovar viciae	*biovar trifolii*	*biovar phaseoli*							
Pisum sativum, *Vicia sativa*, *Vicia hirsuta*	+	+−	+−	−	−			−	−	−
Phaseolus vulgaris	+−−	+−−	+	−	+−			+−	+−	+−−
Trifolium repens	+−	+	+−−	−	−			−	−	−
Lotus corniculatus	−	−	−	−	+			−	−	−
Medicago sativa	−	−	−	+	+−−			−	−	−
Macroptilium atropurpureum	−	−	+−	−	+−			+−	+−	+−−
Glycine max	−	−	−	−	−	+−		+	+−	+−−
Glycine soya★	−	−	−	−	−	+		+−	+−	+−−
Vigna sinensis	−	−	−	−	−			+−	+−	+−−
Ornithopus sativus	−	−	−	−	−			−	−	+
Galega sp.	−	−	−	−	−		+	−	−	−

★ Wildtype of soybean.
+ More than 90% of isolates nodulate with effective (fix^+) nodules.
+− 20–90% of strains nodulate, but often ineffectively (fix^-).
+−− Less than 20% of strains nodulate with mostly ineffective nodules.
− No nodulation.

Taxonomic distinctions within genera are still primarily based on the host plant infection spectrum. This has meant that some species which were earlier divided are now united (e.g. *Rhizobium leguminosarum*) and completely new species have been recognized (e.g. *Rhizobium loti* and *Sinorhizobium fredii*). The nodulation characteristics of *Rhizobium* and *Bradyrhizobium* are summarized in Table 3.1. Here nodulation means the morphologically recognizable formation of nodules on host plants. Nodulation can be highly host specific in the case of *R. meliloti* or relatively unspecific in the case of the strains belonging to the species *Bradyrhizobium japonicum*. Apart from soybeans, these other strains also nodulate a significant proportion of *Phaseolus vulgaris*, *Vigna sinensis* and *Ornithopus sativus* cultivars. *R. trifolii* and *R. phaseoli* are now classified as biovarieties (races) within *R. leguminosarum*. It is also probable that further taxonomic work will give rise to subclassification and races within *B. japonicum*. The new types of *R. loti* and the *Rhizobium* strains isolated from *Galega* are clearly different from other rhizobia.

The previous *Rhizobium leguminosarum* bv. *phaseoli* subgroup II has been separated as a new species, called ***Rhizobium tropici*** (Martinez-Romero *et al.*, 1991). Strains of this new species tolerate temperatures up to 40°C and high levels of acidity. They nodulate besides *Phaseolus* sp. also *Leucaena leucocephala* and *Leucaena esculenta*. **DNA homology** with other *Rhizobium* species was **less than 30%**.

Hybridization experiments between DNA from various Rhizobiaceae and the 23S rRNA from *R. meliloti* show clear clustering into single infection groups (Fig. 3.1). Relatively close to each other are the clusters from *R. meliloti*, *S. fredii* and R. *leguminosarum*. Somewhat further apart are strains of *R. loti*. The distance to *B. japonicum* is even greater. This was one of the most important factors in the separation of the genus *Rhizobium* from *Bradyrhizobium*. These results were confirmed when DNA from various strains of *B. japonicum* and *Rhizobium* spp. were hybridized with 23S rRNA from *B. japonicum* (Fig. 3.2). The *Bradyrhizobium* strains are clustered together and *R. loti* and *R. meliloti* somewhat further removed. The meaning of the Sym-plasmid for the host-specificity of the genus *Rhizobium* is discussed in section 3.1.3(a).

For the identification of a bacterium as *Rhizobium*, a nodulation test is still decisive. This means it is very hard to identify non-nodulating strains. In one experiment, 49 strains of *R. meliloti* were tested for resistance to nine antibiotics. Their combined resistance properties fell into nine different groups. Thus the classification of strains of *R. meliloti* on the basis of their antibiotic resistance is not possible.

The introduction of molecular methods such as transposon mutagenesis stimulated the research in *Rhizobium* to a large extent (Beringer *et al.*, 1979).

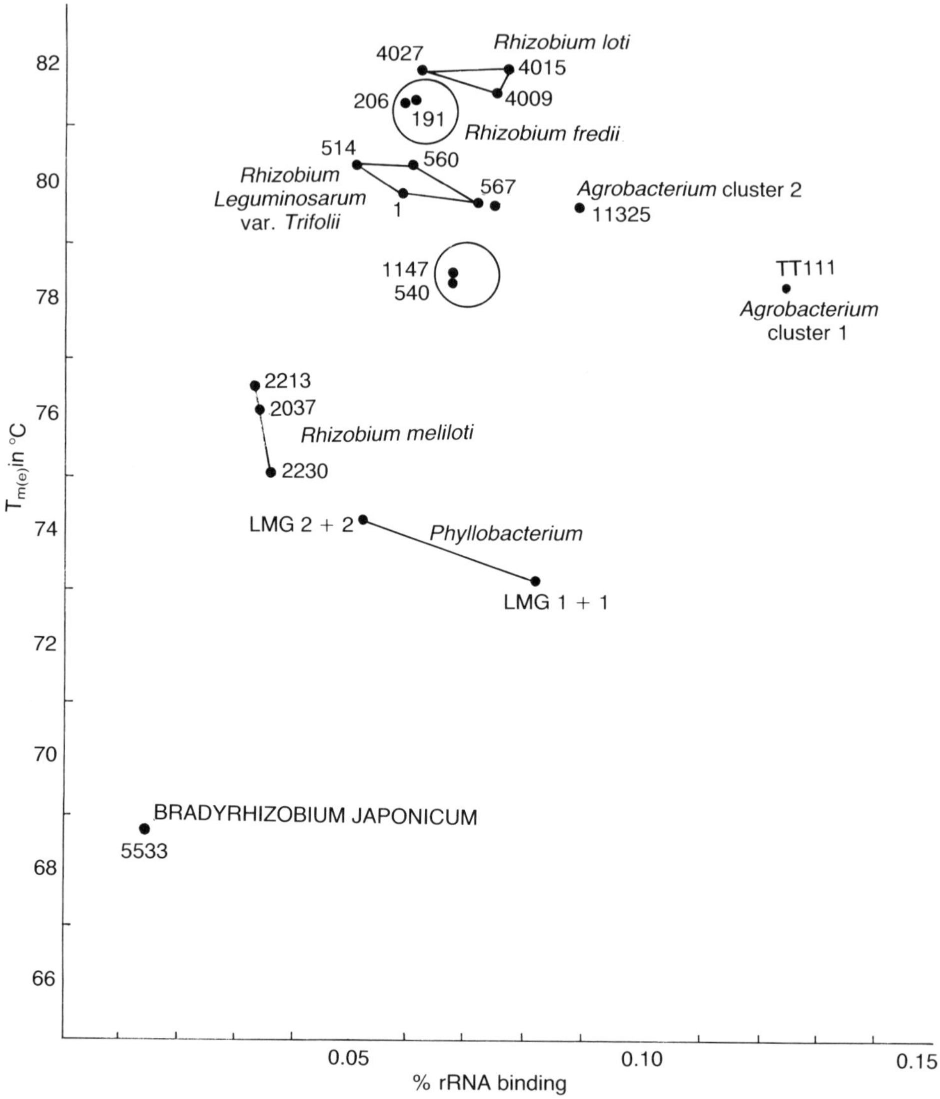

Fig. 3.1. RNA-homologies in *Rhizobium* and *Bradyhizobium*: correlation of T_m (●) (50% denaturation of a hybrid between 23S rRNA of *R. meliloti* SU 47 and DNA of the respective bacterium) with the percentage of rRNA–DNA binding on a filter (Based on Jarvis, *et al.*, 1986.)

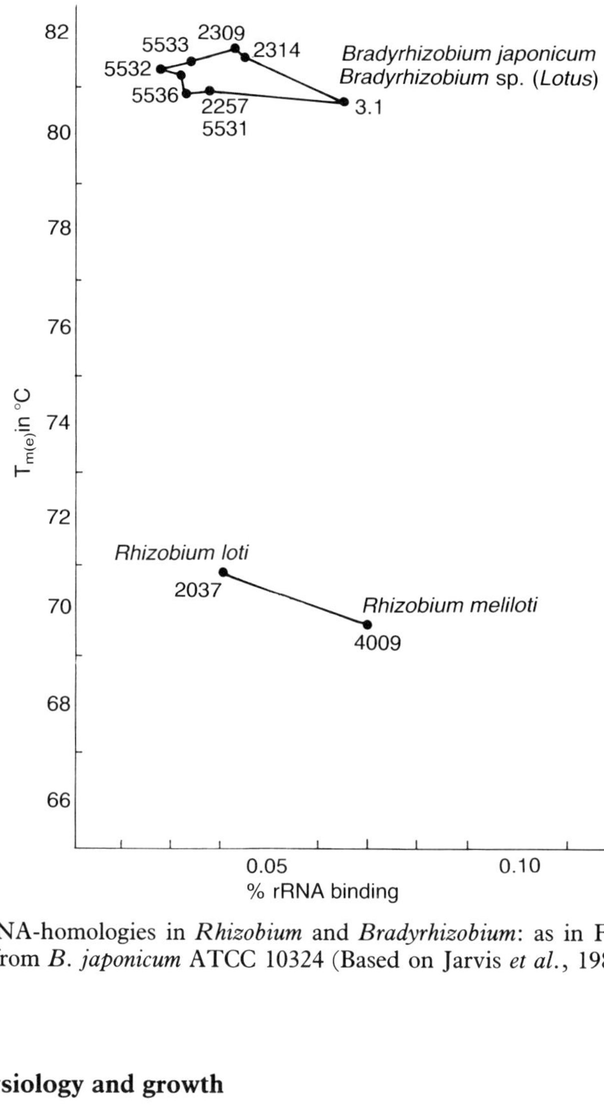

Fig. 3.2. RNA-homologies in *Rhizobium* and *Bradyrhizobium*: as in Fig. 3.1, with 23S rRNA from *B. japonicum* ATCC 10324 (Based on Jarvis *et al.*, 1986.)

3.1.2 Physiology and growth

The genera *Rhizobium* and *Bradyrhizobium* can be differentiated on the basis of their utilization of carbon sources and excretion of metabolites. Almost all strains of *Rhizobium* can grow on sucrose or mannose as single carbon-source. They export 3-ketoglycosides and acidify their growth medium. *Bradyrhizobium* does not grow on sugars such as sucrose or rhamnose (Table 3.2). However, both genera have several characteristics in common, such as growth on organic acids as C-source, metabolism of carbohydrates via the Entner-Doudoroff pathway and the absence of coloration of colonies with Congo Red. Comparisons based on further physiological and biochemical

Table 3.2 Physiological and biochemical characteristics of the genera *Rhizobium* and *Bradyrhizobium*

Characteristic	*Rhizobium*	*Bradyrhizobium*
Growth on yeast extract + mannitol (Generation time)	fast (2–4 h)	slow (5–10 h)
Growth on sucrose as C-source	+	–
Growth on rhamnose as C-source	+	–
Growth on organic acids as C-source	+	+
Increase of pH in the medium with sugars (pentoses) as C-source	–	+
Acidification of the medium with carbohydrates as C-source	+	–
Colonies stained with Congo Red (0.0025%)	–	–
Carbohydrate metabolism via Entner–Doudoroff pathway	+	+
Growth on media with 2% NaCl	some strains of *R. meliloti*	–
H_2S production	some strains of *R. meliloti*	–
Enhanced growth with biotin	+	+
Nitrate respiration	+ (many strains)	+ (many strains)
High affinity uptake of phosphate	+	+
High requirements for Ca, Fe and Co	+	+
Degradation of phenolic compounds via the β-ketoadipate pathway	+	+

+ More than 90% of the analysed strains were positive.
– Negative (with exceptions).

properties reveal that not all strains of one infection- or serological-group are uniform.

Both *Rhizobium* and *Bradyrhizobium* can use nitrate and NH_4^+ as nitrogen sources. Some strains of *Bradyrhizobium* are denitrifiers and use nitrate as the terminal respiratory electron acceptor. Strains of both genera have high affinity phosphate uptake systems which are more efficient than the systems in *Escherichia coli* and *Bacillus subtilis*. In comparison with other soil bacteria, both *Rhizobium* and *Bradyrhizobium* have high requirements for Ca^{2+}, Fe^{2+} and Co^{2+}. Vitamin requirements are very variable. As a general rule, *Bradyrhizobium* is stimulated by biotin, whereas *Rhizobium* requires both thiamine and pantothenate. In general, additional nicotinic acid, pyridoxin, folic acid, *p*-aminobenzoic acid, riboflavin and cyanocobalamin are not required for growth, although auxotrophic mutants requiring these metabolites are known. Strains of both genera can metabolize phenols via the β-ketoadipate pathway. The regulation of this pathway by succinate provides another physiological difference between e.g. *B. japonicum* and *Pseudomonas* strains (Röhm and Werner, 1985). Glucose

Table 3.3 Physiological characterization of strains of *B. japonicum* and *S. fredii* which nodulate soybean. (Based on Sadowsky *et al.* 1983)

Characteristic	*% of isolates with positive reaction*	
	B. japonicum	*S. fredii*
Catalase	100	100
Oxidase	100	100
Urease	100	100
Penicillinase	71	100
Nitrate reduction	100	100
3-Ketolactose production	0	0
H_2S-production	0	0
Citrate utilization	0	0
Gelatinase	0	100
2% NaCl (tolerance)	0	100
pH 4.5 (tolerance)	100	0
pH 9.0 (tolerance)	0	100
pH 9.5 (tolerance)	0	100

transport is also repressed by succinate, as demonstrated for *Rhizobium* strains effective on *Onobrychis vicifolia* and *Astragalus alpinus* (Bigwaneza *et al.*, 1991).

The optimum temperature for growth of *R. leguminosarum* strains is between 32 and 38°C, *R. meliloti* and *B. japonicum* do not grow above 43°C. Strains of both genera grow in a relatively wide pH spectrum between 4.5 and 8.5. The optimum for *R. leguminosarum* biovar *trifolii* is between 6.8 and 7.2, and that for *B. japonicum* between 5.8 and 6.2. Thus one difference between *B. japonicum* and *S. fredii* (both of which nodulate soybean) (Table 3.3) is that 100% of all *B. japonicum* strains still grow at pH 4.5, but not at pH 9.0. *Sinorhizobium fredii* strains on the other hand, tolerate alkaline but not acid conditions.

(a) Energy: C, N and H metabolism

The use of organic acids as C- and energy-sources by *R. meliloti* is especially well studied. Of 46 strains tested 45 grew on citrate, fumarate, malate and succinate as single C-source when they were present at a concentration of 2 mM (Table 3.4). At higher concentrations (20 mM) the same results were obtained except that no strain grew on either 20 mM malate or 20 mM citrate. The exact opposite result was found for glyoxylate, i.e. only eight strains could grow on 2 mM-glyoxylate, but 38 strains could grow on 20 mM. There was no correlation between the growth of strains on various organic acids and their N_2-fixing ability in nodules.

Table 3.4 Growth of 46 strains of *R. meliloti* having different symbiotic effectiveness (N_2-fixation in symbiosis) on organic acids as sole carbon source. (Based on Antoun *et al.*, 1984)

C-source	*Concentration (mM)*	*Number of positive strains (growth)*		
		HE	*E*	*I*
Acetate	2	4	28	3
	20	5	33	4
Citrate	2	5	35	5
	20	0	0	0
Fumarate	2	5	35	5
	20	5	34	5
Glyoxylate	2	0	7	1
	20	3	32	3
α-Ketoglutarate	2	2	18	1
	20	0	0	0
Lactate	2	5	30	4
	20	5	32	4
Malate	2	5	35	5
	20	0	0	0
Oxaloacetate	2	4	18	0
	20	3	12	0
Pyruvate	2	3	24	5
	20	5	32	5
Succinate	2	5	35	5
	20	5	34	5

HE = highly effective, E = effective, I = ineffective strains.

Apart from chemo-organotrophy, some strains of *B. japonicum* also exhibit chemolithotrophy. Following derepression of their hydrogenase, these strains exhibit ribulosebiphosphate carboxylase (Rubisco) activity (EC 4.1.1.39) and can reduce CO_2 using H_2 as an electron source (Simpson *et al.*, 1979). Succinate inhibits the derepression of Rubisco. The hydrogenase of *B. japonicum* is a membrane-bound enzyme which catalyses the reaction of 2 mol of H_2 with 1 mol O_2. The apparent K_m for H_2 is between 0.05 μM and 1.8 μM. The K_m for oxygen for the free enzyme is around 10 nM. This high affinity for O_2 means that the enzyme can also be active at the very low partial pressures of oxygen present in root nodules. The reaction is inhibited by acetylene and carbon monoxide. All C-sources, including succinate, inhibit hydrogenase induction, but this can be reversed by cyclic AMP. It is evident that hydrogenase expression is dependent on carbon nutrition. One important side-effect of hydrogenase activity is the reduction of cellular O_2 partial pressure, allowing increased protection of nitrogenase from O_2.

Rhizobium, *Bradyrhizobium* and *Agrobacterium* strains possess a specific form of glutamine synthetase (GS 2). This enzyme is completely different from GS 1, which is the normal prokaryotic form. GS 2 is absent in the bacteroid forms of the bacteria. Strains of *R. leguminosarum* have high affinity transport systems for the amino acids leucine, isoleucine, valine, alanine, threonine, histidine, serine and methionine. Under conditions of ammonia export the cell probably re-imports nitrogen as amino acids. Apart from the high-affinity transport system described above there is also a passive system for the uptake of amino acids. Ammonium flows freely in and out of cells, the intracellular concentration quickly adjusting to external concentrations. Different transport activities for choline have been found in *R. meliloti* where this metabolite has nutritional and osmoregulatory functions (Pocard *et al.*, 1989).

(b) Cell surface properties (cell wall, periplasmic proteins, EPS, LPS and CPS)

Basically the cell envelope in rhizobia is constructed as in other Gram-negative bacteria. The cytoplasmic inner membrane with carrier proteins is surrounded by a rigid peptidoglycan layer and an outer membrane. The space between the outer membrane and the inner membrane is called the periplasm, where a number of nutrient-binding proteins are located. The outer membrane carries three types of proteins: lipoproteins, porins and specific carriers such as the Fec protein, involved in the uptake of Fe^{3+} citrate (Brass, 1986).

Strains of *R. leguminosarum* do not export extracellular proteins. They do, however, synthesize a range of periplasmic proteins, e.g. alkaline phosphatase (EC 3.1.3.1), cyclic phosphodiesterase (EC 3.1.4.1) and inorganic pyrophosphatase (EC 3.6.1.1). The synthesis of some periplasmic proteins is repressed by phosphate and amino acids. Exopolysaccharides are of special importance for specificity and cell surface characteristics of both *Rhizobium* and *Bradyrhizobium*.

The exopolysaccharides (EPS) of *Rhizobium* are of three different types:

1. the neutral β-1,2-glucans of about 20 glucose units
2. the acidic exopolysaccharides
3. the lipopolysaccharides

The synthesis of the oligosaccharide repeat unit of acidic EPS is assumed to occur at the surface of the inner membrane. Mutants in β-1,2-glucan synthesis are nod$^-$. These exopolysaccharides may have a general role in adaptation to osmotic variations. They are also differentiated on the basis of their carbohydrate composition (Table 3.5). The basic structure of the acidic EPS repeat unit of *R. leguminosarum* is shown in Fig. 3.3. Two glucuronic acids are connected to five glucose units and these to one galactose unit. The last two sugars are linked to pyruvate. Acetate and β-hydroxybutyrate modify these repeat units in different strains

Table 3.5 Carbohydrate composition of exo- and capsular polysaccharides from *Rhizobium* spp. and *Bradyrhizobium japonicum*, as a percentage of total carbohydrates. (Based on Carlson, 1982)

Species/Strain	*Glucose*	*Galactose*	*Mannose*	*Glucuronic acid*	*Galacturonic acid*	*Pyruvate*	*Acetate-residues*	*4-O-Methyl-galactose*	*Rhamnose*	*4-O-Methyl-glucuronic acid*
R. leguminosarum biovar *trifolii*										
Strain 0266	56	11	0	12	0	9	11	0	0	0
Strain Ar 3	48	10	0	13	0	15	14	0	0	0
R. leguminosarum biovar *viciae*										
Strain PRE	48	11	0	19	0	13	9	0	0	0
R. leguminosarum biovar *phaseoli*										
Strain U 453	51	12	0	16	0	12	10	0	0	0
R. meliloti										
Strain U 27	67	14	0	0	0	11	7	0	0	0
B. japonicum										
Strain 110	36	24	18	0	12	0	4	5	0	0
Strain 3I1b	0	0	0	0	0	0	0	0	71	29

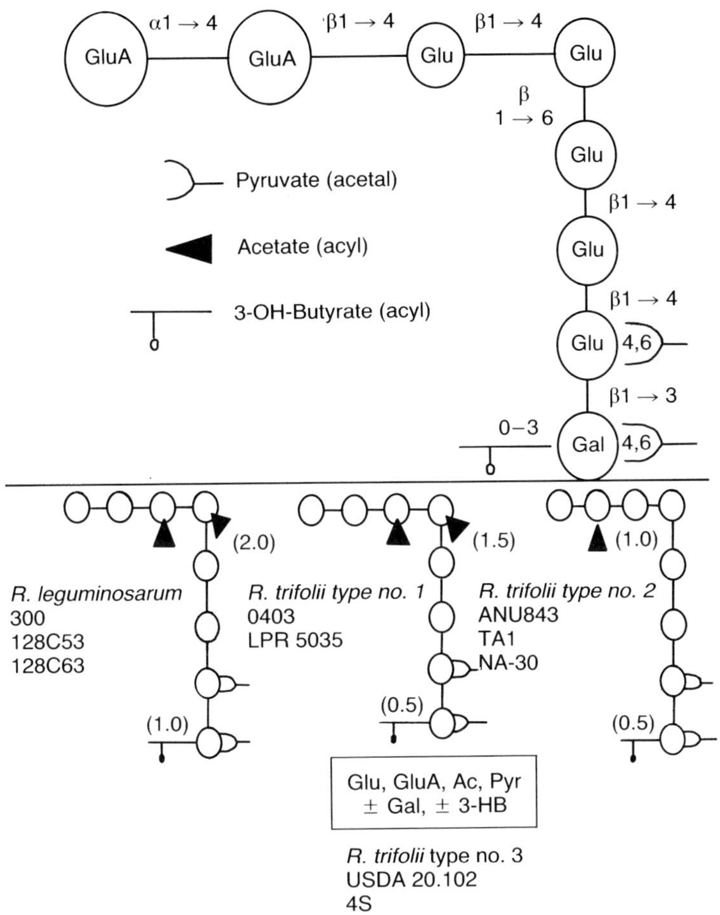

Fig. 3.3. Repeat units of the acidic exopolysaccharides of *R. leguminosarum* bv. *viciae* and bv. *trifolii*. (Based on Philip-Hollingsworth *et al.*, 1989.)

(Hollingsworth *et al.*, 1988; Philip-Hollingsworth *et al.*, 1989). In the genus *Bradyrhizobium* the situation is also not uniform. One group of strains clustered around *B. japonicum* USDA 110 typically have galacturonic acid and mannose present. Other groups around *B. japonicum* 3I1b have only rhamnose and 4-*O*-methylglucuronate. The functions of (EPS) for the infection process have been studied using two types of mutant (Müller *et al.*, 1988; Pühler *et al.*, 1991): strains without EPS and strains over-producing EPS. Both interactions are Inf^{-}, i.e. nodules are formed, but no infection threads and no bacteroids are present. DNA located on the megaplasmid 2 of *R. meliloti* can complement the EPS^{-} mutation (Keller *et al.*, 1988). Mutants in lipopolysaccharide (LPS) production form nodules with differentiated bacteroids but do not fix nitrogen (Priefer, 1989). The genes for LPS production are located on the chromosome in *R. leguminosarum* biovar. *viciae*. Specific nodule induced forms of LPS have been found to be expressed only in mature bacteroids, not in bacteria just released from the infection thread (van den Bosch *et al.*, 1989). A high-molecular-weight form of LPS is a prerequisite for an effective symbiosis. Specific cell surface antigens associated with LPS are expressed only under low oxygen tensions (Kannenberg and Brewin, 1989).

Apart from acid polysaccharides, neutral glucans have been found in *B. japonicum* capsular polysaccharides (CPS). This is particularly interesting in so far as glucans from culture medium or from the cell walls of pathogenic micro-organisms elicit phytoalexin formation in plants. The role of phytoalexins in the legume–*Rhizobium* symbiosis is considered in section 3.3.4(f). The LPS from *Rhizobium* and *Bradyrizobium* are more heterogeneous than EPS and CPS. The basic structure is similar for all Gram-negative bacteria and consists of a repeated oligosaccharide sequence, the so-called O-antigen, and a core oligosaccharide. This core oligosaccharide is bound to lipid A with an acid-labile 2-keto-3-desoxyoctaric acid. In 17 tested strains of *Rhizobium* and *Bradyrhizobium*, 11 different chemotypes of LPS composition were found.

3.1.3 Genetics

Linkage analyses of chromosomal genes in *Rhizobium* have been supplemented by intensive research on plasmid-localized genes in the past few years. In some strains, up to 25% of total DNA is present in the form of plasmids. Through the use of transposon mutants and sequencing of identified genes, knowledge of genes involved in nodule development and symbiotic nitrogen fixation has grown rapidly.

A very promising method to identify DNA sequences different at the strain to strain level is the **subtraction-hybridization method** (Cooper and Bjourson, 1991).

(a) Chromosome and plasmids

The chromosomal order of biosynthetic and metabolic genes in *Rhizobium* is generally quite different from that found for *E. coli*. Genes encoding similar functions are not clustered, but rather randomly distributed in the genome (Johnston *et al.*, 1978). A chromosomal map of *R. meliloti* is shown in Fig. 3.4. The genes for tryptophan biosynthesis are in three different regions. This is also true for *R. leguminosarum*. Within biovarieties with different chromosome types the same *sym*-plasmid is present (with *nod* genes). On the other hand similar chromosome types in different bacteria can harbour different plasmids. They thus belong to different biovarieties. Of all the

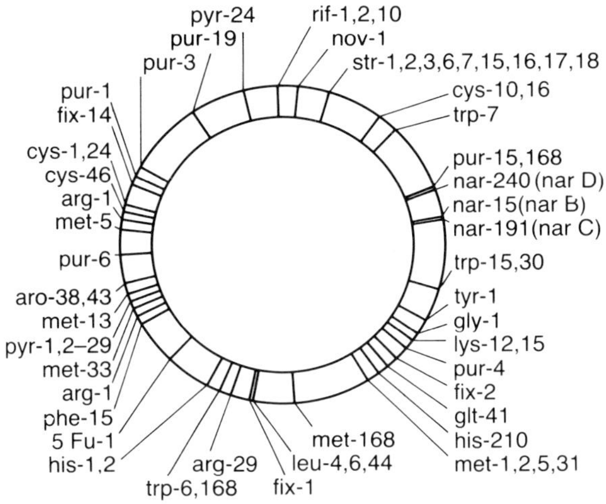

Fig. 3.4. Chromosomal map of *R. meliloti* strain 41 (Based on O'Brien, 1984.)

plasmids known in *Rhizobium* the *sym*-plasmid is especially noteworthy (Prakash and Atherly, 1986) because it carries the information needed for symbiosis, including host specificity and nitrogen fixation. Also present are genes for bacteriocins and pigment production. *Sym*-plasmids of the genus *Rhizobium* have molecular weights of between 100 MDa and over 300 MDa. Some strains of *R. leguminosarum* biovar *trifolii* may contain up to seven further plasmids with molecular weights of between 40 and 800 MDa. In *R. meliloti*, two megaplasmids of 800–1000 MDa are present. One carries the *nod* and *nif* genes, whereas the other carries genes for extracellular polysaccharide production. The products of these genes are also essential for the successful nodulation of alfalfa. Strains of *Sinorhizobium fredii* contain four plasmids of 100 MDa or greater. In *B. japonicum* it is in many cases difficult to show the presence of plasmids. Similar differences are known in the genus *Azotobacter. Azotobacter chroococcum* has up to six plasmids of between 7 and 200 MDa, whereas in *Azotobacter vilandii* no plasmids could be separated. Host-specificity is coded for by *sym*-plasmid genes. The exchange of these plasmids also means an exchange of host specificity. Thus *Agrobacterium* containing *sym*-plasmid can infect legumes. The functions of genes on the other plasmids are largely unknown. However, the difference in their number indicates that they are not of central importance. Some plasmids are incompatible, thus the insertion of one plasmid into *Agrobacterium tumefaciens* can lead to the loss of another. Expression of plasmid-localized genes can be affected by the host cell. *Ti*-plasmids from *Agrobacterium tumefaciens* are expressed after wounding of Liliaceae and Amaryllidaceae (monocots), but without tumour production. The mobilization of plasmids can be increased by the insertion of a so-called *mob*-region. By coupling of the *mob*-region with an antibiotic resistance marker, incorporation can be easily followed. The greater the number of plasmids in *R. leguminosarum* biovar *trifolii*, the less is the symbiotic N_2-fixation. Competitive growth is however improved.

Up to 25% of the total DNA can be present in plasmids in *Rhizobium* strains (Martinez *et al.*, 1990). DNA reiteration seems to be a general characteristic of the Rhizobiaceae (Flores *et al.*, 1987). Reiteration can involve the chromosome, different plasmids or a plasmid and the chromosome. Examples of such reiterated sequences are the promoter regions of *nif* HDK (Better *et al.*, 1983), the *nod* D gene (Göttfert *et al.*, 1986) and insertion elements (IS) (Wheatcroft and Watson, 1987, 1988). A high degree of such reiteration can also favour genomic rearrangements and deletions. A high frequency of deletions has been observed, for example, for the *nif* region of *B. japonicum* 110 (Hahn and Hennecke, 1987).

(b) *Nif*, *fix* and *nod* genes

Work on the genes for nitrogen fixation in *Rhizobium* and *Bradyrhizobium* was based on results of previous work with *Klebsiella*. The 21 *nif* genes (nif = nitrogen fixation) are clustered together in *Klebsiella* on the circular bacterial chromosome. Because the transfer of these genes to other non-fixing prokaryotes leads to nitrogen fixation, it is believed that only these genes code and regulate nitrogen fixation. The order and function of these 21 *nif* genes are given in Fig. 3.5. Four other genes (*nif* C, T, W and Z) have been added to the previous 17-gene *nif* cluster. Their functions are unknown. The structural genes for dinitrogenase reductase and dinitrogenase are genes K, D and H. Two subunits are required to form dinitrogenase reductase, also called the Fe-protein (see Fig. 3.6). The products of the *nif* K and D genes form dinitrogenase, also called the MoFe protein. The comprehensive term nitrogenase is to encompass the functions of both proteins. A further five genes, *nif* Q, B, V, N and E code for products which are involved in the biosynthesis or assembly of the Fe-Mo cofactor. Products of *nif* F and J genes serve in electron transport by way of synthesizing flavodoxin and pyruvate flavodoxin oxidoreductase. The gene products from *nif* S and U take part in the processing of dinitrogenase and that of *nif* M in the processing of dinitrogenase reductase. The gene product of *nif* A is the transcriptional activator for all operons apart from those of *nif* L and *nif* A. The *nif* A gene product is thermolabile (37°C). The *nif* C operon codes for a transcription repressor which inactivates the *nif* A protein in the presence of oxygen or fixed nitrogen. Thus nitrogen fixation is regulated by temperature, O_2 and fixed nitrogen. The functions of *nif* C, W, X, Y and Z are not yet known. The molecular weight of the various *nif* gene products is given in Fig. 3.6. The 21 genes are organized in eight

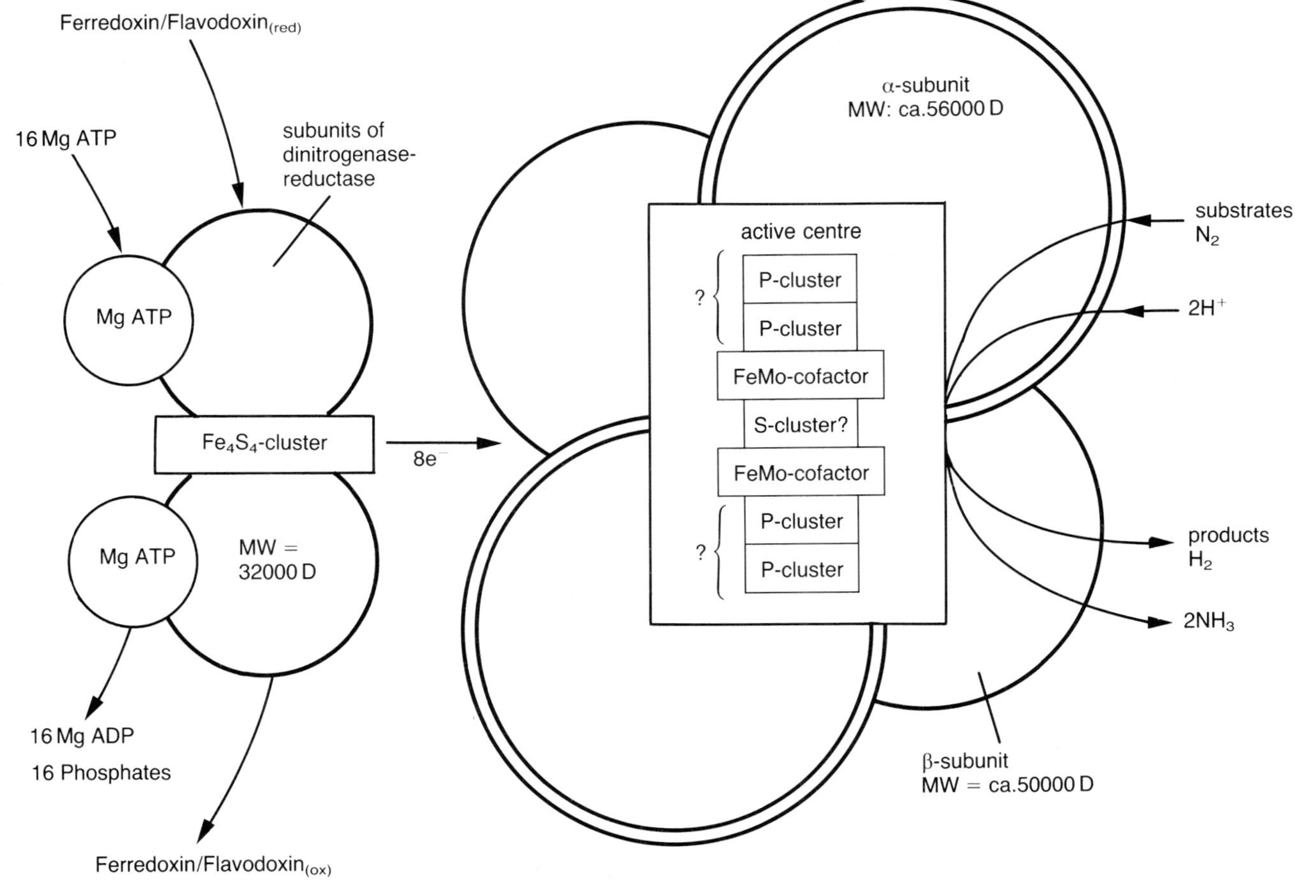

Fig. 3.5. Structures of dinitrogenase reductase and dinitrogenase. (Modified from Leigh, 1990.)

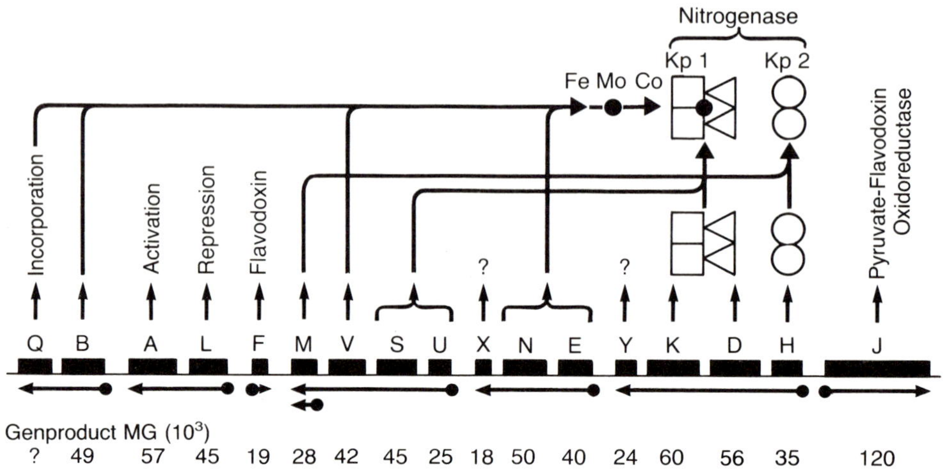

Fig. 3.6. *Nif* gene organization in *Klebsiella pneumoniae* (non-Symbiotic bacterium). Horizontal arrows indicate direction of transcription: −24, −12 promoter sequence, with NifA upstream activator sequence. Vertical arrow heads indicate gene functions. (Based on Cannon *et al.*, 1985 and Merrick, 1988.)

	−23 −10 +1
nifJ	C G A G C T G G C A C A G G C T G T G C T T G A G G C A A C A A
nifH	A C G G C T G G T A T G T T C O C T G C A C T T C T C T G C T G
nifE	G C T T C T G G A G C G C G A A T T G C A T C T T C C C C C T C
nifU	T T C T C T G G T A T C G C A A T T G C T A G T T C G T T A T C
nifM	G T G G C T G G C C G G A A A T T T G C A A T A C A G G G A T A
nifF	C A A C C T G G C A C A G C C T T C G C A A T A C C C C T G C G
nifB	A C C T C T G G T A C A G C A T T T G C A G C A G G A A G G T A
nifL	G A T A A G G G C G C A C G G T T T G C A T G G T T A T C A C C
glnA	A A A G T T G G C A C A G A T T T C G C T T T A T A T T T T T T

Fig. 3.7. *Nif* and *gln*A promoter sequences from *Klebsiella pneumoniae*. (Based on Cannon *et al.*, 1985.)

Table 3.6 Rhizobia *nif* and *fix* genes and gene product functions. (Modified from Stacey, 1990 and Hansmann et al., 1989)

Gene	*Gene product and functions*
nif A	DNA binding regulatory protein required for the transcription of *nif* genes and some *fix* genes
nif H, D, K	structural genes for dinitrogenase reductase (*nif* H) and dinitrogenase (*nif* D, *nif* K)
nif B, E, N	synthesis of the FeMo cofactor of the dinitrogenase
nif S	maturation of dinitrogenase
fix L, J	transcriptional regulation of the *nif* A gene, *fix* L with a sequence homology to a membrane sensor family, *fix* J sequence similarity to transcriptional regulatory proteins
fix A, B, C, X	may function in electron transport to nitrogenase
fix G, H, I, S	proposed to function as a membrane-bound complex coupling a cation pump to redox processes
fix K, N, R	functions unknown

operons. It is noticeable that only those two genes involved in electron transport (*nif* F and J) are transcribed in the reversed direction. The sequence of the eight *nif* promotors shows conserved regions, separated by eight basepairs, the sequence CTGG for basepairs 26–23 and the sequence TGCA for basepairs 13–10 (Fig. 3.7). These sequences are also found in *nif* promotors from *Rhizobium*. In *Rhizobium*, *Bradyrhizobium*, *Azorhizobium* and *Sinorhizobium*, in addition to the *nif* genes with similar functions as in *Klebsiella*, a number of fix genes, which are important for nitrogen fixation in symbiosis or have unknown functions at present, have been identified (Table 3.6).

The *fix* X gene encodes a ferredoxin like protein, essential for nitrogen fixation in the symbiotic state and in free-living cells under microaerobic conditions (Gubler *et al.*, 1989). The inhibition of the *nif* A dependent *nif* gene activation by oxygen is at the *nif* A protein level. The half-life of the *nif* A mRNA from *B. japonicum* is the same under aerobic and anaerobic conditions (Kullik *et al.*, 1989).

The essential role of oxygen concentration in *nif* gene regulation can be summarized by comparing the current models for *nif* regulation in *Klebsiella pneumoniae* (*oxytoca*) (Fig. 3.8a) and *B. japonicum* (Fig. 3.8b). In *Klebsiella*, there are two pairs of regulatory proteins controlling *nif* transcription: the

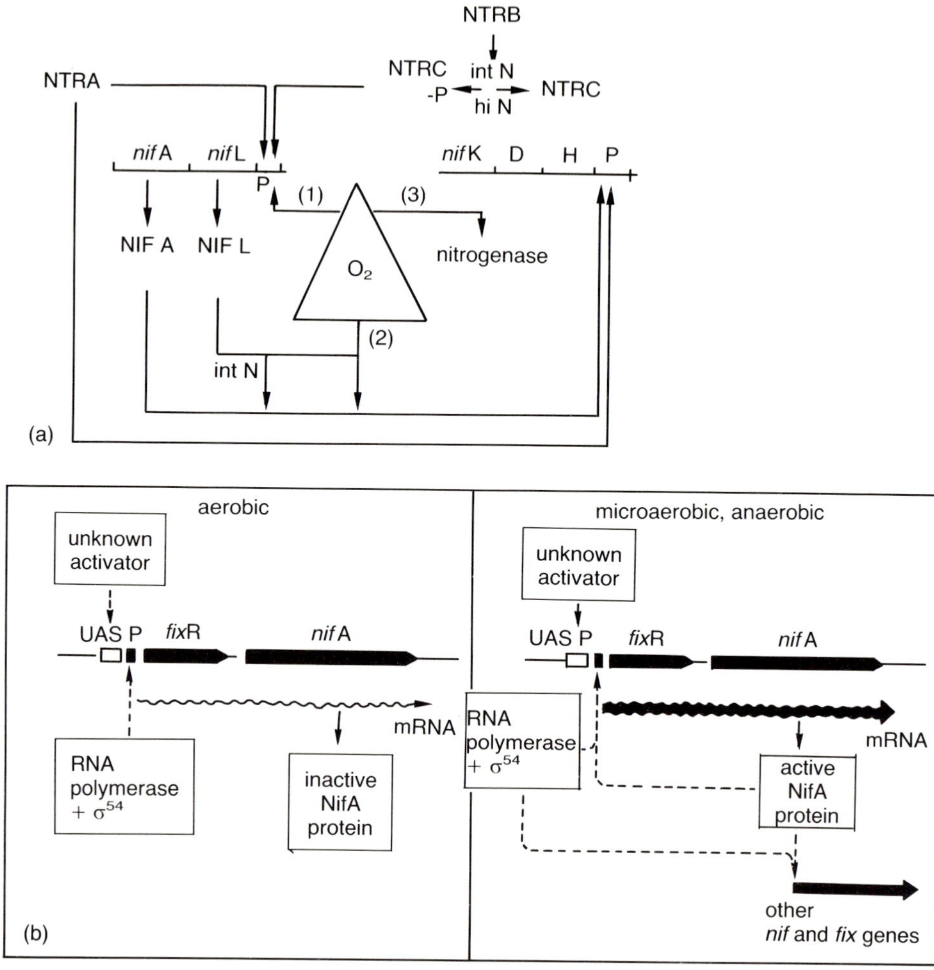

Fig. 3.8. (a) *Nif* gene regulation in *Klebsiella pneumoniae* by high (hi) and intermediate (int) concentrations of reduced N and regulation by three levels of O_2. (1) 60 μM O_2 inhibits derepression of *nif*LA. (2) 0.1 μM O_2 inhibits derepression of the other *nif* operons by the *nif*L gene product. (3) 6 μM O_2 inhibits nitrogenase activity. N-deficiency also affects nitrogenase synthesis, but at a later stage. (Based on Hill, 1988; Hill *et al.*, 1988.) (b) *Nif* and *fix* gene regulation under aerobic and microaerobic and anaerobic conditions in *B. japonicum*. (From Thöny *et al.*, 1989, and H. Hennecke, personal communication.)

NtrBC protein and the NifA and NifL proteins. Transcription of *nif* A and *nif* L is activated by phosphorylated NtrC protein. High levels of oxygen inhibit the expression of the *nif* A and *nif* L promoter. Under low oxygen concentrations and under nitrogen limitation *nif* A activates the other *nif* genes, i.e. *nif* KDH. Derepression of *nif* KDH is inhibited by the *nif* L

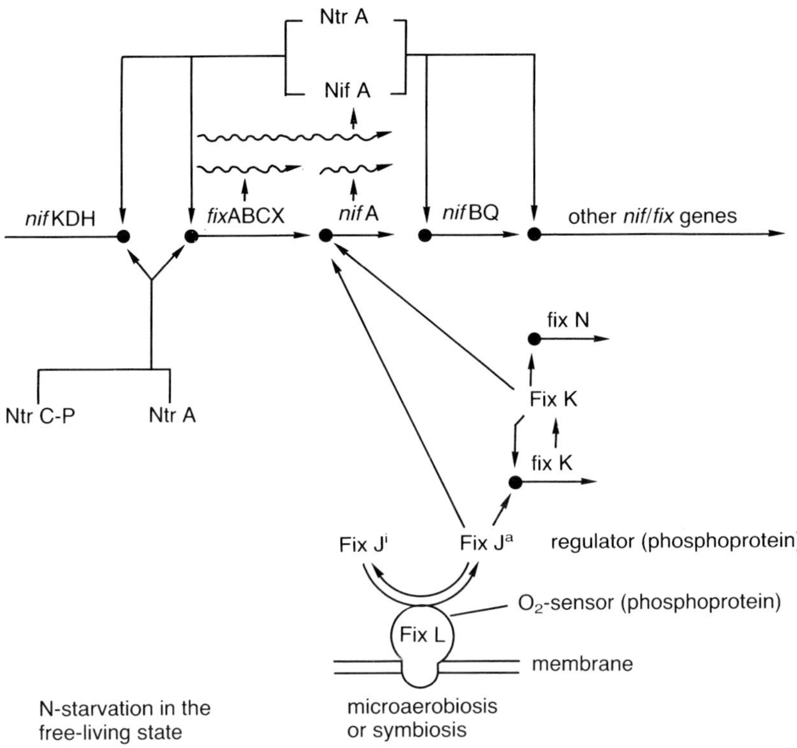

Fig. 3.9. C Regulation of *nif* and *fix* gene expression in *R. meliloti*. (Modified from de Bruijn *et al.*, 1990.)

product at oxygen concentrations as low as 0.1 μM. At 6 μM oxygen, nitrogenase activity *per se* is inhibited. Thus there are at least three discrete levels of oxygen concentration involved in *nif* gene regulation and the regulation of nitrogenase activity (Fischer *et al.*, 1991).

In *B. japonicum*, *nif* gene regulation by oxygen is different from *Klebsiella* and also apparently different from *R. meliloti* (Fig. 3.9). The *fix*A–*nif*A operon is activated by a sigma factor and an additional unknown activator. Under aerobic conditions this operon is expressed at moderate levels. However, the gene product is inactive. Under microaerobic or anaerobic conditions, the NifA protein is active and serves as a transacting factor, enhancing the synthesis of *nif*A mRNA and increasing NifA activity. The *nif* and *fix* genes are activated by NifA protein.

The *nod* genes in *Rhizobium* and *Bradyrhizobium* are the most symbiosis-specific genes of the microsymbionts, allowing the communication with the host plant to build the housing (D. Verma) for a high population density without competition from other soil micro-organisms: 1 g of nodule tissue can harbour more than 10^{11} rhizobial cells. A high titre of a single species of

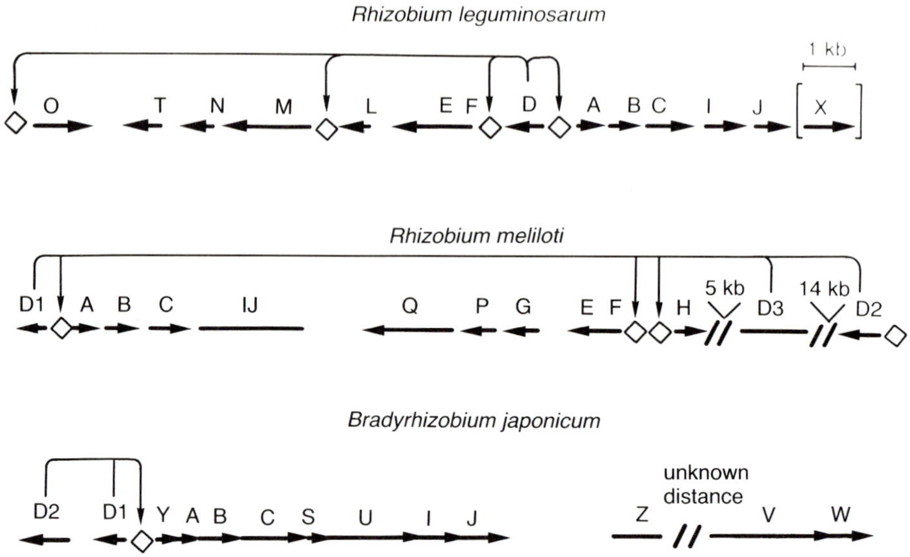

Fig. 3.10. Organization of *nod* genes in *Rhizobium* and *Bradyrhizobium*. (Modified from Hansmann *et al.*, 1989; Thöny *et al.*, 1989.)

free-living soil bacteria is in the range of 10^6–10^7 cells per g of soil. The presence of *nod* genes (together with a number of other genes) results in an advantage of 1 : 10^4–10^5 in population density. The *nod* genes A, B and C are called the common nod genes and are present in all rhizobia (Fig. 3.10). Mutations in these genes lead to a nod$^-$ phenotype. The defective genes can be complemented by *nod*ABC genes from other species. A larger number of *nod* genes determine host specificity and are called *hsn* genes (*h*ost *s*pecificity of *n*odulation). Mutations in these genes lead to reduced or retarded nodulation, e.g. in *nod* E and *nod* F of *R. leguminosarum* (Horvath *et al.*, 1986). Mutations in the *nod* V and *nod* W genes in *B. japonicum* give a nod$^-$ phenotype in mung bean, however only slightly delayed nodulation in soybean (Göttfert, 1990). A summary of all *nod* genes and *nod* gene functions known at present is given in Table 3.7. The *nod* D gene product is a transcriptional activator which binds to the promotors of other nod genes at the so called nod box. It is activated by different flavonoids and flavones, described in more detail in section 3.3.1.

There is a very interesting homology of the ligand binding region of the *nod* D protein of *Rhizobium meliloti* and vertebrate estrogen receptors (Györgpal and Kondorosi, 1991).

A breakthrough in 1990 was the identification of a phytohormone-like substance, *Nod* Rm 1 from *R. meliloti* synthesized by gene products of *nod*ABC, in the laboratory of J. Denarié in Toulouse (Lerouge *et al.*, 1990).

Table 3.7 Nodulation (*nod*) genes in *Rhizobium* (R), *Bradyrhizobium* (B), *Sinorhizobium* (S) and *Azorhizobium* (A). (Modified from Göttfert 1990; Göttfert *et al.*, 1990 and Stacey, 1990)

Gene	*Functions of gene products*
*nod*ABC	Essential for root hair curling and cortical cell division. *Nod*ABC likely synthesize a phytohormone-like substance. *R. meliloti*: Rmnod-1 identified as a *N*-acyl-tri-*N*-acetyl-β-1,4-D-glucosamine tetrasaccharide bearing a sulphate group on carbon 6 of the reducing sugar moiety.
*nod*D1, D2, D3	Transcriptional regulator required for the expression of the other nod genes
*nod*E	Host range, reported sequence similarity to the *fab*B gene of *E. coli*, condensing enzyme of fatty acid synthase. Membrane protein. Mutants lacking NodE are affected in capsular polysaccharide biosynthesis.
*nod*F	Host range, reported sequence similarity to acyl carrier protein. *Nod*F carries 4′-phosphopantetheine as a prosthetic group. At least two different forms of NodF can be isolated from a NodFE-overproducing strain of *R. leguminosarum* bv. *viciae*.
*nod*G	Host range, reported sequence similarity to dehydrogenases.
*nod*H	Host range, may modify factor produced by *Nod*ABC. May function as a sulphate transferase for the synthesis of NodRm-I
*nod*I	Reported sequence similarity to ATP binding transport proteins. *R. leguminosarum* bv. *trifolii* mutants defective in *Nod*I have a decrease in the 3-hydroxybutyrate substitution of the capsular polysaccharide
*nod*J	Membrane protein, may act in conjunction with *nod*I.
*nod*K	Reported only in *B.* sp. (*Parasponia*). Function unknown.
*nod*L	Host range, reported sequence similarity to acetyltransferase.
*nod*M	Host range, reported sequence similarity to amidotransferase. Glucosamine synthase for the synthesis of NodRm-1.
*nod*N	Reported. Involved in biosynthesis of hair deformation factor.
*nod*O	Function unknown, encoded protein is exported. Reported to bind calcium.
*nod*P	Function unknown, gene hybridizes to *E. coli* genomic DNA.
*nod*Q	Host range, reported sequence similarity to GTP binding proteins, EF-tu and EF-1. May modify factor produced by *Nod*ABC.
*nod*R	Reported only in *R. leguminosarum* bv. *trifolii*. Host range.
*nod*S	Function unknown. MW 23 kDa
*nod*T	Host range. Function unknown.
*nod*U	Function unknown. MW 62 kDa
*nod*V	Reported only in *B. japonicum*. Important for host range, sequence similarity to membrane sensor family.
*nod*W	Reported only in *B. japonicum*. Important for host range, sequence similarity to family of transcriptional regulatory proteins.
*nod*X	Reported only in *R. leguminosarum* strains capable of nodulating Afghanistan pea.

Table 3.7 *Continued*

Gene	*Functions of gene products*
*nod*Y	Reported only in *B. japonicum*, but found in all rhizobia by hybridization. Function unknown.
*nod*Z	Host range required for growth under microaerobic conditions.
*nod*1A	Host range. Essential for nodulation of selected soybean cultivars. Reported only from *B. japonicum*. Shows sequence similarity to MerR, a transcriptional regulatory protein.
*nol*D, *nol*E, *nol*F, *nol*G	Host range in *R. meliloti*.

The *hsn* genes *nod* H and *nod* Q are involved in the modification of this signal (Banfalvi and Kondorosi, 1989). The metabolite was identified as a *N*-acyl-tri-*N*-acetyl-β-1,4-glucosamine tetrasaccharide with a sulphate group on C-6 of the reducing sugar (Fig. 3.11). The resemblance to chitin precursors, elicitors in phytopathogenic reactions, is obvious.

The *nod*-factors can be also looked at as a special type of oligosaccharides, called **oligosaccharines**, functioning as elicitors in plant defence reactions or as signal molecules in cell growth and differentiation (Ryan and Farmer, 1991).

This compound is active at the very low concentration of 10^{-11} M in root hair branching. The nodulation factor from *R. leguminosarum* bv. *viciae* (Nod Rlv-V) was identified in the laboratory of B. Lugtenberg in Leiden. Besides the common nod genes ABC, the host specificity genes *nod* FEL are involved in the production of this factor (Spaink *et al.*, 1991). The major differences between the two structures are indicated by arrows in Fig. 3.11. The highly unsaturated fatty acid in Nod Rlv-V is determined by Nod E, and *nod* L is required for the production of the *O*-acetyl substituent. In *R. leguminosarum* bv. *viciae* also a nod factor with 4 sugars instead of 5 can be produced. *Nod* F (the gene product of *nod* F) is homologous to an acyl carrier protein. Nod O is homologous to a Ca^{2+} binding protein of the haemolysin family of proteins. Genes required for *Nod* O secretion are apparently not located on the sym plasmid. The *nod* O protein is also present in the growth medium of a culture of *R. leguminosarum* biovar. *viciae*, induced for *nod* gene expression (Economou *et al.*, 1990), and therefore not dependent on other plant factors. These homologies suggest that the *hsn nod* genes determine host specificity by modifying the root hair curling factor (Downie *et al.*, 1990) or by involvement in the synthesis of other similar factors. Other additional functions are possible, e.g. modification of the bacterial environment in the host cell wall or the prevention of host defence reactions (Long and Ehrhardt, 1989).

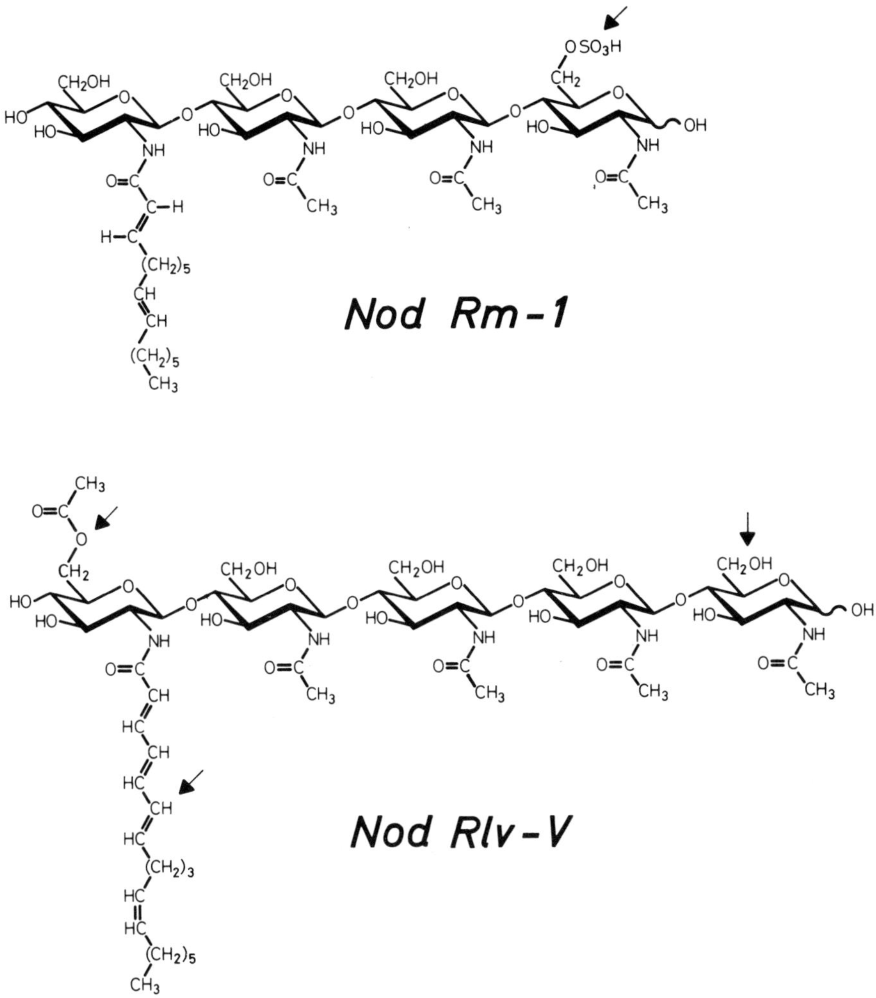

Fig. 3.11. Nodulation factors from *Rhizobium meliloti* (Nod Rm-1) and from *Rhizobium leguminosarum* bv. *viciae* (Nod Rlv-V). Compiled from data by personal communication from J. Denarié (Toulouse) and B. Lugtenberg (Leiden) and from Lerouge *et al.* (1990) and Spaink *et al.* (1991). Arrows indicate major differences between the two structures.

Rhizosphere genes (*rhi* A, B and R) have been identified adjacent to the nod genes (Economou *et al.*, 1989). Mutations in these genes have no effect on nodulation. However, their expression is, together with flavonoids, controlled by nod D. A survey of all symbiosis-related genes in the microsymbiont is presented in Table 3.8.

Table 3.8 Symbiotic genes in *Rhizobium/Bradyrhizobium*. (Modified from Göttfert, 1990)

Gene	*Definition*	*Function*
hac	*ha*ir *c*urling	curling of root hairs
nod	*nod*ulation	triggering of nodule development
hsn	*h*ost *s*pecificity of *n*odulation	determination of the host specificity within the infection groups
efn	*ef*ficiency of *n*odulation	additional genes for efficient nodulation
ndv	*n*odule *d*e*v*elopment	regulation of nodule development, e.g. by cyclic glucans
exop	*ex*tracellular *p*olysaccharides	regulation of infection and release of bacteroids
nif	*ni*trogen *f*ixation	genes involved in nitrogen fixation with homology to the *nif* gene cluster of *Klebsiella*
fix	*fix*ation	additional genes involved in nitrogen fixation in the symbiotic stage (without homology to *Klebsiella nif* genes)
dct	*d*i*c*arboxylic acid *t*ransport	transport of dicarboxylic acids as substrates for bacteroid nitrogen fixation

3.1.4 Ecology

Hardly any natural habitat is without some representative of the Leguminosae. Thus, the question, whether *Bradyrhizobium* and *Rhizobium* can survive over many years in soil without host plants, can only be answered under controlled experimental conditions. The best example of such an investigation is the so-called Broadbalk-Project of the Agricultural Research Station, Rothamstead, where wheat has been grown since 1843. In this soil *R. leguminosarum* biovar. *viciae* is present at a concentration of 10^5 cells per g soil. The concentration of biovar. *trifolii* is 10^2 cells per g soil. The results indicate that the influence of an occasional legume growing as a weed is negligible. *Rhizobium* can therefore survive over many decades in soil without its host. It is also clear that in fields intensively cultivated with legumes the concentrations of bacteria are higher. With monocultures of *Trifolium*, *R. leguminosarum* biovar *trifolii* reached 7×10^5 cells per gram soil. With *Vicia faba*, *R. leguminosarum* biovar. *viciae* reached 2×10^6 cells per g soil and with *Medicago lupulina*, *R. meliloti* was present at a concentration of 5×10^5 cells per g soil.

(a) Competition and the ability to survive

There are three major groups of factors influencing the population of *Bradyrhizobium* and *Rhizobium* strains:

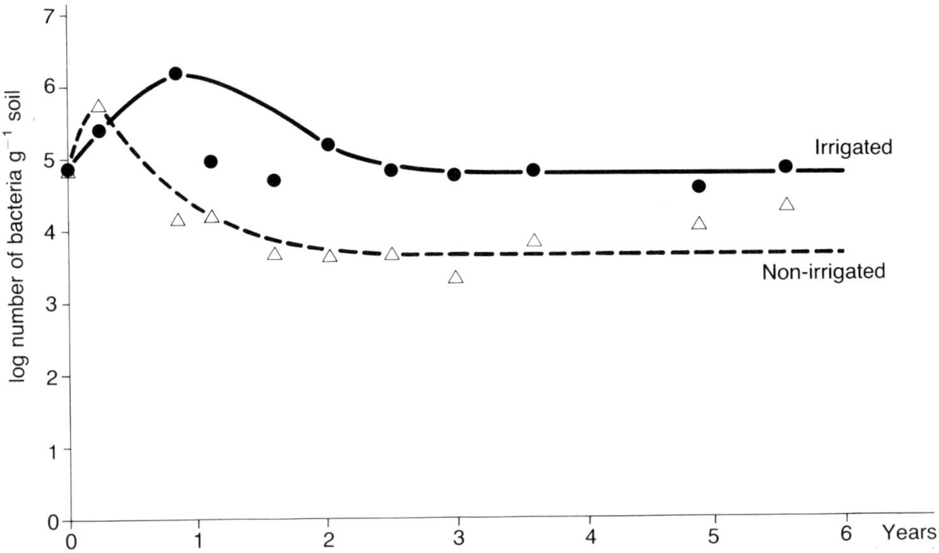

Fig. 3.12. Survival of *B. japonicum* in irrigated and non-irrigated soil, after a single cultivation period of soybean during the first year. (Based on Crozat *et al.*, 1982.)

1. abiotic factors,
2. microbiotic factors,
3. biotic factors from the plant root system.

(i) *Abiotic factors*

A. Water content Experiments in Southern France have shown that after a single inoculation with *B. japonicum* the bacterial population was, over many years, one order of magnitude higher in irrigated fields than in non-irrigated fields (Fig. 3.12). Notice especially the constant level of 10^4 cells per g soil in the dry field and the constant level of 10^5 cells per g soil in the irrigated field over a period of several years.

B. Temperature Above 40°C, the survival of the majority of *Rhizobium* and *Bradyrhizobium* strains is strongly reduced. *Rhizobium* strains isolated from warm and dry areas have, as a general rule, an increased heat tolerance compared to those isolated from temperate zones. High temperature tolerance is often connected with the ability to withstand desiccation. Thus temperature effects are often related to soil humidity. The relative dominance of endogenous *Rhizobium* in soils is clearly dependent on soil temperature. For example, in the USA, the response of legumes to inoculation increases from North to South. Another factor here is the lower nitrogen content of soils in the south.

C. Soil type The growth and survival of *Rhizobium* is also affected by the soil type. Montmorillonite appears to protect *R. meliloti* against temperatures over 50°C; *B. japonicum* on the other hand is not protected. The effect of other soil components is only known through model experiments. The situation in natural soil is almost totally unknown. In soil, *Rhizobium* strains grow very slowly. The population doubling time for *B. japonicum* is 10–15 days in plant-free soil, but 9–10 h in the rhizosphere. Movement of *Rhizobium* is primarily dependent, in agricultural soils, on water movement. Other components counteract this, e.g. *Rhizobium* is strongly absorbed by phytin aggregates, but weakly by humic acid conditions.

D. pH In acid soils particularly, the low pH tolerance of *Rhizobium* plays an important role. At pH 4–5, acid tolerant strains are selected. These strains, when nod^+ and fix^+, are of great practical importance especially in the tropics, where soils are generally more acid.

E. Salt content and heavy metals Although some strains of *B. japonicum* tolerate up to 500 mM-NaCl, others are inhibited at 80 mM-NaCl. Strains of *R. meliloti* can tolerate up to 600 mM-NaCl (Rigaud, 1987). The inhibition of legume nodulation in high salt areas is, however, primarily due to inhibition of the infection process and nodule formation, processes influenced by salt tolerance of the plant host. Selection of salt-tolerant *Rhizobium*–legume associations is of special importance for agriculture due to the rapidly expanding areas with high salt concentrations.

A high concentration of heavy metals in soils favours the survival of ineffective (non-N_2-fixing strains of *R. leguminosarum* biovar. *trifolii* (Giller *et al.*, 1989).

(ii) Microbiotic factors

A. Soil fungi and bacteria Common soil fungi such as *Alternaria*, *Cephalosporium*, *Rhizopus* and *Trichoderma* can inhibit growth of *Rhizobium* strains in soil. Similar results have been found with soil actinomycetes and bacteria, such as *Streptomyces*, *Actinomyces*, *Bacillus*, *Pseudomonas*, *Xanthomonas*, *Flavobacterium*, *Alcaligenes*, *Achromobacter*, *Arthrobacter* and *Corynebacterium*. The inhibition is probably caused either by the excretion of antibiotics or by competition for nutrients. Nodulation can also be inhibited by these fungi and bacteria.

B. Parasites and predators There is no significant influence of bacteriophages or *Bdellovibrio bacteriovorus* on soil rhizobial populations. However, protozoans of the genera *Colpoda* can drastically reduce *Rhizobium* titres in soils from 2×10^8 to less than 1×10^7 per g soil, within 60 days. The protozoans exert no effect when the *Rhizobium* titre in the soil is 10^3 cells

g^{-1}. When protozoans are inhibited by cycloheximide the nodulating *Rhizobium* titre rises again.

C. Competition among strains *R. leguminosarum* biovar. *trifolii* strain T24 excretes an antibiotic, which inhibits growth of strains of the same biovariety as well as *R. phaseoli* and *S. fredii*. However, *Bradyrhizobium* spp., *R. meliloti* and *Agrobacterium* spp. are not inhibited. In general, the properties of *Bradyrhizobium* and *Rhizobium* strains which make them competitive over other strains are not well known.

Incorporation of genes for trifolitoxin production, an antirhizobial peptide, increases significantly the competitiveness for nodule occupancy against sensitive strains of *R. leguminosarum* biovar. *trifolii* (Triplett, 1990). A screening of 360 isolates of *B. japonicum* from soybean nodules from 18 locations in Delaware, USA revealed 12 different serogroups. Nearly one-third of all strains were non-reactive with all antisera tested and about 205 produced rhizobitoxin symptoms (Fuhrmann, 1990).

(iii) ***Biotic factors, effect of plants***

Compared to other plants, Fabales are distinguished by relatively high rates of root exudation. Thus they can enrich *Rhizobium* and *Bradyrhizobium* and other soil bacterial populations in their rhizosphere. Gram-negative bacterial strains have been found to be predominant in the rhizosphere of legumes, even when the surrounding soil is dominated by Gram-positive bacteria. A 10-day-old seedling of *Cicer arietinum* can excrete 80 µg amino acids and 30 µg sugar and uronic acids. This supports the growth of about 3×10^8 cells of *B. japonicum*. *Rhizobium* and *Bradyrhizobium* have been shown to move chemotactically towards root exudate components such as cinnamic acid (Fig. 3.13). Chemicals absolutely specific for *Rhizobium* chemotaxis are, however, unknown. When *Rhizobium* or *Bradyrhizobium* are inoculated into soil they often have difficulty becoming established in the presence of indigenous strains. These indigenous strains have very different serological reactivity, antibiotic resistance, plasmid spectra and degrees of auxotrophy. Nodules growing on parts of the root system in different depths of soil can contain different strains of *Rhizobium*.

Transfer of the symbiotic plasmid from, for example, *R. leguminosarum* to *R. meliloti* can increase the genetic diversity of the species (Broughton *et al.*, 1987b). A natural population of one species is often very heterogenous in terms of plasmid content and antibiotic resistance (Broughton *et al.*, 1987a). In one field of 100 m^2 size, 40 distinct strains of *R. meliloti* were isolated from 100 different plants (Dowling and Broughton, 1986).

In a restricted region, a coevolution of certain lines and cultivars of host plants and the local *Rhizobium* strains can be assumed. This has been shown for peas from Afghanistan and a distinct area of Turkey. These cultivars are not nodulated by *R. leguminosarum* strains from European soil, but form a

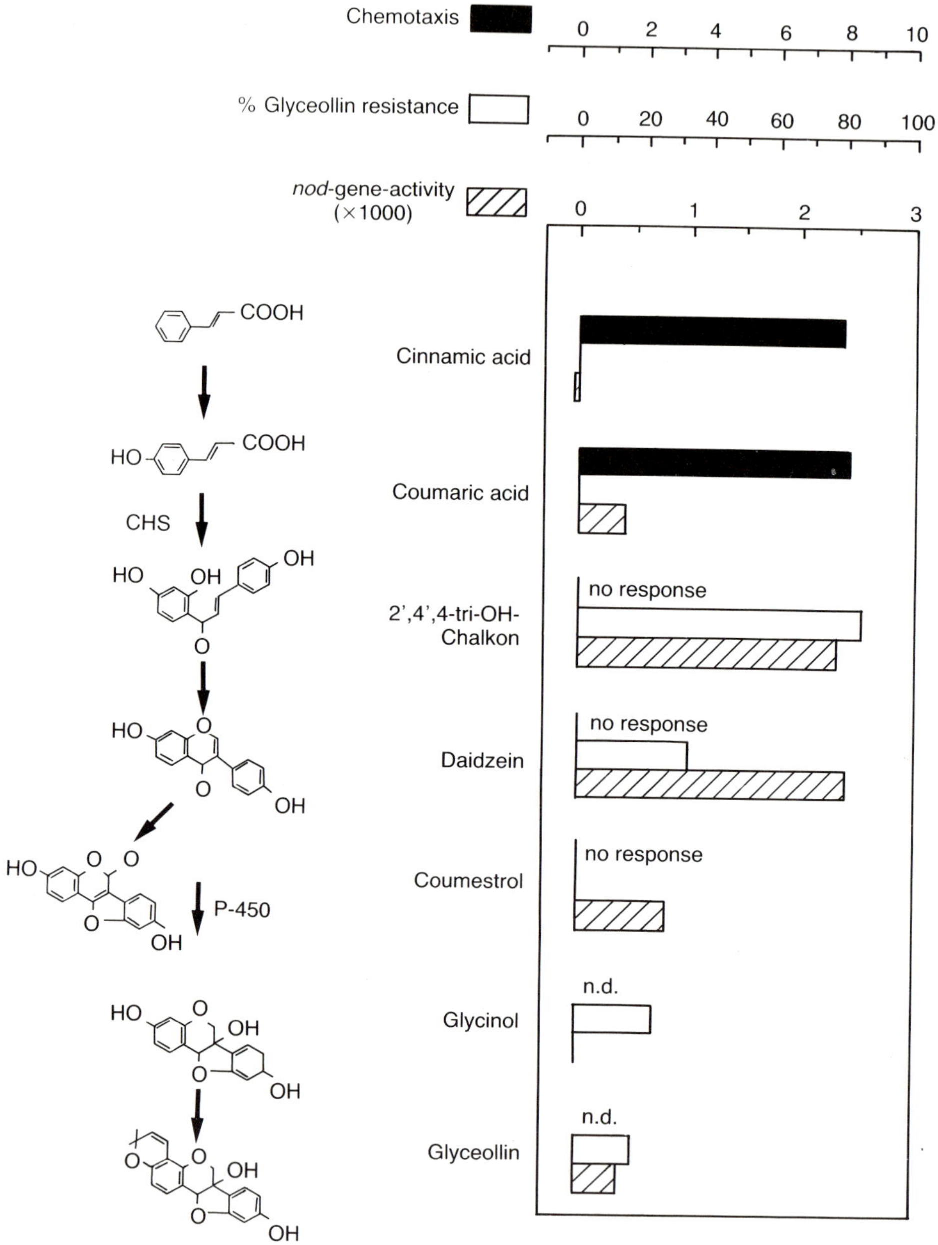

Fig. 3.13. Chemotaxis, nodulation and phytoalexin resistance, effects of intermediates of phenylpropane-metabolism (Developed from Parniske *et al.*, 1991a and Kape *et al.*, 1991.)

symbiosis with local effective strains (Lie *et al.*, 1987, 1988). Also a general coevolution can be assumed (Verma and Stanley, 1989).

More than 30 genes are involved in coding for the structure and function of the bacterial flagella (Lengeler, 1986). Therefore, chemotaxis towards newly identified compounds such as phytoalexin biosynthesis intermediates (Fig. 3.13), which are also effective as signalling substances for nodulation, presumably requires a complicated regulatory systems (Gitte *et al.*, 1978).

(b) Growth, competition and survival rates under defined conditions

Rhizobium and *Bradyrhizobium* are not particularly specialized in the use of C-sources or tolerance of physicochemical soil factors. In comparison with other soil bacteria such as *Pseudomonas* and *Arthrobacter* they grow relatively slowly. Thus it is hard to imagine that they can survive many decades in soil without infecting plants. In this context high affinity substrate uptake systems are important. A dialysis culture system (Fig. 3.14) allows com-

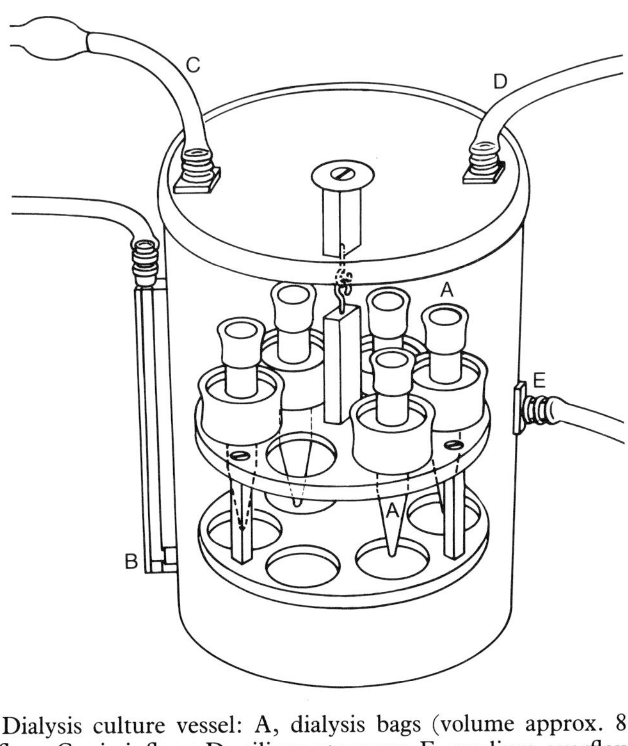

Fig. 3.14. Dialysis culture vessel: A, dialysis bags (volume approx. 8–10 ml); B, medium inflow; C, air inflow; D, silicon stoppers; E, medium overflow. The main vessel has a volume of 1.8 l.

parison of up to eight different strains of bacteria simultaneously at lower concentrations of nutrients than possible with chemostats or fermentor cultures. With such dialysis culture systems it has been shown that *B. japonicum* (generation time under optimal conditions 8–10 h) grows much slower than *Enterobacter aerogenes* (generation time under optimal conditions 30 min) when the succinate concentration is between 10^{-2} and 10^{-3} M (Fig. 3.15). At 10^{-4} M-succinate, *B. japonicum* had a population doubling time of 20 h, whereas that of *Enterobacter aerogenes* had increased to 50 h. Below this substrate concentration the *Enterobacter aerogenes* strain failed to grow whereas *B. japonicum* still had a 40–45 h doubling time. Apparently, carbon sources present in trace amounts in pure water were able to support bacterial growth since some growth still occurred in the absence of any added carbon source. Addition of 10^{-7} M succinate failed to stimulate bacterial growth above this background level. *Bradyrhizobium japonicum* strain 61-A-101 has two succinate uptake systems: the normal bacterial system with average affinity plus a second with higher affinity. The latter is absent in the *Enterobacter* strain tested (Humbeck *et al.*, 1985).

Bradyrhizobium strains can also survive at low concentrations of unknown C-sources for very long periods. During the first days after transfer of *B. japonicum* into distilled water the living cell count increases one order of magnitude (Fig. 3.16). For up to 16 months, the number of colony-forming units remains above that figure. Under optimal conditions; the survival rate for *Rhizobium* and *Bradyrhizobium* is also very high in peat; the survival

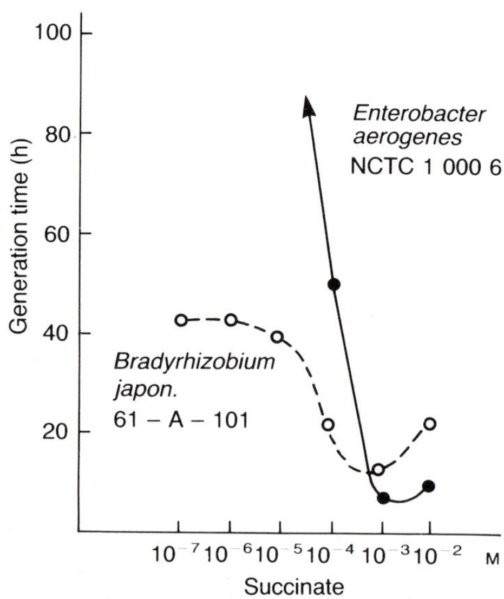

Fig. 3.15. Generation times of *B. japonicum* and *Enterobacter aerogenes* in dialysis culture at various succinate concentrations in continually flowing medium.

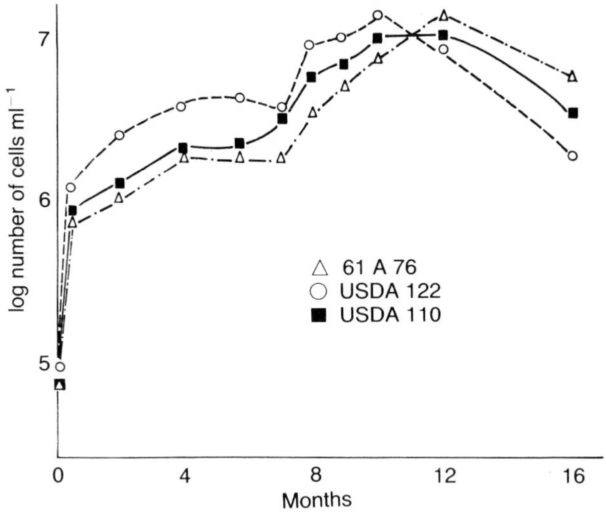

Fig. 3.16. Growth and survival of three strains of *B. japonicum* kept in distilled water at 24°C for 16 months. (Based on Crist *et al.*, 1984.)

rates being higher in sterilized than in non-sterile peat. Humus/peat is used as a carrier for inocula. However, in some areas sterilization is not always possible. Other carriers such as wood charcoal, sugar cane waste or compost have also been used with varying degrees of success. In some countries commercial inocula contain a mixture of strains whereas in others only one strain is present. In humus, counts of colony-forming units (cfu) rapidly increase to 10^{10} cells per g humus, then slowly fall to 10^{8}–10^{9} cells after 6 months (Fig. 3.17). With storage at −135°C the survival rate of *R. leguminosarum* biovar. *viciae* is 100% after one year. For *B. japonicum* (70%) and for *R. meliloti* (60%) the important factor is quick thawing of the cultures. Inoculation of pregerminated seeds with bacterial suspensions has the advantage that cell loss due to sticking to the seed coat is reduced. Even when rhizobia survive inoculation into soils, suppression of endogenous strains presents a further problem. After a legume harvest, an average soil can contain up to 10^{15} rhizobia per hectare in the upper 10 cm. This has to be compared with an inoculation of seeds or seedlings with 10^{12} rhizobia per hectare. For successful competition against the endogenous bacteria, the introduced strain must possess some selective advantage. When only 10^{4}–10^{5} endogenous bacteria per g of soil are present the inoculated strain can be found in 50–65% of the resulting nodules. This situation is more difficult in the tropics where the endogenous population of 10^{2} cells per g soil can seriously affect inoculation even at inoculation titres of up to 10^{9} *Rhizobium* cells per plant.

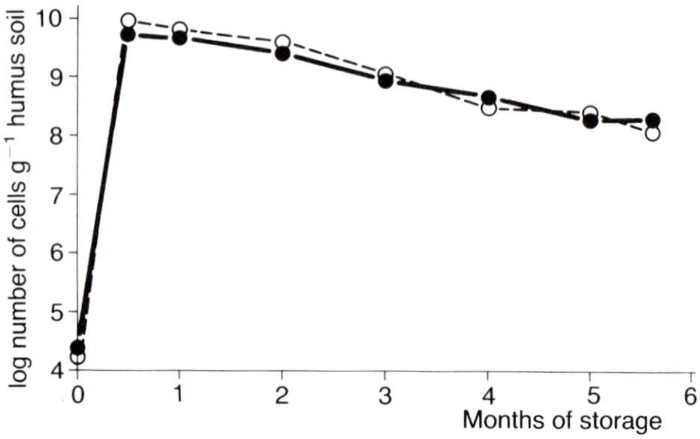

Fig. 3.17. Growth and survival of *B. japonicum* USDA 110 maintained in humic soil at 28°C. Dilutions in distilled water (○) or in yeast extract/mannitol (●). (Based on Somasegaran and Halliday, 1982.)

About 80% of the alfalfa production area in the United States is inoculated, but only 15% of the soybean production acreage.

3.1.5 Nitrogen fixation in pure cultures of *B. japonicum*

In 1975 research groups in several countries reported that *Bradyrhizobium* strains could fix nitrogen in pure culture, and were thus comparable to bacteroids in legume nodules (Kurz and LaRue, 1975; McComb *et al.*, 1975; Pagan *et al.*, 1975; Tjepkema and Evans, 1975). Previously it had been shown that nitrogenase could be expressed in associations with plant tissue culture. Since not only tissue cultures from legumes, but also from other plants were effective, it was clear that the function of the plant cells is the conditioning of the medium and the creation of niches of low oxygen partial pressure. For nitrogen fixation in pure or associative culture, bacteria require the equivalent of 1% O_2 at a precisely regulated rate and the presence of a C-source in the medium which supports the high energy turnover needed for nitrogenase activity. Moreover, the majority of the cells must be in a non-dividing developmental state (Fig. 3.18). In surface cultures on agar nitrogenase may be active under air where colonies contain a microaerobic zone. Here again, an easily metabolizable C-source such as succinate must be present. After transfer to such conditions, nitrogenase is newly synthesized, since the addition of chloramphenicol and rifampicin inhibit N_2-fixation (Fig. 3.19). With higher nitrogenase activity, cells show a reduction of their ATP content and energy charge. The specific activity of nitrogenase in *Bradyrhizobium* grown in pure culture is much lower (up to 150 nmol C_2H_4 h^{-1} mg^{-1} protein) than in free-living nitrogen fixers such as *Azospirillum brasilense* or *Klebsiella oxytoca* (up to 2000 nmol C_2H_4 h^{-1} mg^{-1}

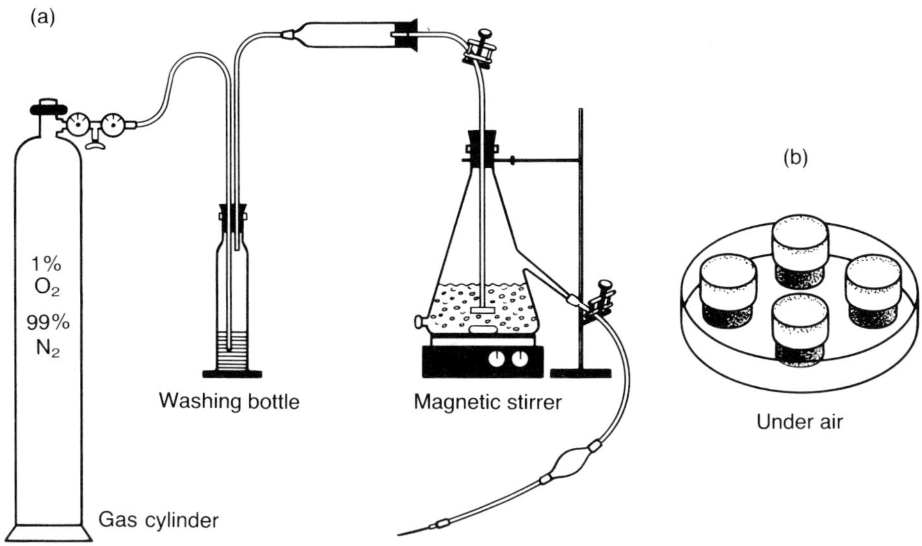

Fig. 3.18. Cultivation methods for derepression of nitrogenase in *B. japonicum*. (a) Suspension culture gassed with 1% O_2 and 99% N_2. (b) Surface culture (colonies) under air. For measurements of acetylene reduction, colonies were covered with glass or plastic rings and air-tight serum stoppers.

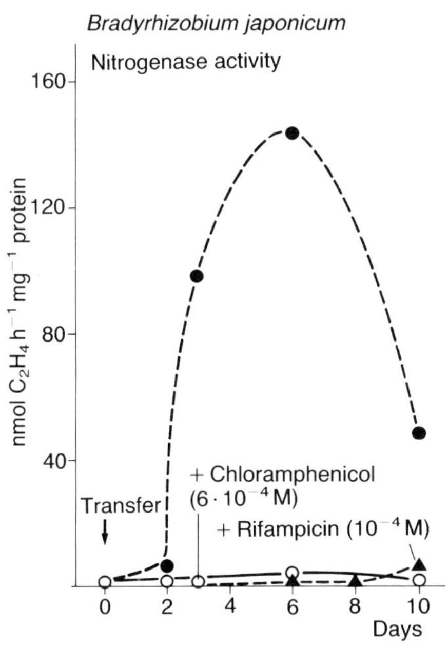

Fig. 3.19. Derepression of nitrogenase in axenic cultures of *B. japonicum* (●). Transfer of bacterial suspension to microaerobic conditions at point 0. Addition of chloramphenicol (○) or rifampicin (▲). (Based on Werner, 1978.)

protein). Nitrogenase activity is, however, in the same order of magnitude as the activity of isolated bacteroids.

3.2 THE MACROSYMBIONT: HOST PLANT; FABALES

The order Fabales (legumes) consists of more than 17 000 species within over 200 genera and is thus one of the largest in the plant kingdom (Allen and Allen, 1989). Characteristic properties of legumes are the synthesis of over 300 different, non-proteinogenic amino acids and specific alkaloids. Also noteworthy is the production of lectins and proteinase inhibitors in seeds. It is not known whether these properties have any connection with the ability to form symbioses with (*Brady*) *rhizobium* bacteria. Legumes existed as long ago as 55 million years, i.e. in the late Tertiary period. Papilionoideae similar to *Trifolium* and Mimosoideae similar to *Acacia* have been found in deposits from that age (Crepet and Taylor, 1985).

3.2.1 Systematics

Members of the Fabales are divided into three families:

1. Mimosaceae
2. Caesalpiniaceae
3. Fabaceae (Papilionaceae)

About 3400 species (20% of all legumes) have been tested for nodulation by rhizobia (Faria *et al.*, 1989). In the three families, the percentage of nodulated species is quite different: 97% of the Papilionaceae, 90% of the Mimosaceae and only 23% of the Caesalpiniaceae were nodulated. Considerable scepticism surrounds reports of nodulation in the genera *Amherstia*, *Bauhinia*, *Caesalpinia*, *Copaifera*, *Dialium*, *Senna* and *Tamarindus*. Most of the infectable Mimosaceae for symbioses with *Bradyrhizobium*. *Rhizobium* is also a possible symbiont of some members of the genera *Leucaena*, *Acacia* and *Mimosa*.

Since the Caesalpiniaceae are the most primitive family of the Fabales, it is assumed that the symbiosis was developed in a relatively late stage during legume evolution and had a polyphyletic origin.

For the Fabaceae species already examined, only the genera *Chaetocalyx* and *Cladrastis* have been convincingly demonstrated to become nodulated. Subfamilies can be differentiated on the basis of infection by either *Rhizobium* or *Bradyrhizobium* (Figs 3.20 and 3.21). Thus Trifolieae, Cicereae and Vicieae are infected, without exception, by *Rhizobium*. The Genisteae and Sophoreae, on the other hand, only by *Bradyrhizobium*. There are also subfamilies, e.g. Loteae and Phaeseoleae, which can be nodulated by either *Rhizobium* or *Bradyrhizobium*.

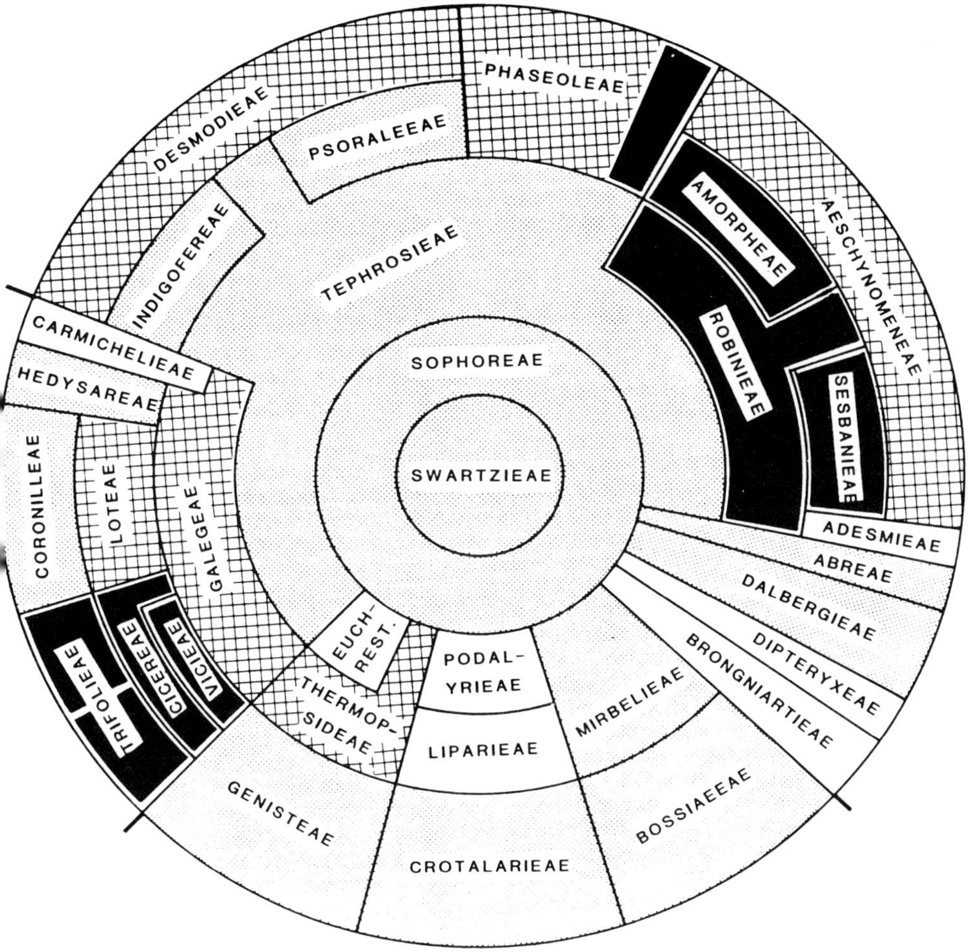

Fig. 3.20. Nodulation in subfamilies of the Fabaceae: exclusively by strains of the genus *Rhizobium* (black); exclusively by strains of the genus *Bradyrhizobium* (grey); by strains of both genera (chequered); no data (white). (Based on Young, 1985.)

The three families can be divided in the karyotype according to their chromosome number: 28% of the Fabaceae possess eight chromosomes (haploid), 15% have $n = 11$, 12% $n = 7$ and 16% $n = 16$ chromosomes. All chromosome counts for haploid cells are between 4 and 27. Polyploidy sometimes leads to counts as high as 80–90 chromosomes. For the other two families, chromosome counts are generally higher. In the Caesalpiniaceae 28% have $n = 14$, 30% $n = 12$ and 8% $n = 8$. The most prominent number in the Mimosaceae is $n = 13$ (57%), 15% have $n = 14$ and 12% $n = 26$.

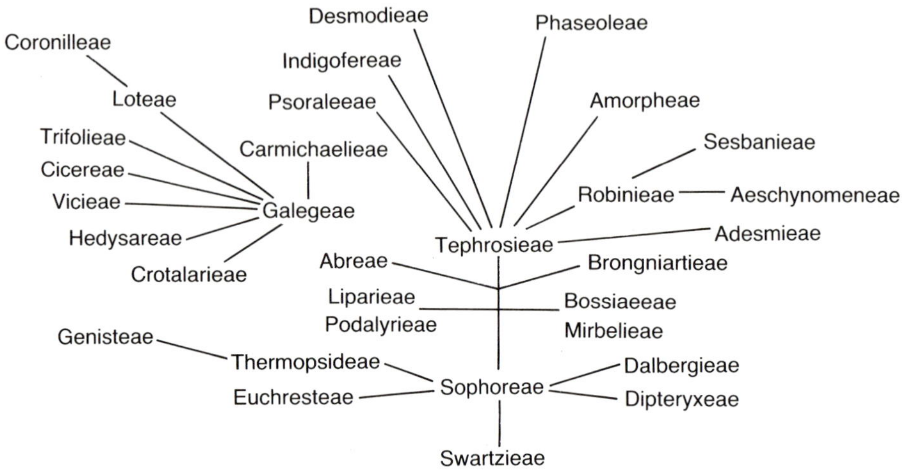

Fig. 3.21. Phylogenetic relationships between subfamilies of the Fabaceae.

3.2.2 Special physiological and biochemical properties

Roots, shoots and seeds of legumes are distinctive with respect to other plants in that they are especially enriched in unusual biochemical components (Bell, 1971).

(a) Non-protein amino acids

The organs of legumes contain more than 300 non-proteinogenic amino acids, such as methylene-glutamine (common to all legumes), canavanine found in the subfamily Lotoideae, and lathyrine, found only in the genus *Lathyrus*. Part of the chemotaxonomy of the legumes is based on the presence or absence of these compounds. Their physiological or ecological functions are mostly unknown. Compounds toxic for animals are thought to defend the plants against consumption. Non-toxic compounds are assumed to play some role in nitrogen storage. Canavanine, resembling arginine, toxic to bacteria, algae, fungi, higher plants, insects and mammals is restricted to occur in Papilionaceae only, verified in less than half of 1200 species (Bell *et al.*, 1978). The possible relationship between biosynthesis of these 'luxury' amino acids and the plant's ability to enter into symbiotic associations is unresearched. In addition to these amino acids, special peptides are also known. For example L-glutamyl-1-D-alanine has been found in pea root exudate.

(b) Alkaloids

To date, around 300 different alkaloids have been found in legumes. Around 10% of there are known to occur in species from other orders. Common to legumes are alkaloids belonging to the tertiary chinolizide alkaloids like α-isosparteine from *Sarothamnus scoparius* or 17-oxalupanine from *Lupinus polyphyllus* (Table 3.9). An example of highly specialized alkaloids is erythratinone in the genus *Erythrina*. Specific functions for such alkaloids in plant growth, development or resistance are not known. High N-content such as in agmatine from *Pisum sativum* or vicine from *Vicia faba* could possibly indicate an N-storage function. Nitrile groups could be substrates for nitrogenase. Thus alkaloids such as 2-aminopropionitrile from *Lathyrus odoratus* could affect nitrogen fixation. The general bacteriostatic effect of legume alkaloids such as spartein and lupanine, is three to four orders of magnitude lower than the effect of known antibiotics (Tyski *et al.*, 1988). The best hypothesis for the function of alkaloids in seeds is that they serve both as protectants against predation and as a nitrogen pool (Kishinevsky *et al.*, 1988).

(c) Phytoalexins

These compounds are synthesized by higher plants following pathogenic infection or imposition of stress. They have an antibiotic function. Their synthesis and chemistry differs from one legume to another. Among the phytoalexins of the Leguminosae, isoflavonoids are the most prominent group, both in distribution and in structural diversity. Other groups with more restricted occurrence are the benzofurans, chromones, flavanones, furanoacetylenes and the stilbenes. Isoflavones are synthesized by the genus *Lotus*, simple pterocarpanes by *Trifolium*, complex pterocarpanes by *Phaseolus* and furanoacetylenes by *Vicia* and *Lens* (Ebel, 1986).

Table 3.9 Alkaloids in the Fabales

Alkaloid	*Distribution (examples)*
Vicine	*Vicia faba*
Homostachydrine	*Medicago sativa*
2-Aminopropionitrile	*Lathyrus odoratus*
Erythratinone	*Erythrina glauca*
17-Oxalupanine	*Lupinus polyphyllus*
α-Isosparteine	*Cytisus scroparius*
Agmatine	*Pisum sativum*

(d) Isoflavonoids

In marked contrast to the flavonoids, the isoflavonoids have a very limited distribution in the plant kingdom, and are almost entirely restricted to the subfamily Papilionoidae. Their oestrogenic effect was discovered following the observation of a decline is birth rate for sheep fed on *Trifolium subterraneum*. Flavonoids and isoflavonoids have recently been identified as essential symbiotic signal compounds for nodulation (section 3.3.3). Changes in the isoflavonoid spectrum have been observed in suspension cultures of soybean challenged with intact bacteria (*Pseudomonas syringae* cv. *glycinea*) (Zacharius and Kalan, 1990). A sharp decline in the levels of daidzein and genistein occurred concurrently with increased glyceollin formation. In intact plants, only small amounts of genistein, the major isoflavonoid in cells from suspension cultures, were found.

Seeds and roots release different flavonoids. (Sathiyamoorthy, 1990). Quercetin-3-O-galactoside has been identified as the dominant flavonoid from alfalfa seeds, roots exude mainly deoxy-molecules (Philipps *et al.*, 1990; Hartwig *et al.*, 1991b). Luteolin 7-O-glucoside from the seed exudate can probably be hydrolysed by the plant as well as by *R. meliloti* to luteolin (Hartwig *et al.*, 1991a). The availability of flavonoids, such as luteolin, in the rhizosphere can actually limit nodulation in the rhizosphere (Philipps *et al.*, 1988).

(e) Lectins

Lectins are proteins which reversibly bind specific sugars or sugar sequences. They are known to occur in half the legume genera. They are

Table 3.10 Lectins in the Fabales

Origin	*Mol. wt. of subunits (kDa)*	*Specificity*
Glycine	30 (4×)	*N*-Acetylgalactosamine, D-galactose
Lotus	120 58 117	α-*L*-Fucose
Phaseolus	250 125	*N*-Acetylgalactosamine
Pisum	53	*N*-Acetylglucosamine D-Mannose
Robinia	100	*N*-Acetylneuraminic acid
Trifolium	50	2-Deoxy-D-glucose
Vicia	5.6 (2×) 20 (2×)	D-Glucose, D-mannose

especially concentrated in seeds, where they form up to 10% of the soluble protein. They are located to a large extent in the protein bodies of cotyledon cells. During germination the lectin content of cotyledons drops rapidly. In *Vicia faba*, lectin levels drop after digestion of reserve proteins whereas in *Dolichos biflorus* the levels of both decrease simultaneously. Seed and root isolectins may have different properties despite their general similarities (Diaz *et al.*, 1990).

The possible role of root lectins in recognition and infection during the development of the symbiosis will be described in section 3.3.1. Other functions of lectins could be in defence against consumption by animals, defence against phytopathogens, cell wall extension, growth rate regulation or carbohydrate transport. The sugar-specificity of important seed lectins from Fabales is given in Table 3.10.

(f) Sugar acids and lactones

More than 25 different sugar-related acids and lactones have been identified from legume leaves (Szczepanowska *et al.*, 1989). Several of these may have additional unknown functions in legume nodules.

3.2.3 Ecological distribution and agricultural importance

Crop legumes are planted annually worldwide on 1.3–1.5 million km^2 of land, resulting in an annual production of 200 million tonnes and an average harvest of 15 dt per hectare. Approximately half this area is planted with soybeans and 20% with various subtropical legumes such as *Cajanus* and *Cicer*. Peanuts grow on 10% and the rest is divided among *Phaseolus* spp., *Vicia faba* and *Pisum* spp. The market value of this production is well over $US 25 billion. Over 31 million km^2 of grassland area on the five continents is partially covered with legumes. In temperate climates *Trifolium*, *Lotus* and *Medicago* are most important. The quantity of legumes present is affected by various factors including soil phosphate content. The regular fertilization of pasture land with superphosphate raises the coverage of clover from 3% total plant mass to over 60% (Table 3.11). This can be reversed by grazing pressure from ruminants. *Medicago sativa* (alfalfa) alone is harvested from more than 300 000 km^2 world wide. A legume that has followed human migration all over the world is white clover, closely associated with the domestication of cattle (Klebesadel, 1986).

In tropical and subtropical open areas *Stylosanthes humilis*, *Macroptilium atropurpureum* and *Mimosa* spp. are very widespread. Since many tropical and subtropical legumes are trees, they constitute a large percentage of forest areas in these regions. This is especially so for the over 700 species of acacia in the bush and forest of Australia and the savannas of East Africa and South America. Fast-growing tree legumes could be very useful for

Table 3.11 Percentage coverage by legumes and non-legumes in grasslands in relationship to phosphate fertilization (after 20 years of fertilization). (Based on Crocker and Tiver, 1948)

	Superphosphate ($kg\ ha^{-1}\ year^{-1}$)		
	0	*100*	*200*
Grasses	80	25	14
Clover	3	45	63
Other herbaceous plants	17	30	23

reafforestation projects. Under good conditions *Leucaena leucocephala* and *Leucaena diversifolia* grow around 4 m per year (Ahmad and Ng, 1981). Their leaves also function as fodder for domestic animals. Work on *Leucaena* as a new agricultural plant has concentrated on the following points:

1. Breeding and selection of cultivars which are specialized for soil and climatic conditions in the various regions of the tropics and subtropics.
2. Selection and characterization of *Rhizobium* strains with effective N_2-fixation and nodulation. *Rhizobium* strain isolated from *Leucaena* nodules belong to the fast-growing, acidifying strains. Some also infect *Medicago sativa*.
3. Properties and uses of *Leucaena* wood.
4. Nutritional value and physiological effects of grazing animals on *Leucaena* leaves.
5. *Leucaena* leaves as green manure for other crops such as rice, maize, coffee and coconuts.

Table 3.12 Effect of legumes and actinorhizal plants on the growth of *Pinus densiflora* in N-free culture solution. Fallen leaves were removed. Plant parameters were recorded after 30 months, with the data from a *Pinus* monoculture taken as 100%. (Based on Uemura, 1961)

Culture	*Plant height (%)*	*Shoot weight (%)*	*Total plant weight (%)*
Pinus densiflora monoculture	100	100	100
+ *Robinia pseudoacacia*	222	1112	748
+ *Acacia mollissima*	154	471	798
+ *Alnus tinctoria*	167	393	314
+ *Elaeagnus umbellatus*	160	363	276

Leguminous trees stimulate the growth of neighbouring plants. This effect is not only due to the N-rich leaves which fall from the legumes. The growth of *Pinus densiflora* in mixed culture with *Robinia pseudoacacia* is stimulated eightfold within 30 months (Table 3.12). This increase is smaller

Table 3.13 The 50 major N_2-fixing trees (Based on Brewbaker, 1990)

Acacia auriculiformis (fuel, pulpwood, timber, ornamental), *A. farnesiana* (perfume tannin, fuel, dye, ornamental), *A. mangium* (pulpwood, timber), *A. mearnsii* (tannin, pulp, charcoal, shelterbelt, poles, fuel), *A. nilotica* (fuel, charcoal, tannin, fodder (pods, leaves), gum), *A. saligna* (fuel, soil stabilization), *A. senegal* (gum arabic), *A. tortilis* (fodder (pods, leaves), fuel)
Aeschynomene elaphroxylon (corky wood, ornamental, fodder)
Albizia lebbeck (timber, ornamental, fuel), *A. procera* (timber, ornamental, tannin, poles, fuel), *A. saman* (timber, craftwood, fodder (pods), ornamental)
Alnus acuminata (fuel, craftwood), *A. glutinosa* (fuel, soil stabilization, craftwood), *A. nepalensis* (furniture)
Baphia nitida (dye, craftwood, medicinal, green manure)
Cajanus cajan (food, fodder, green manure, fuel)
Calliandra calothyrsus (fuel, green manure)
Casuarina equisetifolia (fuel, shelterbelt, ornamental)
Dalbergia sissoo (timber, fuel)
Derris indica (fuel, fodder, oil, medicinal, ornamental)
Elaeagnus angustifolia (shelterbelt, ornamental)
Enterolobium cyclocarpum (ornamental, fuel)
Erythrina poeppigiana (fencerow, ornamental, windbreak, nurse tree, green manure), *E. variegata* (ornamental, nurse tree, windbreak)
Faidherbia albida (fodder, shade, green manure)
Flemingia macrophylla (green manure, dye)
Gliricidia sepium (fuel, nurse tree, ornamental, fodder)
Hardwickia binata (timber, fodder)
Hippophae rhamnoides (food (berry), medicinal)
Inga vera (fuel, timber, shade, honey)
Leucaena diversifolia (nurse tree, poles, fuel), *L. leucocephala* (fuel, fodder, poles, food, pulp, nurse tree, green manure)
Mimosa scabrella (fuel, pulp, nurse tree, ornamental)
Paraserianthes falcataria (timber, pulpwood, ornamental)
Parkia javanica (food (pods), ornamental)
Pithecellobium dulce (poles, ornamental, fodder (pods), fuel)
Prosopis chilensis (fuel, fodder (pods), ornamental), *P. pallida* (fuel, charcoal, fodder (pods), honey), *P. tamarugo* (fuel, fodder (pods), alkaline soil stabilization)
Pterocarpus erinaceus (craftwood, fodder (leaves), dye, tannin), *P. indicus* (timber, ornamental)
Robinia pseudoacacia (fuel, poles, ornamental, nurse tree, honey)
Sesbania grandiflora (fodder (foliage), food (flowers, pods, leaves) ornamental, nurse tree), *S. sesban* (fodder, green manure)
Sophora japonica (medicinal, ornamental)
Tephrosia candida (green manure, nurse tree, ornamental)
Tipuana tipu (timber, ornamental)

when *Pinus* is planted in mixed culture with *Alnus tinctoria* or *Elaeagnus umbellata*.

Assessment of nitrogen fixation by legume trees is more difficult than for annual agricultural crop plants since nodule age and distribution over the root system is more heterogenous. In addition, a diurnal variation with the highest nitrogenase activity at the end of the light period has been observed for *Acacia albida* and *Leucaena leucocephala* (Sellstedt *et al.*, 1989). The annual variation for *Acacia catechu* trees shows a maximum between August and September, i.e. at the end of the rainy season in India (Pokhriyal *et al.*, 1990). A total of 50 different species of nitrogen fixing trees have been designated by the Nitrogen Fixing Trees Association as outstanding in their economic or ecological importance (Table 3.13) (Brewbaker, 1990).

3.3 THE SYMBIOSIS

From the over 17 000 species of legumes, only 10% have been investigated with regard to their ability to form symbioses, their microsymbionts and nitrogen fixation capability. It is more difficult to prove that a certain species does not nodulate than to prove that it does (Sprent, 1990). *Delonix regia* (Royal poinciana), for example, has neither in the field nor in the laboratory (with many different rhizobia) been found to be nodulated. Due to their agricultural importance, the best-studied legumes are soybeans (*Glycine max*), peanut (*Arachis hypogaea*), lucerne (alfalfa) (*Medicago sativa*), pea (*Pisum sativum*), clover (*Trifolium* spp.), the broad bean (*Vicia faba*) and the common bean (Phaseolus vulgaris) (Vance, 1990).

The development of symbiosis can be viewed as an exchange of signals (Fig. 3.22) (Brewin, 1991; Werner, 1992). The first part involves an exchange of signals for recognition between host and microsymbiont during the early stages of infection of root hairs or other epi- or rhizodermal cells. The differentiation of free-living bacterial cells to bacteroids (section 3.3.4(a)) modifies the signals from the microsymbiont. Host reactions are the synthesis of nodule-specific proteins (nodulins), e.g. leghaemoglobin, or sucrose synthase. Evolution of symbiosis seems to be more a gathering of genes than a gene invention (Long, 1989a,b; Sprent and Sprent, 1990). The nodulation of legume rhizobia symbiosis is also a sensitive bioindicator towards toxic chemicals in the soil such as polycyclic aromatic hydrocarbons (Wetzel *et al.*, 1991).

3.3.1 Recognition

The components involved in the recognition process between legume plants and *(Brady)rhizobium* bacteria have not yet been definitively identified. Numerous experiments indicate that lectins (from the plant) and EPS

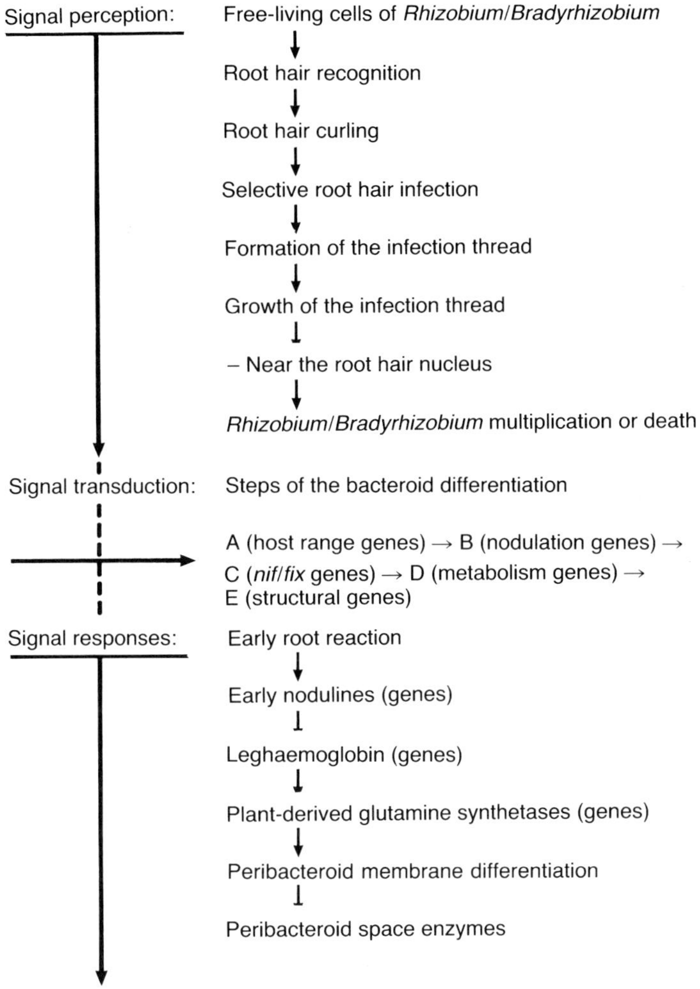

Fig. 3.22. Signal transduction in the Fabales–*Rhizobium/Bradyrhizobium* symbiosis.

and LPS (from the bacteria) are involved in recognition. However, other experiments indicate that these are not the only factors involved.

In extracts of peanut nodules a mannose-specific lectin has been detected, which was not present in root extracts. In addition, a galactose/lactose-specific lectin was present in both roots and nodules (Kishinevsky *et al.*, 1988). The specific nodule lectin differed from the seed lectin in molecular weight and amino acid composition and appeared to be a glycoprotein (Law *et al.*, 1988). As in the binding and recognition between erythrocytes and malaria merozoites, the O-linked tetrasaccharide of the glucophorin, carrying two subunits of neuraminic acid, is involved (Hermentin, 1987).

A correlation between extracellular fibrils and the attachment of microsymbionts to root hair tips was found in *R. leguminosarum* and pea plants (Smit *et al.*, 1986).

3.3.2 Surface interactions between legume root hairs and rhizobia

The interaction starts with an exchange of diffusible factors. Flavonoids from the plant rhizosphere (Fig. 3.23) induce the synthesis of the gene products of the *nod* and *hsn* genes, such as Nod Rm-1 (Fig. 3.11). These compounds are involved in the redirection of apical growth of root hairs following bacterial attachment. Attachment to root hairs is a two step process in the pea–*R. leguminosarum* biovar. *viciae* system (Kijne *et al.*, 1990). In the first step, single rhizobial cells adhere to the surface of the developing root hair using the calcium-binding protein rhicadhesin. This 14 kDa protein also mediates attachment to non-legume root hairs. The further development of a single rhizobial cell to a small colony involves cellulose fibrils from the bacteria and lectins from the host plant. In peas

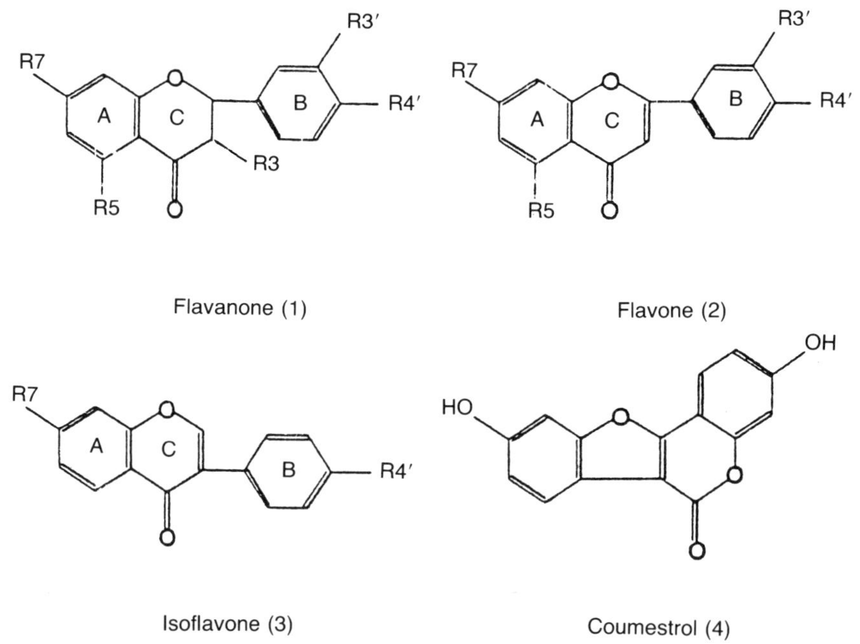

Fig. 3.23. Structures of flavonoids present in root exudates of host plants and inducing *nod* gene expression in rhizobia. (1) as 3,5,7,3′-tetrahydroxy-4′-methoxyflavanone, inducer in *Rhizobium leguminosarum* bv. *viciae*; (2) as 3′,4′,5,7-tetrahydroxyflavone, inducer in *Rhizobium meliloti*; (3) as 4′,7-dihydroxyisoflavone, inducer in *Bradyrhizobium japonicum*; (4) coumestrol: intermediate in phenylpropane metabolism, weak inducer. (Modified from Hansmann *et al.*, 1989.)

Table 3.14 Concentrations of mineral elements in root hairs and in the root system of 8-day-old seedlings of *Glycine max* (soybean) and *Triticum aestivum* (wheat). Germination on mineral agar. (Based on Werner *et al.*, 1985a)

Element	*Concentration (μg g^{-1} dry matter)*			
	Soybean		*Wheat*	
	Root hair	*Root*	*Root hair*	*Root*
Fe	414 ± 138*	31 ± 5	120 ± 35	44 ± 26
Co	7.9 ± 3.8	0.9 ± 0.4	2.6 ± 0.8	1.3 ± 1.1
Ca	2 200 ± 460	287 ± 70	246 ± 60	288 ± 70
Mo	3.1 ± 0.5	5.4 ± 0.7	0.6 ± 0.1	0.5 ± 0.3
K	1 1740 ± 2 450	12 840 ± 2 640	4670 ± 1010	4780 ± 990
S	530 ± 165	560 ± 170	180 ± 55	190 ± 60
P	1 326 ± 410	2 590 ± 790	526 ± 165	1570 ± 475

* Standard deviation

the presence of a 49 kDa lectin with two binding sites for oligosaccharides on the outer surface of the trichoblast plasmalemma has been demonstrated. Some lectins are secreted into the rhizosphere (Diaz, 1989). Some of these lectins may also remain at the cell wall surface, half-way between the plasmalemma and the rhizosphere. Rhizobial *nod* genes are not required for the lectin-mediated surface colony growth. The pea root surface lectin appears at the surface of the epidermal cells when they differentiate into trichoblasts and emerging root hairs. The surface of the mature root hair zone no longer contains detectable lectins and is also not susceptible to *Rhizobium* infection. Involvement of plant lectins in the host specificity of infection has been elegantly shown by transformation of white clover roots with a functional pea lectin gene. As a result, *R. leguminosarum* biovar. *viciae*, the microsymbiont of peas, could induce infection threads in clover (Diaz *et al.*, 1989).

Some other characteristics of root hairs may also be important. By using the differences in mechanical stability of root hairs and roots after freezing, it is possible to obtain sufficient material for studies on a preparative scale (Röhm and Werner, 1987). With such preparations some specific properties of root hairs have been found (Table 3.14). Iron, cobalt and calcium are eightfold enriched in root hairs relative to roots. Legume root hairs contain 3–8 times more of these elements than root hairs of wheat. The calcium concentration in soybean root hairs is 2200 μg g^{-1} dry matter whereas that in wheat root hair is 250 μg g^{-1}. Ca^{2+} contents of the tap roots of soybean and wheat are approximately equal at 280 μg g^{-1}. Potassium and sulphur are equally abundant in root hairs and total roots of soybean (12 000 and 540 μg g^{-1}, respectively) (Werner *et al.*, 1985a).

The enrichment of Fe, Co and Ca in root hairs is interesting because the microsymbiont, *B. japonicum* has a high requirement for these elements. This has led to the hypothesis that the bacteria, after infecting the plant are supplied not only with C-sources, but also with specific minerals. Legume plants relying on symbiotic N_2-fixation have a higher requirement for Fe. Under conditions of iron deficiency, nodule initiation was more severely retarded than asymbiotic growth of the host plant. This indicates that nodule initiation and functioning has a high iron requirement (Tang *et al.*, 1990).

The differentiation of root cells into root hair cells is characterized by root hair-specific proteins. These have been found in both soybean and peas. They occur in both soluble and membrane fractions. Their function is still unknown. Especially noteworthy are the membrane proteins with extracellular domains structurally suggestive of a function as signal or receptor molecules. A membrane protein from peas 32 kDa has been found, using fluorescence labelling techniques *in situ* and subsequent fractionation (Röhm and Werner, 1987). This, however, is not identical to the pea lectin (53 kDa) (Table 3.10).

3.3.3 Infection and nodule development

In most legumes, nodule development begins with the infection of a developing root hair. The first morphologically recognizable reaction of the plant is the curling of the root hair (Fig. 3.24). In the soybean system (Fig. 3.25), the bacteria then penetrate the slime layer and are pressed against the root hair cell wall or the cell wall of a nearby epidermal cell, through the curling action of the root hair. The cell wall is subjected to local disintegration. This constitutes the initiation of the infection thread. The so-called infection sack grows intracellularly within the root hair. At this point the bacteria are embedded in an electron translucent matrix material (Fig. 3.26). Around this is a net of cell wall fibrils which are connected with the regularly ordered fibrils of the cell wall. These are separated from the cell cytoplasm of the root hair cell by a membrane which is continuous with the plasma membrane. In the surrounding cell cytoplasma a massive increase in the amount of Golgi, endoplasmic reticulum (ER) and microtubules is evident. This **infection sack** branches out in the form of tubes or threads (Fig. 3.27). The cortical cells at this stage already have a meristematic character, with a small vacuole, a large cell nucleus and clear nucleoli. Those host cells further away from the **infection thread** have much larger central vacuoles. The root hair cell itself remains vacuolated and the increase in dense cytoplasm and cellular organelles is localized around the infection sack. The tips of the growing infection threads are surrounded by only a very thin layer of cell wall material. Three days after inoculation, successful infection threads enter the cortex. The favoured path is between

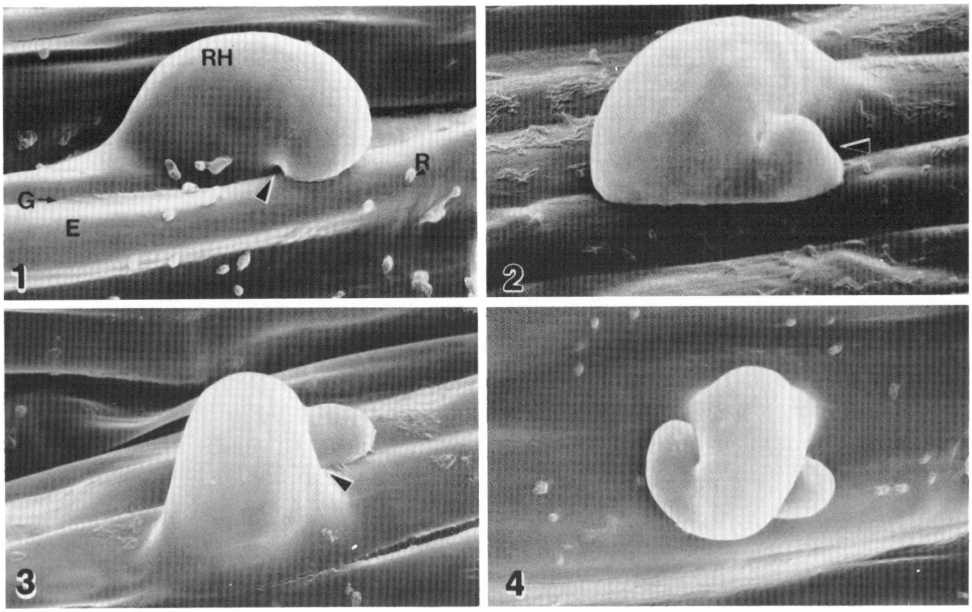

Fig. 3.24. Curling root hairs of soybean after infection with *B. japonicum*. RH, root hair cell; E, neighbouring endodermis cell; R, *Bradyrhizobium* cell; G, gap between curled root hair and epidermal cell. (Based on Turgeon and Bauer, 1985.)

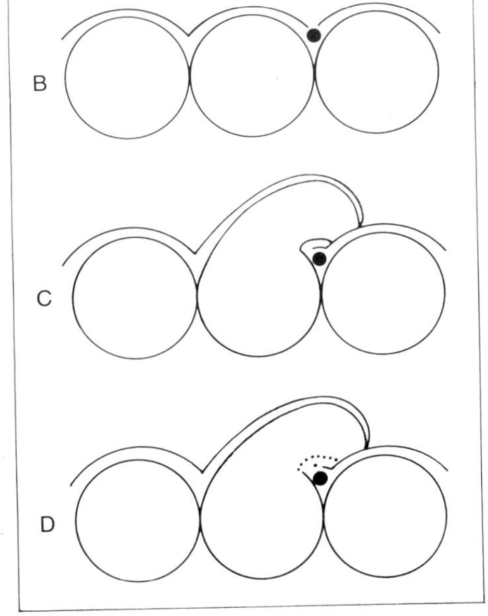

Fig. 3.25. Model for root hair infection in soybean. (A, B) *B. japonicum* cell (●) penetrates the root mucigel layer between two rhizodermis cells. (C) Development of rhizodermal cell into a root hair, deformation of the root hair and entrapped bacterial cell. (D) Local changes of root hair cell wall (. . .) and synthesis of new wall material between bacterium and old cell wall. (Based on Turgeon and Bauer, 1985.)

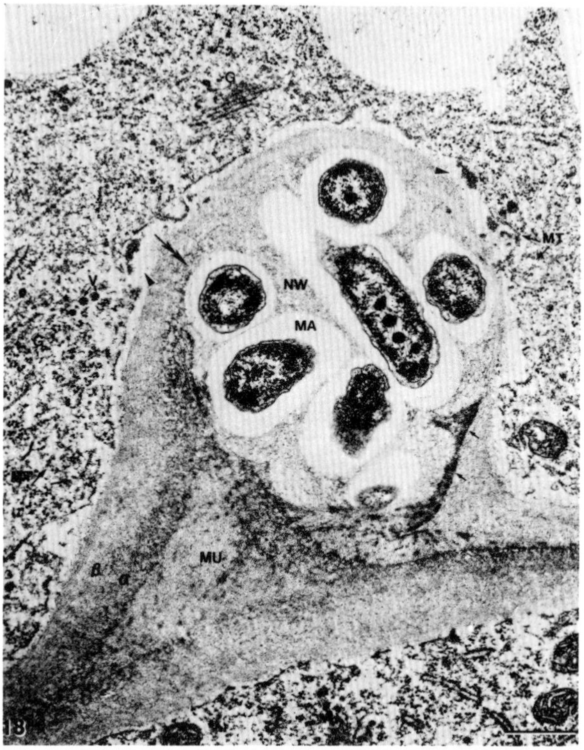

Fig. 3.26. Infection thread at the base of a root hair cell from *Glycine max*. Bacteroids are surrounded by light-coloured matrix (MA) next to a layer of disarranged wall fibrils (NW). The cytoplasm contains abundant Golgi vesicles (G) and microtubules (MT). MU, Mucigel. (Based on Turgeon and Bauer, 1985.)

two newly divided cells. Pretreatment of *R. meliloti* with luteolin also triggers root curling in the non-host plant of white clover and in *Avena sativa* (oat). However, as expected, no infection thread formation is observed in the oat seedlings (Terouchi and Syono, 1990). A treatment of clover root hairs with cell wall-degrading enzymes (cellulase and pectolyase) removes some plant barriers which also determine host specificity. This allows, for example *R. loti*, which otherwise will not infect *Trifolium repens*, to nodulate clover (Al-Mallah *et al.*, 1987).

A light microscopic study of the sequence of cell divisions is presented in Fig. 3.28. After inoculation the first cell divisions can be observed in the hypodermis within 12 h. Four to eight daughter cells are formed through anticline divisions in the cortex. At this time a small group of dark-coloured meristematic cells are observed. At 3–4 days after inoculation the meristematic zone has developed a recognizable swelling on the root surface. The

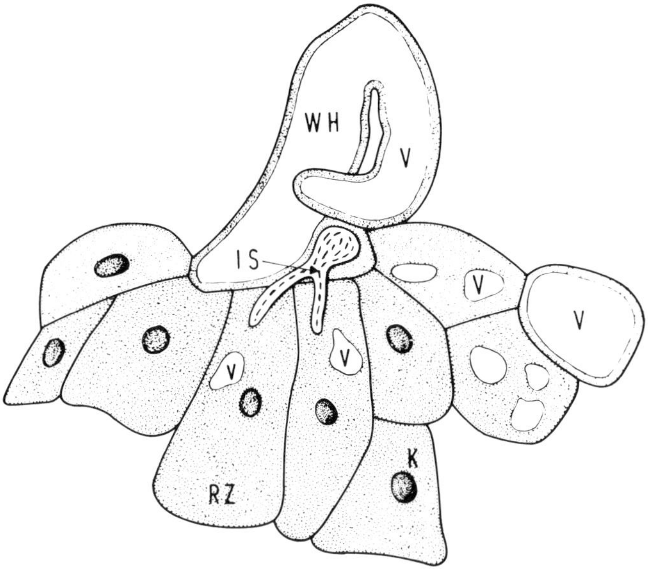

Fig. 3.27. Soybean: development of a pouch-like, then slender and branched infection thread from a root hair cell (WH) towards meristematic cortex cells (RZ) with small vacuoles (V) and large central nuclei (K). Infection thread and bacteria are not reproduced in correct scale (enlarged).

nodule meristem is connected to the vascular system of the root about 6 days after inoculation (Fig. 3.28).

Parallel to this typical sequence of infection events is the occurrence of many so-called **pseudo infections**. In these cases, initiation of cortical cell division is observed without the formation of infection threads. Cell division is restricted to the outermost two or three layers of cells. This does not lead to a macroscopically observable nodule. Sometimes in the case of successful infection thread formation, the growth of the thread precedes cell division. The fact that infection thread growth and cortical cell division are not coupled is also supported by experiments with *sym*-plasmid-containing *Agrobacterium tumefaciens* or *E. coli*, where cell division without infection thread formation is observed. In these cases, cell divisions do result in obvious nodule swellings, even though no bacteria are found inside.

Three days after inoculation with *B. japonicum*, 80 pseudoinfections were found on a 7 cm long piece of root together with 50 root hair infections. Such roots usually develop five nodules. Thus, even where infection threads are initiated, over 90% of these are abortive.

When successful infection threads reach the meristematic cortex, bacteria bud off into these cells. This first invagination is bordered by a membrane *in continuum* with the host cell plasma membrane. The individual bacteria

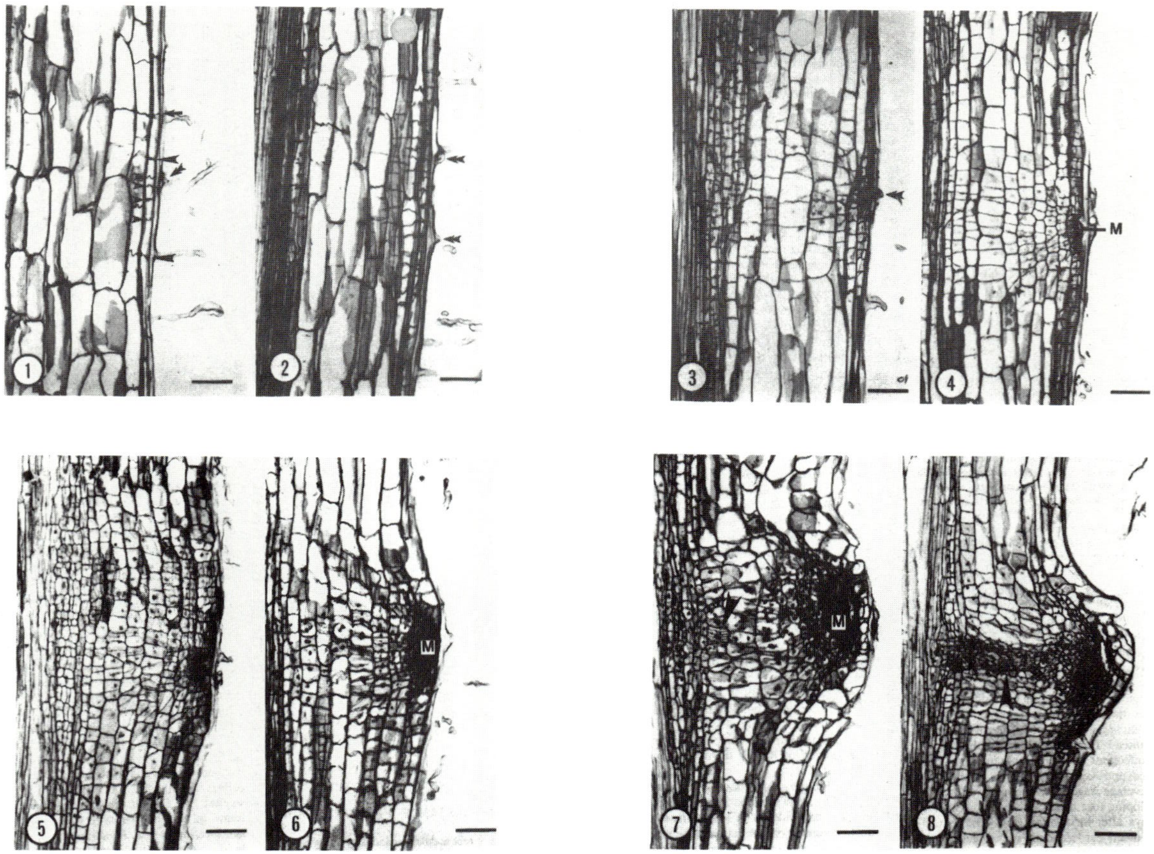

Fig. 3.28. Longitudinal sections of the cortex of a root and a developing soybean nodule, after infection with *B. japonicum*. M, Meristem. Arrows indicate areas of active cell divisions. (Based on Calvert *et al.*, 1984.)

are later enclosed in the so-called symbiosome membrane (section 3.3.4(b)). This process is influenced by bacterial genes. There are mutant strains of *B. japonicum* which are not released from the infection thread. At the time of release the meristematic cells are not yet polyploid. The question arises: is polyploidy needed for infection or do infected cells become polyploid? For soybean and clover the latter alternative may be true. Development of bacteroid-containing cells from diploid as well as from polyploid cells has been observed in lupins (Olszewska and Legocki, 1989). The degree of ploidy is raised in *Lathyrus*, *Medicago*, *Trifolium* and *Vicia* from $2n$ in the cortex cells to tetraploidy in infected nodule cells. In nodule cells of *Pisum sativum* it can be as high as $8-16n$.

Legume nodules contain high concentrations of auxins, kinins and gibberellins. In pea and alfalfa nodules, auxin (indoleacetic acid, IAA) can reach levels 40–60 times higher than in roots. *Rhizobium* can metabolize exogenous tryptophan to auxin. Whether these high auxin levels are produced by the plant or bacteria is not known. IAA oxidase activity in nodules is dramatically reduced after infection. The high IAA concentrations ($10^{-7}-10^{-6}$ g IAA g^{-1} nodule fresh weight) may simply be due to reduced degradation. Auxin can induce rapid changes in phosphatidylinositol metabolites, an effect that is generally involved in signal transduction by auxins (Ettlinger and Lehle, 1988). Several auxin binding proteins (ABP) have been identified. The most conclusive studies indicate that in maize a 22 kDa ABP has a receptor function (Jones, 1990). In nodules, the presence of such receptors still has to be established.

Cytokinin in nodules of *Phaseolus vulgaris* reaches 1.5×10^{-7} g kinetin equivalents per g nodule fresh weight. In *Vicia faba* nodules this figure is 10–15 times higher than in roots. Pure cultures of *R. leguminosarum* produce cytokinin to an extracellular concentration of 10^{-9} g ml^{-1}. LPS from *Rhizobium* also contains cytokinin (up to 2×10^{-7} g g^{-1} LPS). However, it is not known if cytokinin production by bacteria in the infection thread stimulates cell division. The gibberellin concentration in nodules is also 40–100 times higher than in root tissue. *Rhizobium leguminosarum* can export various gibberellins. The involvement of these three phytohormones in the various steps of nodule differentiation can only be explained when the location of the biosynthetic and catabolic pathways is known. The functions of abscissic acid and ethylene in nodules are not known apart from the fact that they inhibit nodule development in concentrations too low to inhibit secondary root formation. The sequence of different tissues in an indeterminate nodule such as from pea or faba bean is summarized in Fig. 3.29, together with the expression of some nodule specific genes. Nodule form and morphology fall basically into two categories. Type A has an apical meristem with non-determinate growth and a cylindrical nodule. These occur for example on *Vicia* and *Trifolium*. These nodules have a branched vascular system, the uppermost end being con-

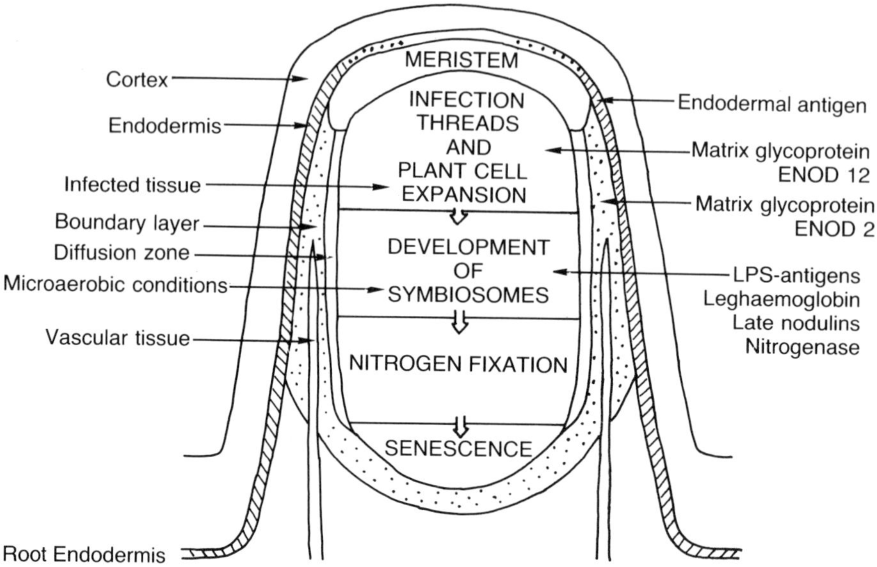

Fig. 3.29. Diagrammatic longitudinal section through a pea nodule. Left: organization of cell layers; centre: tissue invasion by *Rhizobium* sp. and the development of nitrogen-fixing organelles; right: development of gene expression as identified by antibody or cDNA probes. (Based on Nap and Bisseling, 1990; Kannenberg and Brewin, 1989.)

nected to the meristematic zone. These nodules are biochemically characterized by amides, asparagine and also glutamine as N-transport compounds, and have transport cells in the pericyclic tissue. The second type of nodules are spherical, with determinate growth. The vascular branches encircle the infected tissues and are connected to the root vascular bundle at the base. Transfer cells have not been reported here. N-transport in such nodules is mediated via ureides such as allantoin and allantoic acid. This type of nodule is characteristic for *Phaseolus* and *Glycine*. A relationship between nodule type and N-transport form is postulated. Since ureides are less soluble than amino acids they can be better transported in the closed vascular system of a type B nodule. Nodules of other legumes, e.g. *Aotus ericoides*, lie between the two types. Cross-inoculation experiments prove nodule morphology to be plant-determined (Fig. 3.30).

Nodules developing on soybean root fall into several discontinuous size classes (Fig. 3.31). Independent of their age, some size classes are favoured over others (classes 2 mm, 2.6 mm and 3.3 mm diameter). Size classes between these are less abundant. It is evident that nodule size and nodule age are independent factors. Young, big nodules can be present along with old, small ones. The number of nodules developing on a root system is

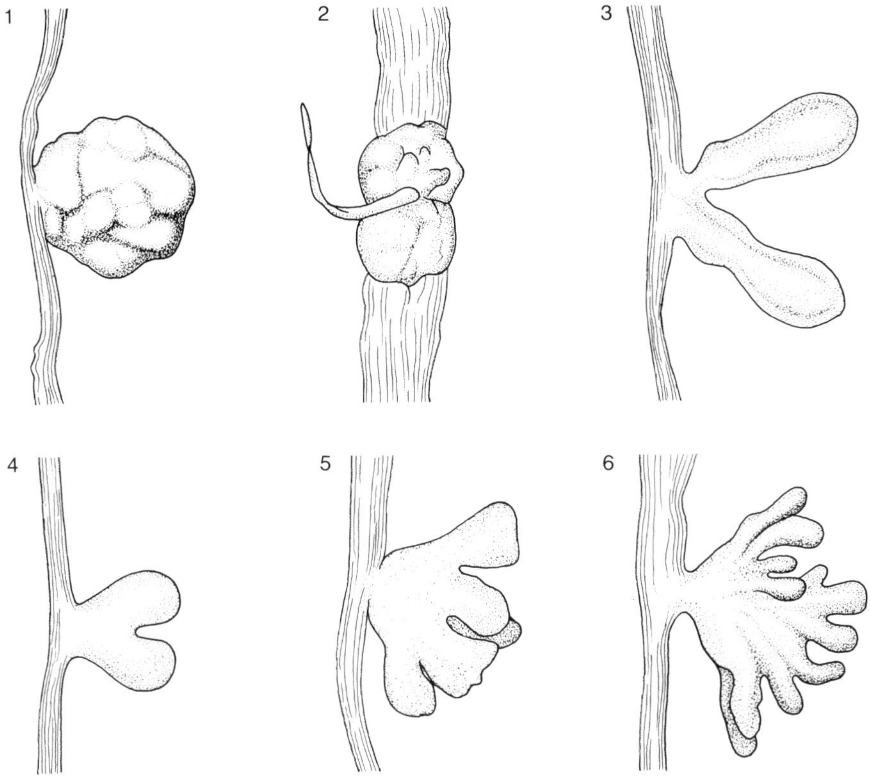

Fig. 3.30. Morphology of Fabales nodules. (1) Spherical/globose, e.g. in *Glycine*, *Erythrina*, *Vigna*; (2) 'Siamese', e.g. in *Arachis*; (3) elongate, e.g. in *Vicia*; (4) bifurcate, e.g. in *Medicago*, *Leucaena*; (5) fan-shaped and palmate, e.g. in *Acacia*, (b) coralloid, e.g. in *Ibiziamollucana*. (Based on Allen and Allen, 1981.)

affected by nitrogen fixation. In ineffective symbioses (non-N_2-fixing), over 100 nodules per root system can be observed. In effective (N_2-fixing) symbioses the average is 20–30 nodules in 50 days (Fig. 3.32).

(a) Nodulins: nodule specific proteins

Over 30 nodule-specific proteins have been visualized using molecular techniques. A further 40 are postulated to exist based on nucleic acid analyses. The limitations of these techniques means that for the majority only their molecular weight is known. After mRNA isolation and translation in the presence of 35S methionine more than 500 labelled polypeptides can be separated on two-dimensional gels. The *in vitro* translation system used translates only eukaryotic mRNA, so bacterial polypeptides are

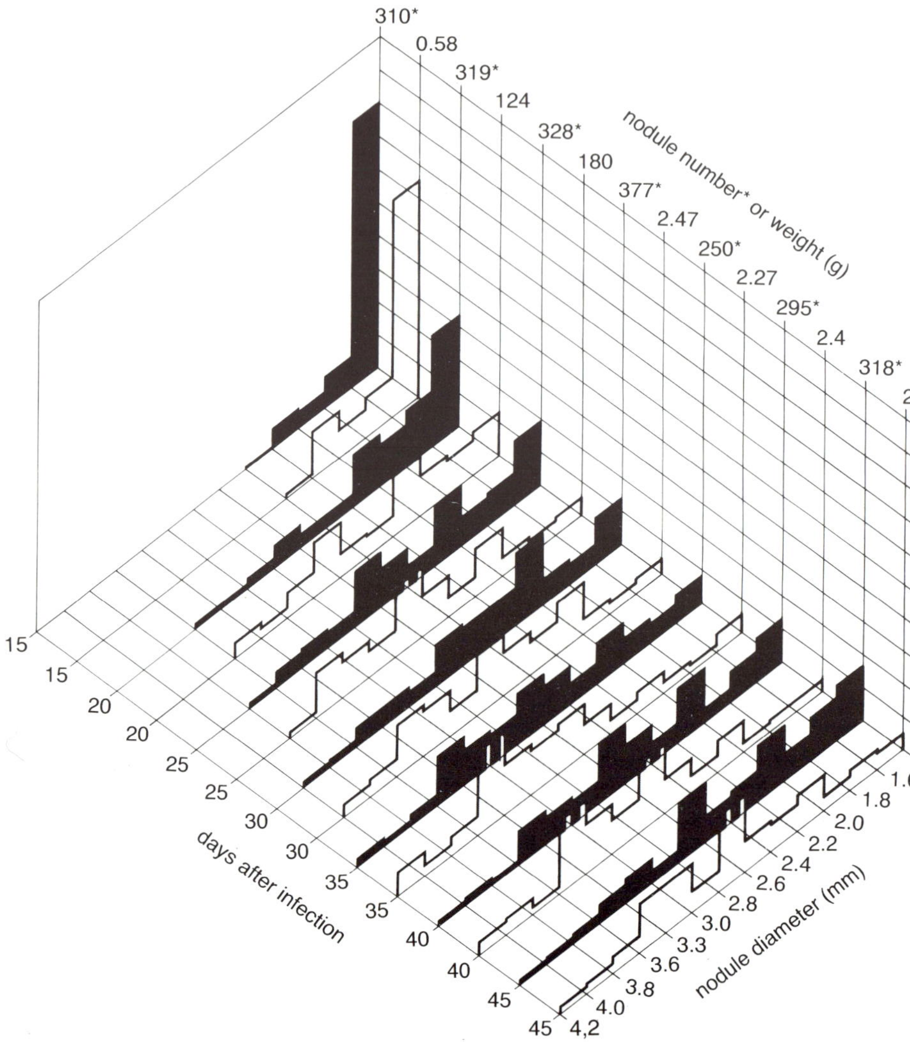

Fig. 3.31. Distribution of different nodule sizes (1.4–4.2 mm diameter) in *Glycine max* cv. Caloria infected with *B. japonicum* 61-A-101, 15–45 days after infection. Black bars: percentage of total nodule number; white bars: percentage of total nodule fresh weight. Data are given for ten plants. (Based on Stripf and Werner, 1980.)

not labelled. Comparisons between gels of root mRNA and nodule mRNA result in the detection of nodulins.

In infected root hairs, one 40 kDa nodulin has been found. At 15 days after infection the four leghaemoglobins are to be seen together with 15 other nodulins. Nodulins N-68 (68 kDa) and N-21 are qualitatively highly expressed (Fig. 3.33). These results have been confirmed using nodulin-

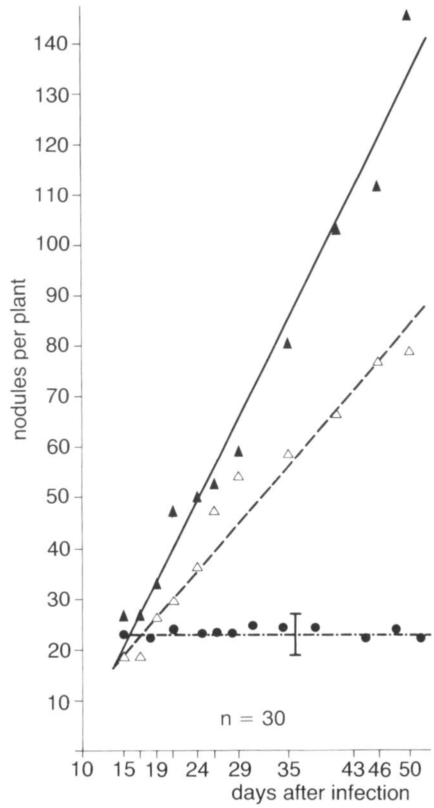

Fig. 3.32. Nodule number in *Glycine max* cv. Caloria (▲) and cv. Mandarin (△) after infection with the ineffective (*fix*$^-$) strain *B. japonicum* 61-A-24. Nodule numbers after infection with the effective (*fix*$^+$) strain 61-A-101 varies only within the range indicated with bars on the horizontal line.

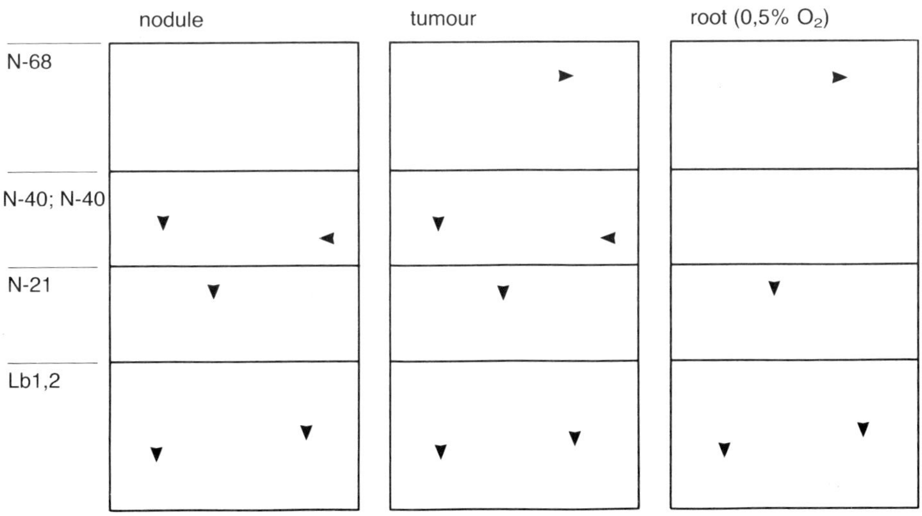

Fig. 3.33. Nodulins from 15-day-old pea nodules, compared to proteins from tumours and from roots grown under 0.5% O_2. *In vitro* tRNA was analysed by fluorography. (Based on Bisseling *et al.*, 1980a.)

Table 3.15 Expression of early nodulins. (Modified from Franssen *et al.*, 1990 and Scheres *et al.*, 1990)

Nodulin	*Protein characteristic*	*Appearance (days after inoculation)*	*Tissue organ specificity*
ENOD 12	hydroxyproline-rich cell wall protein	8–10	Root hairs, cortical cells with infection threads
ENOD 5	arabinogalactan protein	10–13	Early symbiotic zone
ENOD 2	hydroxyproline-rich cell wall protein	10–13	Symbiotic zone, after disappearance of ENOD 12
ENOD 14	Cysteine clustered protein	13–17 just before start of nitrogenase activity	Symbiotic zone peribacteroid space ?
ENOD 3	Cysteine clustered	Just before start of nitrogenase activity	Symbiotic zone peribacteroid space ?
Leghaemoglobin (for comparison)		13–17	Host cell cytoplasm

specific cDNA clones. Gene expression and characteristics of the early nodulin proteins are summarized in Table 3.15. For the induction of ENOD 12, the common *nod* genes and the *hsn* genes of *R. leguminosarum* must be active (Scheres *et al.*, 1990b). Some early nodulin genes could be also induced by auxin transport inhibitors (Hirsch *et al.*, 1989).

The functions of only a few nodulins are known. Glutamine synthetase is 10–100 times more active in nodules than in roots. Besides the three genes (*gln* α, *gln* β, *gln* γ) that encode cytosolic GS polypeptides a fourth gene (*gln* δ) encodes a polypeptide located in the plastid (Fig. 3.34). The *gln* γ gene is expressed at a very low level in roots (150–750-fold lower than in nodules) and also in petioles (50 times lower than in nodules) and in cotyledons (about 10-fold lower than in nodules). Thus the *gln* γ gene does not encode a nodulin, but the expression in nodules is strongly enhanced (Cullimore *et al.*, 1990). The expression of the *gln* γ gene is not due to the presence of ammonia derived from N_2-fixation, since *gln* γ mRNA also increases in nodules infected with a *fix*$^-$ strain. In soybeans nodulin N-35 is a nodule specific form of uricase (urate oxidase, uricase II). This is located exclusively in the peroxisomes of uninfected cells. A nodule-specific form of sucrose synthetase belongs to the enzymes of primary carbon metabolism in the cytoplasm of infected cells (Verma and Stanley, 1989; Verma *et al.*, 1991). A nodulin involved in the architectural restructuring of infected cells

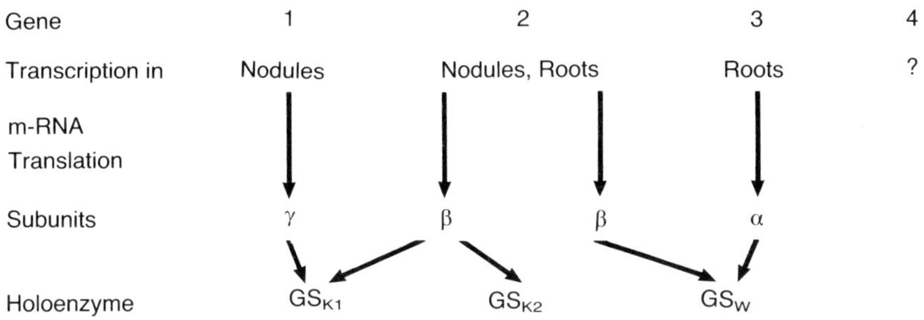

Fig. 3.34. Expression of the plant glutamine synthetase gene in nodules and roots of *Phaseolus vulgaris*. (Based on Cullimore, 1985.)

and with a known function is the nodule-specific form of choline kinase (CK II) (Mellor *et al.*, 1986).

There is some evidence that nodulin 26 might be the dicarboxylate carrier in the symbiosome membrane, regulated by phosphorylation/dephosphorylation by a CA^{2+}-stimulated protein kinase (Day, 1992).

The mechanism of nodulin induction has also been studied by incubating nuclei of uninfected soybean tissues with nodule extracts. Transcription of nodulin 23, nodulin 24 and leghaemoglobin c3 was activated, but not of nodulin 35. Nodulin 35 was induced in uninfected cells in the later stages of nodule development (Mauro and Verma, 1988).

Four isoenzymes of phosphoenolpyruvate (PEP) carboxylase, each with a molecular weight of 100 kDa, have been found in nodules and roots, but nodules contain severalfold more enzyme activity than root tissue (Egli *et al.*, 1991a,b). This means that PEP-carboxylase belongs to the nodule-enhanced proteins.

Besides the synthesis of new proteins during nodule development, the absence of specific proteins compared to uninfected roots has also been observed. One example is an 18 kDa protein, found in roots but not in nodules of lupins (Sikorski *et al.*, 1989).

(b) Leghaemoglobin genes

Soybean leghaemoglobin genes are activated 7 to 8 days after infection. However, the first real increase in transcription becomes apparent 4 days later. At this time bacterial nitrogenase is still repressed. Thus the functional requirements for nitrogen fixation are not necessary for induction of leghaemoglobin. The four leghaemoglobin genes of soybean (Lba, Lbc, Lbc2, Lbc3) all contain three introns in identical positions, at codon 32, codon 68–69 and codon 103–104. These positions are exactly those found in all other plant globin genes. Thus it is believed that the leghaemoglobin gene and other globin genes have a common ancestor (Landsman *et al.*, 1986; Bogusz *et al.*, 1988). A haemoglobin from *Parasponia andersonii* exhibits over 50% homology in nucleotide sequence with legume leghaemoglobin and also shows homology with haemoglobin genes from *Casuarina*. Activation of leghaemoglobin genes is in the order Lbc3, Lbc2, Lbc and Lba. Connected with these four structural genes are at least two others with possible regulatory functions (Marcker *et al.*, 1984; Christensen *et al.*, 1991).

Leghaemoglobin is localized in the cytoplasm of the host cell. Haem biosynthesis is stimulated in rhizobia under microaerobic conditions. But it is still not clearly verified that the haem group for leghaemoglobin in the nodules is supplied by the microsymbionts.

(c) Other plant genes involved in nodulation

Apart from genes coding for proteins such as leghaemoglobin and sucrose synthase, there must be other plant genes that are involved in the regulation of the symbiosis, for example, genes which regulate, directly or indirectly, the level of phytohormones and genes which control the plant's defence mechanisms (Fig. 3.35). The most important element in the control of plant defences is the symbiosome membrane (section 3.3.4c). The cultivar-specific interaction of soybeans with *Sinorhizobium fredii* is regulated by the genotype of the host plant root. By grafting experiments, it was demonstrated that the shoot had no influence (Balatti and Pueppke, 1990).

Using direct genetic analysis of plant mutants, 15 different genes have been identified in pea which are involved in nodule development, function and regulation (Table 3.16). The genes *sym* 5, *sym* 15, *sym* 16 and *sym* 17 affect the number of nodules per plant. The gene *sym* 2 determines the host spectrum in relationship to the nod-genes of *R. leguminosarum*. Mutations in *sym* 9, *sym* 10, *sym* 11 and *sym* 14 result in non-nodulating plants. The complete number of structural and regulatory plant genes involved in the development and function of nodules is very hard to estimate. The low number of nodule genes known at the present must be seen in relation to the up to 100 000 genes of higher plants and up to 70 000 different types of mRNA.

3.3.4 The symbiosome

The term symbiosome proposed by Roth *et al.* (1988) includes the bacteroid, the symbiosome space and the peribacteroid (symbiosome) membrane as a functional unit. It is a very useful term, emphasizing the homology to other cell organelles such as chloroplasts and mitochondria developed in evolution by independent endosymbiotic processes.

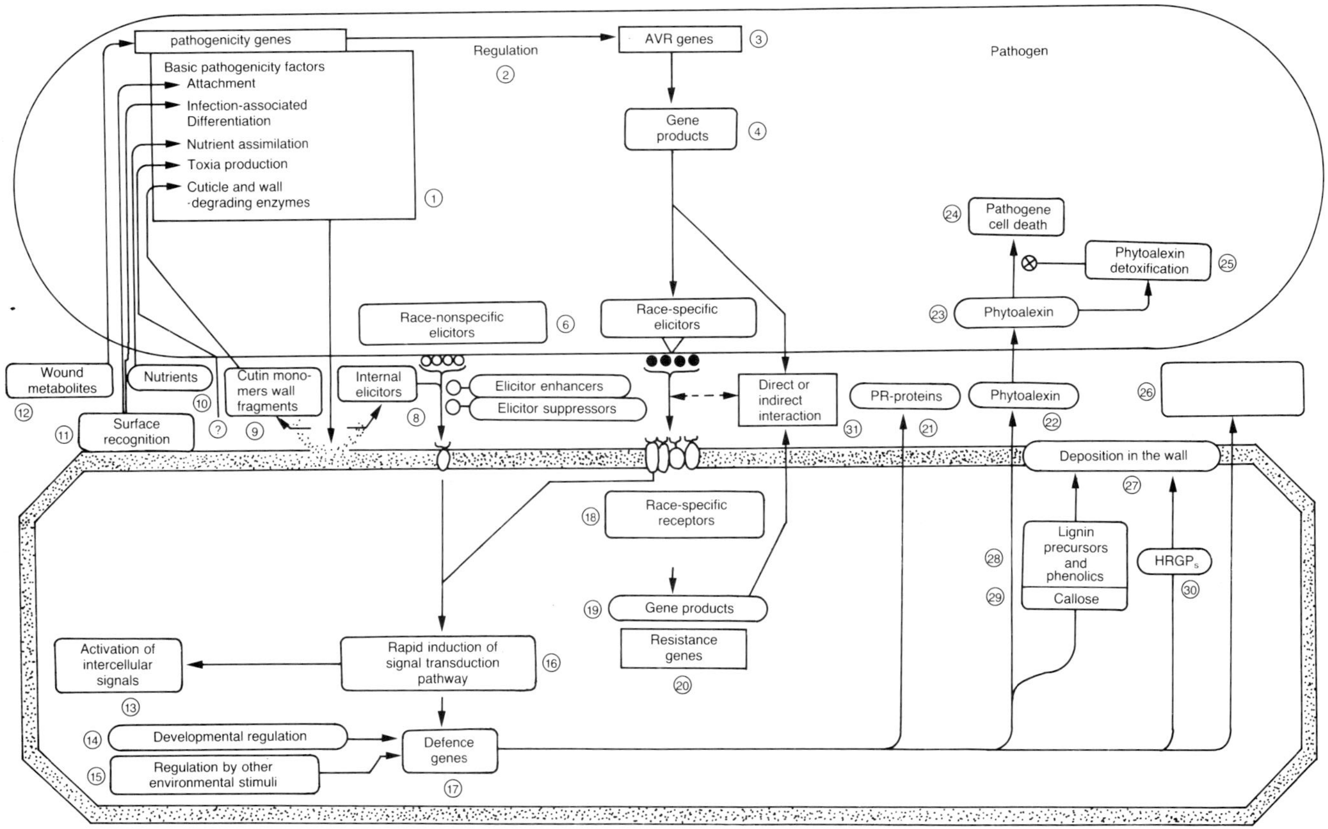

Fig. 3.35. Plant–microbial pathogen interaction: signal molecules and responses. (Modified from Lamb *et al.*, 1989.)

Table 3.16 *Sym* mutants isolated from pea cultivar 'Sparkle'. (Based on Weeden *et al.*, 1990)

Gene	*No. mutants isolated*	*Mutants designations*	*Phenotype of mutants*
sym 2	2	N25, N27	strain specificity
sym 5	7	E2, others	low nodule number
sym 7	2	E69, N12	non-nodulating
sym 8	3	E14, R19, R25	non-nodulating
sym 9	1	R72	non-nodulating
sym 10	1	N15	non-nodulating
sym 11	1	N24	non-nodulating
sym 13	1	E135F	non-fixing
sym 14	1	E135N	non-nodulating
sym 15	1	E151	short lateral roots, few nodules
sym 16	1	R50	short lateral roots, few nodules
sym 17	1	R82	stubby roots and shoots, few nodules
sym 18	1	E54	strain specificity, flowering
sym 19	2	NEU5, NMU1	non-nodulating
brz	1	E107	iron accumulator, non-nodulating

(a) Bacteroid differentiation

In the endosymbiotic form, bacteria of the genera *Bradyrhizobium*, *Rhizobium*, *Azorhizobium* and *Sinorhizobium* are called bacteroids. The differentiation of free-living cells of *Rhizobium* and *Bradyrhizobium* into bacteroids includes the changes summarized in Table 3.17. Functionally bacteroids can be seen as N_2-fixing cell organelles. In size and shape they can be rather similar to the free-living form as in the genus *Bradyrhizobium* or irregular and branched as often found in biovarieties of *Rhizobium*. In bacteroids DNA content per cell can be 4–8 times higher than in free-living rhizobia (Bisseling *et al.*, 1977). However, this is due to the multicellular form of bacteroids which increase their size without undergoing cell division. In *B. japonicum* it has been demonstrated that, per unit cell volume, the DNA content of both forms is about the same (Paau *et al.*, 1979; Wilcokson and Werner, 1979). *In vitro* forms of bacteroids have been produced in soft agar cultures (Pankhurst and Craig, 1978). Between an upper layer and a bottom layer containing rod-shaped cells there is a narrow zone with rather pleomorphic bacteroid-like cells which includes polyphosphate bodies and a polar distribution of PHB (poly-beta-hydroxybutyrate). In nodules, bacteroid form is under host plant control. This was demonstrated for one strain of *Rhizobium* which in *Vigna mungo* formed rod-shaped bacteroids, but in *Arachis hypogaea* formed single spherical bacteroids enclosed in peribacteroid membranes (Dart, 1977). Bacteroids from *Rhizobium* and *Bradyrhizobium* strains can redifferentiate to viable free-living cells. Between

Table 3.17 Differentiation of bacteroids from free-living cells of *Rhizobium* sp. and *Bradyrhizobium* sp. (Modified from Switton, 1981)

Process	*Bacteroid state*
Viability	only slightly reduced
Osmotic sensitivity	enhanced
EPS formation	reduced
DNA content	unchanged, related to the same cell size
Lipid metabolism	PHB accumulation, also in fix$^-$ strains of *Bradyrhizobium*
Electron transport	special branch, without cytochromes *c* and aa_3
Haem synthesis	enhanced (probably for leghaemoglobin)
Nitrogenase	up to 10% of the soluble protein
*Nif*A dependent processes	under oxygen control

20 and 90% colony-forming bacteroids have been isolated from *Glycine max*, *Ornithopus sativus* and *Phaseolus aureus*. Supplementation of the media with 0.2 or 0.3 M-mannitol increases the percentage of colony forming units (Sutton *et al.*, 1977; Tsien *et al.*, 1977).

There is increasing biochemical and genetic evidence that *B. japonicum* contains three different respiratory chains. One chain is similar to the one in mitochondria:

$$2\,H \rightarrow Q \rightarrow FeS/bc_1 \rightarrow c \rightarrow aa_3 \rightarrow O_2$$

In bacteroids there is another specific branched electron transport pathway, leading from the bc_1 complex to the terminal oxidase, since mutations in cytochrome *c* and aa_3 were fix$^+$ (Hennecke *et al.*, 1990). Cytochrome *c* oxidase purified from bacteroid membranes is apparently a glycoprotein (Maier *et al.*, 1990). Other bacteroid specific cytochromes are *P*-450, *P*-420, *c*-552 and *c*-554 (Appleby, 1984). These bacteroid-specific cytochromes are apparently different in *Rhizobium* and *Bradyrhizobium*. In bacteroids of *R. leguminosarum* biovar. *viciae*, there was no cytochrome *P*-450, but instead a protein with a Soret peak at about 424 nm (Kretovich *et al.*, 1973). The somewhat greater O_2-tolerance of pea bacteroids compared to bacteroids from soybeans may be related to this protein (Appleby, 1984).

The activity of bacterial glutamine synthetase is severely reduced and β-hydroxybutyrate dehydrogenase increased in bacteroids. The latter is seen in relationship to the enrichment in the bacteroid of poly-β-hydroxybutyric acid. This is often seen as clear droplets in the bacteroid cytoplasm and

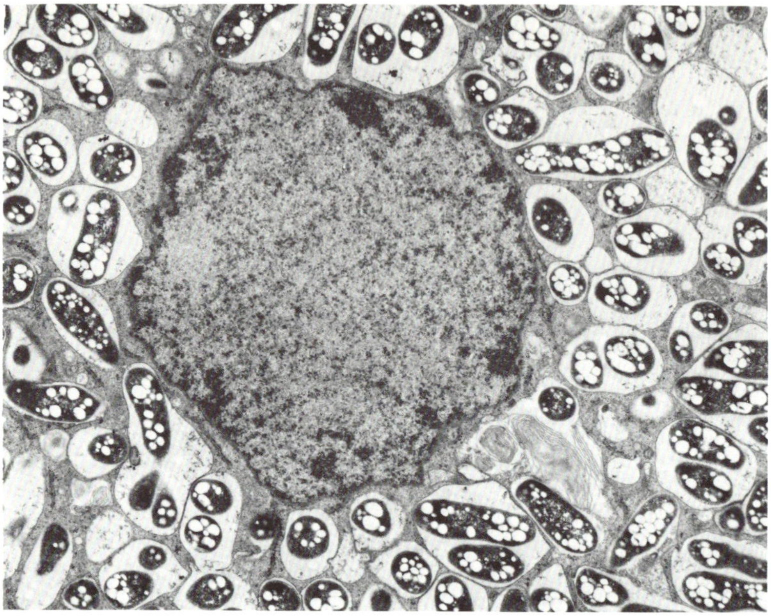

Fig. 3.36. Ultrastructural view of a host cell from a *Glycine max* cv. Mandarin nodule, 29 days after infection with *B. japonicum* 61-A-101 (fix^+). Magnification ×8300.

functions as a C-reserve. In fully derepressed bacteroids nitrogenase can comprise up to 10% of total bacterial soluble protein. Fixed nitrogen is exported passively as ammonia into the plant cytoplasm.

(b) The symbiosome space

The symbiosome space (the space between the bacteroids and the symbiosome membrane) contains proteins and other biochemicals, but at a much lower density than the plant cytoplasm (Fig. 3.36). Convincing biochemical identification of molecules characteristic of this compartment requires that cross-contamination from the host cell cytoplasm, mitochondria, plastids, ER and Golgi as well as contamination from bacteroid periplasm and cytoplasm can be ruled out (Mellor and Werner, 1986; Kinnback *et al.*, 1987). The following components have been identified in this compartment:

1. acid phosphatase (Kinnback and Werner, 1990)
2. α-mannosidase II (Kinnback *et al.*, 1987)
3. α-glucosidase (Kinnback and Werner, 1990)
4. proteases (Mellor *et al.*, 1986)
5. protease inhibitors (Garbers *et al.*, 1988)

6. aspartate aminotransferase (Kinnback and Werner, 1990)
7. trehalase (Mellor, 1988)

The symbiosome space can be considered as a special type of lytic compartment (Mellor, 1989; Mellor and Werner, 1990). This is obvious in soybean nodules infected with *B. japonicum* strain RH 31-Marburg where lysis of bacteroids occurs in the vicinity of the host cell nucleus early after infection (Werner *et al.*, 1984). α-Mannosidase III is known to be a marker for the extracellular space, whereas the isoenzymes I and II are markers for the plant vacuome (one or several vacuoles). Since only isoenzyme II of α-mannosidase has been found in the symbiosome space, this means that the symbiosome space is different from both the plant vacuome and the extracellular space. How bacteroids in an effective symbiosis are stabilized in the symbiosome space in the presence of several hydrolases is not well understood. Proteases could be inhibited by the presence of protease inhibitors (Garbers *et al.*, 1988). Other hydrolases may be regulated by pH. A model for pH balance in symbiosomes has been proposed by Kannenberg and Brewin (1989). As the result of two transport activities located in the peribacteroid membrane, the dicarboxylic acid transport and a vectorial proton ATPase, the space would tend to be acidified. On the other hand, actively N_2-fixing bacteroids would increase the pH by uptake and utilization of malate/succinate and by the excretion of ammonia into the symbiosome space. A consequence of this model would be a somewhat lower pH in fix$^-$ symbiosomes.

(c) The symbiosome membrane synonymous to the peribacteroid membrane

The symbiosome membrane (SM) is produced by the host plant via membrane flow from the Golgi and/or directly from the ER. The fatty acid composition of the SM in nodules of *Glycine max* is more similar to the ER than to the Golgi (Bassarab *et al.*, 1989) (Table 3.18). In this well-established pathway, it was shown by a number of independent methods that the microsymbiont has a decisive role in formation, stability, protein composition and particle density of this membrane. This evidence is as follows:

1. The SM from soybean nodules infected with the mutant RH 31-Marburg of *B. japonicum* (Fig. 3.37) has in the PF face a particle density of only 1200–1300 particles μm^{-2} compared to the wild type with about 220 particles in the PF face (Werner *et al.*, 1988).
2. A discrete set of particles in the size range 13–14 nm are missing in the EF face of the mutant-affected SM compared to the wild type (Werner *et al.*, 1988) (Fig. 3.38).

Table 3.18 Fatty acid composition of membrane systems from *Glycine max* root nodules, harvested 21–26 days after infection with different strains of *B. japonicum* (Modified from Bassarab *et al.*, 1989)

Membrane	*Fatty acid (% of total area)*					
	16:0	*18:1*	*18:2*	*18:3*	*18:0*	*18:n/16:0*
B. japonicum 61-A-101						
ER	36.1 + 4.6*	17.0 + 1.5	27.0 + 3.3	9.7 + 1.1	11.8 + 1.4	1.4
Golgi	24.2 + 2.6	13.1 + 1.8	38.7 + 2.7	17.3 + 1.6	4.3 + 2.0	2.9
M	34.5 + 1.8	15.9 + 2.3	29.3 + 2.0	10.1 + 0.7	7.0 + 0.7	1.6
B. japonicum 110 spc 4						
ER	28.8 + 1.8	8.6 + 1.4	42.1 + 1.9	14.0 + 0.8	6.1 + 0.4	2.4
Golgi	25.7 + 0.9	11.0 + 1.7	37.2 + 3.1	16.9 + 1.5	7.7 + 2.3	2.5
M	30.0 + 0.7	11.4 + 1.1	37.7 + 0.8	14.0 + 1.5	6.5 + 0.3	2.1
B. japonicum RH-31-M						
ER	37.1 + 3.5	8.7 + 3.7	36.4 + 3.2	16.5 + 1.4	7.2 + 0.5	1.9
Golgi	25.8 + 3.2	9.8 + 0.4	39.2 + 4.0	17.7 + 1.3	8.2 + 1.1	2.6
M	33.9 + 0.7	12.4 + 1.4	29.0 + 1.7	11.5 + 2.2	13.0 + 2.4	1.6

* mean values and standard deviation from four determinations

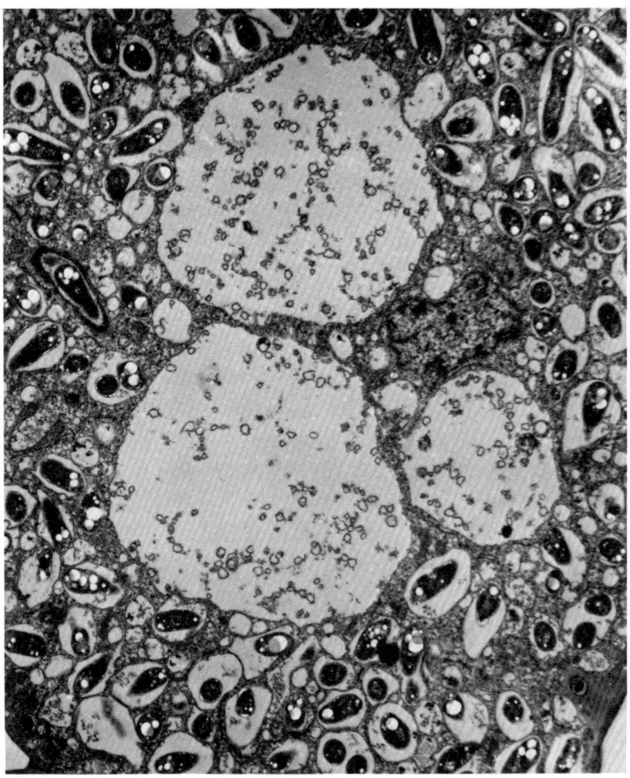

Fig. 3.37. Ultrastructural view of a host cell from a *Glycine max* cv. Mandarin nodule, 34 days after infection with *B. japonicum* RH 31-Marburg. Magnification ×4700. (From Werner and Mörschel, 1978.)

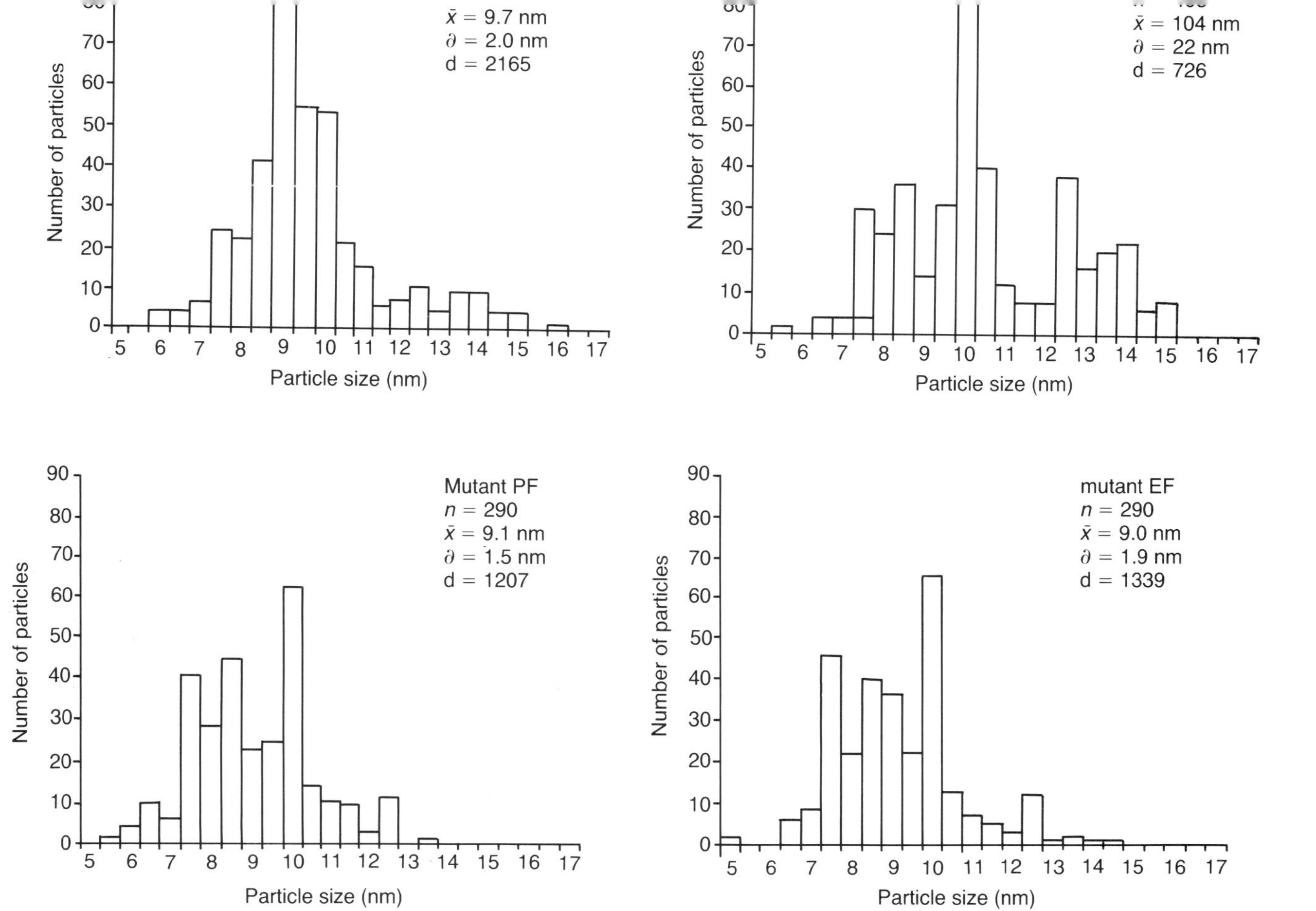

Fig. 3.38. Size distribution of freeze-fracture particles from peribacteroid membranes from *Glycine max* nodules infected with *B. japonicum* 61-A-101 (wild type) and *B. japonicum* RH 31-Marburg (mutant). n = number of particles counted; $\bar{x}$ = average particle size; ∂ = standard deviation; d = particles μm^{-2}. EF: exoplasmic fracture face; PF: protoplasmic fracture face. (From Werner *et al.*, 1988.)

3. Glycosylation of several SM proteins is affected by the microsymbiont (Werner *et al.*, 1988).
4. Three other mutants of *B. japonicum* are responsible for the loss of different specific SM proteins (Mellor *et al.*, 1989).
5. Different wild type strains of *B. japonicum* (strain 61-A-101 and strain 110*spc*4) change the ratio of 18:*n*/16:0 unsaturated to saturated fatty acids in the SM significantly (Bassarab *et al.*, 1989).

In summary it appears that there are at least four different types of signal (and at least as many genes) in the microsymbiont which affect production, stability and protein composition of the peribacteroid membrane.

Nodulin-24 and nodulin-26 have been identified to be SM associated (Fortin *et al.*, 1987). A specific form of choline kinase (choline kinase II) is involved in the biosynthesis of the SM. In soybean it is present at significant activities only in nodules producing a stable peribacteroid membrane (Mellor *et al.*, 1986). The occurrence of this isoenzyme is not related to N_2 fixation since fix$^-$ nodules with a stable peribacteroid membrane (such as produced by strain 110*spc*4/A3) also have choline kinase II (Mellor *et al.*, 1987).

The following functions have been identified in the SM (PBM):

1. A H^+-ATPase of the host plasma membrane type (Blumwald *et al.*, 1985) and probably a second ATPase of the vacuolar type (Bassarab *et al.*, 1986)
2. A calcium-dependent protein kinase activity (Bassarab and Werner, 1987)
3. A dicarboxylate carrier (Day *et al.*, 1990)
4. A magnesium-dependent pyrophosphatase (Bassarab and Werner, 1989)

The interiors of endocytic vesicles, lysosomes, part of the trans-Golgi apparatus and some secretory vesicles, are acidic (Anderson and Orci, 1988). Tonoplast membranes also show pyrophosphatase activity and PP_i dependent H^+-uptake (Marquardt and Lüttge, 1987).

Various membrane building enzymes show increased activities in nodules infected with different strains of *B. japonicum* (Table 3.19). In infected nodules choline kinase and choline phospotransferase are stimulated 350% and 200% relative to roots. The stimulation is even greater in the case of glycosyltransferases. For example, the enzyme of terminal glycoprotein glycosylation UDP-ASGF *N*-acetylgalactosamine transferase, is 16 times more active in effective nodules than in roots. Thus it has been postulated that specific protein or lipid glycosylation is involved in the re-routing of membranes.

It is interesting to compare the biosynthetic pathway for the symbiosome with that for the formation of protein bodies in seeds or cotyledons from

Table 3.19 Increased activity of enzymes involved in membrane biosynthesis of soybean root nodules, after infection with different strains of *B. japonicum*. Figures represent percentage stimulation in comparison with uninfected root tissue. (Based on Mellor *et al.*, 1986)

Microsymbiont strain	*Choline kinase*	*Cholin phospho-transferase*	*GDP-DMP mannosyl-transferase*	*UDP-ASGF galactosyl-transferase*	*UDP-ASGF* N-*acetyl-galactosamine transferase*
B. japonicum					
61-A-101	350	200	300	780	1600
RH 31-Marburg	80	120	280	800	800
61-A-24	50	90	180	550	570

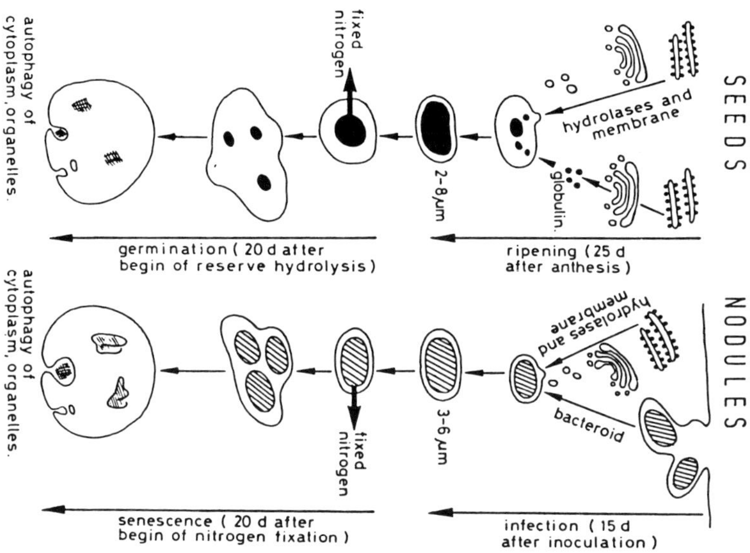

Fig. 3.39. Schematic presentation of the developmental cycle of protein bodies in soybean seeds/cotyledons, compared to the formation of endosymbiotic *Bradyrhizobium*-containing organelles in soybean root nodules. The time scale is approximate and depends strongly upon external factors. (Based on Mellor, 1988.)

seed legumes (Fig. 3.39). The symbiosome can be considered to be homologous to protein bodies, where, N in the form of proteins is stored and finally exported, whereas ammonia is exported from symbiosomes. Membrane fusion, digestion and autophagy in senescence may be rather similar.

(d) Cytology and ultrastructure of infected cells

From ultrastructural studies it is clear that in fully effective symbioses the microsymbiont is always surrounded by the symbiosome (peribacteroid) membrane (Fig. 3.36). However, ineffective symbioses may also have this property. Mutations in the genome of the microsymbiont can modify this phenotype considerably. Examples are given in Figs 3.39 and 3.40. In one phenotype the symbiosome membrane is lost early in development but the bacteroids remain stable and enzymatically active in a slowly degenerating plant cytoplasm (Fig. 3.40). A second phenotype (Fig. 3.37) involves fusion of peribacteroid membranes to larger lytic compartments, mostly centred around the cell nucleus, where bacteroids are digested. Thus apart from the previously described *nod*, *nif* and *fix* genes the bacteria must contain further genes which affect the biosynthesis and stability of the symbiosome membrane.

During nodule development, cellular organelles accumulate at the cell periphery (Fig. 3.41). Early on in the process, mitochondria are noticeable

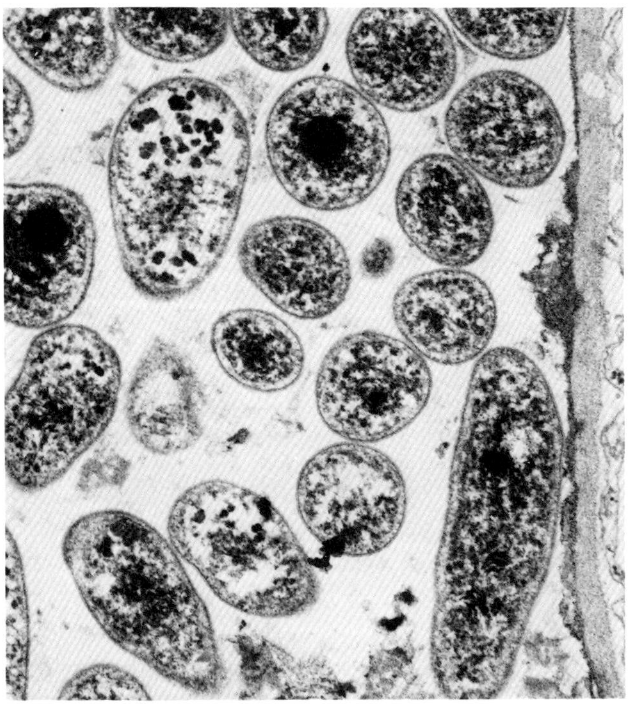

Fig. 3.40. Ultrastructural view of a host cell from a *Glycine max* cv. Mandarin nodule, 29 days after infection with *B. japonicum* 61-A-24 (fix^-). Magnification ×27 000

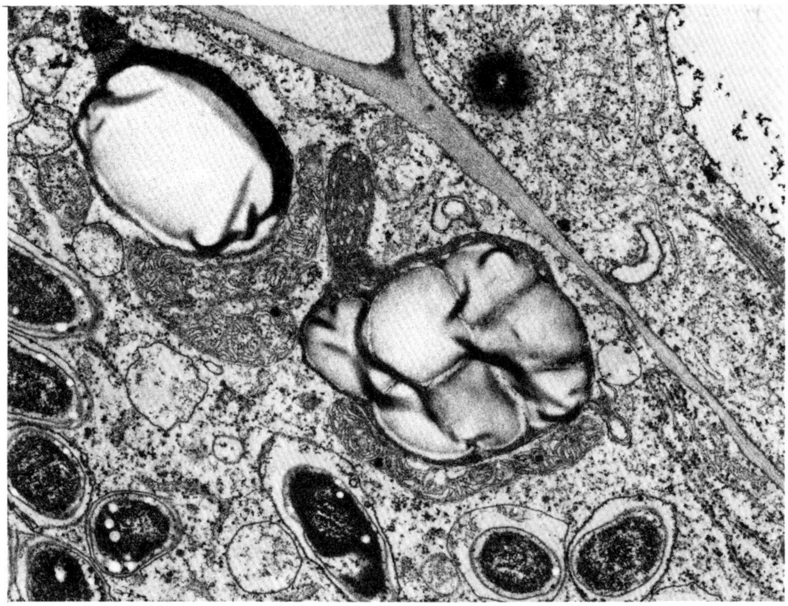

Fig. 3.41. Association of mitochondria and amyloplasts in a host cell from a *Glycine max* cv. Caloria nodule, 15 days after infection with *B. japonicum* 3I1b85 (fix^+). Magnification ×15 000.

in close contact with amyloplasts. The mitochondria are much richer in cristae than is normal for roots. Starch in the amyloplasts is reduced drastically during the first half of the nodule development (Fig. 3.42). Plant genotype also influences nodule fine structure, for example, *Medicago sativa* genotype Mn Pl-480 produces tumour-like nodules following infection with wild type *R. meliloti*. In these nodules very few infection threads are seen and very few infected cells. These nodule cells retain a high starch content for a long time. Plant determined ineffective lucerne nodules have much reduced activities of glutamine synthetase, glutamate synthase (GOGAT) and PEP-carboxylase.

Plant cells filled with bacteroids have either a very small central vacuole (peas) or none at all (soybean). With the onset of nodule senescence, lytic compartments form and fuse with the symbiosomes. Bacteroids are lysed. The volume of the nucleus is reduced, as is the number of ribosomes. The ER becomes fragmented. This process can be accelerated by removal of leaves. However, the plant is able to recover when new foliage is established. Under these conditions the intact infection threads simply infect new cells. These newly infected cells are primarily derived from the inner nodule cortex in peas and from the outer layer of the cortex in soybeans.

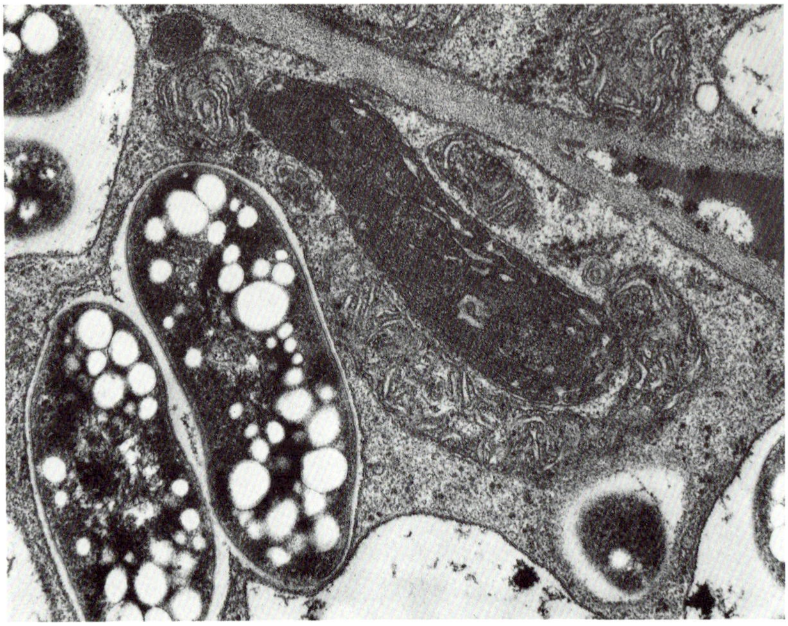

Fig. 3.42. Association of plastids and mitochondria in close proximity to a bacteroid in a host cell from a *Glycine max* cv. Caloria nodule, 29 days after infection with *B. japonicum* 3I1b85 (fix^+). Magnification ×30 000.

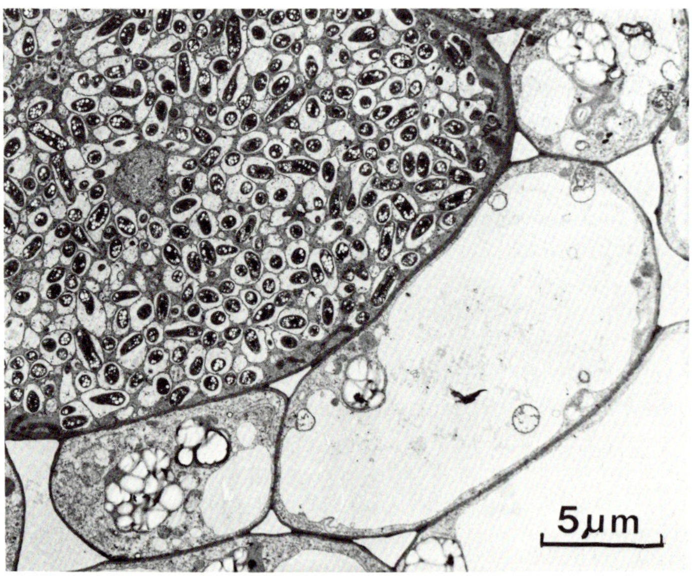

Fig. 3.43. Infected host cell from a *Glycine max* cv. Caloria nodule filled with bacteroids inside peribacteroid membranes. Nearby are several smaller uninfected cells with large central vacuoles and amyloplasts abundant in starch.

(e) Non-infected host cells

Between the large infected cells filled with bacteroids there are often many smaller, vacuolated non-infected cells (Fig. 3.43). Bending of the cell wall between the cell types implies a high osmotic pressure in the infected cells. Plasmodesmata are often seen between infected and non-infected cells, but only sometimes between two infected cells, and practically never between two non-infected cells. Peroxisomes, which accumulate on the periphery of the cell, are prominent in non-infected cells (see Fig. 3.44). Immunogold localization has shown these organelles to be the site of uricase II (EC 1.7.3.3), a nodulin involved in purine degradation (van den Bosch and Newcomb, 1986). In addition, long strands of sER and tER can be found in these cells. Non-infected cells are involved in the production of allantoin and allantoic acid, important N-transport forms in many tropical legumes (Fig. 3.57). For ureide-exporting nodules, it is assumed that N-transport is

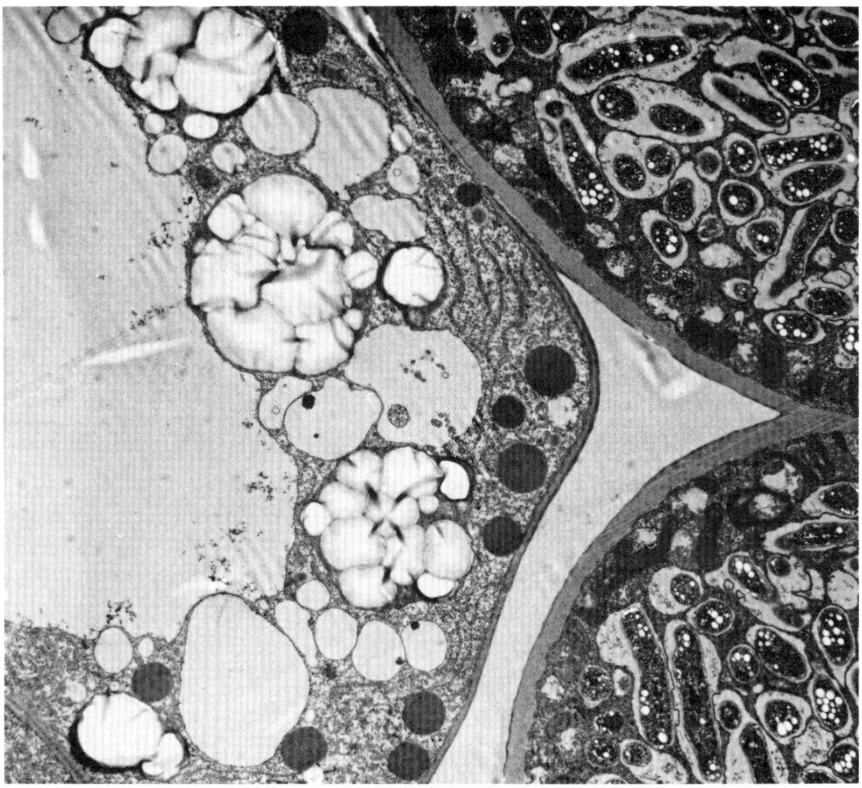

Fig. 3.44. Left: section of an uninfected cell in nodule tissue of *Glycine max* with enlarged dark peroxisomes at the periphery. Magnification approx. ×12 000. (Original photograph E. Mörschel, Marburg.)

more efficient, because the N/C ratio is higher for ureides than for amino acids such as asparagine.

(f) The symbiosis viewed as evolution from chemical defence to structural defence

The symbiosome membrane is the decisive physiological barrier between microsymbiont and host cytoplasm. Using *B. japonicum* strains and mutants (Figs 3.36, 3.37 and 3.40) which give rise to various peribacteroid membrane phenotypes, a strong correlation has been found between symbiosome membrane disintegration and the onset of plant defence responses. In other words, the bacterium is recognized as a pathogen. In soybean, this recognition is apparent as the production of the phytoalexins, glyceollin I (Table 3.20). Isomer I is quantitatively the most important of the three glyceollin isomers (Fig. 3.45). At 20 days after infection with *B. japonicum* 61-A-24 glyceollin I levels had reached 6 $\mu mol\,g^{-1}$ nodule dry weight. In comparison, treatment of oat leaves with various elicitors resulted in a concentration for the phytoalexin avenalumin of 70–100 $\mu g\,g^{-1}$ (Mayama *et al.*, 1989). The mutation of the *nif* A gene of *B. japonicum* leads also to a pathogenic-like response of the host plant. Soybean nodules induced by the *nif* A mutant A9 exhibited symptoms of a hypersensitive reaction (HR) normally observed in plant pathogen interactions as localized death of infected cells and the accumulation of the phytoalexin glyceollin (Parniske *et al.*, 1991b). *Bradyrhizobium* reacts to the production of glyceollin with an induced resistance, which is also induced in the flavonoids daidzein and genistein, strong inducers of *nod* genes in cooperation with the *nod* D gene products. Resistance against this phytoalexin was also induced in a nod D_1D_2Y ABC deletion mutant of *B. japonicum*, indicating another recog-

Table 3.20 Concentrations of glyceollin I in nodules of *Glycine max* cv. Mandarin, infected with N_2-fixing (fix^+) and ineffective (fix^-) strains of *B. japonicum*. (Based on Werner *et al.*, 1985b)

Tissue	B. japonicum *Strain*		*Days after infection*	*Glyceollin I (pmol mg^{-1} dry matter)*
Nodule	61-A-101	(fix^+)	20	140
Nodule	61-A-101	(fix^+)	34	<50
Nodule	110 USDA	(fix^+)	20	630
Nodule	110 USDA	(fix^+)	34	<50
Nodule	RH 31-Marburg	(fix^-)	20	<50
Nodule	RH 31-Marburg	(fix^-)	34	<50
Nodule	61-A-24	(fix^-)	20	6250
Nodule	61-A-24	(fix^-)	34	1600

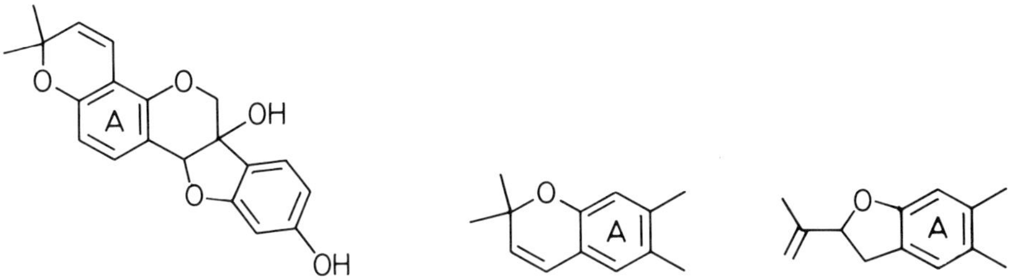

I II III

Fig. 3.45. Isomers of the phytoalexin glyceollin (I–III) which accumulate in soybean after infection with *Phytophthora megasperma* pv. *glycinea*.

nition site for flavonoids besides the *nod* D genes (Parniske *et al.*, 1991a). By varying the host plant genotype, three *Glycine* genotypes, *G. max* cv. Williams, *G. soja* PI 468379 and *G. soja* PI 342434 in combination with the two rhizobial strains *B. japonicum* USDA 123 and *R. fredii* USDA 193, were analysed for phytoalexin concentration in the nodules. In the nodules of Pl 468397/*B. japonicum* USDA 123 there was very strong glyceollin I accumulation around 30 days post infection. Ultrastructural analysis of these nodules showed several symptoms of a severe plant defence response associated with plant cell death (hypersensitive reaction). The cytoplasm of the infected cells was degraded and organelles had vanished. The cell walls of the infected cells showed remarkable thickening. This plant defence response could only be observed in this strain/genotype interaction. The same strain did not elicit a phytoalexin accumulation in the other plant genotypes tested, indicating that this response occurs at the genotype-specific level. This special character of *G. soja* Pl 468397 is heritable as indicated by glyceollin I analysis of the nodules formed by F1 hybrids of Pl 468397 × Williams inoculated with *B. japonicum* USDA 123. The genotype/strain-specific occurrence of the hypersensitive response in root nodules resembles the race/cultivar-specific incompatibility of several plant pathogen interactions (Parniske *et al.*, 1990).

High phytoalexin levels were not found in nodules where the peribacteroid membrane remained intact, regardless of whether the symbiosis was fix$^+$ or fix$^-$. Thus in wild type nodules, the host plant does not recognize the more than 10 000 endosymbionts per cell as non-self. This is apparently due to the presence of an intact peribacteroid membrane. At the low level of phytoalexins in fully effective nodules, the plant genotype is more important than the bacterial strain (Wolff and Werner, 1990).

Structural defence against pathogens includes a number of different strategies (Ride, 1986):

1. lignification of cell walls
2. deposition of phenolic acids in cell walls
3. suberization
4. accumulation of hydroxyproline-rich protein in cell walls
5. formation of papillae during deposition of new cell wall formation
6. vascular occlusion
7. callose deposition

Callose synthesis proceeds via stimulation of callose synthase. Other defence reactions involve transcriptional activation of defence genes (Dixon and Lamb, 1990).

No induction of **chalcone synthase (CHS)** mRNA synthesis is observed after inoculation of axenically grown seedlings with *B. japonicum*. However, inoculation with *Agrobacterium tumefaciens* produces enhanced expression within this time period. After 19–23 days CHS mRNA synthesis is also enhanced in nodules (Wingender *et al.*, 1989). In plant cells in general, some defence reactions are triggered by non-specific perturbations of the plasma membrane (Kauss, 1990). Several phytoalexin elicitors from fungal or plant cell walls are non-specific compared to specific host plant–cultivar–pathogen biovar. infections. Ca^{2+} is an important, but not the only, signal during the induction of callose synthesis by 1,3-β-glucan synthase (GS II) as a general structural defence mechanism (Kauss *et al.*, 1989). In addition, free polyunsaturated fatty acids and polycations can activate GS II *in vitro*.

Symbiosis specific phenylalanine ammonia-lyase and chalcone synthase genes are not induced by pathogen interaction or by stress (Estabrook and Sengupta-Gopalan, 1991).

An interesting model for defence reactions is the '**ion channel defence model**' (Gabriel *et al.*, 1988). This model proposes that resistance genes are structural genes encoding membrane-bound proteins, acting as ionophores following binding of specific trigger molecules. Binding produces depolarization of the membrane leakage of ions and finally cell death in a hypersensitive reaction. For example, avirulent strains of the bacterial pathogen *Pseudomonas solanacearum* cause this hypersensitive reaction whereas virulent strains do not. The latter multiply in the intercellular spaces of the invaded organ and finally cause death of the whole plant (Halverson and Stacey, 1986).

A model of signals and responses in plant–pathogen interactions is summarized in Fig. 3.35. A number of these mechanisms may also be present during certain stages of development of the symbiosis (Lamb *et al.*, 1989; Rolfe *et al.*, 1982).

3.3.5 Nitrogen fixation

A physiological result of the symbiosis is the derepression of dinitrogenase. Differentiation of nodule cells can be seen as a strategy for optimizing this process. Thus the energy invested by the host plant in the restructuring of carbon metabolic pathways and the development of C_2-protection mechanisms is repayed by the bacteria by NH_4^+ supplied to the plant. Nitrogen compounds synthesized by the plant from NH_4^+ supplied are exported from the nodule and used in biosynthetic pathways in roots and leaves, but most importantly, in the seeds of the host plant. It is estimated that biological N_2-fixation represents 50%, industrial N-reduction 32% and atmospheric N-fixation about 17–18% of the total N_2-fixation of about 290 million t year^{-1} (Isermann, 1987).

N_2-fixation in legume nodules follows the general equation:

$$N_2 + 8e^- + 8H^+ + 16\,MgATP \rightarrow 2NH_3 + H_2 + 16\,MgADP + 16\,P_i$$

(a) Nitrogenase: structure, function and regulation

In the old terminology, nitrogenase (dinitrogen oxidoreductase: ATP hydrolysing: EC 1.18.6.1) is a multisubunit enzyme. It is formed from two components, a Mo-Fe-protein of 245 kDa and a Fe-protein of 64 kDa. The Mo-Fe-protein is a tetramer consisting of two non-identical subunits, i.e., it is a protein of the ααββ-type. The Fe-protein has two identical subunits. In the new nomenclature, the Mo-Fe-protein is called dinitrogenase and the Fe-protein dinitrogenase reductase. A model of the dinitrogenase reductase/dinitrogenase reaction is shown in Fig. 3.46. It is apparent that the high ATP requirement (16 mol ATP per mol N_2 reduced) is a function of the dinitrogenase reductase reaction and not the actual reduction of N_2. The high ATP requirement (16 mol ATP per mol N_2 reduced) is a function of the dinitrogenase reductase reaction and not the actual reduction of N_2 (see Fig. 3.5). Both proteins are highly conserved in all N_2-fixing organisms, irrespective of whether they are aerobic, microaerobic, anaerobic or symbiotic. Subunits can be exchanged between most diazotrophic organisms without affecting enzyme activity. Both nitrogenase components are highly labile in the presence of O_2. In *Azotobacter chroococcum* and *Azotobacter vinelandii*, an **alternative nitrogenase** has been found in which the Mo-Fe-protein is replaced by a vanadium-protein and is active under conditions of molybdenum limitation (Bishop and Premakumar, 1992).

Apart from the natural substrates, N_2 and H^+, nitrogenase is able to reduce other compounds containing triple bonds $C \equiv N$ or $N \equiv N$ (Table 3.21). Azide (N_3) is reduced to N_2 and ammonia with a K_m value of 0.2–1.0 mM-N_3. Nitrous oxide (N_2O) is reduced to N_2 and H_2O. The reduction of cyanide products to 90% methane and ammonia plus 10% methylamine. Acetylene (C_2H_2) is reduced to ethylene (C_2H_4). Ethane formation from acetylene has been used as a test for the presence of a vanadium nitrogenase (Dilworth *et al.*, 1987). Cyanamide $N \equiv C\text{-}NH_2$ is also a substrate for nitrogenase and is reduced to CH_3NH_2 and NH_3. The ability of nitrogenase to utilize such a broad spectrum of substrates has

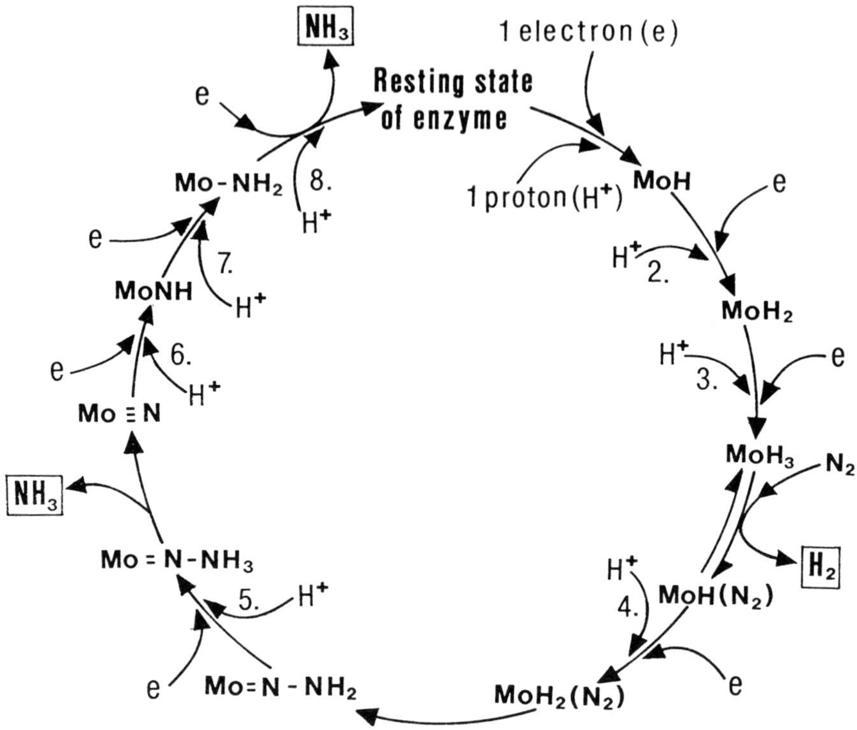

Fig. 3.46. Model of N_2-reduction and H_2-production by molybdenum nitrogenase. (Modified from Erfkamp and Müller, 1990.)

Table 3.21 Substrates of the nitrogenase

Substrate	*Products*
N_2*	2 NH_3
2 H^+*	H_2
N_3^-	N_2, NH_3
N_2O	N_2, H_2O (little NH_3)
HCN	CH_4, NH_3, CH_3NH_2 (traces of C_2H_4 and C_2H_6)
CH_3CN	C_2H_6, NH_3
$CH_2{=}CH{-}C{\equiv}N$	$CH_3{-}CH{=}CH_2$, C_3H_8, NH_3
CH_3NC	CH_3NH_2, CH_4, C_2H_6, C_3H_8, C_3H_6
CH_2CHCN	C_3H_6, NH_3, C_3H_8
C_2H_2	C_2H_4

* Natural substrate

Table 3.22 Energetics of N_2-reduction, nitrate reduction and ammonia assimilation. (Based on Schubert, 1982 and Albrecht and Gaskins, 1982)

Physiological process	*mol glucose consumed / mol N_2 (NO_3^-/NH_4^+) fixed*	*ATP equivalents*
1. *Reduction of N_2*		
Supply of reduction equivalents	0.25	9
ATP supply	0.50	12
2. *H_2 production by nitrogenase*		
supply of reduction equivalents	0.08	3
ATP supply	0.17	4
Sum of 1 and 2	1.00	28
3. *Nitrate reduction*		
supply of reduction equivalents	0.33	12
ATP supply	–	–
4. *Ammonia assimilation*		
supply of reduction equivalents	0.08	3
ATP supply	0.11	4

formed the basis for a theory of its evolution. It is postulated that in prokaryote evolution, the main function of nitrogenase was the detoxification of cyanides and nitriles. At this time N_2-fixation was an unimportant by-product. N_2-fixation became more important with time as cyanide levels fell below the K_m value of nitrogenase for this substrate and reduced nitrogen became limiting. It should be noted here that cyanide inhibits CO-dehydrogenase from *Clostridium* under anaerobic conditions.

The energy requirements for N_2-fixation, nitrate reduction and NH_4 assimilation are compared in Table 3.22. It is apparent that for every mol of N_2 fixed, 1 mol of glucose is required. Thus N_2-fixation uses much more energy than the assimilation of other biological macroelements. ATP requirements, for example, for CO_2 fixation are 3 mol ATP mol^{-1} CO_2 fixed in the Calvin cycle or 5 mol ATP mol^{-1} CO_2 for CAM plants.

(i) Regulation by oxygen and ammonia

The optimal O_2 concentration for free-living N_2-fixation in *B. japonicum* is 1–6 μM, depending on the carbon source. Higher O_2 concentrations repress nitrogenase synthesis and inactivate enzyme activity. Oxygen regulation is controlled by the *nif*LA operon (Figs 3.8a and b). Regulation by NH_4 is difficult to study in *B. japonicum* since the ammonia produced cannot be used by the cells without the addition of exogenous amino acids. Contrary results were obtained for strains of *Azorhizobium sesbania* from stem nodules of *Sesbania*. The turnover rates of nitrogenase, dinitrogenase reductase and leghaemoglobin are the same as the average rate obtained for plant cyto-

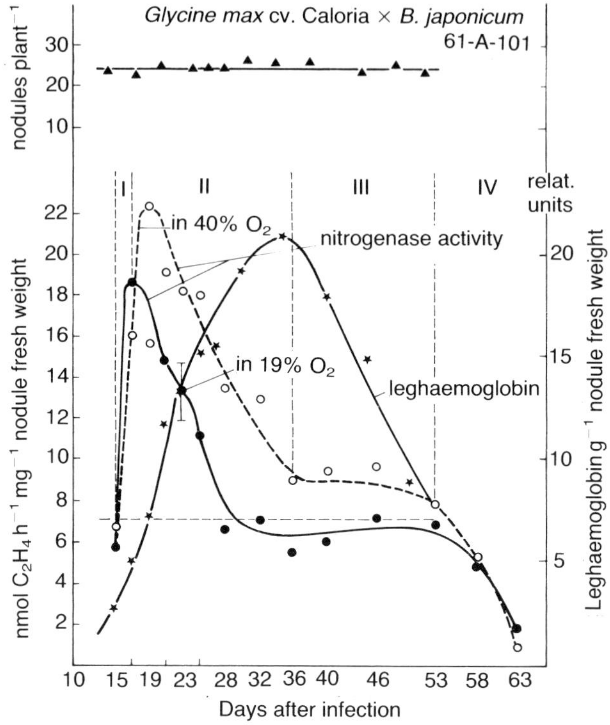

Fig. 3.47. Nitrogenase activity under 19% and 40% O_2, and leghaemoglobin concentration in synchronous nodules of *Glycine max* infected with *B. japonicum* 61-A-101. (From Werner, 1992.)

plasmic and bacteroid proteins. The half-life in all these cases is about 2 days (Bisseling *et al.*, 1980b).

(b) Oxygen protection, cork layers, water-filled diffusion barriers, leghaemoglobin, respiration

Maintenance of low O_2 partial pressure in infected host cells of nodules is achieved by four major, independent factors:

1. the outer cork layer of the nodule;
2. water-filled diffusion barriers (intercellular spaces);
3. leghaemoglobin;
4. cellular respiration (bacteroids and mitochondria).

The bacteroid zone in nodules contains an anastomosing net of intercellular air spaces. In one soybean nodule, approximately 100 000 such air spaces

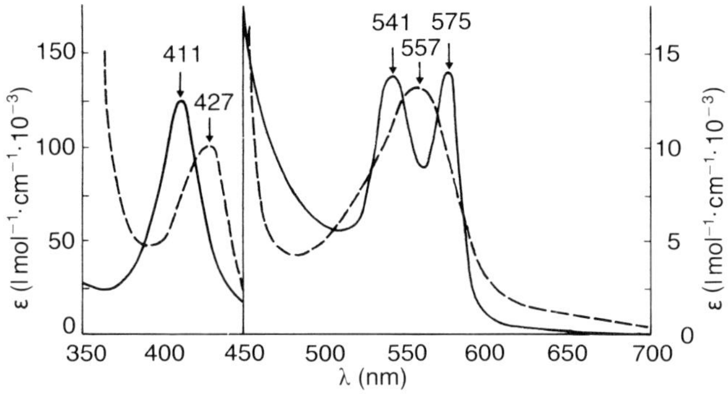

Fig. 3.48. Absorption spectra of leghaemoglobin in crude extracts from soybean nodules. Fe(II)-oxyleghaemoglobin (——); Fe(II) deoxygenated leghaemoglobin (–––) (Based on Appleby and Bergersen, 1980.)

can be found. In the nitrogen-fixing phases (phases II and III, Fig. 3.47) oxygen concentration is suboptimal despite this intercellular network of air spaces. Nitrogen fixation may be significantly increased at increased oxygen levels. Only during early development and senescence (phases I and IV) when leghaemoglobin content is below a critical value, does this oxygen enhancement effect disappear (Fig. 3.47). In pea nodules, the appearance of leghaemoglobin precedes that of nitrogenase and nitrogenase reductase activities (Bisseling *et al.*, 1980a).

Leghaemoglobin cannot be detected in leaves, stems or non-infected roots of symbiotic plants, even when sensitive immunological methods are used. In mature nodules, however, leghaemoglobin can constitute up to 40% of the total soluble protein. The concentration thus reaches 0.5–1-mM. There are four isoforms of soybean leghaemoglobin, *a*, C_1, C_2, C_3. They differ in molecular weight, pI and amino acid sequence. They are, however, antigenically cross-reactive. Unfractionated Fe(II) oxyleghaemoglobin shows an absorption maximum at 411 nm and two other smaller maxima at 541 and 575 nm (Fig. 3.48). Deoxygenated leghaemoglobin has maxima at 427 nm and 557 nm. An experiment showing the transfer of oxygen from oxyleghaemoglobin to *Bradyrhizobium* cells is shown in Fig. 3.49.

Four different mechanisms have been proposed for the conversion of leghaemoglobin into the active, O_2-binding ferrous Lb^{2+} from the inactive Lb^{3+} form (Becana and Klucas, 1990):

1. reduction of the Lb^{3+} form by flavoprotein reductase, with molecular weight 100 kDa and a K_m of for Lb^{3+} 9 μM
2. direct reduction of Lb^{3+} by NAD(P)H, ascorbate and cysteine
3. reduction via free flavin
4. reduction by an unidentified heat-stable compound

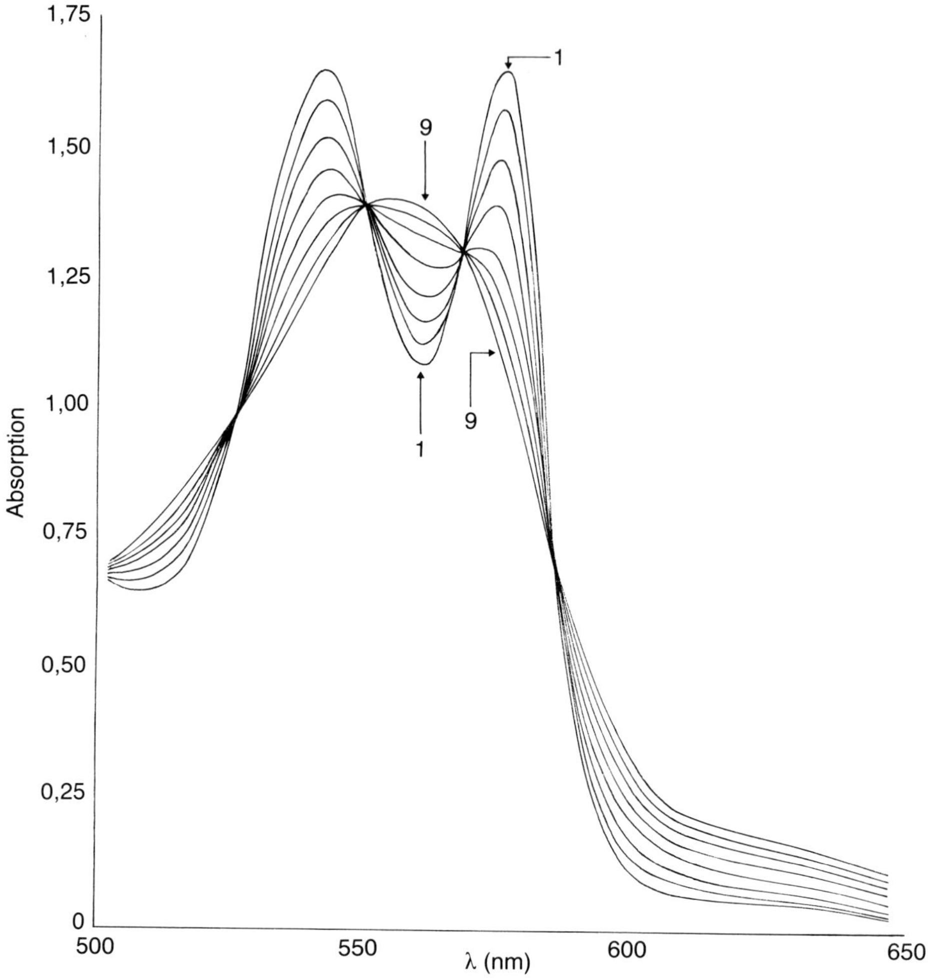

Fig. 3.49. Deoxygenation of leghaemoglobin in the presence of *Bradyrhizobium* sp. (cowpea). Nine spectra were recorded at time intervals of 3.5 min. (Based on Appleby and Bergersen, 1980.)

Protection of nitrogenase from oxygen damage is achieved not only by means of physical barriers to gaseous diffusion (both fixed and variable), but also by active removal of O_2. The latter is achieved to rapid bacterial respiration. Due to the high energy demands of nitrogen fixation the respiration rate of bacteroids is around 6 nmol CO_2 min^{-1} g^{-1} nodule dry weight. This is six times higher than in ineffective nodules (Fig. 3.50). When the partial pressure of O_2 is reduced, the respiration rate of effective bacteroids is also reduced whereas for ineffective bacteroids it remains at its

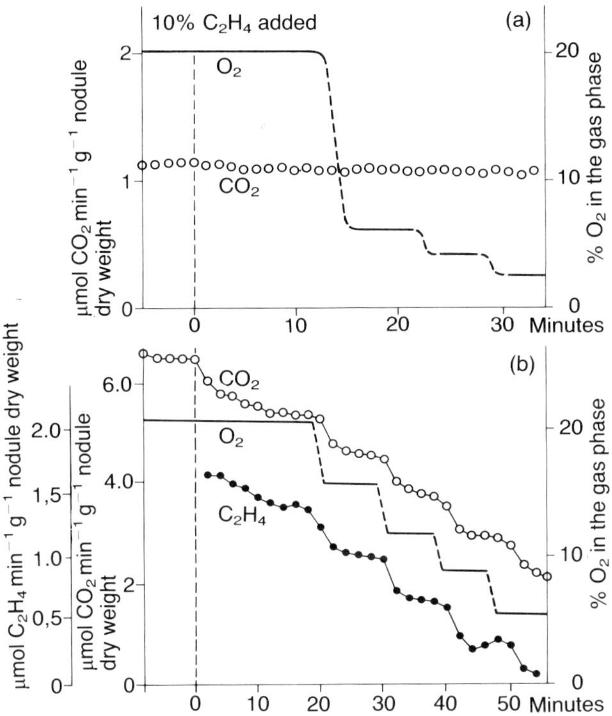

Fig. 3.50. Time dependency of acetylene reduction and respiration during stepwise reduction of O_2 partial pressure in isolated nodules of *Pisum sativum*. (a) ineffective (fix$^-$) nodules; (b) effective (fix$^+$) nodules. (Based on Witty *et al.*, 1983.)

previous low level. Using leghaemoglobin spectroscopy and with microelectrodes, the concentration of free O_2 in active bacteroid-infected cells has been estimated to be on average only 10 nM (Rawsthorne and LaRue, 1986a). In ineffective nodules this value was 60–100 µM. Mitochondria isolated from nodules and hypocotyls of cowpea have similar O_2-uptake rates and ADP/O ratios. However, NAD-malic enzyme activity is 12-fold lower in nodule mitochondria compared with hypocotyl mitochondria (Rawsthorne and LaRue 1986b). By using variables such as respiration rate, concentration of free oxygen and the buffering capacity of leghaemoglobin, mathematical models have been developed to explain the response of N_2-fixation rate to variations in oxygen availability. It has become clear from such computer modelling that a variable barrier to oxygen diffusion must be present in nodules. It has been proposed that this is a water-filled barrier, which increases its resistance when nodule respiration rate falls due to, for example, restricted supply of photosynthate (Sheehy *et al.*, 1985). This is especially important in crop legumes where photosynthate availability

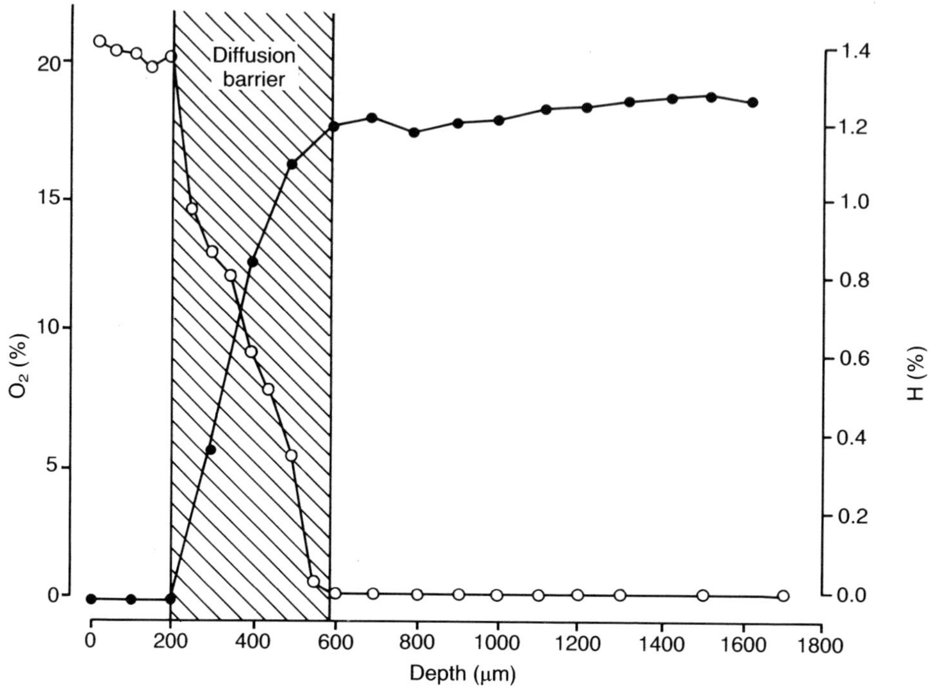

Fig. 3.51. Traces from O_2 and H_2 specific microelectrodes as they were advanced into a soyabean nodule (Clarke/RCR 3442). (From Witty and Minchin, 1990.)

fluctuates due to shoot harvest or grazing. The effect of the water-filled diffusion barrier on O_2 and H_2 concentrations in nodules is shown in Fig. 3.51. The oxygen diffusion resistance of nodules from a supernodulating mutant of soybeans is considerably larger than that from wild-type nodules (Schuller *et al.*, 1988).

(c) Nitrogen fixation per g nodule, per plant, per hectare

Improvements in the development and functioning of legume nodules are dependent on a thorough understanding of the vegetative morphology of root systems. The interaction of roots and nodules with stems and leaves is also important (Lersten and Carlson, 1987; Heichel *et al.*, 1989).

Nitrogen fixation rates obtained for nodules from soyabean, pea, clover and bean are between 10 and 50 μmol $C_2H_4\,h^{-1}\,g^{-1}$ fresh weight. In phases I and IV of nodule development, these rates are obviously much smaller. The ratio of acetylene reduction to N_2 reduction is not 3:1. The factor 3:1 is based on the stoichiometry of electron transfer, N_2 reduction to ammonia involves the transfer of six electrons whereas the reduction of C_2H_2 to C_2H_4

Table 3.23 Acetylene reduction, $^{15}N_2$-reduction and H_2-evolution in legume nodules in the presence of C_2H_2 and N_2. (Based on Van Kessel and Burris, 1983)

Species	*μmol h^{-1} g^{-1} nodule weight*			
	C_2H_4	*H_2 in the presence of C_2H_2*	*H_2 in the presence of N_2*	*$^{15}N_2$*
Trifolium pratense	145	9	69	33
Trifolium repens	99	9	59	19
Medicago lupulina	209	21	94	58

involves only two. This ratio fails to take into account the loss of electrons due to hydrogen production by nitrogenase and the partial repression of hydrogen production in the presence of acetylene. With N_2 as substrate, half the electrons are used to reduce protons to hydrogen (Table 3.23). In the presence of acetylene, only 10% of the electrons are used in H_2 synthesis. Thus the actual conversion factor for the calculation of N_2-reduction rate based on C_2H_2 reduction is between 5 and 6. Measurements of acetylene reduction in a closed system normally underestimate nitrogenase activity. As is shown in Fig. 3.50, falling rates of C_2H_2-reduction occur even when oxygen concentration is held constant. Within a 30 min incubation period, the measured rate can decrease to between 30% and 50% of the real rate. Since incubation times of 60 min are often used in the acetylene reduction assay, errors of at least 50% can be expected.

Table 3.24 N_2-fixation in legumes. (Based on Quispel, 1982)

Species	*N_2-fixation (kg N ha^{-1} $year^{-1}$)*	
	Threshold values	*Average values*
Clover	45–670	250
Pea	50–500	150
Alfalfa	90–340	250
Lupin	140–200	150
Soybean	60–300	100
Groundnut	50–150	100
Lentil	50–150	80
Broad bean	100–300	200

Fixation rates for whole plants are heavily dependent on the age of the plant since age has an effect on both the number and the size of nodules. Eight-week-old alfalfa plants produce 1.5–2 μmol C_2H_2 plant^{-1} h^{-1} whereas 5–9-week-old peas produce 1.5–12.5 μmol C_2H_4. Rates per hectare and per year for the most important crop legumes are between 50 and over 600 kg N ha^{-1} year^{-1} (Table 3.24). This wide range is due to soil and climatic factors, choice of cultivar and intensity of cultivation.

(d) Dependence of N_2-fixation on climatic and soil conditions and on plant development

(i) Temperature

The temperature optimum for N_2-fixation in *Vicia sativa* and *Medicago trunculata* is 20°C. For *Trifolium subterraneum* and *Trifolium pratense* it is between 25 and 30°C, and for *Medicago sativa* 35°C. For plants native to temperate zones, the lower temperature limit is around 2°C, for tropical species, around 10°C. Low and high temperatures affect symbiotic nitrogen fixation severely, leading to abrupt cut-offs at temperatures where the bacterial cells can still grow and metabolize. For example, *Vigna unguiculata* (Fig. 3.52) cannot fix nitrogen below 15°C. However, the rate of increase in nodule dry weight at this temperature is only reduced by 25% compared

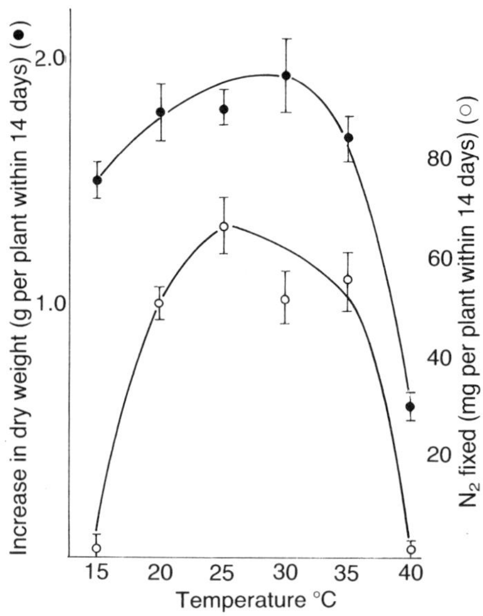

Fig. 3.52. Temperature dependency of growth and N_2-fixation in *Vigna unguiculata*. Variation of root system temperature at constant shoot temperature of 30°C/20°C (day/night). (Based on Rainbird *et al.*, 1983.)

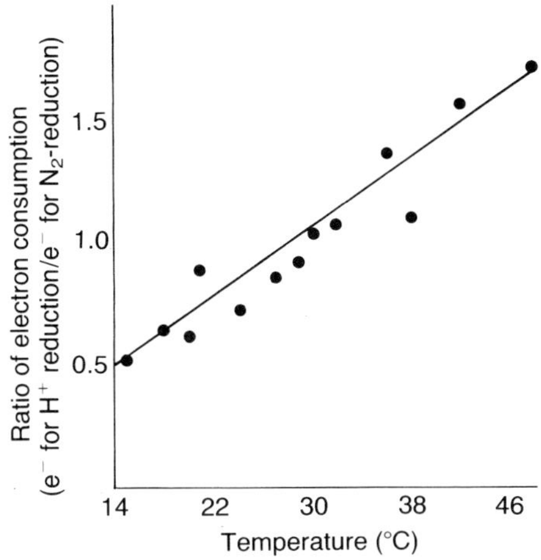

Fig. 3.53. Temperature dependency of proton- and N_2-reduction in nodules of *Vigna unguiculata*. (Based on Rainbird *et al.*, 1983.)

with optimal conditions. Root hair infection is much more temperature sensitive than nodule development. This gives legumes, such as clover, a high growth rate in summer, and ecological dominance over grasses. In autumn, winter and spring, however, the grasses can quickly re-assert their dominance. The relative distribution of electrons between N_2-reduction and proton reduction is also temperature-dependent. Figure 3.53 shows that at 14°C the ratio is 0.5, but at 38°C it is 1.3. Thus at higher temperatures, nitrogenase favours H^+ over N_2-reduction.

(ii) ***Water content***

Water content changes total N_2-fixation by affecting the number of nodules per root (Fig. 3.54). Eleven weeks after inoculation, water-limited *Vicia faba* plants had less than 200 nodules per plant whereas control plants had 500 and waterlogged plants 800. Nitrogenase activity per plant was 20 μmol h^{-1} for water-limited plants, 40 μmol h^{-1} for average watered plants and 60 μmol h^{-1} for waterlogged plants. The number of nodules is also affected by the size of the plants. After 11 weeks, dry shoot mass for water-limited plants was about 2 g, whereas control plants had 7 g shoot dry mass and waterlogged plants 10 g. It should be noted, however, that root development showed the opposite trend. Root dry masses were 2.7 g, 2.6 g and 2.0 g for water-limited, control and waterlogged plants, respectively. Thus nodule numbers and N_2-fixation rates do not reflect the size of the root system, but rather the above ground parts, especially the leaf-area. Prolonged waterlogging leading to oxygen limitation results in reduced N_2-fixation.

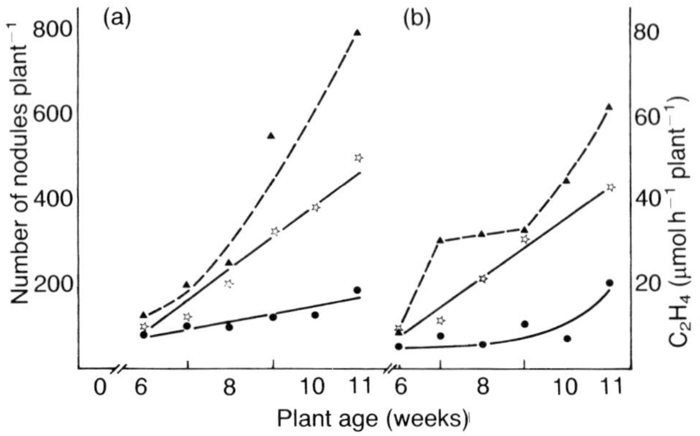

Fig. 3.54. Relationship between (a) nodule number and (b) nitrogenase activity per plant and soil water content in *Vicia faba*. Substrate: humus/sand (1 : 1). Control (☆); water deficiency (●); water logging (▲). (Based on Gallagher and Sprent, 1978.)

(iii) O_2 and CO_2 content of soil

Since atmospheric concentrations of O_2 support only suboptimal N_2-fixation rates, it is obvious that further reduction of oxygen levels will lead to even more reduced activities. Thus the majority of Fabales grow better on sandy, aerated soil. Due to the intensive respiration in nodules and roots relatively high CO_2 concentrations are found. This helps nitrogen fixation by way of the PEP-carboxylase reaction.

(iv) Diseases

All pathogenic fungi, bacteria, viruses and mycoplasmas which affect stem, leaf or root development in Fabales, also inhibit N_2-fixation. Legume nodules are particularly susceptible to larvae of *Sitona lineatus*, insects that are common on clover.

(e) Effects of NH_4^+ and NO_3^- fertilizer: nitrate-tolerant mutant plants

In muture nodules both nodule development and N_2-fixation are inhibited by nitrate concentrations in the range 3–10 mM. Nitrate inhibits the synthesis of the clover lectin trifoliin A, root hair curling, infection thread development and nodule morphogenesis. The effect of nitrate on N_2-fixation in mature nodules is biphasic (Fig. 3.55). Following a rapid decrease during the first 1–2 days after nitrate application, N_2-fixation rates stabilize at around 30% of the original rate. A second decrease corresponding to 100% inhibition occurs 4–6 days later. The first decrease corresponds to an

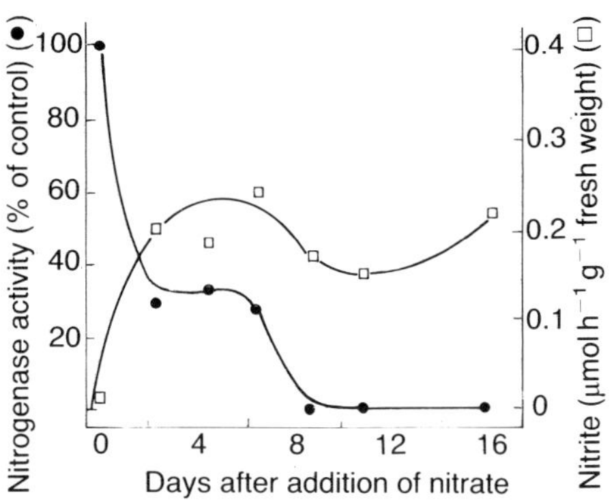

Fig. 3.55. Effect of 10 mM nitrate on nitrogenase activity and nitrate reduction in *Trifolium repens*. (Based on Carroll and Gresshoff, 1983.)

increase in the activity of nitrate reductase, but there is no further increase during the second phase. The initial inhibition of nitrogenase can be partially reversed by increasing photosynthate supply, however, the second does not respond to this treatment. A comparison between soybeans, dependent solely on N_2 fixation as a nitrogen source and nitrogen-fixing plants supplemented with NH_4NO_3, gave a similar net CO_2 fixation rate per unit leaf area but a significantly lower protein/carbohydrate ratio in the leaves, indicating a difference in partitioning (Veau *et al.*, 1990).

In comparing low N_2-fixation rates, we have to realize that different soil types have a significant effect also on the **variability of nitrate analysis**, which is in the order of ± 20 kg N per hectare (Heyland, 1991).

Supplying ammonia to plants with already developed nodules results in an initial decline in nitrogenase activity similar to that seen with nitrate. Over the first 1–2 days, N_2-fixation is reduced to 50% of the original rate (Fig. 3.56). After a stable period the level declines again. Nitrogenase activity in isolated bacteroids is not inhibited by NH_4^+. It is therefore apparent that inhibition of N_2-fixation is not due to effects on nitrogenase. It has been proposed that availability of carbon skeletons needed for N assimilation could explain inhibition by NO_3^- and NH_4^+. This is supported by the finding that methionine sulphoximine (MSX), an inhibitor of glutamine synthetase, overcomes the repressive effect of NH_4^+ on N_2-fixation (Fig. 3.56).

Through classical mutagenetic techniques using soybean seeds and ethyl-methyl-sulphonic acid, it has been shown that nitrate inhibition of nodule development is controlled by the plant rather the bacterial partner in the symbiosis. In several pure-bred mutants, nodulation is no longer inhibited by nitrate, but rather stimulated. These plants, grown in the absence of nitrate, form a very large number of nodules and are called supernodulating

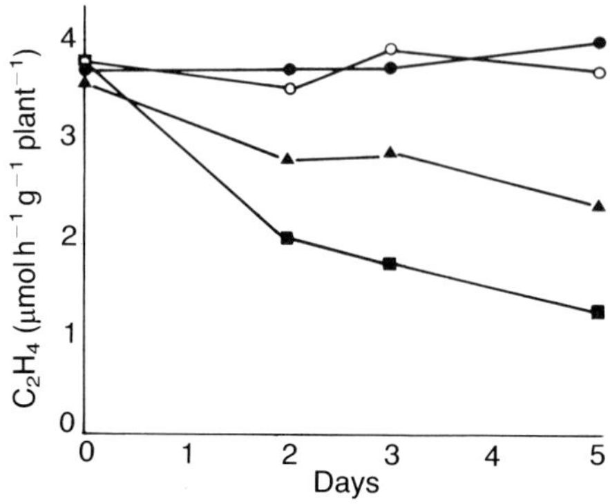

Fig. 3.56. Effect of NH_4^+ and methionine sulploximine (MSX) on nitrogenase activity in intact plants of *Pisum sativum*. At point 0 the nodulated plants were transferred into culture solutions with (○) 20 mM-KCl; (■) 20 mM-NH_4Cl; (●) 20 mM-KCl + 10 μM-MSX; (▲) 20 mM-NH_4Cl + 10 μM-MSX. (Based on Houwaard, 1978.)

mutants. Nodule numbers on these mutants are between 5 and 20 times higher than usually found on wild-type plants and thus reach the numbers found on *Vicia* plants (Table 3.25). Supernodulating mutants of *Glycine max* have a similar number of subepidermal cell divisions to the wild type after inoculation with *B. japonicum*. However, a much larger percentage of these divisions develops further to advanced stages of nodule development (Mathews *et al.*, 1989). Supernodulation is apparently the consequence of reduced transport of a shoot-derived inhibitor. This inhibitor is a water-soluble, low-molecular-weight compound, synthesized in the leaves. Its synthesis is apparently induced by early nodulation events and controlled by a single mendelian gene. Recent studies found a molecular RFLP

Table 3.25 Characteristics of nitrate-tolerant, supernodulating mutants of soybean. The plants were 26-days old. (Based on Gresshoff and Delves, 1986)

Genotype	*Nodules*				*Nitrogenase activity (nmol C_2H_4 min^{-1} plant^{-1})*	
	Number per plant		*Dry weight (mg plant^{-1})*			
	−NO_3	*+NO_3*	*−NO_3*	*+NO_3*	*−NO_3*	*+NO_3*
Wild type (cv. Bragg)	26	19	34	5	71	1
Mutant nts 382	576	1007	166	193	119	69
Mutant nts 1007	334	991	92	179	98	88
Mutant nts 183	351	712	101	141	62	66
Mutant nts 2264	478	907	124	188	99	55
Mutant nts 1116	101	74	66	30	85	23
Mutant nts 733	457	797	115	172	88	54

Table 3.26 Contribution of N_2-fixation to total N-assimilation in legumes under field conditions

Species	*Percentage*
Soybean	50
Bush bean	50
Lentil	70
Pea	70
Broad bean	80
Alfalfa	90
Clover	90

marker which is closely linked (showing no recombination) to the *nts* locus (Landau–Ellis *et al.*, 1991). It is directly involved in nodulation arrest and facilitates nitrate inhibition of nodulation (Gresshoff, 1990). Nitrogenase activity in these mutant plants is barely affected by nitrate concentrations which are completely inhibitory for wild-type plants. N_2-fixation in these plants is increased by more than 50% over wild-type levels. Thus these soybeans can almost support their own nitrogen demand. Clover and alfalfa nodules can supply up to 80–90% of the plant's nitrogen requirements (Table 3.26).

Excision of root nodules on primary roots stimulates the formation of other nodules on lateral roots. The same result is produced by bacteria free nodules, induced by exopolysaccharide deficient mutants (Caetano-Anollés and Gresshoff, 1991). These results seem to indicate that **autoregulation of nodule numbers** is dependent on the host plant structures, perhaps their meristematic activity. On the other side, the much larger number of ineffective (fix^-) nodules compared to fix^+ nodules on the same host (Fig. 3.32) indicates also an effect of the genotype of the microsymbiont or the symbiotic functions. Only 10% of total plant nitrate reduction capaity in soybeans and *Phaseolus* beans is located in the roots, whereas in pea and broad bean the figure is 30–60% (Wallace, 1986).

Plants dependent solely on N_2-fixation as a nitrogen source take up an excess of cations over anions. This is normally compensated for by an efflux of protons into the rhizosphere. The uptake of 100 kg of NH_4^+ means the excretion of 6 kg H^+. To neutralize these protons 300 kg $CaCO_3$ are needed.

(f) C and N metabolism: the relationship between N_2-fixation, respiration, photosynthesis and H_2-production

Nitrogen fixed by bacteroids is exported to the host cytoplasm as ammonia. This is incorporated into organic compounds via the glutamine synthetase/glutamate synthase pathway. Aspartate is formed directly by the amination of oxaloacetate and asparagine through amidation of aspartate (Fig. 3.57).

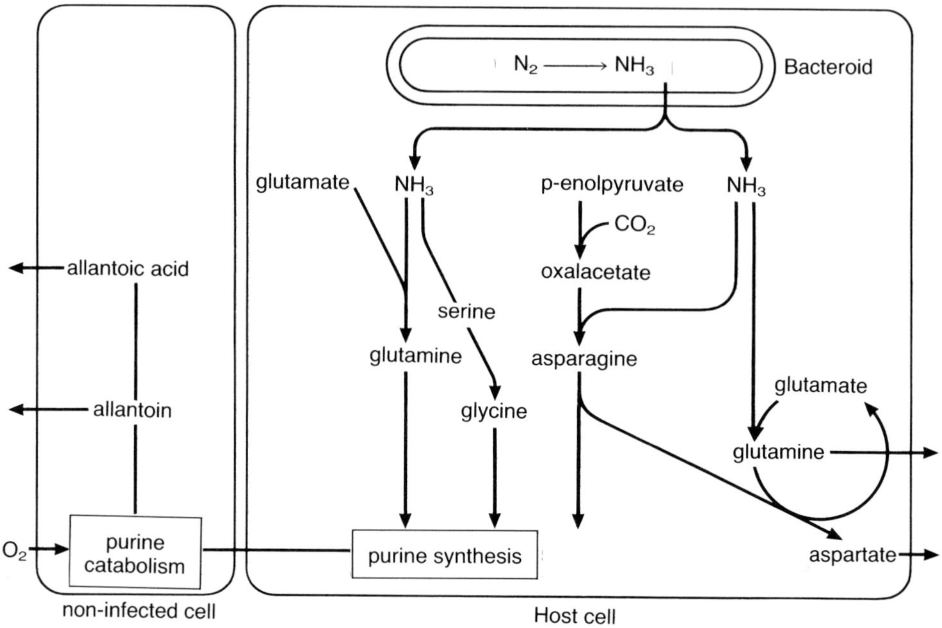

Fig. 3.57. Carbon and nitrogen metabolism in infected and non-infected cells of legume nodules. (Based on Dilworth and Glenn, 1984.)

Glutamine synthetase is also a central enzyme for NH_4^+ assimilation in chloroplasts (Fig. 3.58). Glutamine and asparagine are the most important transport forms of nitrogen in nodules with non-determinate growth (e.g. pea nodules). The most important transport form for N in the determinate (or spherical) nodules of, for example, soybeans are allantoin and allantoic acid. The synthesis of these components occurs in non-infected cells and involves oxidative degradation of purines. Purine synthesis takes place in the infected cells (Fig. 3.57). The concentration of organic nitrogen in nodules, and xylem can be very different. With lower O_2 partial pressure in nodules the incorporation of NH_4^+ into organic forms is much more strongly reduced than N_2-fixation (Table 3.27). Atomic $^{15}N_2$-enrichment is reduced at 0 kPa O_2 to 29% of the value at 20 kPa O_2. This change in O_2 level results in almost complete inhibition of N incorporation into amino acids. Worthy of note is the observation that at 10 or 5 kPa O_2, N incorporation into serine and valine is less reduced than that into glutamine or aspartate. The explanation for a strong reduction in allantoin synthesis below 10 kPa O_2 is that oxygen is required for purine biosynthesis. α-Oxoglutarate, oxaloacetate and glyoxylate, used for asparagine and glycine synthesis, probably become limiting as the tricarboxylic acid cycle becomes oxygen-limited.

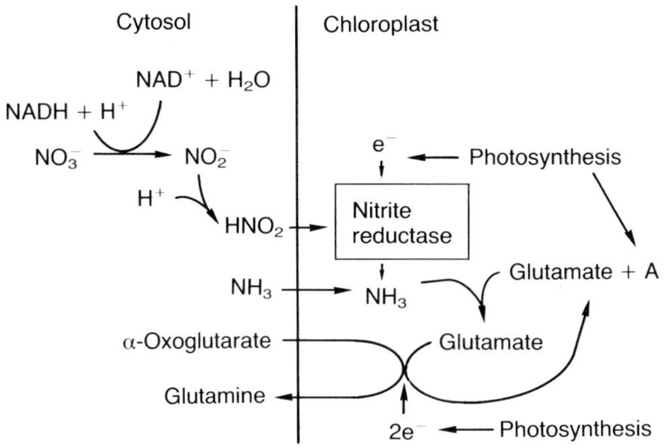

Fig. 3.58. Reduction of NO_3^- in the cytosol and HNO_2^--reduction and NH_3-assimilation in chloroplasts of plant cells. (Based on Mengel, 1992.)

Table 3.27 Effect of oxygen partial pressure on the incorporation of ^{15}N into N-metabolites. $^{15}N_2$ was applied to intact nodules of *Glycine max*. (Based on Ohyama and Kumazawa, 1980)

Metabolite	% *Atomic accumulation*						
	P_{O_2} (*kPa*)						
	20	*10*		*5*		*0*	
Ammonia	0.76	0.37	(57)*	0.20	(31)	0.22	(29)
Glutamine	0.79	0.65	(49)	0.28	(37)	0.01	(1.3)
Asparagine	0.16	0.08	(50)	0.05	(31)	0.00	(0)
Glutamate	3.98	2.68	(67)	1.03	(26)	0.00	(0)
Aspartate	1.68	–		0.12	(7)	0.00	(0)
Alanine	3.26	2.30	(71)	1.10	(34)	0.00	(0)
Serine	1.50	1.73	(115)	1.04	(69)	0.00	(0)
Glycine	1.58	1.09	(69)	0.13	(8)	0.00	(0)
γ-Aminobutyrate	1.27	1.14	(90)	0.51	(40)	0.00	(0)
Isoleucine, Leucine	0.56	0.54	(96)	–		0.00	(0)
Valine	0.38	0.45	(118)	0.14	(37)	0.01	(2.6)
Allantoin	2.03	0.73	(36)	0.43	(21)	0.01	(0.5)

* Figures in brackets: % of ^{15}N-incorporation at 20 kPa P_{O_2}

The biosynthesis of glutamine in the host cytoplasm is catalysed by glutamine synthetase as follows:

$$\text{L-Glutamate} + \text{ATP} + NH_4^+ \xrightarrow{Mg^{2+}} \text{Glutamine} + \text{ADP} + P_i$$

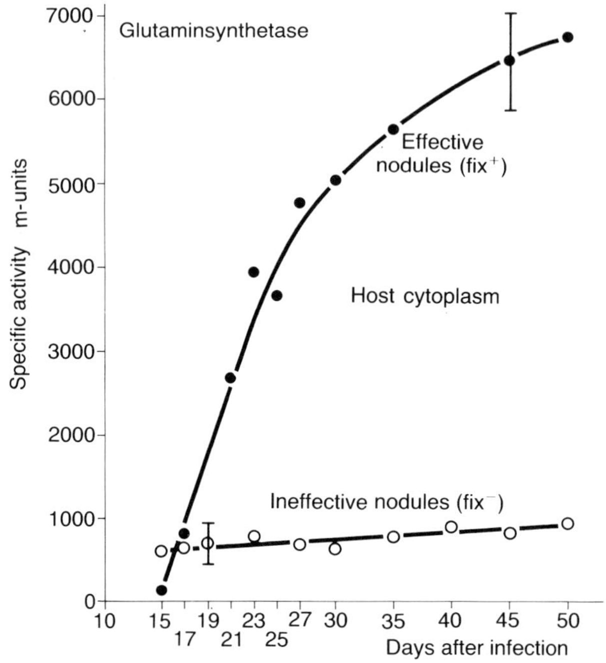

Fig. 3.59. Glutamine synthetase activity in the host cytoplasm of effective (fix^+) and ineffective (fix^-) soybean nodules, 15–50 days after infection with fix^+ or fix^- strains of *B. japonicum*. (Based on Werner *et al.*, 1980.)

This enzyme, which can constitute up to 2% of the soluble protein in the cell, has a molecular weight of 380 kDa and is composed of eight subunits. ADP and AMP inhibit the enzyme, indicating that it is responsive to cellular energy charge. The level of glutamine synthetase activity in nodules is positively correlated with the level of N_2-fixation (Fig. 3.59). Activity in effective nodules is often more than 10 times higher than in ineffective nodules. Bacterial glutamine synthetase activity responds in the opposite way. In ineffective bacteroids the activity is high, whereas in effective bacteroids it is low (Fig. 3.60).

In cells from the infected zone, asparagine synthetase (EC 6.3.1.1) is enriched 150 times relative to roots. The apparent K_m for aspartate is 1.24 mM, for Mg-ATP 0.08 mM and for glutamine 0.16 mM. The K_m value for NH_4^+ as N-donor is over 40 times higher than that for glutamine. Glutamine synthetase (GS), glutamine 2-oxyglutarate aminotransferase (GOGAT), aspartate aminotransferase (AAT), asparagine synthase (AS), phosphoenolpyruvate carboxylase (PEPC) in effective nodules of alfalfa declines following defoliation (Egli *et al.*, 1989).

Carbohydrate consumption by symbiotic N_2-fixation in soybean nodules

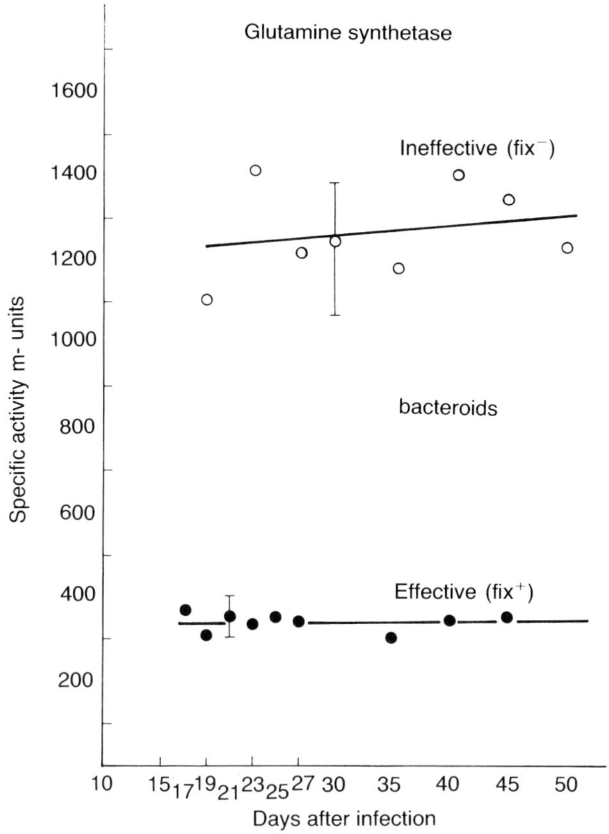

Fig. 3.60. Glutamine synthetase activity in bacteroids from effective (fix^+) and ineffective (fix^-) soybean nodules 15–50 days after infection with fix^+ or fix^- strains of *B. japonicum*. (Based on Werner *et al.*, 1980.)

is about 12 g glucose g^{-1} N fixed (Table 3.28). Of this, 7.3 g carbohydrate is required to support N_2-fixation and H_2 evolution, 2.7 g are needed for nodule metabolism, 0.3 g for nodule growth and 1.9 g for assimilation and transport. The latter is also needed when NH_4^+ or NO_3^- uptake replaces N_2-fixation.

Dependence of nitrogenase activity in the nodules from the photosynthate partitioning depends to a large extent on the condition of the plants. In alfalfa a decrease of N_2-fixation rate by 80% after removal of the shoot has been observed (Cralle and Heichel, 1981) as well as a steady rate for 12 h (Cralle and Heichel, 1982). Photosynthate partitioning was not altered in populations of alfalfa, differing by 40% in nodule mass (Cralle *et al.*, 1987).

The form of the carbon taken up by the bacteroids to be used as an energy source to support N_2-fixation is still debatable. Nitrogenase activity

Table 3.28 Carbohydrate consumption by symbiotic N_2-fixation in nodules of *Glycine max*. (Based on Streeter and Salminen, 1985)

Physiological process	*Carbohydrate (g g^{-1} N fixed)*
N_2-fixation and H_2-evolution	7.3
Maintenance metabolism of nodules	2.7
Nodule growth	0.3
Assimilation and transport of NH_4	1.9
Total	12.2

Table 3.29 Organic acid contents in various organs of 33-day-old soybeans, infected with *B. japonicum* strain USDA 110, in relationship to additional N-fertilization. (Based on Stumpf and Burris, 1981)

Tissue	*N-nutrition*	*Organic acid content (µmol g^{-1} fresh weight)*			
		Malonate	*Succinate*	*Fumarate*	*Malate*
Leaf	N_2	3.51	0.38	0.68	0.79
	$+NO_3^-$	3.26	0.26	0.64	0.94
	$+NH_4^+$	3.40	0.30	0.54	0.91
Stem	N_2	3.38	0.50	8.81	0.84
	$+NO_3^-$	3.04	0.41	8.51	0.79
	$+NH_4^+$	2.15	0.41	2.16	0.86
Root	N_2	4.99	0.26	0.20	1.79
	$+NO_3^-$	3.30	0.17	0.14	1.40
	$+NH_4^+$	2.20	0.17	0.23	1.58
Nodule	N_2	5.40	0.72	1.27	4.18
	$+NO_3^-$	2.70	0.77	0.91	4.00
	$+NH_4^+$	1.72	0.45	0.78	3.80

of isolated bacteroids can be supported by organic acids, but not by sugars. Dicarboxylic acid transport (dct^-) mutants cannot fix N_2. It thus appears that malate and/or succinate are used as the energy source for N_2-fixation. The *dct* genes in *R. meliloti* are located on the same megaplasmid as the genes for EPS production (Watson *et al.*, 1988). In intact nodules of soybeans the addition of malate increased nitrogenase activity (Heckmann and Drevon, 1988). Malonate, malate and fumarate are found in higher concentrations than succinate in nodules (Table 3.29). This enrichment is not nodule-specific but is found in roots, stems and leaves of soybean. Malonate is the dominant organic acid, except in stem tissue, where very high fumarate concentrations are found. The malonate content of nodules and roots is raised considerably when N_2-fixation instead of nitrate or NH_4^+ is the plant's N-source. In roots, malonate is synthesized by acetyl-CoA-carboxylase. It has been proposed that malate functions as an alternative

CO_2 acceptor to PEP in legume nodules (Schramm, 1982). An interesting mechanism postulated for the transport of ammonia out of bacteroids involves coupling with the synthesis of malonamate from malonate (Kim and Chae, 1990). This model is supported by the presence of two types of malonamidases in bacteroids and one type in the host cell cytosol.

Free-living and bacteroid forms of *B. japonicum* can oxidize malonate at a rate of 1.5 μmol h^{-1} mg $protein^{-1}$. Malonate turnover is almost totally inhibited by succinate (0.7 mM) (Fig. 3.61). Similar concentrations of arabinose do not disturb malonate utilization. Succinate utilization by either free-living or bacteroid forms of *B. japonicum* is hardly affected by even high concentrations of malonate (Fig. 3.62). Measurements of total pools of

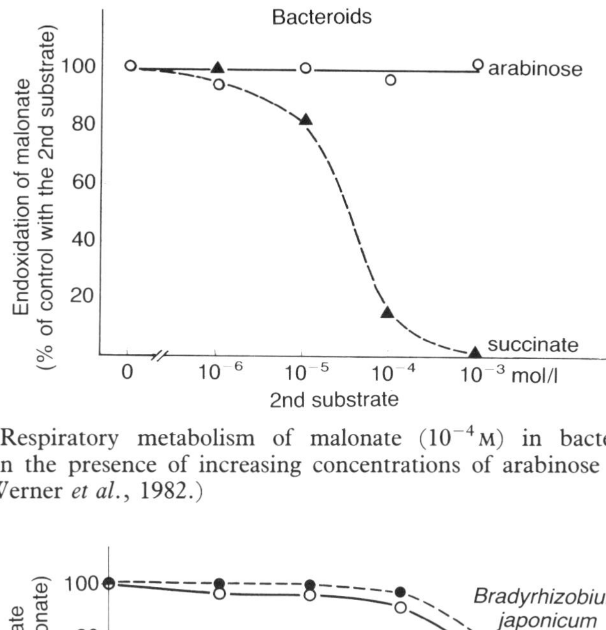

Fig. 3.61. Respiratory metabolism of malonate (10^{-4} M) in bacteroids of *B. japonicum*, in the presence of increasing concentrations of arabinose or succinate. (Based on Werner *et al.*, 1982.)

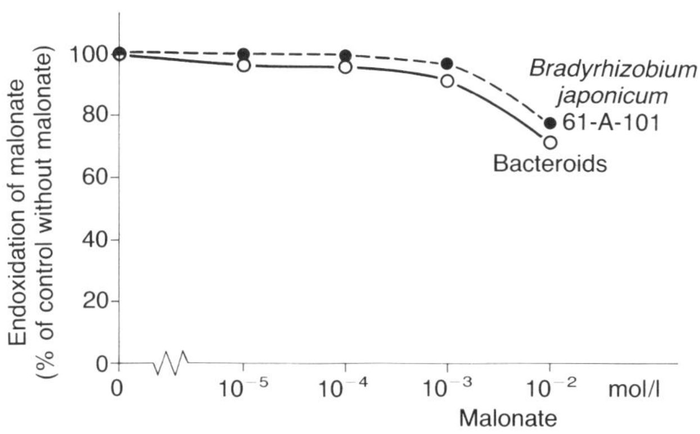

Fig. 3.62. Respiratory metabolism of succinate (10^{-4} M) in *B. japonicum* (free-living or as bacteroids), in the presence of increasing concentrations of malonate. (Based on Werner *et al.*, 1982.)

organic acids can be misleading because they fail to take into account possible subcellular partitioning. As mentioned earlier, bacteroids are surrounded by the symbiosome membrane. Measurements of metabolites in the peribacteroid space show that no malonate is detectable there. On the other hand, malate is at least five times more concentrated in the peribacteroid space than in the host cell cytoplasm (Humbeck and Werner, 1987).

A dicarboxylate transport system has also been characterized in the symbiosome membrane, with a K_m of 0.1–0.2 mM for malate or succinate (Day *et al.*, 1990). This carrier has only weak affinity for malonate. Glutamate and glucose uptake by intact soybean symbiosomes proceeds at a rate of only 5% of that of malate uptake.

All these results indicate that malate is the *in situ* carbon source for nitrogen-fixing bacteroids and that only after disturbance of its supply, amino acids or malonate may be of significance. During their differentiation (i.e. before the induction of nitrogenase) bacteroids may use a wide range of other carbon and also nitrogen sources.

With mitochondria from lupin nodules, the respiration rate with succinate was more than four times the rate with malate as a substrate. The PBU from the same source had twice the V_{max} with succinate (about 40 nmol O_2 min^{-1} mg $protein^{-1}$) than with malate (Tomaszewska *et al.*, 1991).

Carbohydrate is transported into nodules in the form of sucrose. Invertase (EC 3.2.1.26), an exclusively plant protein, is mostly found in the cells of the nodule cortex. Bacteroids from soybeans have only very little glycolytic activity (Tajima and Kouchi, 1990).

Nodules from Fabales contain relatively high concentrations of phenolic compounds. Their physiological functions are mostly unknown. *Bradyrhizobium japonicum* can metabolize phenol via the β-ketoadipate pathway (Table 3.30). Contrary to the situation in *Pseudomonas putida*, phenol degradation in *B. japonicum* is totally inhibited by succinate and stimulated by pyruvate. The specific activity of phenylalanine ammonium lyase (PAL) and O-methyltransferase is much higher in nodules than in roots. In the degradation pathway from sugars to organic acids, two other nodule enzymes have noticeably high activities. PEP carboxylase (EC 4.1.1.31) is 20 times more active in nodules than in root tissue. The apparent K_m for PEP is 9.4×10^{-2} mM and for HCO_3^- 4.1×10^{-1} mM. This enzyme cytalyses CO_2 fixation in nodules. The resulting oxaloacetate can then be used as a substrate for aspartate amino transferase which is involved in the synthesis of aspartic acid. The PEP carboxylase from soybean nodules is a 440 kDa tetramer. At pH 7 it is activated by glucose 6-phosphate, fructose-6-phosphate and dihydroxyacetone phosphate, whereas malate, aspartate, glutamate, citrate and 2-oxoglutarate are potent inhibitors (Schuller *et al.*, 1991). Pyruvate kinase (PK) (EC 2.7.1.40) also has high activity in nodules.

Table 3.30 Degradation of phenol by *B. japonicum* 61-A-101 and its bacteroid form and by *Pseudomonas putida* 548, in the presence of succinate and pyruvate. (Based on Röhm and Werner, 1985)

^{14}*C-labelled C-source*	*Concn. (mM)*	*Additional C-source*	*Concn. (mM)*	*CO_2 produced (μmol h^{-1} 10^{-9} cells)*		
				B. japonicum		*Ps. putida*
				free-living	*bacteroids*	
Succinate	10	Phenol	0.1	11.8	17.4	76.0
Pyruvate	10	Phenol	0.1	1.6	0.9	6.8
Phenol	0.1	–	–	0.3	0.03	5.6
Phenol	0.1	Succinate	10	0.0	0.00	5.0
Phenol	0.1	Pyruvate	10	0.8	0.04	5.6

PK plays a central role in the regulation of carbon metabolism. PK is activated by AMP and fructose-1,6-bisphosphate and inhibited by organic acids and ATP. The energy charge (ATP + 0.5 ADP / ATP + ADP + AMP) in nodules frozen before detachment from the root, was high with a value of 0.94. An ATP concentration of 340 nmol g^{-1} fresh weight was also measured (Tajima and Kouchi, 1990). It has been proposed that N_2-fixation is not limited by photosynthate availability but rather by restrictions on the utilization of the carbon compounds in an oxygen-limited environment (Vance and Heichel, 1991).

(g) Hydrogen uptake and utilization

Implicit in the mechanism of N_2-reduction by nitrogenase (Fig. 3.46) is the 25% of the electrons used the evolution of H_2 (Evans *et al.*, 1987). The reutilization of this hydrogen by an effective uptake hydrogenase avoids a potential waste of energy. H_2-dependent formation of ATP has been demonstrated with bacteroid preparations of *B. japonicum* (Emerich *et al.*, 1979) and *R. leguminosarum* (Nelson and Salminen, 1982). The percentage of strains with an active uptake hydrogenase (hup$^+$) varies with about 10% of *R. leguminosarum* strains, 20% of *B. japonicum* strains and up to 90% of *Bradyrhizobium* – cowpea group strains (Brewin, 1984). Chemolithotrophic growth, using H_2 as sole energy source and CO_2 as sole carbon source is even more restricted. Coincident expression of hydrogenase and ribulose-1,5-bisphosphate carboxylase/oxygenase has been observed in some strains of *B. japonicum* (Simpson *et al.*, 1979). The physiological electron acceptor for the membrane-bound hydrogenase of bacteroids and chemolithotrophically growing cells of *B. japonicum* is not yet known. Cytochrome b_{559} may be involved, since it is reduced to a maximum extent only with H_2 as

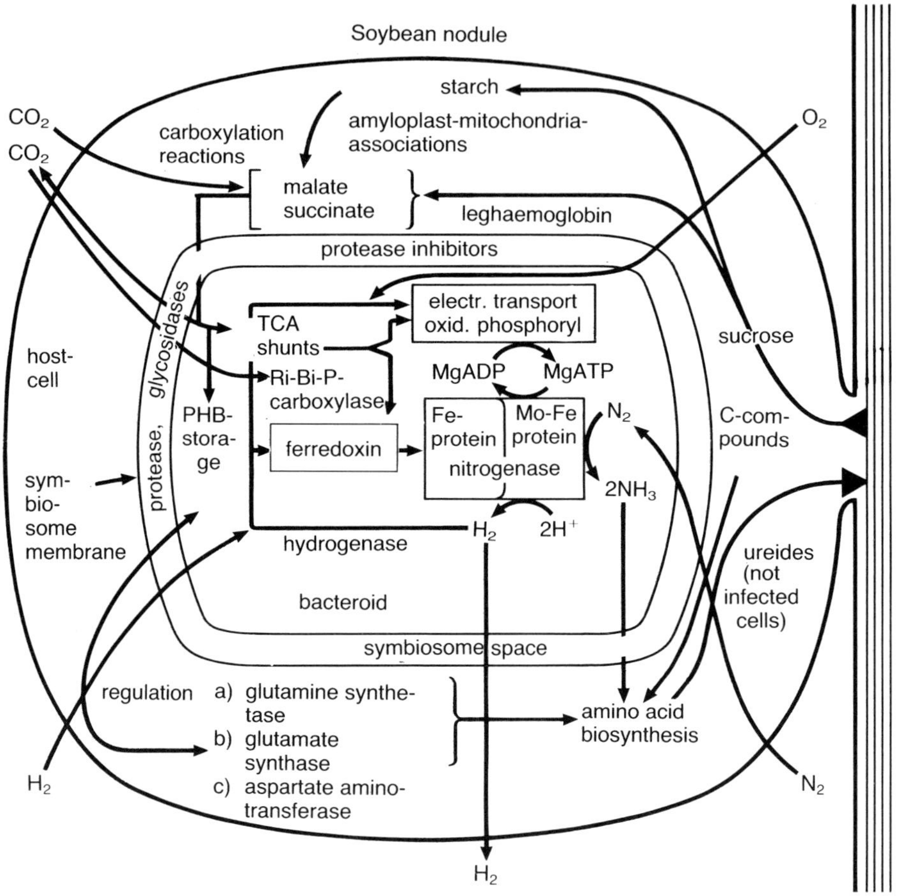

Fig. 3.63. Scheme of the functions in a soybean nodule. (Based on Werner, 1992.)

the electron donor. With NADH or succinate only 40% of maximum reduction is obtained (Evans *et al.*, 1987).

The hydrogenase from *B. japonicum* contains two subunits with molecular weights of 60 and 30 kDa, and is immunologically homologous to the same proteins from *Escherichia coli* (Harker *et al.*, 1985, 1986). *Hup* genes from *R. leguminosarum* are organized in six transcriptional units (*hup* I to *hup* VI). *Hup* I codes for the hydrogenase protein. All six regions are transcribed only in the bacteroid state, not in vegetative cells grown under aerobic conditions (Palacios *et al.*, 1990). Hydrogenase from *B. japonicum* is a seleno-protein (Boursier *et al.*, 1988).

The energy loss due to H_2-production differs greatly between the various members of the Fabaceae (Table 3.31). The efficiency is given by the equation:

Table 3.31 Hydrogen evolution and nitrogenase activity in nodules of legumes and actinorhizal plants. (Based on Schubert and Evans, 1976; Schubert, 1982)

Species	*H_2-evolution under*		*C_2H_2-reduction*	*Effiency H_2-evolution in air*
	Air	*Argon*		*C_2H_2-reduction*
	($\mu mol\ h^{-1}\ g^{-1}$ nodule fresh weight)			
Legumes				
Medicago sativa	7.1	16.4	15.0	0.51
Melilotus albus and *M. officinalis*	4.0	10.2	9.8	0.59
Trifolium fragiferum	4.5	11.1	10.0	0.55
Trifolium hybridum	2.6	5.7	8.8	0.68
Trifolium repens	3.8	7.5	6.9	0.45
Trifolium pratense	2.0	4.1	4.0	0.49
Trifolium subterraneum	1.6	2.7	2.1	0.20
Lotus corniculatus	2.4	6.0	4.3	0.44
Lupinus sp.	1.5	3.8	4.9	0.69
Glycine max	3.7	9.1	7.3	0.52
Pisum sativum	6.3	12.5	17.3	0.63
Cytisus scoparius	0.5	1.0	2.2	0.78
Vigna sinensis	0.03	10.5	14.2	0.99
Phaseolus aureus	1.8	6.3	12.3	0.82
Actinorhiza plants				
Alnus rubra	0.03	0.5	6.9	0.99
Purshia tridentata	0.03	0.03	1.2	0.97
Elaeagnus angustifolia	0.3	2.9	14.8	0.97
Ceanothus velutinus	0.3	0.8	1.8	0.77
Myrica californica	0.001	0.1	1.4	0.99

$$1 - \frac{H_2\text{-production in air}}{C_2H_2 \text{ reduction}}$$

For *Medicago*, *Trifolium*, *Lotus* and *Glycine* this value lies between 0.5 and 0.6. *Sarothamnus scoparius*, *Vigna sinensis* and *Phaseolus aureus* have efficiencies between 0.8 and 1.0. Efficiency is also high for most actinorhiza plants. Carbohydrate and O_2 repress the H_2 uptake system in *B. japonicum*. Reducing equivalents and ATP can be made available using this uptake system. Root respiration of nitrate-fed plants uses 11–16% of the total carbon assimilated, whereas respiration of nodulated roots uses 22–33%. Symbioses with an H_2-uptake system show a reduced requirement for additional photosynthate. Figure 3.63 summarizes the interactions between C-metabolism, amino acid metabolism, N_2-fixation and H_2 production in nodules subdivided into the three symbiotic compartments: host cell cytoplasm, bacterial cytoplasm and symbiosome (peribacteroid) space.

3.3.6 Stem nodules: *Sesbania* spp., *Aeschynomene* spp. and *Neptunia* spp.

Stem nodules were first detected as long ago as 1928 by Hagerup (Hagerup, 1928). A total of 18 species of *Aeschynomene*, four species of *Sesbania* and one *Neptunia* species have been found to form stem nodules (Lahda *et al.*, 1990). In addition, some cultivars of *Vicia faba* (Fyson and Sprent, 1980) and *Arachis hypogaea* can develop nodules on the lower part of the stems.

The development of stem nodules is somewhat different from that of root nodules. The *Azorhizobium* strains able to infect such stems have a doubling-time of 3–4 h. They can grow on sucrose and lactose, but not on citrate, malate or fumarate. Unlike all other rhizobia, they can grow on their own fixed nitrogen in derepressed pure cultures. Bacteroids isolated from *Sesbania* nodules can utilize glucose, succinate and, surprisingly, also lactate to support nitrogenase activity (Trinchant and Rigaud, 1987). The nodules develop only from preformed sites on the shoot axis, the so-called mammillae, which may generate side roots. Nodule morphogenesis (Fig. 3.64) begins in the degenerated or dead rind cells of the stalk. The bacteria divide in the intercellular spaces and induce growth of meristematic

Fig. 3.64. Stem nodules of *Sesbania rostrata*.

tissue. Infection threads develop initially in the intercellular area and later inside cells Infection of host cells is achieved by a process similar to endocytosis. A high population of *Azorhizobium caulinodans*, the microsymbiont of *Sesbania rostrata* has been found in seeds of these plants, indicating that some type of continuous symbiosis exists in this system, which is not found in other legumes. However, it might be worth checking other legume seeds for the presence of rhizobia. *Sesbania rostrata*, which can grow as tall as 3 m, can fix nitrogen at a rate of 17 kg N ha^{-1} day^{-1}. *Sesbania* cultivation has, therefore, a high potential as green manure for rice paddy fields in the tropics.

3.3.7 Gene technology and N_2-fixing symbioses

Eukaryotic cells cannot fix N_2. This property is restricted to approximately 200 prokaryotes. Therefore, atmospheric N_2 can only become available for the 500 000 known eukaryotic plants (340 000 flowering plants and at least 100 000 fungi) and over 1 000 000 species of animals, through symbioses or other close associations. It is noteworthy that neither direct evolution nor convergent evolution, has resulted in the 21 *nif* genes being stably incorporated into any eukaryotic genome. The principle of symbiosis, where phylogenetically distinct organisms have developed associations with diazotrophic genera such as *Rhizobium*, *Bradyrhizobium*, *Frankia*, *Anabaena* or *Nostoc* has been much more successful in terms of evolution. Why was it more efficient to develop and express several dozen symbiosis-related genes in plants and microsymbionts rather than integrate the *nif* genes into the eukaryotic genome?

In principle *nif* genes could be integrated into three different genomes: (1) into the cell nucleus; (2) into the genome of the plastid (chloroplast), or (3) into the mitochondrial genome. Mitochondria in higher plants, fungi and animals have a somewhat unique genetic code. Thus the successful translation of *nif* gene DNA in mitochondria is improbable. On the other hand, translation and transcription in plastids is very similar to that in enterobacteria. RNA polymerase from *Escherichia coli* recognizes the initiation codons in chloroplasts. The chloroplast gene coding for the large subunit of Rubisco can be correctly transcribed and translated in *Escherichia coli* and chloroplast rRNA and cRNA genes exhibit high homology. Chloroplast DNA-dependent RNA polymerase is different from bacterial polymerases. Initiation codons in chloroplast genes in the region −10 to −35 are, however, very similar to those in bacteria. It appears likely that plastid genomes are probably the best targets for the integration of *nif* genes. The insertion of the *nif* gene clusters alone would not lead to nitrogen fixation since gene products of *ntr*C and *ntr*A are needed for transcription. However, if *nif*A gene expression were constitutive, then the *ntr*C gene product would be redundant. *Nif*A and *ntr*A would need to be

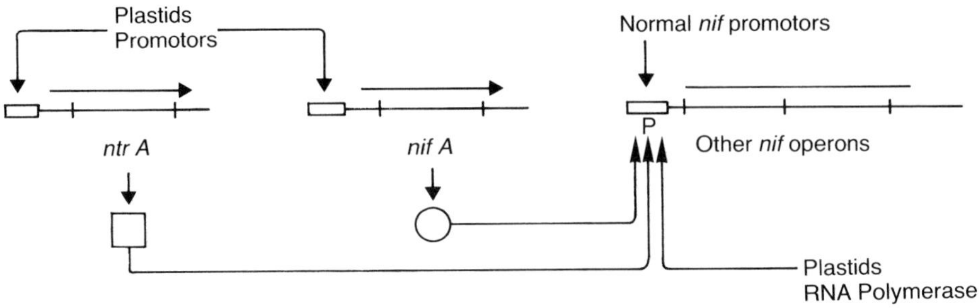

Fig. 3.65. Scheme for genetic engineering of regulatory genes for the expression of *nif* genes in plastids. (Based on Merrick and Dixon, 1984.)

coupled behind chloroplast promoter sequences if they were to be expressed (Fig. 3.65). A further requirement would be that plastid RNA polymerase could transcribe *nif* genes. If this were not the case, eight *nif* gene promotors would need to be swapped with plastid promoters. This route, although complicated, is still simpler than the modifications needed if the 21 *nif* genes were to be expressed in the cell nucleus. Assuming methods such as microinjection of modified *nif* genes into plastids could be successfully applied, the following problems would still remain:

1. The sufficient production of reducing equivalents (reduced ferredoxin) needed for the *nif*-specific electron transport chain;
2. The production of an efficient ATP-production system;
3. The fixation of the resulting NH_4^+ by GS/GOGAT;
4. The protection of the system against O_2 damage.

Although points 1–3 may be covered by the enzymatic machinery in plastids, point 4 is a problem, especially in photosynthetically active plastids. In cyanobacteria lacking heterocysts, such as *Gloeocapsa*, this problem is solved by the chronological separation of photosynthesis (day) and N_2-fixation (night). Incorporation of *nif* genes into amyloplasts would avoid the conflict between photosynthetic O_2-production and the needs cited above.

Other possibilities with more immediate chance of success include the broadening of the host-spectrum in the symbioses. *Frankia* and cyanobacterial symbioses have relatively wide host spectra. *Rhizobium* also forms an effective symbiosis with one non-legume, *Parasponia* (Ulmaceae) (Chapter 4).

3.3.8 Nodulation in the absence of *Rhizobium*

Strictly speaking, nodulation in the absence of a microsymbiont cannot be classified as a symbiotic phenomenon. However, experiments with some cultivars of alfalfa were interesting in several respects. In strictly axenic cultures, root-derived structures developed (1–4 per plant in about 1 to 5% of the plants), which also expressed ENOD 2, but no leghemoglobin (Truchet *et al.*, 1989). The development of these often multilobed structures was inhibited by the addition of 15 mM nitrate and the nodules are capable of inducing autoregulation of nodulation (Caetano-Anollés *et al.*, 1991). The structures revealed some characteristics of indeterminate nodules with an apical meristem, transfer cell, a central tissue and peripheral vascular bundles. Of course no bacteria were present in these tissues. The ontogeny of spontanenous nodules follow that of *Rhizobium*-induced nodules (Joshi *et al.*, 1991). The ability to form nodules in the absence of *Rhizobium* appears to be controlled by a single dominant genetic element (Caetano-Anollés and Gresshoff, 1992).

3.4 REFERENCES

Ahmad, N. and Ng, F.S.P. (1981) Growth of *Leucaena leucocephala* in relation to soil pH nutrient levels and *Rhizobium* concentration. *Leucaena Res Rep.*, **2**, 5–7.

Albrecht, S.L. and Gaskins, M.H. (1982) in *Energy Conservation and Use of Renewable Energies in Bioindustries*. Pergamon Press, Oxford.

Allen, O.N. and Allen, E.K. (1981) *The Leguminosae*. Macmillan, London.

Al-Mallah, M.K., Davey, M.R. and Cocking, E.C. (1987) Enzymatic treatment of clover root hairs removes a barrier to *Rhizobium*-host specificity. *Bio/Technology*, 5, 1319–22.

Anderson, R.G. and Orci, L. (1988) A view of acidic intracellular compartments. *J. Cell Biol.*, **106**, 539–43.

Antoun, H., Bordeleau, L.M. and Sauvageau, R. (1984) Utilization of the tricarboxylic acid cycle intermediates and symbiotic effectiveness in *Rhizobium meliloti*. *Plant Soil*, **77**, 29–38.

Appleby, C.A. (1984) Leghemoglobin and *Rhizobium* respiration. *Ann. Rev. Plant Physiol.*, **35**, 443–78.

Appleby, C.A. and Bergersen, F.J. (1980) Preparation and experimental use of leghaemoglobin, in *Methods for Evaluating Biological Nitrogen Fixation* (ed. F.J. Bergersen), Wiley, Chichester, New York, pp. 315–35.

Balatti, P.A. and Pueppke, S.G. (1990) Cultivar-specific interactions of soybean with *Rhizobium fredii* are regulated by the genotype of the root. *Plant Physiol.*, **94**, 1907–09.

Banfalvi, Z. and Kondorosi, A. (1989) Production of root hair deformation factors by *Rhizobium meliloti* nodulation genes in *Escherichia coli*: HsnD (NodH) is involved in the plant host-specific modification of the NodABS-factor. *Plant Mol. Biol.*, **13**, 1–12.

Bassarab, S., Mellor, R.B. and Werner, D. (1986) Evidence for two types of Mg^{++}-ATPase in the peribacteroid membrane from *Glycine max* root nodules. *Endocyt. Cell Res.*, **3**, 189–96.

Bassarab, S., Schenk, S.U. and Werner, D. (1989) Fatty acid composition of the peribacteroid membrane and the ER in nodules of *Glycine max* varies after infection by different strains of the microsymbiont *Bradyrhizobium japonicum*. *Bot. Acta*, **102**, 196–201.

Bassarab, S. and Werner, D. (1987) Ca^{2+}-dependent protein kinase activity in the peribacteroid membrane from soybean root nodules. *J. Plant Physiol.*, **130**, 233–41.

Bassarab, S. and Werner, D. (1989) Mg^{2+}-dependent pyrophosphatase, a tonoplast enzyme in the peribacteroid membrane of *Glycine max* root nodules. *Symbiosis*, **7**, 81–94.

Becana, M. and Klucas, R.V. (1990) Enzymatic and nonenzymatic mechanisms for ferric leghemoglobin reduction in legume root nodules. *Proc. Natl. Acad. Sci. USA*, **87**, 7295–9.

Bell, E.A. (1971) Toxins in seeds, in *Chemotaxonomy of the Leguminosae* (eds J.B. Harbourne, D. Boulter and B.L. Turner), Academic Press, London, pp. 144–61.

Bell, E.A., Lacey, J.A. and Polhill, R.M. (1978) Systematic significance of canavanine in the Papilionoideae (Faboideae). *Biochem. Syst. Ecol.*, **6**, 201–12.

Beringer, J.E., Brewin, N., Johnston, A.W.B. *et al.* (1979) The *Rhizobium*-legume symbiosis. *Proc. Roy. Soc. London, Series B*, **204**, 219–33.

Better, M., Lewis, B., Corbin, D. *et al.* (1983) Structural relationships among *Rhizobium meliloti* symbiotic promoters. *Cell*, **35**, 479–85.

Bigwaneza, P.C., Prévost, D., Bordeleau, L.M. and Antoun, H. (1991) Glucose transport by an arctic and a temperate strain of rhizobia. *Can. J. Microbiol.* **37**, 105–9.

Bishop, P.E. and Premakumar, R. (1992) Alternative nitrogen fixation systems, in *Biological Nitrogen Fixation*, (eds G. Stacey, R.H. Burris, H.J. Evans), Chapman & Hall, New York, London, pp. 736–62.

Bisseling, T., van den Bos, R.C., van Kammen, A. *et al.* (1977) Cytofluorometrical determination of the DNA contents of bacteroids and corresponding broth-cultured *Rhizobium* bacteria. *J. Gen. Microbiol.*, **101**, 79–84.

Bisseling, T., Moen, A.A., von den Bos, R.C. and van Kammen, A. (1980a) The sequence of appearance of leghaemoglobin and nitrogenase components I and II in root nodules of *Pisum sativum*. *J. Gen. Microbiol.*, **118**, 377–81.

Bisseling, T., van Straten, J. and Houwaard, F. (1980b) Turnover of nitrogenase and leghemoglobin in root nodules of *Pisum sativum*. *Biochim. Biophys. Acta*, **610**, 360–70.

Blumwald, E., Fortic, M.C., Rea, P.A. *et al.* (1985) Presence of host-plasma membrane type H^+-ATPase in the membrane envelope enclosing the bacteroids in soybean root nodules. *Plant Physiol.*, **78**, 665–72.

Bogusz, D., Appleby, C.A., Landsmann, J. *et al.* (1988) Functioning haemoglobin genes in non-nodulating plants. *Nature*, **331**, 178–80.

Boursier, P., Hanus, F.J., Papen, H. *et al.* (1988) Selenium increases hydrogenase expression in autotrophically cultured *Bradyrhizobium japonicum* and is a constituent of the purified enzyme. *J. Bacteriol.*, **170**, 5594–600.

Brass, J.M. (1986) The cell envelope of Gram-negative bacteria: new aspects of its function in transport and chemotaxis. *Curr. Top. Microbiol. Immunol.*, **129**, 1–92.

Brewbaker, J.L. (1990) Nitrogen fixing trees, in *Fast Growing Trees and Nitrogen Fixing Trees* (eds D. Werner and P. Müller) G. Fischer Verlag, Stuttgart, New York, pp. 253–61.

Brewin, N.J. (1984) Hydrogenase and energy efficiency in nitrogen-fixing symbionts, in *Genes Involved in Plant Microbe Interactions* (eds D.P.S. Verma and T. Hohn) Springer Verlag, Wien, New York, pp. 179–203.

Brewin, N.J. (1991) Development of the legume root nodule. *Ann. Rev. Cell Biol.*, **7**, 191–236.

Broughton, W.J., Heycke, N., Priefer, U. *et al.* (1987a) Ecological genetics of *Rhizobium meliloti*: diversity and competitive dominance. *FEMS Microbiol. Lett.*, **40**, 245–9.

Broughton, W.J., Samrey, U. and Stanley, J. (1987b) Ecological genetics of *Rhizobium meliloti*: symbiotic plasmid transfer in the *Medicago sativa* rhizosphere. *FEMS Microbiol. Lett.*, **40**, 251–5.

Caetano-Anollés, G. and Gresshoff, P.M. (1991) Excision of nodules induced by *Rhizobium meliloti* exopolysaccharide mutants releases autoregulation in alfalfa. *J. Plant Physiol.*, **138**, 765–7.

Caetano-Anollés, G. and Gresshoff, P.M. (1992) Plant genetic suppression of the non-nodulation phenotype of *Rhizobium meliloti* host-range *nodH* mutants: gene-for-gene interaction in the alfalfa-*Rhizobium* symbiosis? TAG (in press).

Caetano-Anollés, Joshi, G. and Gresshoff, P.M. (1991) Spontaneous nodules induce feedback suppression of nodulation in alfalfa. *Planta*, **183**, 77–82.

Calvert, H.E., Pence, M.K., Pierce, M. *et al.* (1984) Anatomical analysis of the development and distribution of *Rhizobium* infections in soybean roots. *Can. J. Bot.*, **62**, 2375–84.

Cannon, F., Beynon, J., Buchanan-Wollaston, V. *et al.* (1985) Progress in understanding organization and expression of *nif* genes in *Klebsiella*, in *Nitrogen Fixation Research Progress* (eds H.J. Evans, P.J. Bottomley and W.E. Newton) Martinus Nijhoff Dordrecht, pp. 453–60.

Carlson, R.W. (1982) Surface chemistry, in *Nitrogen Fixation*, Vol. 2: *Rhizobium*, (ed. W.J. Broughton) Clarendon Press, Oxford, pp. 199–234.

Carroll, B.J. and Gresshoff, P.M. (1983) Nitrate inhibition of nodulation and nitrogen fixation in white clover. *Z. Pflanzenphysiol.*, **110**, 77–88.

Chen, W.X., Yan, G.H. and Li, J.L. (1988) Numerical taxonomic study of fast-growing soybean rhizobia and a proposal that *Rhizobium fredii* be assigned to *Sinorhizobium* gen nov. *Int. J. Syst. Bacteriol.*, **38**, 392–7.

Christensen, T., Dennis, E.S., Peacock, J.W. *et al.* (1991) Hemoglobin genes in non-legumes: cloning and characterization of a *Casuarina glauca* hemoglobin gene. *Plant Mol. Biol.*, **16**, 339–44.

Cooper, J.E. and Bjourson, A.J. (1991) Simplified subtraction-hybridization system for isolation of strain-specific *Rhizobium* DNA probes, in *Proc. 2nd Int. Conf. on the Release of Genetically Engineered Microorganisms*, (ed. D.E.S. Stewart-Tull), Nottingham, September 1991.

Cralle, H.T. and Heichel, G.H. (1981) Nitrogen fixation and vegetative regrowth of alfalfa and birdsfoot trefoil after successive harvests or floral debudding. *Plant Physiol.*, **67**, 898–905.

Cralle, H.T. and Heichel, G.H. (1982) Temperature and chilling sensitivity of nodule nitrogenase activity of unhardened alfalfa. *Crop Sci.*, **22**, 300–4.

Cralle, H.T., Heichel, G.H. and Barnes, D.K. (1987) Photosynthate partitioning in plants of alfalfa populations selected for high and low nodule mass. *Crop Sci.*, **27**, 96–100.

Crepet, W.L. and Taylor, D.W. (1985) The diversification of the Leguminosae: first fossil evidence of the Mimosoidae and Papilionoideae. *Science*, **228**, 1087.

Crist, D.K., Wyza, R.E., Mills, K.K. *et al.* (1984) Preservation of *Rhizobium* viability and symbiotic infectivity by suspension in water. *Appl. Environ. Microbiol.*, **47**, 895–900.

Crocker, R.L. and Tiver, N.S. (1948) Survey methods in grassland ecology. *J. Br. Grassland Soc.*, **3**, 1–26.

Crozat, Y., Cleyet-Marel, K.C., Giraud, J.J. and Obaton, M. (1982) Survival rates of *Rhizobium japonicum* populations introduced into different soils. *Soil Biol. Biochem.*, **14**, 401–5.

Cullimore, J.V. (1985) Glutamine synthetase of *Phaseolus vulgaris*: expression of a nodule specific isoenzyme, in *Analysis of the Plant Genes Involved in the Rhizobium–Legume Symbiosis*, OECD, Paris, pp. 121–7.

Cullimore, J.V., Cock, J.M., Farnden, K.J.F. and Bennett, M.J. (1990) Regulation of expression of the GLN-τ gene of *Phaseolus vulgaris* L, in *Nitrogen Fixation: Achievements and Objectives* (eds P.M. Gresshoff, L.E. Roth, G. Stacey and W.E. Newton), Chapman and Hall, New York, London, pp. 713–14.

Dart, P.J. (1977) Infection and development of leguminous nodules, in *A Treatise on Dinitrogen Fixation.* Section III: *Biology*, (eds R.W.F. Hardy and W.S. Silver) Wiley, New York, pp. 367–472.

Day, D.A., Ou Yang, L.-J. and Udvardi, M.K. (1990) Nutrient exchange across the peribacteroid membrane of isolated symbiosomes, in *Nitrogen Fixation: Achievements and Objectives* (eds P.M. Gresshoff, L.E. Roth, G. Stacey and W.E. Newton) Chapman and Hall, New York, London, pp. 219–26.

Day, D.A. (1992) Metabolite transport between endosymbiotic partners. *Symbiosis*, 12, Lectures from the International Symbiosis Congress, Jerusalem, Nov. 1991.

deBruijn, F.J., Hilgert, U., Stigter, J. *et al.* (1990) Regulation of nitrogen fixation and assimilation genes in the free-living versus symbiotic state, in *Nitrogen Fixation: Achievements and Objectives* (eds P.M. Gresshoff, L.E. Roth, G. Stacey and W.E. Newton) Chapman and Hall, New York, London, pp. 33–44.

Diaz, C.L. (1989) Root lectin as a determinant of host-plant specificity in the *Rhizobium*–legume symbiosis. PhD Thesis, Leiden University, The Netherlands.

Diaz, C.L., Hosselet, M., Logman, G.J.J. *et al.* (1990) Distribution of glucose/mannose-specific isolectins in pea (*Pisum sativum* L.) seedlings. *Planta*, **181**, 451–61.

Diaz, C.L., Melchers, L.S., Hooykaas, P.J.J. *et al.* (1989) Root lectin as a determinant of host-plant specificity in the *Rhizobium*–legume symbiosis. *Nature*, **338**, 579–81.

Dilworth, M.J., Eady, R.R., Robson, R.L. and Miller, R.W. (1987) Ethane formation from acetylene as a potential test for vanadium nitrogenase *in vivo*. *Nature*, **327**, 167–8.

Dilworth, M. and Glenn, A. (1984) How does a legume nodule work? *TIBS*, **9**, 519–23.

Dixon, R.A. and Lamb, C.J. (1990) Molecular communication in interactions between plants and microbial pathogens. *Ann. Rev. Plant Physiol. Plant Mol. Biol.*, **41**, 339–67.

Dowling, D.N. and Broughton, W.J. (1986) Competition for nodulation of legumes. *Ann. Rev. Microbiol.*, **40**, 131–57.

Downie, J.A., Economou, A., Scheu, A.K. *et al.* (1990) The *Rhizobium leguminosarum* bv. *viciae* NodO protein compensates for the exported signal made by the host-specific nodulation genes, in *Nitrogen Fixation: Achievements and*

Objectives, (eds P.M. Gresshoff, L.E. Roth, G. Stacey and W.E. Newton), Chapman and Hall, New York, London, pp. 201–6.

Dreyfus, B., Garcia, J.L. and Gillis, M. (1988) Characterization of *Azorhizobium caulinodans* gen. nov., sp. nov., a stem-nodulating nitrogen-fixing bacterium isolated from *Sesbania rostrata. Int. J. Syst. Bacteriol.*, **38**, 89–98.

Eaglesham, A.R.J., Ellis, J.M., Evans, W.R. *et al.* (1990) The first photosynthetic N_2-fixing *Rhizobium*: characteristics, in *Nitrogen Fixation: Achievements and Objectives* (eds P.M. Gresshoff, L.E. Roth, G. Stacey and W.E. Newton) Chapman and Hall, New York, London, pp. 805–11.

Ebel, J. (1986) Phytoalexin synthesis: the biochemical analysis of the induction process. *Ann. Rev. Phytopathol.*, **24**, 235–64.

Economou, A., Hamilton, W.D.O., Johnston, A.W.B. and Downie, J.A. (1990) The *Rhizobium* nodulation gene *nod*O encodes a Ca^{2+}-binding protein that is exported without N-terminal cleavage and is homologous to haemolysin and related proteins. *EMBO J.*, **9**, 349–54.

Economou, A., Hawkins, F.K.L., Downie, A.J. and Johnston, A.W.B. (1989) Transcription of *rhiA*, a gene on a *Rhizobium leguminosarum* bv. *viciae* Symplasmid, requires *rhiR* and is repressed by flavonoids that induce *nod* genes, *Mol. Microbiol.*, **3**, 87–93.

Egli, M.A., Griffith, S.M., Miller, S.S. *et al.* (1989) Nitrogen assimilating enzyme activities and enzyme protein during development and senescence of effective and plant gene-controlled ineffective alfalfa nodules. *Plant Physiol.*, **91**, 898–904.

Egli, M.A., Larson, R.J., Hruschka, W.R. and Vance, C.P. (1991b) Synthesis of nodulins and nodule enhanced polypeptides by plant gene-controlled ineffective alfalfa. *J. Exp. Bot.*, **43**, 441.

Egli, M.A., Vance, C.P. and Larson, R.J. (1991a) Regulation of nodule gene expression in plant-controlled ineffective alfalfa, in *Molecular Strategies of Pathogens and Host Plants* (eds S.S. Patil, S. Ouchi, D. Mills, C. Vance) Springer-Verlag, New York, pp. 215–26.

Emerich, D.W., Ruiz-Argueso, T., Ching, T.M. and Evans, H.J. (1979) Hydrogen-dependent nitrogenase activity and ATP formation in *Rhizobium japonicum* bacteroids. *J. Bacteriol.*, **137**, 153–60.

Erfkamp, J. and Müller, A. (1990) Die Stickstoff-Fixierung. Aktuelle chemische und biologische Aspekte. *Chemie in unserer Zeit*, **24**, 267–79.

Estabrook, E.M. and Sengupta-Gopalan, C. (1991) Differential expression of phenylalanine ammonia-lyase and chalcone synthase during soybean nodule development, *The Plant Cell*, **3**, 299–308.

Ettlinger, C. and Lehle, L. (1988) Auxin induces rapid changes in phosphatidylinositol metabolites. *Nature*, **331**, 176–8.

Evans, H.J., Bottomley, P.J. and Newton, W.E. (1985) *Nitrogen Fixation Research Progress*, Nijhoff/Junk Publ., Dordrecht.

Evans, H.J., Harker, A.R., Papen, H. *et al.* (1987) Physiology, biochemistry and genetics of the uptake hydrogenase in rhizobia. *Ann. Rev. Microbiol.*, **41**, 335–61.

Faria, S.M. de, Lewis, G.P., Sprent, J.I. and Sutherland, J.M. (1989) Occurrence of nodulation in the Leguminosae. *New Phytol.*, **111**, 607–19.

Fischer, H.M., Anthamatten, D., Kullik, I. *et al.* (1991) Complex regulatory network for *nif* and *fix* gene expression in *Bradyrhizobium japonicum*, in *Advances in Molecular Genetics of Plant–Microbe Interaction* (ed. H. Hennecke and D.P.S. Verma) Vol. 1, Kluwer Academic, Dordrecht, pp. 203–10.

Flores, M., Gonzalez, V., Brom, S. *et al.* (1987) Reiterated DNA sequences in *Rhizobium* and *Agrobacterium* spp. *J. Bacteriol.*, **169**, 5782–8.

Fortin, M.G., Morrison, N.A. and Verma, D.P.S. (1987) Nodulin 26, a PBM nodulin, is expressed independently of the development of the peribacteroid compartment. *Nucleic Acids Res.*, **15**, 813–27.

Franssen, H.J., Scheres, B., van de Wiel, C. *et al.* (1990) Nodulins and nodule development, in *Nitrogen Fixation: Achievements and Objectives* (eds P.M. Gresshoff, L.E. Roth, G. Stacey and W.E. Newton) Chapman and Hall, New York, London, pp. 709–12.

Fuhrmann, J. (1990) Symbiotic effectiveness of indigenous soybean *Bradyrhizobia* as related to serological, morphological, rhizobitoxine, and hydrogenase phenotypes. *Appl. Environ. Microbiol.*, **56**, 224–9.

Fyson, A. and Sprent, J.I. (1980) A light and scanning electron microscope study of stem nodules in *Vicia faba* L. *J. Exp. Bot.*, **31**, 1101–6.

Gabriel, D.W., Loschke, D.C. and Rolfe, B.G. (1988) Gene-for-gene recognition: the ion channel defense model, in *Molecular Plant–Microbe Interactions* (eds D.P. Kerma and R. Palacios) APS Press, St Paul, MN, pp. 3–14.

Gallagher, A.E. and Sprent, J.I. (1978) The effect of different water regimes on growth and nodule development of greenhouse-grown *Vicia faba. J. Exp. Bot.*, **29**, 413–23.

Garbers, C., Meckbach, R., Mellor, R.B. and Werner, D. (1988) Protease (thermolysin) inhibition activity in the peribacteroid space of *Glycine max* root nodules. *J. Plant Physiol.*, **132**, 442–5.

Giller, K.E., McGrath, S.P. and Hirsch, P.R. (1989) Absence of nitrogen fixation in clover grown on soil subject to long-term contamination with heavy metals is due to survival of only ineffective *Rhizobium. Soil Biol. Biochem.*, **21**, 841–8.

Gitte, R.R., Rai, P.V. and Patil, R.B. (1978) Chemotaxis of *Rhizobium* sp. towards root exudate of *Cicer arietinum* L. *Plant Soil*, **50**, 553–66.

Göttfert, M. (1990) Signalmoleküle in der *Rhizobium*-Leguminosen Interaktion. *Forum Mikrobiol.*, **13**, 546–51.

Göttfert, M., Grob, P. and Hennecke, H. (1990a) Proposed regulatory pathway encoded by the *nod*V and *nod*W genes, determinants of host specifity in *Bradyrhizobium japonicum. Proc. Natl. Acad. Sci. USA*, **87**, 2680–4.

Göttfert, M., Hitz, S. and Hennecke, H. (1990b) Identification of nodS and nodU, two inducible genes inserted between the *Bradyrhizobium japonicum* nod YABC and nodIJ genes. *Mol. Plant–Microbe Interactions*, **3**, 308–16.

Göttfert, M., Horvath, B., Kondorosi, E. *et al.* (1986) At least two *nod*D genes are necessary for efficient nodulation of alfalfa by *Rhizobium meliloti. J. Mol. Biol.*, **191**, 411–20.

Gresshoff, P.M. (1990) The importance of biological nitrogen fixation to new crop development, in *Advances in New Crops*. Proc. First Nat. Symp. New Crops: Research, Development Economics. Indianapolis, Indiana, October 23–26, 1988 (eds. J. Janick and J.E. Simon), Timber Press Inc., Portland, Oregon, pp. 113–19.

Gresshoff, P.M. and Delves A.C. (1986) Plant genetic approaches to symbiotic nodulation and nitrogen fixation in legumes, in *Plant Gene Research* (eds A.D. Blonstein and P.J. King), 3rd edn, Springer-Verlag, Wien, pp. 159–206.

Gubler, M., Zürcher, T. and Hennecke, H. (1989) The *Bradyrhizobium japonicum fix*BCX operon: identification of *fix*X and of a 5′ mRNA region affecting the level of the *fix*BCX transcript. *Mol. Microbiol.*, **3**, 141–8.

Györgypal, Z. and Kondorosi, A. (1991) Homology of the ligand-binding regions of *Rhizobium* symbiotic regulatory protein NodD and vertebrate nuclear receptors. *Mol. Gen. Genet.*, **226**, 337–40.

Hagerup, O. (1928) Morphological and cytological studies of *Bicornes. Dansk Bot. Arkiv*, **6**(2), 1–26.

Hahn, M. and Hennecke, H. (1987) Mapping of a *Bradyrhizobium japonicum* DNA region carrying genes for symbiosis and an asymmetric accumulation of reiterated sequences. *Appl. Environ. Microbiol.*, **53**, 2247–52.

Halverson, L.J. and Stacey, G. (1986) Signal exchange in plant–microbe interactions. *Microbiol. Rev.*, **50**, 193–225.

Hansmann, P., Maerz, M. and Sitte P. (1989) Cytosymbiosis, in *Progress in Botany*, Vol. 51, Springer-Verlag, Berlin, Heidelberg, pp. 21–47.

Harker, A.R., Lambert, G.R., Hanus, F.J. and Evans, H.J. (1985) Further evidence that two unique subunits are essential for expression of hydrogenase activity in *Rhizobium japonicum. J. Bacteriol.*, **164**, 187–91.

Harker, A.R., Zuber, M. and Evans, H.J. (1986) Immunological homology between the membrane-bound uptake hydrogenases of *Rhizobium japonicum* and *Escherichia coli. J. Bacteriol.*, **165**, 579–84.

Hartwig, U.A. and Philipps, D.A. (1991a) Release and modification of nod-gene-inducing flavonoids from Alfalfa seeds. *Plant Physiol.*, **95**, 804–7.

Hartwig, U.A., Joseph, C.M. and Philipps, D.A. (1991b) Flavonoids released naturally from Alfalfa seeds enhance growth rate of *Rhizobium meliloti. Plant Physiol.*, **95**, 797–803.

Heckmann, M.-O. and Drevon, J.-J. (1988) Effet du malate sur la réponse de l'activité nitrogénase (réduction de l'acétylène) des nodosités de soja à l'oxygène. *C. R. Acad. Sci. Paris*, **306**, Série III, 559–64.

Heichel, G.H., Barnes, D.K., Vance, C.P. and Sheaffer, C.C. (1989) Dinitrogen fixation technologies for alfalfa improvement. *J. Prod. Agric.*, **2**, 24–32.

Hennecke, H., Bott, M., Ramseier, T. *et al.* (1990) A genetic approach to analyze the critical role of oxygen in bacteroid metabolism, in *Nitrogen Fixation: Achievements and Objectives* (eds P.M. Gresshoff, L.E. Roth, G. Stacey and W.E. Newton), Chapman and Hall, New York, London, pp. 293–300.

Hermentin, P. (1987) Malaria invasion of human erythrocytes. *Parasitol Today*, **3**, 52–5.

Heyland, K.U. (1991) Pflanzenproduktion in der Landwirtschaft, in *Angewandte Biologie* (eds. D. Klämbt, H. Kreiskott and B. Streit), Verlag Chemie, Weinheim, pp. 37–50.

Hill, S. (1988) How is nitrogenase regulated by oxygen? *FEMS Microbiol. Rev.*, **54**, 111–30.

Hill, S., Kavanagh, E.P., Arnott, M. *et al.* (1988) Annual Report, 114, AFRC Institute of Plant Science Research, Norwich.

Hirsch, A.M., Bhuvaneswari, T.V., Torrey, J.G. and Bisseling, T. (1989) Early nodulin genes are induced in alfalfa root outgrowth elicited by auxin transport inhibitors. *Proc. Natl. Acad. Sci. USA*, **86**, 1244–8.

Hollingsworth, R.I., Dazzo, F.B., Hallenga, K. and Musselman, B. (1988) The complete structure of the trifoliin a lectin-binding capsular polysaccharide of *Rhizobium trifolii* 843. *Carbohydr. Res.*, **172**, 97–112.

Horvath, B., Kondorosi, E., John, M. *et al.* (1986) Organization, structure and symbiotic function of *Rhizobium meliloti* nodulation genes determining host specificity for alfalfa. *Cell*, **46**, 335–43.

Houwaard, F. (1978) Influence of ammonium chloride on the nitrogenase activity of nodulated pea plants (*Pisum sativum*). *Appl. Environ. Microbiol.*, **35**, 1061–5.

Humbeck, C., Thierfelder, H., Gresshoff, P.M. and Werner, D. (1985) Competitive growth of slow growing *Rhizobium japonicum* against fast growing *Enterobacter* and *Pseudomonas* species at low concentrations of succinate and other substrates in dialysis culture. *Arch. Microbiol.*, **142**, 223–8.

Humbeck, C. and Werner, D. (1987) Separation of malate and malonate pools by the peribacteroid membrane in soybean nodules. *Endocyt. C. Res.*, **4**, 185–96.

Isermann, K. (1987) Quantitative und qualitative Aspekte des Energiebedarfes bei Ernährung der Kulturpflanzen mit technisch, atmosphärisch und biologisch fixiertem Stickstoff. Jahrestagung, Sektion Physiologie und Biochemie der Pflanzen der Biologischen Gesellschaft der DDR, Kühlungsborn.

Jarvis, B.D.W., Gillis, M. and De Ley, J. (1986) Intrageneric and intergeneric similarities between the ribosomal RNA cistrons of *Rhizobium* and *Bradyrhizobium* species and some related bacteria. *Int. J. Syst. Bacteriol.*, **36**, 129–38.

Johnston, A.W.B., Bibb, M.J. and Beringer, J.E. (1978) Tryptophan genes in *Rhizobium*: their organization and their transfer to other bacterial genera. *Mol. Gen. Genet.*, **165**, 323.

Jones, A.M. (1990) Do we have the auxin receptor yet? *Physiol. Plant.*, **80**, 154–8.

Jordan, D.C. (1984) Family III. Rhizobiaceae Conn 1938, 321, in *Bergey's Manual of Systematic Bacteriology*, Vol. 1 (eds N.R. Krieg and J.G. Nolt), William & Wilkins, Baltimore, London, pp. 234–44.

Joshi, P.A., Caetano-Anollés, G., Graham, E.T. and Gresshoff, P.M. (1991) Ontogeny and ultrastructure of spontaneous nodules in alfalfa (*Medicago sativa*). *Protoplasma*, **162**, 1–11.

Kannenberg, E.J. and Brewin, N.J. (1989) Expression of a cell surface antigen is regulated by oxygen and pH for *Rhizobium leguminosarum* strain 3841. *J. Bacteriol.*, **171**, 4543–8.

Kape, R., Parniske, M. and Werner, D. (1991) Chemotaxis and *nod* gene activity of *Bradyrhizobium japonicum* in response to hydroxycinnamic acids and isoflavonoids. *Appl. Environ. Microbiol.*, **57**, 316–19.

Kauss, H. (1990) Role of the plasma membrane in host–pathogen interactions, in *The Plant Plasma Membrane* (eds C. Larsson and I.M. Moller), Springer-Verlag, Berlin, Heidelberg, pp. 320–50.

Kauss, H., Waldmann, T., Jeblick, W. *et al.* (1989) Ca^{2+} is an important but not the only signal in callose synthesis induced by chitosan, saponins and polyene antibiotics, in *Signal Molecules in Plants and Plant–Microbe Interactions* (ed. B.J.J. Lugtenberg), NATO ASI Series, Vol. H 36, Springer-Verlag, Berlin, Heidelberg, pp. 107–16.

Keller, M., Müller, P., Simon, R. and Pühler, A. (1988) *Rhizobium meliloti* genes for exopolysaccharide synthesis and nodule infection located on megaplasmid 2 are actively transcribed during symbiosis. *Mol. Plant–Microbe Interactions*, **1**, 267–74.

Kessel, van C. and Burris, R.H. (1983) Effect of H_2 evolution on $^{15}N_2$ fixation, C_2H_2 reduction and relative efficiency of leguminous symbionts. *Physiol. Plant.*, **59**, 329–34.

Kijne, J.W., Diaz, C.L., de Pater, B.S. *et al.* (1990) Surface interactions between rhizobia and legume root hairs, in *Nitrogen Fixation: Achievements and Objectives* (eds P.M. Gresshoff, L.E. Roth, G. Stacey and W.E. Newton), Chapman and Hall, New York, London, pp. 187–92.

Kim, Y.S. and Chae, H.Z. (1990) A model of nitrogen flow by malonamate in *Rhizobium japonicum*–soybean symbiosis. *Biochem. Biophys. Res. Commun.*, **169**, 692–9.

Kinnback, A., Mellor, R.B. and Werner, D. (1987) Alpha-mannosidase II isoenzyme in the peribacteroid space of *Glycine max* root nodules. *J. Exp. Bot.*, **38**, 1373–7.

Kinnback, A. and Werner, D. (1990) α-Glucosidase and other enzyme activities in the peribacteroid space of soybean root nodules, in *Nitrogen Fixation: Achievements and Objectives* (eds P.M. Gresshoff, L.E. Roth, G. Stacey and W.E. Newton), Chapman and Hall, New York, London, p. 746.

Kinnback, A. and Werner, D. (1991) Glucosidases (α, β) and trehalase (α) in the peribacteroid space and the bacteroid periplasm of *Glycine max* root nodules. *Plant Science*, **77**, 47–55.

Kishinevsky, B.D., Law, I.J. and Strijdom, B.W. (1988) Detection of lectins in nodulated peanut and soybean plants. *Planta*, **176**, 10–18.

Klebesadel, L.J. (1986) Natural selection may modify introduced white clover toward superior winterhardiness. *Agroborealis*, **18**, 13–19.

Kretovich, W.L., Romanov, V.I. and Korolyov, A.V. (1973) *Rhizobium leguminosarum* cytochromes (*Vicia faba*). *Plant and Soil*, **39**, 619–34.

Kullik, I., Hennecke, H. and Fischer, H.-M. (1989) Inhibition of *Bradyrhizobium japonicum nif*A-dependent *nif* gene activation by oxygen occurs at the NifA protein level and is irreversible. *Arch. Microbiol.*, **151**, 191–7.

Kurz, W.G.W. and LaRue, T.A. (1975) Nitrogenase activity in Rhizobia in absence of plant host. *Nature*, **256**, 407–8.

Lahda, J.K., Pareek, R.P., So, R. and Becker, M. (1990) Stem-nodule symbiosis and its unusual properties, in *Nitrogen Fixation: Achievements and Objectives* (eds P.M. Gresshoff, L.E. Roth, G. Stacey and W.E. Newton), Chapman and Hall, New York, London, pp. 633–40.

Lamb, C.J., Lawton, M.A., Dron, M. and Dixon, R.A. (1989) Signals and transduction mechanisms for activation of plant defenses against microbial attack. *Cell*, **56**, 215–24.

Landau-Ellis, D., Angermüller, S., Shoemaker, R. and Gresshoff, P.M. (1991) The genetic locus controlling supernodulation in soybean (*Glycine max* L.) co-segregates tightly with a cloned molecular marker. *Mol. Gen. Genet.*, **228**, 221–6.

Landsmann, J., Dennis, E.S., Higgins, T.J.V. *et al.* (1986) Common evolutionary origin of legume and non-legume plant haemoglobins. *Nature*, **324**, 166–8.

Law, I.J., Haylett, T. and Strijdom, B.W. (1988) Differences in properties of peanut seed lectin and purified galactose- and mannose-binding lectins from nodules of peanut. *Planta*, **176**, 19–27.

Leigh, J. (1990) Metals set a trap for nitrogen. *New* Scientist, 55–58.

Lengeler, J.W. (1986) Molecular mechanisms of chemotaxis of Enterobacteria. *Forum Mikrobiol.*, **9**, 323–32.

Lerouge, P., Roche, P., Faucher, C. *et al.* (1990) Symbiotic host-specificity of *Rhizobium meliloti* is determined by a sulphated and acylated glucosamine oligosaccharide signal. *Nature*, **344**, 781–4.

Lersten, N.R. and Carlson, J.B. (1987) Vegetative morphology, in *Soybeans: Improvement, Production, and Uses*, Agronomy Monograph **16**, 49–94.

Lie, T.A., Göktan, D. and Engin, M. (1988) *Rhizobium* strains from wild and primitive legumes: a nuisance or a valuable gene pool? in *Nitrogen Fixation by Legumes in Mediterranean Agriculture* (eds D.P. Beck and L.A. Materon), ICARDA, pp. 121–9.

Lie, T.A., Göktan, D., Engin, M. *et al.* (1987) Co-evolution of the legume–*Rhizobium* association. *Plant Soil*, **100**, 171–81.

Long, S.R. (1989a) *Rhizobium* genetics. *Ann. Rev. Genet.*, **23**, 483–506.

Long, S.R. (1989b) *Rhizobium*–legume nodulation: life together in the underground. *Cell*, **56**, 203–14.

Long, S.R. and Ehrhardt, D.W. (1989) New route to a sticky subject. *Nature*, **338**, 545–6.

Maier, R.J., Moshiri, F., Keefe, R.G. and Gabel, C. (1990) Molecular analysis of terminal oxidases in electron-transport pathways of *Bradyrhizobium japonicum* and *Azotobacter vinelandii*, in *Nitrogen Fixation: Achievements and Objectives* (eds P.M. Gresshoff, L.E. Roth, G. Stacey and W.E. Newton), Chapman and Hall, New York, London, pp. 301–8.

Marcker, A., Lund, M., Jensen, E.O. and Marcker, K.A. (1984) Transcription of the soybean leghemoglobin genes during nodule development. *EMBO J.*, **3**, 1691–5.

Marquardt, G. and Lüttge, U. (1987) Proton transporting enzymes at the tonoplast of leaf cells of the CAM plant *Kalanchoe daigremontiana*. II. The pyrophosphatase. *J. Plant Physiol.*, **129**, 269–86.

Martinez, E., Romero, D. and Palacios, R. (1990) The *Rhizobium* genome. *Crit. Rev. Plant Sci.*, **9**, 59–93.

Martinez-Romero, E., Segovia, L., Mercante F.M. *et al.* (1991) *Rhizobium tropici*, a novel species nodulating *Phaseolus vulgaris* L. beans and *Leucaena* sp. trees. *Int. J. Syst. Bacterial.*, **41**, 417–26.

Mathews, A., Carroll, B.J. and Gresshoff, P.M. (1989) Development of *Bradyrhizobium* infections in supernodulating and non-nodulating mutants of soybean (*Glycine max* [L.] Merrill). *Protoplasma*, **150**, 40–7.

Mauro, V.P. and Verma, D.P.S. (1988) Transcriptional activation in nuclei from uninfected soybean of a set of genes involved in symbiosis with *Rhizobium. Mol. Plant–Microbe Interactions*, **1**, 46–51.

Mayama, S., Bordin, A.P.A., Morikawa, T. and Tani, T. (1989) Role of phytoalexins in host defence reactions, in *Molecular Strategies of Pathogens and Host Plants* (eds S.S. Patil, S. Ouchi, D. Mills and C.P. Vance), Springer Verlag, New York, pp. 203–13.

McComb, J.A., Elliot, J. and Dilworth, M.J. (1975) Acetylene reduction by pure *Rhizobium* in culture. *Nature*, **256**, 409–10.

Mellor, R.B. (1988) Distribution of trehalase in soybean root nodule cells: implications for trehalose metabolism. *J. Plant Physiol.*, **133**, 173–7.

Mellor, R.B. (1989) Bacteroids in the *Rhizobium*–legume symbiosis inhabit a plant internal lytic compartment: implications for other microbial endosymbioses. *J. Exp. Bot.*, **40**, 831–9.

Mellor, R.B., Christensen, T.M.I.E. and Werner, D. (1986) Choline kinase II is present only in nodules that synthesize stable peribacteroid membranes. *Proc. Natl. Acad. Sci. USA*, **83**, 659–63.

Mellor, R.B., Garbers, C. and Werner, D. (1989) Peribacteroid membrane nodulin gene induction by *Bradyrhizobium japonicum* mutants. *Plant Mol. Biol.*, **12**, 307–15.

Mellor, R.B., Thierfelder, H., Pausch, G. and Werner, D. (1987) The occurrence of choline kinase II in the cytoplasm of soybean root nodules infected with various strains of *Bradyrhizobium japonicum. J. Plant Physiol.*, **128**, 169–72.

Mellor, R.B. and Werner, D. (1986) The fractionation of *Glycine max* root nodule cells: a methodological overview. *Endocyt. Cell Res.*, **3**, 317–36.

Mellor, R.B. and Werner, D. (1990) Legume nodule biochemistry and function, in *Molecular Biology of Symbiotic Nitrogen Fixation* (ed. P.M. Gresshoff), CRC Press, Boca Raton, pp. 111–29.

Mellor, R.B. and Wiemken, A. (1988) Peribacteroid organelles as organ-specific forms of lysosomes, in *Nitrogen Fixation: Hundred Years After* (eds. H. Bothe, F.J. de Bruijn and W.E. Newton), Gustav Fischer Verlag, Stuttgart, New York 1988, p. 528.

Mengel, K. (1992) Nitrogen: Agricultural productivity and environmental problems, in *Nitrogen Metabolism of Plants* (ed. K. Mengel), Oxford University Press, Oxford (in press).

Merrick, M.J. (1988) Organisation and regulation of nitrogen fixation genes in *Klebsiella* and *Azotobacter*, in *Nitrogen Fixation: Hundred years after*, (eds H. Bothe, F.J. de Bruijn, W.E. Newton, G.), Fischer Verlag, Stuttgart, New York, pp. 293–302.

Merrick, M. and Dixon, R. (1984) Why don't plants fix nitrogen? *Trends Biotechnol.*, **2**, 162–6.

Müller, P., Hynes, M. Kapp, D. *et al.* (1988) Two classes of *Rhizobium meliloti* infection mutants differ in exopolysaccharide production and in coinoculation properties with nodulation mutants. *Mol. Gen. Genet.*, **211**, 17–26.

Nap. J.-P. and Bisseling, T. (1990) Nodulin function and nodulin gene regulation in root nodule development, in *Molecular Biology of Symbiotic Nitrogen Fixation* (ed. P.M. Gresshoff), CRC Press, Boca Raton, pp. 181–229.

Nelson, L.M. and Salminen, S.O. (1982) Uptake hydrogenase activity and ATP formation in *Rhizobium leguminosarum* bacteroids. *J. Bacteriol.*, **151**, 989–95.

O'Brien, S.D. (ed.) (1984) *Genetic Maps 1984*. Cold Spring Harbor Laboratory.

Ohyama, T. and Kumazawa, K. (1980) Nitrogen assimilation in soybean nodules. III. Effects of rhizosphere pO_2 on the assimilation of $^{15}N_2$ in nodules attached to intact plants. *Soil Sci. Plant Nutr.*, **26**, 321–4.

Ohmiya, K., Oshino, C. and Shimizu, S. (1989) Cellulose-dependent and penicillin resistant plasmids isolated from *Ruminococcus albus*. *Jpn. J. Zootech. Sci.*, **S 8-3**, 501–2.

Olszewska, M.J. and Legocki, A.B. (1989) Changes in DNA content during rhizobial nodule development in *Lupinus luteus* L. I. Cytophotometry and autoradiography. *Biol. Zentralbl.*, **108**, 221–30.

Paau, A.S., Oro, J. and Cowles, J.R. (1979) DNA content of free-living rhizobia and bacteroids of various *Rhizobium*–legume associations. *Plant Physiol.*, **63**, 402–5.

Pagan, J.D., Child, J.J., Snowcroft, W.R. and Gibson A.H. (1975) Nitrogen fixation by *Rhizobium* cultured on a defined medium. *Nature*, **256**, 406–7.

Palacios, J.M., Murillo, J., Leyva, A. *et al.* (1990) Differential expression of hydrogen uptake (hup) genes in vegetative and symbiotic cells of *Rhizobium leguminosarum*. *Mol. Gen. Genet.*, **221**, 363–70.

Pankhurst, C.E. and Craig A.S. (1978) Effect of oxygen concentration, temperature and combined nitrogen on the morphology and nitrogenase activity of *Rhizobium* sp. strain 32H1 in agar culture. *J. Gen. Microbiol.*, **106**, 207–19.

Parniske, M., Ahlborn, B. and Werner, D. (1991a) Isoflavonoid-inducible resistance to the phytoalexin glyceollin in soybean rhizobia. *J. Bacteriol.*, **173**, 3432–9.

Parniske, M., Fischer, H.-M., Hennecke, H. and Werner, D. (1991b) Accumulation of the phytoalexin glyceollin I in soybean nodules infected by a *Bradyrhizobium japonicum nif*A mutant. *Z. Naturforsch.*, **46c**, 318–20.

Parniske, M., Zimmermann, C., Cregan, P.B. and Werner, D. (1990) Hypersensitive reaction of nodule cells in the *Glycine max* sp./*Bradyrhizobium japonicum* symbiosis occurs at the genotype-specific level. *Bot. Acta*, **103**, 143–8.

Philip-Hollingsworth, S., Hollingsworth, R.I. and Dazzo, F.B. (1989) Host-range related structural features of the acidic extracellular polysaccharides of *Rhizobium trifolii* and *Rhizobium leguminosarum*. *J. Biol. Chem.*, **264**, 1461–6.

Philipps, D.A., Hartwig, U.A., Maxwell, C.A. *et al.* (1990) Host legume control of nodulation in flavonoids, in *Nitrogen Fixation: Achievements and Objectives* (eds P.M. Gresshoff, L.E. Roth, G. Stacey and W.E. Newton), Chapman and Hall, New York, London, pp. 331–8.

Philipps, D.A., Maxwell, C.A., Joseph, C.M. and Hartwig U.A. (1988) Flavonoid nodulation signals in alfalfa, in *Nitrogen Fixation: Hundred Years After* (eds H. Bothe, F.J. de Bruijn and W.E. Newton), G. Fischer Verlag, Stuttgart, New York, pp. 411–15.

Pocard, J.-A., Bernard, T., Tombras Smith, L. and Le Rudulier, D. (1989) Characterization of three choline transport activities in *Rhizobium meliloti*: modulation by choline and osmotic stress. *J. Bacteriol.*, **171**, 531–7.

Pokhriyal, T.C., Negi, D.S., Chaukiyal, S.P. *et al.* (1990) Changes in nitrogenase and nitrate reductase activity in *Acacia catechu*. *Nitrogen Fixing Tree Res. Rep.*, **8**, 108–10.

Prakash, R.K. and Atherly, A.G. (1986) Plasmids of *Rhizobium* and their role in symbiotic nitrogen fixation. *Int. Rev. Cytol.*, **104**, 1–24.

Priefer, U.B. (1989) Genes involved in lipopolysaccharide production and symbiosis are clustered on the chromosome of *Rhizobium leguminosarum* biovar *viciae* VF39. *J. Bacteriol.*, **171**, 6161–8.

Pühler, A., Arnold, W., Buendia-Claveria, A. *et al.* (1991) The role of *Rhizobium meliloti* exopolysaccharides EPS I and EPS II in the infection process of alfalfa nodules, in *Advances in Molecular Genetics of Plant–Microbe Interaction* (eds H. Hennecke and D.P.S. Verma), Vol. 1, Kluwer Academic Dordrecht, pp. 189–94; 203–10.

Quispel, A. (1982) Frontiers of research in symbiotic nitrogen fixation, in *Frontiers of Research in Agriculture* (ed. S.K. Roy), Indian Statistical Institute, Calcutta, pp. 206–23.

Rainbird, R.M., Atkins, C.A., Pate, J.S. and Sanford, P. (1983) Significance of hydrogen evolution in the carbon and nitrogen economy of nodulated cowpea. *Plant Physiol.*, **71**, 122–7.

Rawsthorne, S. and LaRue, T.A. (1986a) Metabolism under microaerobic conditions of mitochondria from cowpea nodules. *Plant Physiol.*, **81**, 1097–102.

Rawsthorne, S. and LaRue, T.A., (1986b) Preparation and properties of mitochondria from cowpea nodules. *Plant Physiol.*, **81**, 1092–6.

Ride, J.P. (1986) Induced structural defences in plants, in *Natural Antimicrobial Systems*. Part I. *Antimicrobial Systems in Plants and Animals* (eds G.W. Gould, A.K. Chamley M.E. Rhodes-Roberts *et al.*) Bath University Press, Bath, pp. 159–75.

Rigaud, J. (1987) Salt tolerance of *Medicago* nodules and bacteroids, in *Plant Genes Involved in Nitrogen Fixation and Productivity of Alfafa*, USDA-INRA Workshop, Auzeville.

Röhm, M. and Werner, D. (1985) Regulation of the β-ketoadipate pathway in *Rhizobium japonicum* and bacteroids by succinate. *Arch. Microbiol.*, **140**, 373–9.

Röhm, M. and Werner, D. (1987) Isolation of root hairs from seedlings of *Pisum sativum*. Identification of root hair specific proteins by *in situ* labelling. *Physiol. Plant.*, **69**, 129–36.

Rolfe, B.G., Shine, J., Gresshoff, P.M. *et al.* (1982) *Rhizobium* and the geneti of a controlled disease. *Aust. Microbiol.*, **3**, 33–7.

Roth, L.E., Jean, K. and Stacey, G. (1988) Homology in endosymbiotic systems: the term 'symbiosome'. In *Molecular Genetics of Plant–Microbe Interactions* (eds R. Palacios and D.P.S. Verma), The American Phytopathological Society Press, St Paul, MN, p. 220.

Ryan, C.A. and Farmer, E.E. (1991) Oligosaccharide signals in plants: A current assessment. *Ann. Rev. Plant Phys. Plant Mol. Biol.*, **42**, 651–74.

Sadowsky, M.J., Keyser, H.H. and Bohlool, B.B. (1983) Biochemical characterization of fast- and slow-growing rhizobia that nodulate soybeans. *Int. J. Syst. Bacteriol.*, **33**, 716–22.

Sathiyamoorthy, P. (1990) Identification of vanillic and *p*-coumaric acid as endogenous inhibitors of soybean seeds and their inhibitory effect on germination. *J. Plant Physiol.*, **136**, 120–1.

Scheres, B., van Engelen, F., van der Knaap, E. *et al.* (1990a) Sequential induction of nodulin gene expression in the developing pea nodule. *Plant Cell*, **2**, 687–700.

Scheres, B., van de Wiel, C., Zalensky, A. *et al.* (1990b) The ENOD12 gene product is involved in the infection process during the pea–*Rhizobium* interaction. *Cell*, **60**, 281–94.

Schramm, R.W. (1982) Carbon assimilation and flux connected with nitrogen fixation in legumes. *Isr. J. Bot.*, **31**, 131–9.

Schubert, K. (1982) *Energetics of Biological Nitrogen Fixation.* Business Office American Society of Plant Physiology, Rockville, Md.

Schubert, K.R. and Evans, H.J. (1976) Hydrogen evolution: a major factor affecting the efficiency of nitrogen fixation in nodulated symbionts. *Proc. Natl. Acad. Sci. USA*, **73**, 1207–11.

Schuller, K.A., Minchin, F.R. and Gresshoff, P.M. (1988) Nitrogenase activity and oxygen diffusion in nodules of soyabean cv. Bragg and a supernodulating mutant: effects of nitrate. *J. Exp. Bot.*, **39**, 865–977.

Schuller, K.A., Turpin, D.H. and Plaxton, W.C. (1991) Metabolite regulation of partially purified soybean nodule phosphoenolpyruvate carboxylase. *Plant Physiol.*, **94**, 1429–35.

Sellstedt, A., Högberg, P. and Jonsson, K. (1989) Diurnal variation in acetylene reduction and net hydrogen evolution in five tropical and subtropical nitrogen-fixing tree symbioses. *J. Exp. Bot.*, **40**, 1163–8.

Sheehy, J.E., Minchin, F.R. and Witty, J.F. (1985) Control of nitrogen fixation in a legume nodule: an analysis of the role of oxygen diffusion in relation to nodule structure. *Ann. Bot.*, **55**, 549–62.

Sikorski, M.M., Szybiak-Strozycka, U., Strozycki, P. *et al.* (1989) Coordinated synthesis of leghemoglobin and root protein R18 in yellow lupin. *Acta Biochim. Pol.*, **36**, 63–72.

Simpson, F.B., Maier, R.J. and Evans, H.J. (1979) Hydrogen-stimulated CO_2-fixation and coordinate induction of hydrogenase and ribulose-bisphosphate carboxylase in a H_2-uptake positive strain of *Rhizobium japonicum. Arch. Microbiol.*, **123**, 1–8.

Smit, G., Kijne, J.W. and Lugtenberg, B.J.J. (1986) Correlation between extra-cellular fibrils and attachment of *Rhizobium leguminosarum* to pea root hair tips. *J. Bacteriol.*, **168**, 821–7.

Somasegaran, P. and Halliday, J. (1982) Dilution of liquid *Rhizobium* cultures to increase production capacity of inoculant plants. *Appl. Environ. Microbiol.*, **44**, 330–3.

Spaink, H.P., Sheeley, D.M., van Brussel, A.A.N. *et al.* (1991) A novel highly unsaturated fatty acid moiety of lipooligosaccharide signals determines host specificity of *Rhizobium. Nature*, **354**, 125–30.

Sprent, J.I. (1990) Evolution, structure and function of nitrogen-fixing root nodules: confessions of ignorance, in *Nitrogen Fixation: Achievements and Objectives* (eds P.M. Gresshoff, L.E. Roth, G. Stacey and W.E. Newton), Chapman and Hall, New York, London, pp. 45–54.

Sprent, J.I. and Sprent, P. (1990) *Nitrogen Fixing Organisms: Pure and applied aspects*, Chapman & Hall, London.

Stacey, G. (1990) Workshop summary: compilation of the *nod*, *fix* and *nif* genes of Rhizobia and information concerning their function, in *Nitrogen Fixation: Achievements and Objectives* (eds P.M. Gresshoff, L.E. Roth, G. Stacey and W.E. Newton), Chapman and Hall, New York, London, pp. 239–44.

Streeter, S. and Salminen, O. (1985) Carbon metabolism in legume nodules, in *Nitrogen Fixation Research Progress* (eds H.J. Evans, P.J. Bottomley and W.E. Newton), Martinus Nijhoff Publ., Dordrecht, pp. 277–83.

Stripf, R. and Werner, D. (1980) Development of discontinuous size classes of nodules of *Glycine max. Z. Naturforsch.*, **35c**, 776–82.

Stumpf, D.K. and Burris, R.H. (1981) Organic acid contents of soybean: age and source of nitrogen. *Plant Physiol.*, **68**, 989–91.

Sutton, W.D., Jepsen, N.M. and Shaw, B.D. (1977) Changes in the number, viability, and amino-acid-incorporating activity of *Rhizobium* bacteroids during lupin nodule development. *Plant Physiol.*, **59**, 741–4.

Sutton, W.D., Pankhurst, C.E. and Craig, A.S. (1981) The *Rhizobium* Bacteroid State, in *Biology of the Rhizobiaceae* (eds K.L. Giles and A.G. Atherly), *Int. Rev. Cytol.*, Suppl. **13**, 149–77.

Szczepanowska, E., Kusmierz, J., Szafranek, J. and Schramm, R.W. (1989) Gas chromatography and mass spectrometry identification of sugar-related hydroxy acids from kidney bean, alfalfa and clover. *Acta Physiol. Plant.*, **11**, 97–109.

Tajima, S. and Kouchi, H. (1990) Metabolism of C_4-dicarboxylates and amino acids in soybean nodule bacteroids, and their role in energy supply to nitrogenase, in *Nitrogen Fixation: Achievements and Objectives* (eds P.M. Gresshoff, L.E. Roth, G. Stacey and W.E. Newton), Chapman and Hall, New York, London, pp. 309–14.

Tang, C., Robson, A.D. and Dilworth, M.J. (1990) The role of iron in nodulation and nitrogen fixation in *Lupinus angustifolius* L. New Phytol., **114**, 173–82.

Terouchi, N. and Syono, K. (1990) Hair curling induced in heterologous legumes and monocots by flavonoid-treated rhizobia. *Plant Cell Physiol.*, **31**, 113–18.

Thöny, B., Anthamatten, D. and Hennecke, H. (1989) Dual control of the *Bradyrhizobium japonicum* symbiotic nitrogen fixation regulatory operon *fix*R *nif*A: analysis of cis- and trans-acting elements. *J. Bacteriol.*, **171**, 4162–9.

Tjepkema, J. and Evans, H.J. (1975) Nitrogen fixation by free-living *Rhizobium* in a defined liquid medium. *Biochem. Biophys. Res. Commun.*, **65**, 625–8.

Tomaszewska, B., Jarmuszkiewicz, W. and Schramm, R.W. (1991) The role of malate in the metabolism of mitochondria and symbiosomes from lupin root nodules. *Plant Physiol. Biochem.*, **29**, 489–96.

Trinchant, J.C. and Rigaud, J. (1987) Acetylene reduction by bacteroids isolated from stem nodules of *Sesbania rostrata*. Specific role of lactate as an energy-yielding substrate. *J. Gen. Microbiol.*, **133**, 37–43.

Triplett, E.W. (1990) Construction of a symbiotically effective strain of *Rhizobium leguminosarum* bv. *trifolii* with increased nodulation competitiveness. *Appl. Environ. Microbiol.*, **56**, 98–103.

Truchet, G., Barker, D.G., Camut, S. *et al.* (1989) Alfalfa nodulation in the absence of *Rhizobium. Mol. Gen. Genet.*, **219**, 65–8.

Tsien, H.C., Cain, P.S. and Schmidt, E.L. (1977) Viability of *Rhizobium* bacteroids. *Appl. Environ. Microbiol.*, **34**, 854–6.

Turgeon, B.G. and Bauer, W.D. (1985) Ultrastructure of infection-thread development during the infection of soybean by *Rhizobium japonicum. Planta*, **163**, 328–49.

Tyski, S., Markiewicz, M., Gulewicz, K. and Twardowski, T. (1988) The effect of lupin alkaloids and ethanol extracts from seeds of *Lupinus angustifolius* on selected bacterial strains. *J. Plant Physiol.*, **133**, 240–2.

Uemura, S. (1961) Reciprocal influence of forest tree species by root excretions. International Union Forestry Research Organizations, IUFRO-Kongress 13, Section 21–4/2, Wien.

Vance, C.P. (1990) Symbiotic nitrogen fixation: recent genetic advances, in *Bio-*

chemistry of Plants: An Advanced Treatise (eds P.K. Stumpf and E.E. Conn), Vol. 16, Academic Press, New York, pp. 43–80.

Vance, C.P. and Heichel, G.H. (1991) Carbon in N_2 fixation: limitation or exquisite adaptation. *Ann. Rev. Plant Physiol. Plant Mol. Biol.*, **42**, 373–92.

van den Bosch, K.A., Brewin, N.J. and Kannenberg, E.L. (1989) Developmental regulation of a *Rhizobium* cell surface antigen during the growth of pea root nodules. *J. Bacteriol.*, **171**, 4537–42.

van den Bosch, K.A. and Newcomb, E.H. (1986) Immunogold localization of nodule-specific uricase in developing soybean root nodules. *Planta*, **167**, 425–36.

Veau, E.J. de, Robinson, J.M., Warmbrodt, R.D. and van Berkum, P. (1990) Photosynthesis and photosynthate partitioning in N_2-fixing soybeans. *Plant Physiol.*, **94**, 259–67.

Verma, D.P.S. and Stanley, J. (1989) The legume–*Rhizobium* equation: a coevolution of two genomes, in *Advances in Legume Biology* (eds C.H. Stirton and J.L. Zarucchi), Monogr. Syst. Bot. Missouri Bot. Gard., **29**, 545–57.

Verma, D.P.S., Miao, G.-H., Cheon, C.-I. and Suzuki, H. (1991) Genesis of root nodules and function of nodulins, in *Advances in Molecular Genetics of Plant–Microbe Interactions* (eds H. Hennecke and D.P.S. Verma), Vol. 1, Kluwer Academic, Dordrecht, pp. 291–9.

Wallace, W. (1986) Distribution of nitrate assimilation between the root and shoot of legumes and a comparison with wheat. *Physiol. Plant.*, **66**, 630–6.

Watson, R.J., Chan, Y.-K., Wheatcroft, R. *et al.* (1988) *Rhizobium meliloti* genes required for C_4-dicarboxylate transport and symbiotic nitrogen fixation are located on a megaplasmid. *J. Bacteriol.*, **170**, 927–34.

Weeden, N.F., Kneen, B.E. and LaRue, T.A. (1990) Genetic analysis of *sym* genes and other nodule related genes in *Pisum sativum*, in *Nitrogen Fixation: Achievements and Objectives* (eds P.M. Gresshoff, L.E. Roth, G. Stacey and W.E. Newton), Chapman and Hall, New York, London, pp. 323–30.

Werner, D. (1978) Differentiation of *Rhizobium japonicum*, III. Inhibition of nitrogenase derepression by chloramphenicol and rifampicin concentrations, not inhibiting growth. *Z. Naturforsch.*, **33c**, 859–62.

Werner, D. (1991) Physiology of nitrogen fixing legume nodules – compartments and functions, in *Biological Nitrogen Fixation* (eds R. Burris, H. Evans and G. Stacey, Chapman and Hall, New York, London (in press).

Werner, D. (1992) Physiology of nitrogen fixing legume nodules: Compartments and functions, in *Biological Nitrogen Fixation* (eds G. Stacey, R. Burris and H.J. Evans), Chapman Hall, New York, London, pp. 399–431.

Werner, D., Dittrich, W. and Thierfelder, H. (1982) Malonate and Krebs cycle intermediates utilization in the presence of other carbon sources by *Rhizobium japonicum* and soybean bacteroids. *Z. Naturforsch.*, **37c**, 921–6.

Werner, D., Kuhlmann, K.-P., Gloystein, F. and Richter, F.-W. (1985a) Calcium, iron and cobalt accumulation in root hairs of soybean. *Z. Naturforsch.*, **40c**, 912–13.

Werner, D., Mellor, R.B., Hahn, M.G. and Grisebach, H. (1985b) Soybean root response to symbiotic infection. Glyceollin I accumulation in an ineffective type of soybean nodules with an early loss of the peribacteroid membrane. *Z. Naturforsch.*, **40c**, 179–81.

Werner, D. and Mörschel, E. (1978) Differentiation of nodules of *Glycine max*. Ultrastructural studies of plant cells and bacteroids. *Planta*, **141**, 169–77.

Werner, D., Mörschel, E., Garbers, C. *et al.* (1988) Particle density and protein composition of the peribacteroid membrane from soybean root nodules is affected by mutation in the microsymbiont *Bradyrhizobium japonicum*. *Planta*, **174**, 263–79.

Werner, D., Mörschel, E., Kort, R. *et al.* (1984) Lysis of bacteroids in the vicinity of the host cell nucleus in an ineffective (fix^-) root nodule of soybean (*Glycine max*). *Planta*, **162**, 8–16.

Werner, D., Mörschel, E., Stripf, R. and Winchenbach, B. (1980) Development of nodules of *Glycine max* infected with an ineffective strain of *Rhizobium japonicum*. *Planta*, **147**, 320–9.

Wetzel, A., Brandt, S., Görge, E. and Werner, D. (1991) A new plant test system for ecotoxicological evaluation of polycyclic aromatic hydrocarbons – The legume–root–nodule symbiosis as effect parameter, in *International Symposium Environmental Biotechnology II*, Technological Institute – Royal Flemish Society of Engineers, Oostende, pp. 687–9.

Wheatcroft, R. and Watson, R.J. (1987) Identification and characterization of insertion sequence ISRml in *Rhizobium meliloti* JJlc10. *Can. J. Microbiol.*, **33**, 314–21.

Wheatcroft, R. and Watson, R.J. (1988) Distribution of insertion sequence ISRml in *Rhizobium meliloti* and other Gram-negative bacteria. *J. Gen. Microbiol.*, **134**, 113–21.

Wilcockson, J. and Werner, D. (1979) On the DNA content of bacteroids of *Rhizobium japonicum*. *Z. Naturforsch.*, **34c**, 793–6.

Wimpee, C., Garvin, D., Wrobel, R. *et al.* (1992) An old symbiosis and a new one: The impact of the abandonment of photosynthesis on the plastid genome of parasitic plants. *Symbiosis*, **12**, Lectures from the International Symbiosis Congress, Jerusalem, Nov. 1991.

Wingender, R., Röhrig, H., Höricke, C. *et al.* (1989) Differential regulation of soybean chalcone synthase genes in plant defence, symbiosis and upon environmental stimuli. *Mol. Gen. Genet.*, **218**, 315–22.

Witty, J.F. and Minchin, F.R. (1990) Oxygen diffusion in the legume root nodule, in: *Nitrogen Fixation: Achievements and Objectives* (eds P.M. Gresshoff, L.E. Roth, G. Stacey and W.E. Newton), Chapman and Hall, New York, London, pp. 285–92.

Witty, J.F., Minchin, F.R. and Sheehy, J.E. (1983) Carbon costs of nitrogenase activity in legume root nodules determined using acetylene and oxygen. *J. Exp. Bot.*, **34**, 951–63.

Wolff, A.B. and Werner, D. (1990) Defence reactions in *Rhizobium*–legume symbiosis: phytoalexin concentration in *Vicia faba* nodules is affected by the host plant genotype. *Z. Naturforsch.*, **45c**, 958–62.

Young, J.P.W. (1985) The phylogeny of the *Rhizobium* symbiosis, in: *Analysis of the Plant Genes Involved in the Rhizobium–Legume Symbiosis*, OECD, Paris, pp. 12–19.

Zacharius, R.M. and Kalan, E.B. (1990) Isoflavonoid changes in soybean cell suspensions when challenged with intact bacteria or fungal elicitors. *J. Plant Physiol.*, **135**, 732–6.

4 The *Bradyrhizobium* sp.–*Parasponia* Symbiosis

Root nodules of *Parasponia* (Ulmaceae) represent the only confirmed symbiosis between *Bradyrhizobium*/*Rhizobium* and a non-leguminous species. As early as 1909, nodules were detected on roots of *Parasponia* trees in Indonesia (Becking, 1992). However, the microsymbiont was not identified until 1973 when it was assigned to the genus *Bradyrhizobium* by Trinick who proved its identity by cross-infection tests with legumes (*Vigna*). *Parasponia* sp., which are trees of up to 15 m in height, grow as pioneer plants on volcanic soils in Indonesia, New Guinea and some parts of Malaysia. Root nodules have so far been found on four species: *P. rugosa*, *P. parviflora*, *P. andersonii* and *P. rigida*.

Bacterial strains isolated from *Parasponia* root nodules are physiologically and serologically closely related to strains of the so-called 'cowpea' group of *Bradyrhizobium*. However, most *Parasponia* isolates nodulate the corresponding tropical legumes only poorly or induce ineffective nodules. Only a few strains have been isolated which produce effective nodules on both *Parasponia* and, for example, *Macroptilium atropurpureum*. In some strains nitrogenase can be derepressed in pure culture (section 3.1.5).

Two to four weeks after infection, seedlings of *Parasponia andersonii* were found to produce nodules with significant nitrogenase activity. The infection of *Parasponia* root hairs takes place at the cell base and not at the root hair tip. During further nodule development, the most striking difference compared to legumes is the retention of bacteroids in the infection thread (with a few exceptions). In the host cells the infection thread branches strongly and is sometimes called 'fixation thread' during this stage. It is not yet clear whether the maintenance of the infection thread is primarily due to a continual synthesis of wall material or by lack of degrading enzymes for this structure. *Parasponia* nodules are modified side roots with a central vascular system. Nodule aggregates of 1–2 cm diameter are formed by dichotomous branching in meristematic zones. Three-month-old plants of *Parasponia* can thus produce up to 15 g of nodule fresh weight. The average number of bacteroids in a defined volume of infected nodule tissue in *Parasponia* is about half that found in *Macroptilium*. This is also

Table 4.1 Comparison of nodules from *Parasponia rugosa* (Ulmaceae) and *Macroptilium atropurpureum* (Fabaceae), both infected with *Bradyrhizobium* sp. ANU 289. (Based on Mohapatra *et al.*, 1985)

	Parasponia	*Macroptilium*
Nodule type	Indeterminate, modified side roots, 5–10 mm in length	Determinate, spherical 1–4 mm in diameter
N_2-fixation rate (μmol $C_2H_4\,h^{-1}\,mg^{-1}$ nodule fresh weight)	9 ± 2	16 ± 4
Number of bacteroids 1 mm^{-3} of fully infected nodule tissue	300	650
Vascular system	Central, with lateral ramification	Cortical
Morphology of infected cells	Bacteroids remain in the infection thread, mitochondria scattered in the cell	Bacteroids in a peribacteroid membrane, mitochondria concentrated close to the cell wall
O_2-transport and O_2-protection	Haemoprotein	Leghaemoglobin

true for specific nitrogenase activity (Table 4.1). Haemoglobin isolated from *Parasponia* has 39% homology in amino acid sequence to soybean leghaemoglobin. The rate of oxygen release for *Parasponia* oxyhaemoglobin is approx. one-third of the value found for soybean leghaemoglobin.

Leghaemoglobin genes and *Parasponia* haemoglobin genes each have three introns at identical positions (Landsmann *et al.*, 1986). Two of these introns are also identical to those in vertebrates. *Parasponia* haemoglobin genes are also transcribed in roots of non-nodulated plants, indicating a role of haemoglobin in the oxygen-related metabolism of all plants (Bogusz *et al.*, 1988).

The induction of nodular structures on non-legumes other than *Parasponia* by rhizobia such as *Brassica napus* and also rice (Al-Mallah *et al.*, 1990) has so far not produced the essential characters of effective legume nodules such as:

1. a significant nitrogenase activity
2. leghaemoglobin induction
3. significant compartmentation such as the symbiosome membranes
4. induction of a large number of other essential nodulines

REFERENCES

Al-Mallah, M.K., Davey, M.R. and Cocking, E.C. (1990) Nodulation of oilseed rape (*Brassica napus*) by rhizobia. *J. Exp. Bot.*, **41**, 1567–72.

Becking, J.-H. (1992) The *Rhizobium* symbiosis of the nonlegume *Parasponia*, in *Biological Nitrogen Fixation*, (eds G. Stacey, R. Burris and H.J. Evans), Chapman & Hall, London, pp. 497–559.

Bogusz, D., Appleby, C.A., Landsmann, J. *et al.* (1988) Functioning haemoglobin genes in non-nodulating plants. *Nature*, **331**, 178–80.

Landsmann, J., Dennis, E.S., Higgins, T.J.V. *et al.* (1986) Common evolutionary origin of legume and non-legume plant haemoglobins. *Nature*, **324**, 166–8.

Mohapatra, S.S., Weinmann, J.W., Scott, K.F. *et al.* (1985) Comparative aspects of *Rhizobium* symbiosis with the non-legume *Parasponia* and the legumes, in *Analysis of the Plant Genes Involved in the Rhizobium–Legume Symbiosis*, OECD, Paris, pp. 90–102.

5
Actinorhiza

5.1 THE MICROSYMBIONT: *FRANKIA*

Frankia is a slow-growing filamentous actinomycete which can form root nodules, called **actinorhiza**, on about 280 species (Table 5.1) of plants from eight different families (Schwintzer and Tjepkema, 1990) (Fig. 5.1). Progress in our understanding of the genus *Frankia* has been made with the development of culture techniques and the fulfilment of Kochs postulates by reinfection (Callaham *et al.*, 1978). Under appropriate physiological conditions, *Frankia* strains can fix N_2 in pure culture. There is a correlation between the formation of vesicles and the nitrogenase activity, which is seen as being similar to the heterocyst formation in cyanobacteria (Akkermans and van Dijk, 1981).

5.1.1 Taxonomy and systematics

Bergey's Manual of Determinative Bacteriology, 8th edition (1974) defined the division of the genus *Frankia* into 10 types (*Frankia allnii*, *F. casuarinae* etc.). In the later edition this has been changed. One *Frankia* species, isolated from *Elaeagnus*, can also infect *Alnus*. Thus the new taxonomy (Becking, 1981) tries to reflect the biochemical and cell physiological properties of the bacteria as well as specific plant reactions and symbiotic specificity.

The GC-content of DNA from the genus *Frankia* is between 68 and 72 mol %. DNA hybridization studies show a high degree of homology between isolates from *Alnus* and *Comptonia* nodules (An *et al.*, 1985). These strains are combined into the so-called 'B-group' whereas other strains, which show similarity to the isolates from *Elaeagnus*, form the 'A-group'. A third group of strains infects *Casuarina* species (Nittayajarn *et al.*, 1990; section 5.3.3). A very stable plasmid has been found in strains belonging to the B-group whereas a plasmid in A-group strains tends to be labile. Lectins with specificities for *N*-acetylglucosamine and sialic acid bind to hyphae, vesicles and sporangia from strains of 'B (*Alnus*) group'-bacteria. The production of pigments, metabolism of carbohydrates at low concentrations

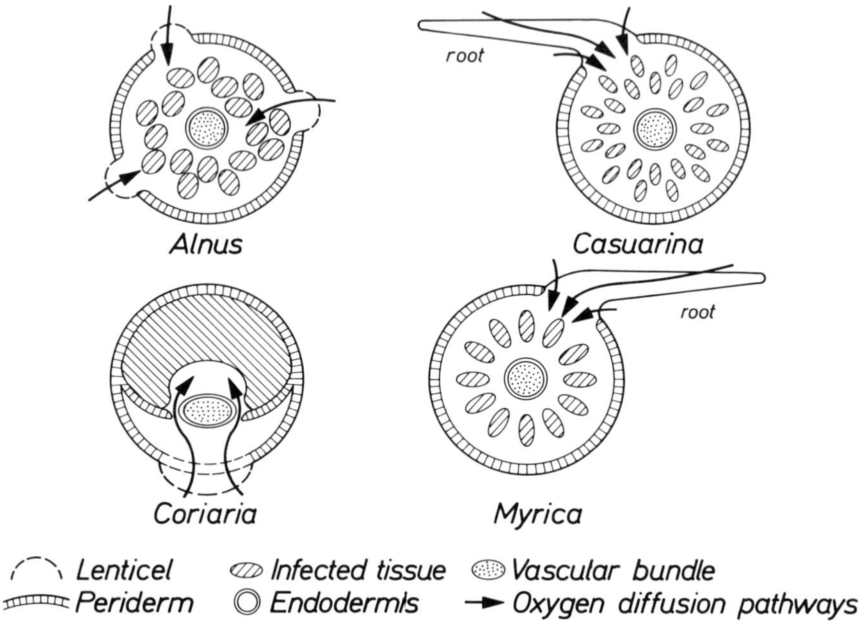

Fig. 5.1. Distribution of infected tissue in different actinorhizal nodules. Modified from Silvester *et al.*, 1990

(0.5%) and saccharide composition are all factors which divide the two groups. These properties are not correlated with morphological characteristics such as the setting of spores or vesicles (Becking, 1981). DNA–DNA hybridization did not reveal a significant relatedness between *Frankia* and other actinomycetes (An *et al.*, 1987). However, on the basis of 16S rRNA, a high degree of phylogenetic relationship has been found between some *Frankia* strains (B-type) and *Geothermatophylus* (Becking, 1970). Oligonucleotide probes with 20–22 nucleotides have been used successfully to identify specific strains from a group of 23 *Frankia* strains with different compatibility groups (Hahn *et al.*, 1989).

By using **restriction fragment length polymorphism (RFLP)** with *nif DH* probes, a large number of *Frankia* strains have been separated into three groups, identical with cross-inoculation studies (Nittayajarn *et al.*, 1990).

5.1.2 Physiology, biochemistry and genetics

Frankia strains from *Comptonia peregrina* grow in defined medium using succinate and glutamine, as C-source, vitamins and the normal mineral salts. *Frankia* isolates from *Alnus incana* nodules grow using acetate and propionate as C-source, but are incapable of using succinate, fumarate,

Table 5.1 Actinorhizal families, genera and species (modified from Baker and Schwintzer and Baker, 1990)

Family	*Genus*	*Number of species*
Betulaceae	*Alnus*	47
Casuarinaceae	*Allocasuarina*	54
	Casuarina	16
	Ceuthostoma	2
	Gymnostoma	18
Coriariaceae	*Coriaria*	16
Datiscaceae	*Datisca*	2
Elaeagnaceae	*Elaeagnus*	38
	Hippophae	2
	Shepherdia	2
Myricaceae	*Comptonia*	1
	Myrica	28
Rhamnaceae	*Ceanothus*	31
	Colletia	4
	Discaria	5
	Kentrothamnus	1
	Retanilla	2
	Talguenea	1
	Trevoa	2
Rosaceae	*Cercocarpus*	4
	Charmaebatia	1
	Cowania	1
	Dryas	1
	Purshia	2

malate or glucose. A general simple medium with all macronutrients, trace elements and propionic acid as C-source can be used for all *Frankia* cultures. Without NH_4^+ *Frankia* can differentiate into vesicles. Vesicles, isolated from *Alnus incana* and *Alnus rubra* nodules, can utilize glucose, glucose-6-phosphate and 6-phosphogluconate. Hexokinase (EC 2.7.1.1), NADP–glucose-6-phosphate dehydrogenase (EC 1.1.1.49) and NAD–6-phosphogluconate dehydrogenase (EC 1.1.1.44) have been found to be active in extracts from vesicles (Lopez and Torrey, 1985; Vikman and Huss-Danell, 1987). Optimal growth temperature is 30°C. No growth is observed under 14°C or over 36°C. Differences in C-metabolism are known for *Actinomyces*, thus it is not surprising that similar differences are also found among the members of *Frankia*. C-reserves include trehalose, which can form up to 20% of dry weight, and glycogen, up to 10% of the cell dry weight. Both of these reserves are rapidly used when nitrogenase is de-repressed (Lopez *et al.*, 1984).

In order to differentiate non-infective and non-spore forming *Frankia* from background actinomyces, biochemical properties of symbiotic strains have been sought. *Frankia* from *Alnus*, *Myrica*, *Comptonia* and *Elaeagnus*, all contain 2-*O*-methylmannose. Phenolic compounds are found in actinorhizal tissue in high concentrations. They influence growth and differentiation in the microsymbiont. Ferulic acid and paracoumaric acid inhibit growth, increase branching of the hyphae and reduce the size and number of sporangia. Benzoic acid and *p*-hydroxybenzoic acid on the other hand do not affect growth but stimulate the formation of vesicles. Dipterocarpol has been identified as a specific growth stimulating factor from *Alnus* roots (Quispel *et al.*, 1989), but not all *Frankia* strains seem to require this compound for growth (Lechevalier and Lechevalier, 1990). A catalogue of *Frankia* strains has improved the identification of available and also new *Frankia* strains (Lechevalier, 1986).

The genome of *Frankia* has been determined with a molecular weight of 6×10^9–8×10^9 Da, equivalent to 10×10^6–12×10^6 base pairs. This is about twice the coding capacity of *E. coli*. In more than 200 different *Frankia* strains, 14 different types of plasmids have been identified, ranging in size from 8 to 60 kb (Normand and Lalonde, 1986). However, only a relatively small number of strains harbour plasmids and no specific functions have so far been assigned to these plasmids (Mullin and An, 1990). An exception is that in one *Frankia* strain *nif H*, *D* and *K* hybridizing sequences which have been found in plasmid DNA as well as in chromosomal DNA (Normand *et al.*, 1988). A large sequence diversity in the *nif H*, *D* and *K* genes has been observed in different *Frankia* strains by RFLP analysis (Mullin and An, 1990). A *nif*-like regulatory gene also seems to be present in *Frankia* (Seguin and Lalonde, 1990).

5.2 THE MACROSYMBIONT: THE HOST PLANT

Compared to the specificity exhibited in the (*Brady*)-*Rhizobium*–legume symbiosis, the host spectrum of *Frankia* is very wide. Host plants are members of 20 different genera from eight families and seven different orders. Newly recognized host plants include *Datisca*, *Rubus*, *Cowania* and *Chamaebatia*.

5.2.1 Systematic overview

At present about 160 species are reported to develop an actinorhiza. The genera *Allocasuarina*, *Alnus*, *Elaeagnus*, *Ceanothus*, *Myrica*, *Gymnostoma*, *Casuarina* and *Coriaria* are the most important ones (Table 5.1). Five subclasses of dicots are actinorhiza-free as are all subclasses of monocots. Previous reports on actinorhiza formation in a *Rubus* species have not been confirmed. More species can possibly be found in the Rhamnaceae and Rosaceae.

It is an interesting hypothesis that during evolution some species and genera may have lost symbiosis genes. It is thus assumed that the genus *Betula* has lost the genes for a symbiosis with *Frankia* (Bousquet and Lalonde, 1990). Proposals for a transfer of these genes from *Alnus* to *Betula* by genetic engineering have been suggested. In forest soils devoid of actinorhizal plants, a very high *Alnus* nodulation capacity was found with *Betula* stands (Smolander and Sundman, 1987). By using RFLP methods and isoenzyme patterns, some *Alnus* species have been found to be less divergent from *Betula* species (*B. papyrifera*) than from other alder species.

5.2.2 Ecological and agricultural aspects

The geographical distribution of actinorhizal plants is normally in the boreal latitudes (Table 5.1). Those found in tropical latitudes are mostly confined to higher, mountainous, cooler regions. An exception to this rule is the genus *Casuarina*, which is found in almost all climatic zones of Australia.

Economically most important is the genus *Alnus*, for example, *Alnus glutinosa* and *A. incana* in European woods and *A. rubra* and *A. crispa* in the forests of North America (Baker, 1990). A 60–90-year-old stand of *A. glutinosa* yields 500–700 m^3 of wood per hectare. It has been estimated that in the US States of Oregon and Washington there are 500 million m^3 of usable timber from *A. rubra* alone. This is about one half of the total hardwood in these states. The importance of alder is not, however, due to the production of alder wood alone. Growth of poplar in mixed stands of alder and poplar, measured over 3 years, was stimulated by 220% whereas only a 55% stimulation occurred when poplar was grown in mixed poplar/plum stands (Table 5.2). Thus *Alnus crispa* has a stimulatory effect on *Populus trichocarpa*. A similar stimulation of *Pinus radiata* grown near alder is documented in Table 5.3. These effects may be specific for a few trees such as pine and poplar since the lack of stimulation with other types of tree has also been reported.

Table 5.2 Growth of poplar (*Populus deltoides* × *canadensis* cv. *marilandica*) interplanted with *Alnus glutinosa* and *Prunus* sp. (Based on Becking, 1970.)

Plantation	*Average height (cm)*			*% Increment*
	1956	*1958*	*1959*	
Poplar monoculture	141	186	237	68
Poplar with *Alnus*	144	289	460	219
Poplar with *Prunus*	142	181	220	55

Table 5.3 Yield and N-content of shoots from *Pinus radiata* seedlings grown on granite soil, after addition of N-fertilizer and litter from *Alnus rubra* or *Pseudotsuga mentziesii*. (Based on Wollum and Youngberg, 1964)

Supplement	*Yield* (mg dry matter per plant)*	*N-content (mg per plant)*
0 ppm N	240	0.82
Pseudotsuga litter	200	0.68
25 ppm N (as NH_4NO_3)	470	1.87
Alnus litter	570	8.06

* Means of 48 plants

5.2.3 Nitrogen-fixation per hectare and year

Values between 50 and 200 kg N fixed $ha^{-1} yr^{-1}$ have been reported for *Alnus glutinosa* and *Alnus rubra*. *Hippophae rhamnoides* actinorhiza are severely affected by the habitat and their fixation rate varies between 2 and 58 kg N $ha^{-1} yr^{-1}$ (Akkermans and van Dijk, 1981). Similar variations between 1 and 60 kg N $ha^{-1} yr^{-1}$ have been reported for *Ceanothus* and *Purshia* species. *Alnus* in mountainous regions of the Alps fixed around 8 kg N $ha^{-1} yr^{-1}$. This represents only 10% of the nitrogen needed for growth. New plantations of *Casuarina equisetifolia* on the Cape Verde Islands fixed 58 kg N $ha^{-1} yr^{-1}$ over a period of 13 years. This N was reported not only to be incorporated into tree tissue, but also enriched in the soil. Soil and climatic factors affect N_2-fixation considerably. Age is one other important factor; 3-year-old *Hippophae* trees fix 27 kg N yr^{-1} whereas 13-year-old trees fix 180 kg N yr^{-1}.

5.3 INFECTION AND THE DEVELOPMENT OF ACTINORHIZA

Actinorhizas are perennial but have a shorter life than the plant. Thus on older plants they tend to be younger than the rest of the root tissue. A 20-year-old *Alnus glutinosa* tree for example has actinorhiza around 3–4 years old, sometimes as old as 8 years. It has been shown with both *Alnus* and *Ceanothus* that when the host plant is absent for long periods the *Frankia* in the soil practically disappear, and no reinfection of young host plants occurs. Actinorhizas develop only when the soil around the roots is well aerated. This is also the case when alder grow in wet habitats, in which actinorhizas only develop above the water level. Dryness and high soil

Fig. 5.2. Growth of *Alnus glutinosa* in N-free nutrient solution. Right: with actinorhiza; left: without inoculation. (Photograph courtesy of J.H. Becking, Wageningen.)

temperatures inhibit infection. Under N-limiting conditions the effect of symbiotic N_2-fixation on growth is marked (Fig. 5.2). The following scheme of infection and development of the symbiosis has mostly been researched in *Alnus*.

5.3.1 Infection of the root hair and rind cells: the formation of primary actinorhiza

As in legumes, the root hair is the site of primary infection in *Alnus*. The interaction between the host and the endophytic microsymbiont begins in the rhizosphere. The actinomycete induces the host to deform root hairs and export a polysaccharide-containing capsular material. This capsular material can be excreted by either roots or root hairs. Outside the root hair cells this is called exocapsule. The endophyte enters the plant via a deformed root hair. The success rate is very low, only one in several thousand. During the entry and subsequently, the hyphae are enclosed by more polysaccharide-containing material, now called the endocapsule.

Hyphae and associated capsular material are surrounded by a membrane derived from the host plasma membrane. The nucleus of the infected root hair moves with the hyphae to the base of the root hair cell. There the hypha crosses the cell wall and is released into neighbouring cells. Those hyphae in infected cells near cell walls differentiate into vesicles. This stage is defined as primary actinorhiza, and is morphologically seen as localized swellings on the root. This can be confirmed by exposing the root to light, whereupon a red pigment is produced. Infection can be supported by the presence of *Pseudomonas* species. This is believed to be due to increased root hair curling.

Significantly different is the infection of Elaeagnaceae. Here primary infection is achieved by direct penetration of the rhizodermis by hyphae and a subsequent colonization of the intercellular spaces of the root.

5.3.2 Secondary actinorhiza: the development of hyphae, vesicles and spores

Further development (Fig. 5.3) begins with the formation of side roots from the pericycle. This is apparently induced by the microsymbiont. These side roots develop into secondary actinorhizas. At 2–3 weeks after infection they can be seen with the naked eye and at this stage N_2-fixation was begun. The various zones of a secondary actinorhiza are shown in Fig. 5.4. The tip contains a meristematic zone consisting of non-infected cells (A). The following zone shows cells filled with symbiotic hyphae (B). Connected to this are zones of cells containing active *Frankia* vesicles (C), followed by a

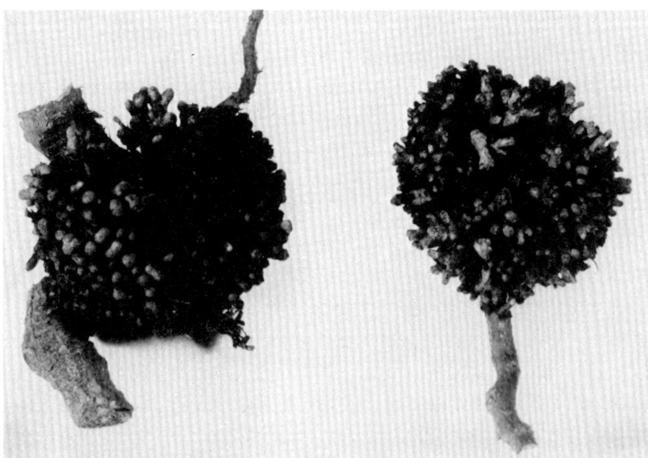

Fig. 5.3. Actinorhiza from *Alnus glutinosa*. (Original photograph A. Wolff, Marburg.)

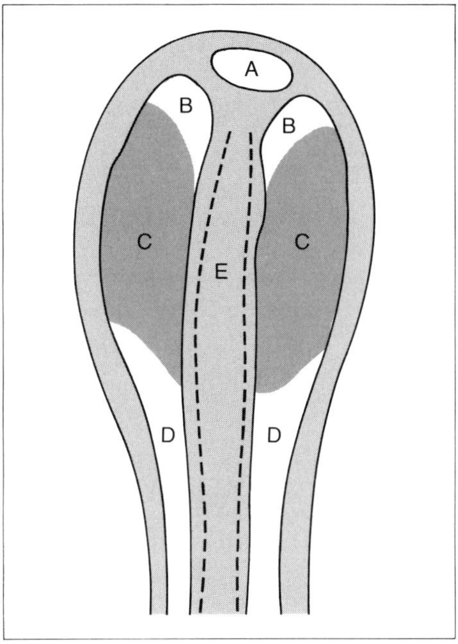

Fig. 5.4. Longitudinal view of the endophyte zone of an actinorhizal lobe from *Alnus glutinosa*. A, meristematic zone with cortex cells that remain uninfected; B, cells filled with hyphae; C, cells with active *Frankia* vesicles; D, zone of senescence; E, vascular bundle. (Based on Akkermans and Van Dijk, 1981.)

senescent zone (D). In the middle, forming an axis of symmetry, is a vascular zone (E). Although this morphology is typical for *Alnus*, other host plants may be quite different. In actinorhiza from *Coriaria* cells only one side of the vascular bundle becomes infected. In *Myrica* only one or two layers of cells are infected whereas up to five cell layers can be infected in *Alnus glutinosa* (Fig. 5.5). Following the different developmental stages along the long axis of an actinorhiza it becomes clear that those cells exclusively filled with hyphae are followed by a much larger number of cells filled with vesicles (Fig. 5.6). At a distance of 0.5–0.6 mm from the nodule tip cells filled with spores can be observed (Fig. 5.7). It is still not established whether the ability of *Frankia* to form spores is connected with the presence of small plasmids, which have been found in *Frankia* strains.

Frankia hyphae are highly branched. The cytoplasm is especially rich in ribosomes and lipid droplets. The development of vesicles generally starts in the proximity of the cell wall. The hyphae swell at the tips and when ripe vesicles have been formed these are cut off from the hyphae by a septum. In pure *Frankia* cultures it can easily be demonstrated that a correlation exists between the number of vesicles and N_2-fixation (Fig. 5.8). Five days after subculturing N_2-fixation is around 5 μmol C_2H_4 mg protein^{-1} h^{-1}. Simultaneously the number of vesicles is raised to 6×10^8 per mg protein. After the addition of 5 mM-propionate both parameters increase about 50%, then fall dramatically. Also with clones of different *Alnus glutinosa* ecotypes propagated *in vitro*, ineffectivity of certain *Frankia* strains was correlated

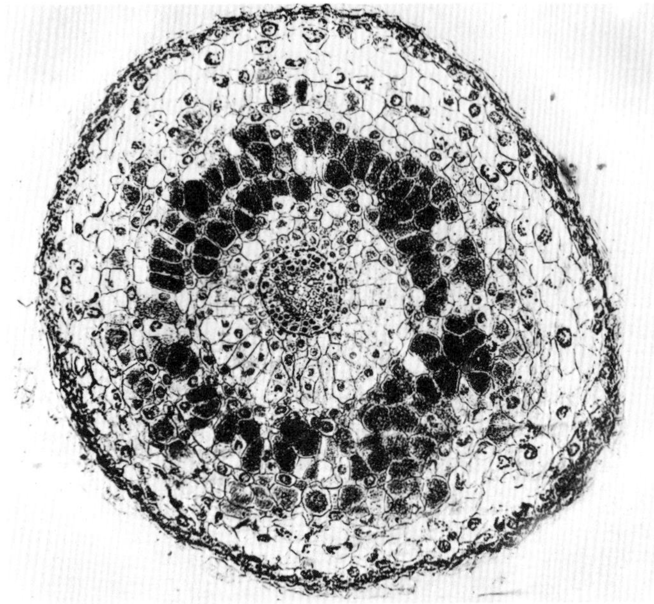

Fig. 5.5. Cross-section through actinorhiza from *Alnus glutinosa*. Cells of the inner cortex layers are densely packed with endophytes. (Photograph courtesy of J.H. Becking, Wageningen.)

Fig. 5.6. Vesicles and spores in infected cortex cells of *Ceanothus* sp. actinorhiza. (Original photograph K. Fischbeck and J. Kummerow.)

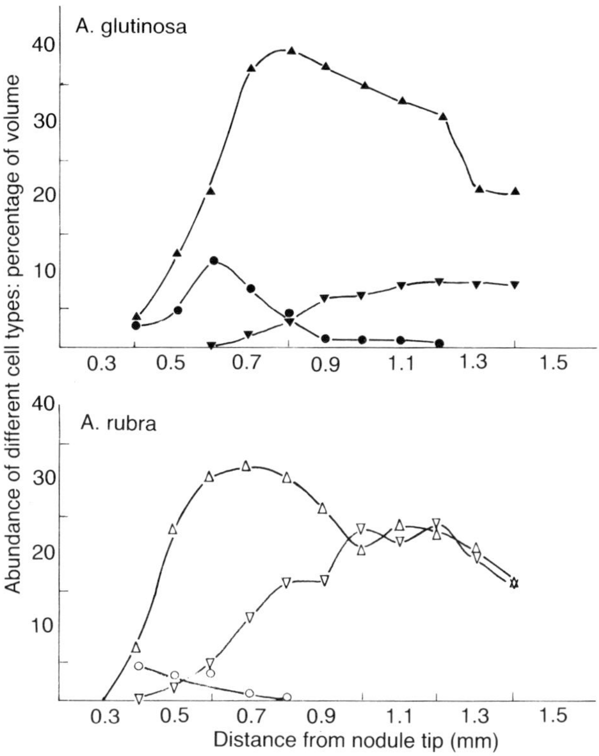

Fig. 5.7. Distribution of cells infected with different types of endophyte along actinorhizal lobes from *Alnus glutinosa* and *A. rubra*. ●, ○, Cells with hyphae; ▲, △, cells with vesicles; ▼, ▽, cells with spores. (Based on Wheeler *et al.*, 1981.)

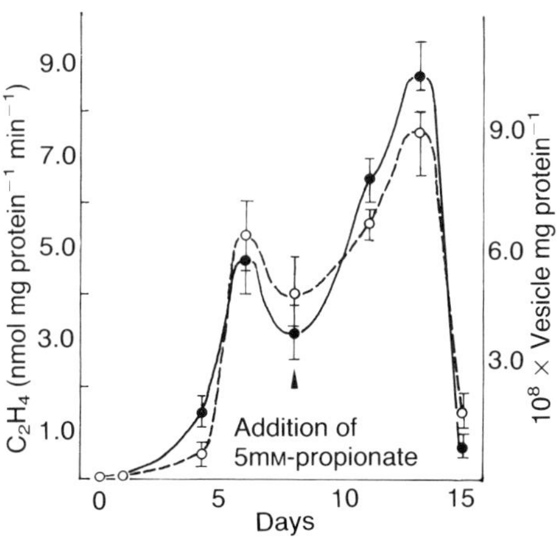

Fig. 5.8. Correlation between vesicle differentiation (○) and nitrogenase activity (●) in pure cultures of *Frankia* strain ArI3. (Based on Murry *et al.*, 1984.)

with the absence of vesicles in the infected host cells (Hahn *et al.*, 1988). By immunogold-labelling, nitrogenase was also only detectable in vesicles, but not in hyphae of the microsymbiont developmental stages (Torrey, 1985; Meesters *et al.*, 1987). The third stage of the life cycle, the setting of spores, has been observed in several *Frankia* spp. and also in the nodules of many host plants from the genera *Alnus*, *Myrica*, *Ceanothus*, *Elaeagnus*, *Hippophae*, *Casuarina* and *Purshia*. Besides the genotype of the microsymbiont, the physiological state of the host plant also seems to be important for sporulation (Torrey, 1987). Glutamine inhibits the development of spores and vesicles. Spores are contained in a thick pectinic capsule, which, as with all developmental stages of the symbiosis, is partitioned from the host cytoplasm by a host-derived membrane. The formation of the 0.5–1.0 μm diameter spores is easily achieved in pure culture using strains that do not form spores in nodules. Thus, spore formation in these actinorhizas is suppressed by the host plants, but this response is genetically determined in the *Frankia* strains. There are no strains in pure culture available that consistently produce spores in the nodules (Schwintzer, 1990). In the degenerating cells of the scenescent zone hyphae and vesicles appear to be lysed and digested by the host. Spores are able to withstand this procedure. Spore formation is, however, not essential for survival, as non-sporulating strains remain infectious.

5.3.3 Inoculation groups

Incompatibility between various *Frankia* isolates is a subject which still needs more research. From the few confirmed experiments it is known that endophytes from *Alnus glutinosa*, *Alnus incarna*, *Alnus crispa* and *Alnus rubra* cannot infect *Casuarina equisetifolia*, *Hippophae rhamnoides*, *Elaeagnus angustifolia*, *Ceanothus velutinus* or *Coriaria myrtifolia*. Conversely isolates from *Casuarina equisetifolia* cannot infect the *Alnus* species given above nor *Myrica gale*. Within the family of the Elaeagnaceae cross infections are possible. *Frankia* isolated from either *Hippophae rhamnoides* or *Elaeagnus angustifolia* may infect the other host. *Frankia* gained from *Hippophae rhamnoides* may also infect *Shepherdia canadensis*. Another successful cross-infection group are the microsymbionts and hosts from *Casuarina equisetifolia*, *Casuarina cristata* and *Casuarina glauca*.

Exact examinations show that within the same genera different degrees of incompatibility can be found. Thus inoculum from *Alnus glutinosa* actinorhiza can infect *Alnus crispa* and *Alnus incarna* to 95%, but only 28% of the plants were infected when *Alnus rubra* was used as host. Furthermore, these actinorhizas on *Alnus rubra* were atypical in that they exhibited negative geotropism, a property associated with actinorhizas from *Myrica* and *Casuarina*. Similar degrees of incompatibility are observed using endophytes from *Coriaria myrtifolia* and *Coriaria japonica*. These are

Table 5.4 Compatibility between host plant species and strains of *Frankia* as endophyte. (Based on Normand and Lalonde, 1986)

Host plant	*Frankia strain from*												
	AGl	*ACr*	*ARg*	*ASi*	*ARb*	*CP*	*MGa*	*MPe*	*HRh*	*EAn*	*EUm*	*CaE*	*CoS*
Alnus glutinosa (AGl)	E	E	E		E	E	E/I				I		
Alnus crispa (ACr)	E	E	E	I	E	E	E	E			O		
Alnus rugosa (ARg)	E	E	E	O	E	E		E					
Alnus sinuata (ASi)	E			E									
Alnus rubra (ARb)	E	E		I	E	E		E			O		
Comptonia peregrina (CP)	E	E		I	E	E		E			O		
Myrica gale (MGa)	E	E		E	E	E	E/I	E			O	I	
Myrica penpsylvanica (MPe)	E							E					
Hippophae rhamnoides (HRh)	O	O	O		O		E		E		E	E	E
Elaeagnus angustifolia (EAn)	O				O				E	E	E/I	E	E
Elaeagnus umbellata (EUm)	O				O					E	E/I		
Casuarina equisetifolia (CaE)									O			E	O
Colletia spinosissima (CoS)		O				O	O		E			E	E

E: Formation of effective actinorhiza; I: formation of ineffective actinorhiza (no N_2-fixation); O: no formation of actinorhiza.

summarized in Table 5.4. Some actinorhizal species, such as *Ceanothus americanus* and *Purshia tridentata*, fail to be nodulated by any of the available pure culture strains of *Frankia*, indicating that the appropriate strains have not been isolated (Torrey, 1990).

In another study, using 50 different strains of *Frankia* also from pure cultures, three major host infectivity groups were defined (Baker, 1987).

1. Strains which nodulate *Alnus* and *Comptonia*;
2. Strains which nodulate *Casuarina* and *Myrica*;
3. Strains which nodulate Elaeagnaceae and *Myrica*.

Strains which nodulate only Elaeagnaceae may be a subgroup of 3.

5.3.4 The influence of soil factors on infection and actinorhizal development

Air-dried actinorhiza, kept at 6°C, retains infectiousness ($10^5 - 10^6$ infection units g^{-1}) after 3 months. After 6 years this value is reduced to 0.01% of the original. It is thus postulated that most of the infecting cells in the soil come from dying actinorhiza, even though in pure culture the cells may replicate. The low infectiousness of soil samples grown for many years in the absence of host cells implies that *Frankia* replication in soil is not ecologically important. Release of endophytes from actinorhiza of *Alnus glutinosa* under field conditions showed that 10 g dry weight actinorhiza m $soil^{-2}$ released 10–100 infectious particles per year when the actinorhiza contained no spores. Actinorhiza containing a sporulating strain released $10^3 - 10^5$ infectious particles per year. These values were confirmed using direct measurements of population density of sporulating and non-sporulating *Frankia* strains in the soil. For a large-scale artificial inoculation of alder samplings the use of *Frankia* strains grown in pure culture is, however, more feasible than using actinorhiza homogenates. *Frankia* strains were best preserved for long periods of time by freeze drying and less efficiently by glycerol preservation (Fontaine *et al.*, 1986).

5.4 STRUCTURE AND FUNCTION OF ACTINORHIZA

The function of the actinorhiza is in the development of an (intracellular) microhabitat where multiplication and differentiation into vesicles and spores is possible without competition from other organisms. The nitrogenase activity induced during vesicle formation can in many cases cover the N demand of the plant.

5.4.1 Nitrogenase specific activity

Nitrogenase activities, normally given as μmol C_2H_4 formed h^{-1} per g actinorhiza fresh weight, is between 1 (or less) up to more than 20 μmol

Table 5.5 Nitrogenase activity of actinorhiza from greenhouse plants (G) or from the field (F) (data from Dalton and Naylor, 1975; Dalton and Zobel, 1977; Heisey *et al.*, 1980; Lalonde, 1979; Schubert and Evans, 1976; van Straten *et al.*, 1977; other references detailed in Werner, 1987)

Host plant	*Nitrogenase activity ($\mu mol\ C_2H_4\,h^{-1}\,g^{-1}$ actinorhiza fresh weight)*
Alnus glutinosa (G)	2–20
Alnus glutinosa (F)	20
Alnus crispa (F)	5
Elaeagnus angustifolia (F)	15
Casuarina cunninghamiana (G)	1.5–3.7
Comptonia peregrina (G)	9.6–22
Myrica gale (G)	6.5–14
Ceanothus velutinus (F)	1.8
Colletia spinosa (F)	2.1
Purshia tridentata (F)	0–5.1
Chamaebatia foliosa (G)	0.7

(Table 5.5). Thus the maximum value is about 30–50% of that found in well-functioning legume nodules. However, it must be taken into account that the zone where active fixation takes place is small. In this infected zone specific activity is thus very high. Factors affecting N_2-fixation are on the one hand species specificity and on the other hand diurnal variation (Wheeler, 1969; Bond and Mackintosh, 1975) and seasonal variations (Fig. 5.9). In field trials in France activity maxima were found in April and again in October/November. The annual minimum was found in June and July. This fluctuation can only be measured *in vivo*. Activity in cell-free extracts taken between April and November gave no differences. It is thus postulated that enzyme levels remain constant and/or energy-sources provided by the plant undergo an annual rhythm. Fixation is on average higher in actinorhiza from *Alnus*, *Comptonia* and *Myrica* as in those from *Ceanothus*, *Colletia*, *Purshia* or *Shepherdia*.

The temperature optimum for fixation in actinorhiza from *Alnus* and *Myrica* is between 20 and 25°C. Over 25°C the activity drops dramatically. This is not so the case with *Casuarina* which, as expected from its ecology and tropical distribution, can tolerate temperatures above 30°C.

Water limitation is an especially important factor in plants from dry areas. N_2-fixation is completely inhibited in, for example, *Purshia tridentata* when the water potential in the xylem is reduced to less than −25 bar. Moderate water stress reduces in alder nitrogen fixation to one half, and severe stress (−1MPa) to almost zero (Sundström and Huss-Danell, 1987).

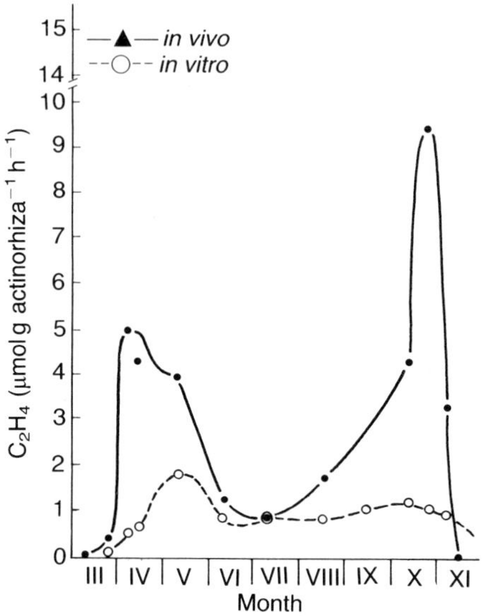

Fig. 5.9. Annual fluctuation of *in vivo* and *in vitro* nitrogenase activity in actinorhiza from *Alnus glutinosa*. (Based on Pizelle, 1984.)

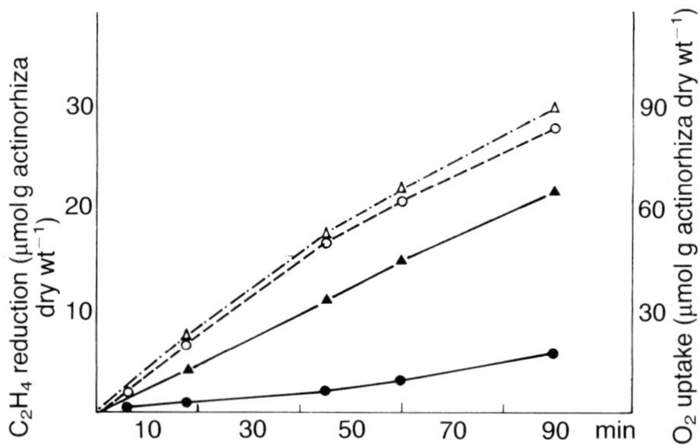

Fig. 5.10. N_2-fixation (acetylene reduction) in actinorhiza from *Alnus glutinosa* before (○) and after (●) infiltration with water, in comparison with respiration before (△) and after (▲) imbibition. (Based on Akkermans and Van Dijk, 1981.)

Oxygen limitation, normally related to the water content of the soil, reduces N_2-fixation by limiting respiration. The gas volume of *Alnus* actinorhiza is around 5–10% of the total volume. Infiltration with water reduces respiration (Fig. 5.10) and strongly inhibits nitrogenase activity.

5.4.2 Nitrogenase in pure cultures of symbionts

Direct relationships between C_2H_4-reduction, $^{15}N_2$-fixation and oxygen consumption can only be studied using pure cultures of the microsymbionts since in actinorhiza the host tissue also consumes oxygen. N_2-fixation has a much narrower oxygen concentration optimum than respiration (Fig. 5.11). *Frankia* CCl.17 has an optimum at 5% O_2 in the gas phase. When the O_2 concentration is raised to 20%, N_2-fixation is reduced to less than 50% whereas respiration is hardly affected. The maximum specific nitrogenase activity found in pure culture, which is about 2 μmol C_2H_4 mg protein^{-1} h^{-1}, is comparable with the values calculated for active vesicles isolated from actinorhiza. Nitrogenase has also a narrower pH optimum than respiration (Fig. 5.12). At pH 7.5 N_2-fixation is only 10% of the maximum whereas total respiration is reduced only to 50%.

5.4.3 Oxygen protection mechanisms

Oxygen tolerance in *Frankia* is significantly higher than in *Bradyrhizobium japonicum*. At 5% O_2 in the gas phase nitrogenase activity in *Bradyrhizobium japonicum* is totally inhibited. This concentration is however around the optimum for *Frankia* (Fig. 5.11). However, pure nitrogenase enzyme from both sources has similar oxygen sensitivity. The difference is due to the fact that in *Frankia* N_2-fixation is protected by the thick vesicle cell wall, in a similar way to that of cyanobacterial heterocysts. The vesicle cells respond to the Po_2, having very thin walls at 2 kPa O_2 and rather thick walls at

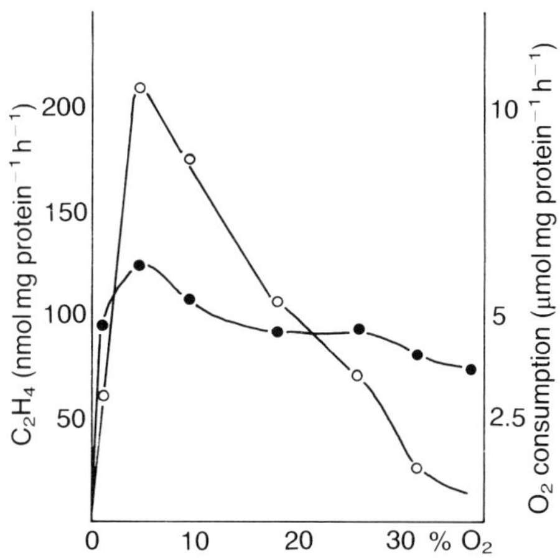

Fig. 5.11. Nitrogenase activity (○) and oxygen uptake (●) in *Frankia* strain Cc 1.17, in relationship to the oxygen content in the gas phase. (Based on Akkermans *et al.*, 1984.)

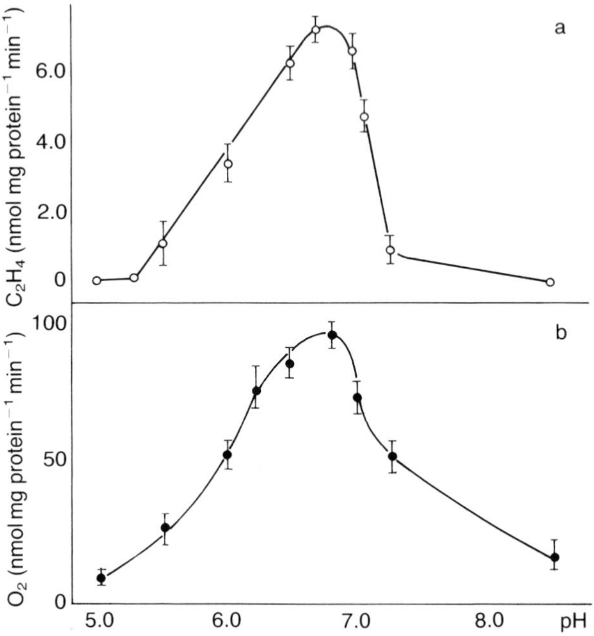

Fig. 5.12. Acetylene reduction (a) and O_2 uptake (b) in relation to the pH. Measurements were made in an atmosphere of argon: O_2:CO_2 (79:20:1). (Based on Akkermans and Van Dijk, 1981.)

70 kPa O_2 (Parsons *et al.*, 1987). The wall includes more layers at higher O_2 concentrations. In addition, the type of available carbon source affects the optimum O_2 concentration for nitrogenase activity. With an organic acid such as propionate, the optimum is at 15–20 kPa O_2, with glucose it is less than 10 kPa O_2 (Lopez *et al.*, 1986). In *Frankia* species which do not make vesicles, e.g. those from *Casuarina* and *Myrica gale*, the nitrogenase is protected from O_2 by two further mechanisms: first, by the presence of a

Table 5.6 Haem content, nitrogenase activity and suberin deposition in actinorhizal cells. (Based on Tjepkema and Murry, 1985)

Species	*Total haem (pmol g^{-1} fresh weight)*	*CO-reactive haem (nmol g^{-1} fresh weight)*	*C_2H_4 production (µmol h^{-1} g^{-1} fresh weight)*	*Suberin deposition*
Casuarina cunninghamiana	97 ± 3	44 ± 5	24 ± 2	+
Myrica gale	107 ± 19	26 ± 7	39 ± 3	+
Alnus rubra	28 ± 2	3 ± 0.3	46 ± 3	–
Datisca glomerata	15 ± 1	0	19 ± 2	–

(CO-sensitive) haemoglobin (Tjepkema and Asa, 1987) and second, by the thickening of host cell walls with suberin (Table 5.6). The function of the actinorhizal haemoglobin is similar to that of leghaemoglobin in legume nodules in facilitating oxygen transport (Silvester *et al.*, 1990).

The main mechanisms which supply oxygen effectively to the microsymbiont and the host plant and also protect nitrogenase can be summarized as follows:

1. the diffusion barrier of the vesicle cell wall or the suberin incorporation in the host cell wall and the presence of haemoglobin;
2. the respiration of the microsymbiont and the host nodule cells.

5.4.4 Ammonium assimilation

Ammonium is commonly assimilated in legume nodules via the GS/GOGAT pathways. In actinorhiza, however, the glutamate dehydrogenase (GDH) pathway is also active. Generally, in most of the higher plants this pathway is of minor importance as suggested by the high apparent K_m values of

Table 5.7 Free amino acids in actinorhiza of *Alnus nitida* (A) and *Datisca cannabina* (D). (Based on Hafeez *et al.*, 1984)

Amino acid	*μmol g fresh weight^{-1}*		*mol %*	
	A	*D*	*A*	*D*
Glutamate	0.95	2.30	16.6	34.4
Arginine	0.62	1.26	10.8	18.9
Citrulline	1.78	0	31.1	0
Aspartate	0.19	0.85	3.4	12.7
Glutamine	0.17	0.69	3.1	10.3
Valine	0.25	0.41	4.3	6.2
α-Aminobutyric acid	0.08	0.29	1.3	4.3
Glycine	0.38	0.29	6.6	4.3
Alanine	0.36	0.21	6.3	3.2
Serine	0.38	0.18	6.6	2.6
Threonine	0.09	0.07	1.6	1.1
Lysine	0.10	0.05	1.7	0.7
Ornithine	0.17	0.04	3.0	0.6
Histidine	0.06	0.03	1.0	0.4
Phenylalanine	0.04	0.02	0.8	0.3
Leucine	0.05	0	0.8	0
Isoleucine	0.04	0	0.7	0
Tyrosine	0.02	0	0.3	0
Total	5.72	6.68	100	100

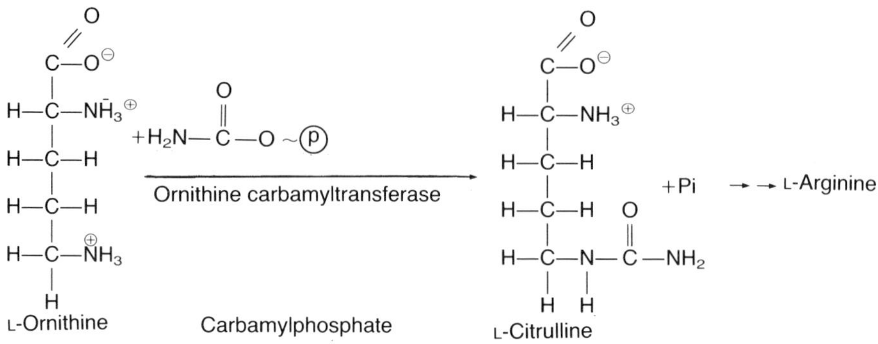

Fig. 5.13. Biosynthesis of citrulline via the arginine synthesis pathway.

glutamate dehydrogenase for NH_4^+. Such values are between 5mM and 70 mM. In some green algae the values are appreciably lower, 0.5–0.7 mM.

Frankia strains also synthesize two different glutamine synthetases (GS I and GS II). The GS II has a high degree of similarity to the GS II from *Bradyrhizobium japonicum* (Rochefort and Benson, 1990).

Glutamine synthetase from *Alnus* has a molecular weight of 360 000 Da and consists of eight subunits each of about 43 kDa. The K_m value for glutamate is around 0.9 mM and for ATP around 0.5 mM. Both GS and GDH are located in the host cell cytoplasm. It is therefore assumed that N_2 fixed in the vesicles is exported as NH_4^+ and assimilated in the host cytoplasm.

In *Alnus* fixed nitrogen is transferred as citrulline, the most dominant amino acid in this species (Table 5.7). In *Datisca cannabina* and other actinorhizas N-transport is accomplished using glutamine, glutamic acid, arginine and aspartic acid. Citrulline is totally absent in *Datisca* actinorhizas. Since citrulline is known as an intermediate in arginine synthesis (Fig. 5.13), it is assumed therefore that there are differences between *Datisca* and *Alnus* in arginine biosynthesis, somewhere between ornithine and arginine.

5.4.5 Other biochemical properties of actinorhiza

A comparison of actinorhiza and roots with respect to various elements is shown in Table 5.8. Clear differences exist in N concentration in both winter and summer, but not in spring or autumn. In contrast to roots, Mg, Mo and Co are present in actinorhiza in high concentrations in spring. Ion and nutrient uptake by roots also has a role in the infection. Exactly as in the legumes, infection is reduced severely by NO_3^- and less by NH_4^+.

The high level of nitrogen supply is shown by the nitrogen content of the leaves at their fall. Alder leaves contain 2–4 times more nitrogen per gram

Table 5.8 Concentrations of elements in roots and actinorhiza from *Alnus glutinosa* (Means of four determinations). (Based on Rodriguez-Barrueco *et al.*, 1984)

Element	*Winter*	*Spring*	*Summer*	*Autumn*
Roots				
N (%)	0.94	1.58	1.01	1.34
P (%)	0.15	0.16	0.17	0.11
K (%)	0.42	0.50	0.37	0.26
Ca (%)	0.52	0.55	0.63	0.40
Mg (%)	0.23	0.20	0.23	0.18
Fe (p.p.m)	5110	3370	1685	2936
Mn (p.p.m)	223	154	165	137
Mo (p.p.m)	0.48	1.50	1.25	0.73
Co (p.p.m)	15.25	10.75	8.88	8.50
Actinorhiza				
N (%)	1.59	1.87	1.44	1.50
P (%)	0.15	0.16	0.11	0.13
K (%)	0.50	0.52	0.30	0.47
Ca (%)	0.39	0.49	0.39	0.26
Mg (%)	0.34	0.43	0.34	0.21
Fe (p.p.m)	2368	3504	1713	2341
Mn (p.p.m)	77	87	48	58
Mo (p.p.m)	1.00	2.90	1.90	0.60
Co (p.p.m)	17.30	20.80	13.30	10.40

dry weight than those from birch, lime or poplar (Table 5.9). Thus the annual demand for nitrogen is higher than in other trees.

Actinorhiza have a very active 'uptake hydrogenase', and therefore produce no free H_2. In autumn, however, when the hydrogenase activity is reduced, H_2-formation may be detected in *Alnus* actinorhiza. This can be reproduced in plants which are temperature-stressed (30–33°C).

A 72 kDa protein from symbiotic *Frankia* and also from free-living strains is immunologically related to the large subunit of hydrogenase from *Alcaligenes latus*. By immunogold-labelling it was demonstrated that the protein is present in vesicles and in hyphae (Sellstedt and Lindblad, 1990).

5.4.6 Cytology and fine structure of infected cells

Hyphae and vesicles, of *Alnus* and *Elaeagnus* endophytes are surrounded by a very noticeable polysaccharide capsule. This capsule is formed by host cell dictyosomes, transported to the plasma membrane and deposited between the plasma membrane and the endophyte. In the third differentiation state, sporangia with spores are surrounded by a thick pectinic capsule (Lalonde

Table 5.9 Ash content and N-content of leaves of different tree species immediately after leaf fall. (Based on Mikola, 1958)

Species	*Ash content (% of dry matter)*	*N-content (% of dry matter)*
Alnus incana	5.9	2.73
Alnus glutinosa	4.5	2.57
Corylus avellana	7.6	1.30
Betula sp.	4.5	1.27
Tilia cordata	8.4	1.16
Sorbus aucuparia	8.4	0.83
Populus tremula	6.4	0.72
Acer platanoides	11.9	0.63
Larix sibirica	3.8	0.54
Pinus sylvestris	1.5	0.49

and Quispel, 1977). The cell wall of the vesicles has three layers, an outer layer continuous with the hyphal cell wall, a middle layer with several thin laminae and an electron-dense inner layer (Newcomb *et al.*, 1987).

5.5 REFERENCES

Akkermans, A.D.L. and Dijk, C.van (1981) Non-leguminous root-nodule symbioses with actinomycetes and *Rhizobium*, in *Nitrogen Fixation*, Vol. I *Ecology*, (ed. W.J. Broughton) Clarendon Press, Oxford, pp. 57–103.

Akkermans, A.D.L., Hafeez, F. Roelofsen, W. *et al.* (1984) Ultrastructure and nitrogenase activity of *Frankia* grown in pure culture and in actinorhizas of *Alnus*, *Colletia* and *Datisca* spp., in *Advances in Nitrogen Fixation Research* (eds C. Veeger, and W.E. Newton) Martinus Nijhoff/Dr W. Junk, The Hague, pp. 311–19.

An, C.S., Riggsby, W.S. and Mullin, B.C. (1985) Relationships of *Frankia* isolates based on deoxyribonucleic acid homology studies. *Int. J. Syst. Bacteriol.*, **35**, 140–6.

An, C.S., Riggsby, W.S. and Mullin, B.C. (1987) DNA relatedness of *Frankia* isolates ArI4 and EuI1 to other actinomycetes of cell wall type III. *Actinomycetes*, **20**, 50–9.

Baker, D.D. (1987) Relationships among pure cultured strains of *Frankia* based on host specificity. *Physiol. Plant.*, **70**, 245–8.

Baker, D.D. (1990) Actinorhizal plants: underexploited trees and shrubs for forestry and agroforestry, in *Nitrogen Fixing Tree Research Reports*, **8**, 3–7.

Baker, D.D. and Schwintzer, C.R. (1990) Introduction in *The Biology of Frankia and Actinorhizal Plants*, (eds C.R. Schwintzer, and J.D. Tjepkema), Academic Press, San Diego, pp. 3–13.

Becking, J.H. (1970) Plant–endophyte symbiosis in non-leguminous plants. *Plant Soil*, **32**, 611–54.

Becking, J.H. (1981) The genus *Frankia*, in *The Prokaryotes*, Vol. II (eds M.P. Starr, H. Stolp, H.G. Trüper *et al.*), Springer Verlag, Berlin, pp. 1991–2003.

Bond, G. and Mackintosh, A.H. (1975) Diurnal changes in nitrogen fixation in the root nodules of *Casuarina*. *Proc. R. Soc. Lond. B*, **192**, 1–12.

Bousquet, J. and Lalonde, M. (1990) The genetics of actinorhizal Betulaceae, in *The Biology of Frankia and Actinorhizal Plants* (eds C.R. Schwitzer and J.D. Tjepkema) Academic Press, San Diego, London, pp. 239–61.

Callaham, D., del Tredici, P. and Torrey, J.G. (1978) Isolation and cultivation *in vitro* of the actinomycete causing root nodulation in *Comptonia*. *Science*, **199**, 899–902.

Dalton, D.A. and Naylor, A.W. (1975) Studies on nitrogen fixation by *Alnus crispa*. *Am. J. Bot.*, **62**, 76–80.

Dalton, D.A. and Zobel, D.B. (1977) Ecological aspects of nitrogen fixation by *Purshia tridentata*. *Plant Soil*, **48**, 57–80.

Fontaine, M.S., Young, P.H. and Torrey, J.G. (1986) Effects of long-term preservation of *Frankia* strains on infectivity, effectivity, and *in vitro* nitrogenase activity. *Appl. Environ. Microbiol.*, **51**, 694–8.

Hafeez, F., Chaudhary, A.H. and Akkermans, A.D.L. (1984) Physiological studies on N_2-fixing root nodules of *Datisca cannabina* L. and *Alnus nitida* Endl. from Himalaya region in Pakistan. *Plant Soil*, **78**, 129–46.

Hahn, D., Dorsch, M., Stackebrandt, E. and Akkermans, A.D.L. (1989) Synthetic oligonucleotide probes for identification of *Frankia* strains. *Plant Soil*, **118**, 210–19.

Hahn, D., Lechevalier, M.P., Fischer, A. and Stackebrandt, E. (1989) Evidence for a close phylogenetic relationship between members of the genera *Frankia*, *Geodermatophilus*, and *Blastococcus* and emendation of the family Frankiaceae. *Syst. Appl. Microbiol.*, **11**, 236–42.

Hahn, D., Starrenburg, M.J.C. and Akkermans, A.D.L. (1988) Variable compatibility of cloned *Alnus glutinosa* ecotypes against ineffective *Frankia* strains. *Plant Soil*, **107**, 233–43.

Heisey, R.M., Delwiche, C.C., Virginia, R.A. *et al.* (1980) A new nitrogen-fixing non-legume: *Chamaebatia foliolosa* (Rosaceae). *Am. J. Bot.*, **67**, 429–31.

Lalonde, M. (1979) Immunological and ultrastructural demonstration of nodulation of the European *Alnus glutinosa* (L.) Gaertn. host plant by an actinomycetal isolate from the North American *Comptonia peregrina* (L.) Coult. root nodule. *Bot. Gaz.*, **140**, (Suppl.), 35–43.

Lalonde, M. and Quispel, A. (1977) Ultrastructural and immunological demonstration of the nodulation of the European *Alnus glutinosa* (L.) Gaertn. host plant by the North American *Alnus crispa* var. *mollis* Fern. root nodule endophyte. *Can. J. Microbiol.*, **23**, 1529–47.

Lechevalier, M.P. (1986) Catalogue of *Frankia* strains. *Actinomycetes*, **19**, 131–62.

Lechevalier, M.P. and Lechevalier, H.A. (1990) Systematics, isolation and culture of *Frankia*, in *The Biology of Frankia and Actinorhizal Plants* (eds C.R. Schwintzer and J.D. Tjepkema) Academic Press, San Diego, London, pp. 35–60.

Lopez, M.F., Fontaine, M.S. and Torrey, J.G. (1984) Levels of trehalose and glycogen in *Frankia* sp. HFPArI3 (Actinomycetales). *Can. J. Microbiol.*, **30**, 746–52.

Lopez, M.F., Murry, M.A. and Torrey, J.G. (1986) Effect of oxygen on substrate utilization for nitrogen fixation and growth in *Frankia* spp. *Arch. Microbiol.*, **145**, 209–14.

Lopez, M.F. and Torrey, J.G. (1985) Enzymes of glucose metabolism in *Frankia* sp. *J. Bacteriol.*, **162**, 110–16.

Meesters, T.M., van Vliet, W.M. and Akkermans, A.D.L. (1987) Nitrogenase is restricted to the vesicles in *Frankia* strain EAN1pec. *Physiol. Plant.*, **70**, 267–71.

Mikola, P. (1958) Liberation of nitrogen from alder leaf litter. *Acta Forest. Fenn.*, **67**, 1–9.

Mullin, B.C. and An, C.S. (1990) The molecular genetics of *Frankia*, in *The Biology of Frankia and Actinorhizal Plants* (eds C.R. Schwintzer, and J.D. Tjepkema) Academic Press, San Diego, London, pp. 195–214.

Murry, M.A., Fontaine, M.S. and Torrey, J.G. (1984) Growth kinetics and nitrogenase induction in *Frankia* sp. HFPArI3 grown in batch culture. *Plant Soil*, **78**, 61–78.

Newcomb, W., Baker, D. and Torrey, J.G. (1987) Ontogeny and fine structure of effective root nodules of autumn olive (*Elaeagnus umbellata*). *Can. J. Bot.*, **65**, 80–94.

Nittayajarn, A., Mullin, B.C. and Baker, D.D. (1990) Screening of symbiotic Frankiae for host specificity by restriction fragment length polymorphism analysis. *Appl. Environ. Microbiol.*, **56**, 1172–4.

Normand, P. and Lalonde, M. (1986) The genetics of actinorhizal *Frankia*: a review. *Plant Soil*, **90**, 429–53.

Normand, P., Simonet, P. and Bardin R. (1988) Conservation of *nif* sequences in *Frankia. Mol. Gen. Genet.*, **213**, 238–46.

Parsons, R., Silvester, W.B., Harris, S. *et al.* (1987) *Frankia* vesicles provide inducible and absolute protection for nitrogenase. *Plant Physiol.*, **83**, 728–31.

Pizelle, G. (1984) Seasonal variations of the sexual reproductive growth and nitrogenase activity (C_2H_2) in mature *Alnus glutinosa. Plant Soil*, **78**, 181–8.

Quispel, A., Svemdsem, A.B., Schripsema, K. *et al.* (1989) Identification of dipterocarpol as isolation factor for the induction of primary isolation of *Frankia* from root nodules of *Alnus glutinosa* (L.) Gaertner. *Mol. Plant Microbe Interact.*, **2**, 107–12.

Rochefort, D.A. and Benson, D.R. (1990) Molecular cloning, sequencing, and expression of the glutamine synthetase II (*gln*II) gene from the actinomycete root nodule symbiont *Frankia* sp. strain CpI1. *J. Bacteriol.*, **172**, 5335–42.

Rodriguez-Barrueco, C., Miguel, C. and Subramaniam, P. (1984) Seasonal fluctuations of the mineral concentration of alder (*Alnus glutinosa* (L.) Gaertn.) from the field. *Plant Soil*, **78**, 201–8.

Schubert, K.R. and Evans, H.J. (1976) Hydrogen evolution: a major factor affecting the efficiency of nitrogen fixation in nodulated symbionts. *Proc. Natl. Acad. Sci. USA*, **73**, 1207–11.

Schwintzer, C.R. (1990) Spore-positive and spore-negative nodules. in *The Biology of Frankia and Actinorhizal Plants* (eds C.R. Schwintzer and J.D. Tjepkema) Academic Press, San Diego, London, pp. 177–93.

Schwintzer, C.R. and Tjepkema, J.D. (1990) *The Biology of Frankia and Actinorhizal Plants*. Academic Press, San Diego, London.

Séguin, A. and Lalonde, M. (1990) Micropropagation, tissue culture, and genetic transformation of actinorhizal plants and *Betula*, in *The Biology of Frankia and*

Actinorhizal Plants (eds C.R. Schwintzer and J.D. Tjepkema) Academic Press, San Diego, London, pp. 215–38.

Sellstedt, A. and Lindblad, P. (1990) Activities, occurrence, and localization of hydrogenase in free-living and symbiotic *Frankia. Plant Physiol.*, **92**, 809–15.

Silvester, W.B., Harris, S.L. and Tjepkema, J.D. (1990) Oxygen regulation and hemoglobin, in *The Biology of Frankia and Actinorhizal Plants* (eds C.R. Schwintzer and J.D. Tjepkema) Academic Press, San Diego, London, pp. 157–76.

Smolander, A. and Sundman, V. (1987) *Frankia* in acid soils of forests devoid of actinorhizal plants. *Physiol. Plant.*, **70**, 297–303.

Sundström, K.-R. and Huss-Danell, K. (1987) Effects of water stress on nitrogenase activity in *Alnus incana. Physiol. Plant.*, **70**, 342–8.

Straten, van J., Akkermans, A.D.L. and Roelofsen, W. (1977) Nitrogenase activity of endophyte suspensions derived from root nodules of *Alnus*, *Hippophae*, *Shepherdia* and *Myrica* spp. *Nature*, **266**, 257–8.

Tjepkema, J.D. and Asa, D.J. (1987) Total and CO-reactive heme content of actinorhizal nodules and the roots of some non-nodulated plants. *Plant Soil*, **100**, 225–36.

Tjepkema, J.D. and Murry, M.A. (1985) Heme content and diffusion limitation of respiration and nitrogenase activity in nodules of *Casuarina cunninghamiana*, in *Nitrogen Fixation Research Progress*, (eds H.J. Evans, P.J. Bottomley and W.E. Newton) Martinus Nijhoff, Dordrecht, p. 370.

Torrey, J.G. (1985) The site of nitrogenase in *Frankia* in free living culture and in symbiosis, in *Nitrogen Fixation Research Progress* (eds H.J. Evans, P.J. Bottomley and W.E. Newton) Martinus Nijhoff, Dordrecht, pp. 293–9.

Torrey, J.G. (1987) Endophyte sporulation in root nodules of actinorhizal plants. *Physiol. Plant.*, **70**, 279–88.

Torrey, J.G. (1990) Cross-inoculation groups within *Frankia* and host – endosymbiont associations, in *The Biology of Frankia and Actinorhizal Plants* (eds C.R. Schwintzer and J.D. Tjepkema) Academic Press, San Diego, London, pp. 83–106.

Vikman, P.-A. and Huss-Danell, K. (1987) Capacity for hexose respiration in symbiotic *Frankia* from *Alnus incana. Physiol. Plant.*, **70**, 349–54.

Werner, D. (1987) *Pflanzliche und mikrobielle Symbiosen.* G. Thieme Verlag, Stuttgart, New York .

Wheeler, C.T. (1969) The diurnal fluctuation in nitrogen fixation in the nodules of *Alnus glutinosa* and *Myrica gale. New phytol.*, **68**, 675–82.

Wheeler, C.T., McLaughlin, M.E. and Steele, P. (1981) A comparison of symbiotic nitrogen fixation in Scotland in *Alnus glutinosa* and *Alnus rubra. Plant Soil*, **61**, 169–88.

Wollum, A.G. II and Youngberg, C.T. (1964) The influence of nitrogen fixation by nonleguminous woody plants on the growth of pine seedlings. *J. Forestry*, **62**, 316–21.

6
Other Bacterial Symbioses

6.1 ALGAE AS HOST CELLS

Strains of algae from widely separated classes live with intracellular bacteria. These include *Volvox* and *Udotea* (Chlorophyceae), *Vaucheria* (Xanthophyceae), *Katodinium* (Dinophyceae) and *Euglena* (Euglenophyceae) (Colombo, 1978; Surek and Melkonian, 1983). In *Euglena mutabilis* rod-shaped bacteria of uncertain identity have been found, of which 50–70 are present per *Euglena* cell and which remain stable in the alga after several years of subculturing. Thus algal and bacterial replication must be synchronized. A function cannot be assigned to these bacteria as growth rates of infected cells are identical with those of uninfected *Euglena*. One strain of *Volvox carteri* contains similar-sized bacteria ($1.5 \times 0.5\,\mu m$) in all developmental stages, i.e. vegetative cells, gonidia and sperm cells, which, in contrast to other bacterial symbioses (e.g. *Paramecium*, Fig. 6.1), are not surrounded by a phagosomal or lysosomal membrane. In the dinophyte *Katodinium glandulum*, symbionts are surrounded by a further membrane (Brown and Williams, 1985). In *Cryptomonas* species some of the endocytic bacteria have been found to contain bacteriophage-like particles which may have a function in maintaining a stable number of bacteria inside the algal cells (Schnepf and Melkonian, 1990).

6.2 PROTOZOA AS HOST CELLS

Many protozoans take up nutrients by phagocytosis. It thus appears reasonable that through such mechanisms many bacteria or unicellular eukaryotes may be taken up and eventually established as endocytobionts. These favourable cytological conditions result in a very wide spectrum of emprisoned prokaryotes ('particles') particularly in *Paramecium aurelia* and *Euplotes* spp. (Table 6.1) (Preer and Preer, 1984; Görtz, 1986). These 'particles', designated alpha, gamma, kappa, lambda or sigma were taxonomically studied. Alpha-'particles' are *Holospora caryophila* and kappa 'particles', *Caedibacter taeniospiralis*. Some properties of these 'particles'

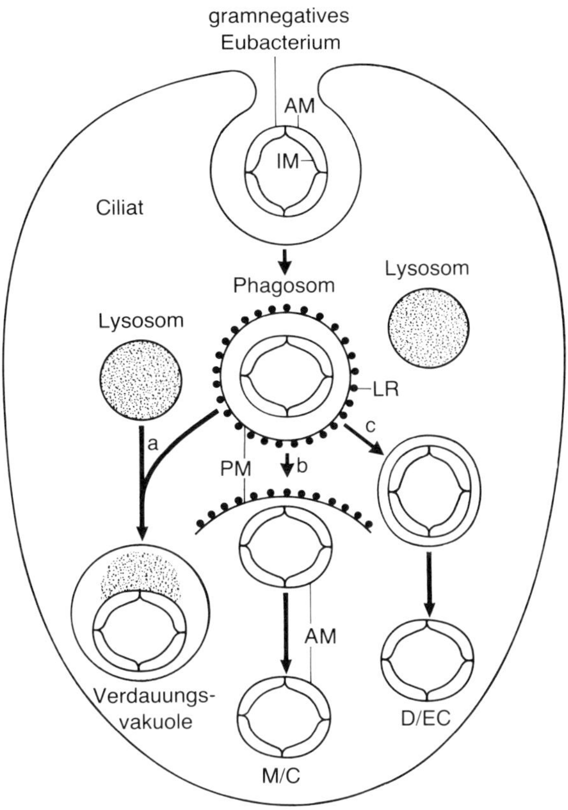

Fig. 6.1. Phagocytosis and development of endocytobionts from bacteria in ciliates. OM, outer bacterial membrane; IM, inner bacterial membrane; LR, lysosomal receptor; PM, phagosomal membrane; M/C, mitochondria–chloroplast-type of cell organelle; D/EC, *Euglena*–chloroplast-type of cell organelle. a, Fusion between lysosome and phagosome; b, rupture of phagosomal membrane; c, loss of lysosomal receptors. (Based on Cavalier-Smith and Lee, 1985.)

include the so-called 'killer function'. The 'particle' produces a toxin and the *Paramecium* can kill other, non-endocytobiotic, paramecia using this toxin. Direct cell contact appears to be needed for this effect. The toxins are temperature-labile, which is in some conditions unfavourable for the host. In the endocytobiont-containing *Paramecium* subcultured in bacteria-free medium kappa-'particles' divide faster than the host, resulting in host cell death. Thus, these 'particles' resemble lytic phages more than endosymbionts, although they have been confirmed to be bacteria. The endocytobionts may be localized in the host cytoplasm, macronucleus or micronucleus. 'Particles' have been shown to be essential for host cell growth in a few cases (Table 6.1). *Lyticum flagellatum* receives essential biopterin and folic

Table 6.1 Endocytobionts in ciliates. (Based on Heckmann *et al.*, 1983; Preer and Preer, 1984)

Endocytobiont (Particle)	*Host species*	*Compartment*	*Function and characteristics*
Caedibacter taeniospiralis (kappa)	*Paramecium aurelia* complex	Cytoplasm	Killer function
Caedibacter paraconjugatus (mu)	*Paramecium aurelia* complex	Cytoplasm	Killer function after cell contact
Holospora undulata (omega)	*Paramecium caudatum*	Micronucleus	
Holospora caryophila (alpha)	*Paramecium aurelia* complex	Macronucleus	
Holospora obtusa (iota)	*Paramecium caudatum*	Macronucleus	
Lyticum flagellatum (lambda)	*Paramecium aurelia* complex	Cytoplasm	Killer function
Lyticum sinuosum (sigma)	*Paramecium aurelia* complex	Cytoplasm	Killer function
Pseudocaedibacter minutus (gamma)	*Paramecium aurelia* complex	Cytoplasm	Killer function
Tectibacter vulgaris (delta)	*Paramecium aurelia* complex	Cytoplasm	
N.N. (omikron)	*Euplotes aediculatus*	Cytoplasm	Essential for the growth of *Euplotes*
N.N. (omikron-like)	*Euplotes eurystomus*	Cytoplasm	Essential for the growth of *Euplotes*
N.N. (epsilon)	*Euplotes minuta*	Cytoplasm	Killer function
N.N. (eta)	*Euplotes crassus*	Cytoplasm	Killer function
N.N. (xenosomes)	*Parauronema acutum*	Cytoplasm	Killer function in some strains

acid from its lambda-'particles'. Omicron-'particles' are obligate symbionts in *Euplotes* strains which cannot divide when the omicron-'particles' are removed with penicillin treatment (Heckmann, 1983). Omicron symbionts are different from other 'particles' as they contain nucleoids, which vary in number between 2 and 20. This species, therefore, has been classified in the new genus *Polynucleobacter* as *P. necessarius* (Heckmann and Schmidt, 1987). There are between 900 and 1000 *Polynucleobacter* per *Euplotes* cell. Total DNA content of a cell is around 6×10^{-3} pg. This is equivalent to 3.5×10^9 Da. Thus it can be calculated that one *Polynucleobacter* cell carries seven copies of its DNA and 0.5×10^9 Da per nucleoid. This is about 25% of the genome of *E. coli*. Lambda-symbionts from *Lyticum flagellatum* contain 10–19 copies of a $4–7 \times 10^8$ Da long genome (Soldo and Godoy, 1973; Soldo, 1987). In protozoans the development resulting in endocytobiosis begins with phagocytosis and the capture of the phagocytosed bac-

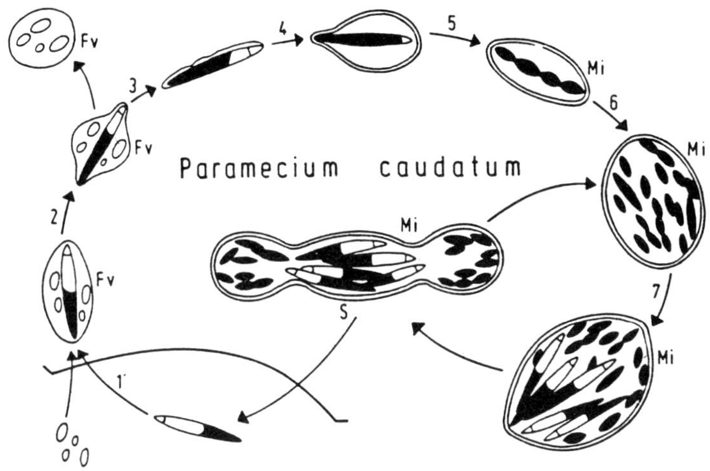

Fig. 6.2. Infection and developmental cycle of *Holospora elegans*. 1, From the culture medium the long infectious form is ingested into a food (= digestive) vacuole (Fv) together with prey bacteria. 2, The digestive vacuole shrinks and, due to the length of the *Holospora*, the bacterium evaginates. The bacterium starts a differentiation process finally leading to the development of the short reproductive form. 3, The symbiont separates from the digestive vacuole. In the host cytoplasm it is surrounded by host membranes. 4, The symbiont enters the micronucleus (Mi). 5, Inside the micronucleus the symbiont constricts at several points, divides, and so develops into the reproductive form. 6, The reproductive form multiplies by binary fission. 7, Some bacteria grow longer and develop into the infectious form again. When the host micronucleus divides, the reproductive forms are transported into the daughter nuclei, while the infectious forms remain within the connecting piece (S). Later the infectious forms are released into the external medium. (Based on Reisser *et al.*, 1985.)

terium in a phagosome. Fusion of this membrane with a lysosome results in a digestive vacuole. This fusion needs the participation of specific lysosome receptors on the phagosome. If the phagosome is destroyed before this fusion, the bacteria can escape into the host cytoplasm and, lacking lysosome receptors, cannot be transported into the digestive vacuole. The bacterium can multiply and establish itself as symbiont or 'particle'. A further possibility is that the lysosome receptors on the phagosome are lost, or defective, or are overcome by a bacterial factor. These bacteria can then also replicate as symbionts, but enclosed in a membrane, comparable to the symbiosome membrane of legume nodules (Chapter 3) or the perialgal membrane (Chapter 8). In *Holospora* the particles colonizing macro- and micronucleus exhibit a 2-part developmental cycle (Fig. 6.2) (Görtz, 1986). A longer infection form, up to 20 μm long, is taken up together with bacterial prey, into the phagosome. This phagosome becomes smaller and lets the particle bacteria out, while the prey remains in the vesicle and is digested. The symbiont then appears to be more or less specifically (depending on *Holospora* sp.) taken up into the micro-nucleus (Görtz and Dieckmann, 1980; Mikami and Görtz, 1985) or macro-nucleus (Fuhishima and Görtz, 1983). Thus receptors have been postulated to exist in the nucleus membrane. The entry of the bacteria into the nuclei may use two different mechanisms. A fusion of the symbiosome membranes with the nuclear membranes, or bacteria leaving first their additional envelope and penetrating the nuclear membrane directly have been observed (Görtz and Wiemann, 1989). Microtubules may be involved in the entry of the bacteria into the nuclei, since in the area of the entry many intranuclear microtubules have been observed (Görtz, 1986). The 20 μm infectious form of the 'particles' divides into 5–7 cells of 2–3 μm, the reproductive form. Some of these differentiate again into the long form whereupon they escape into the surrounding medium and the infection process is repeated. During the differentiation of *Holospora obtusa* from the infectious form to the reproductive form the periplasm is reduced and the cytoplasm expands. This development is correlated with the gradual disappearance of nine major soluble proteins (Görtz *et al.*, 1990). In the macronucleus as well as in the micronucleus, *Holospora* as other endonucleobionts are not enveloped by a host membrane, whereas the majority of endocytobionts in the host

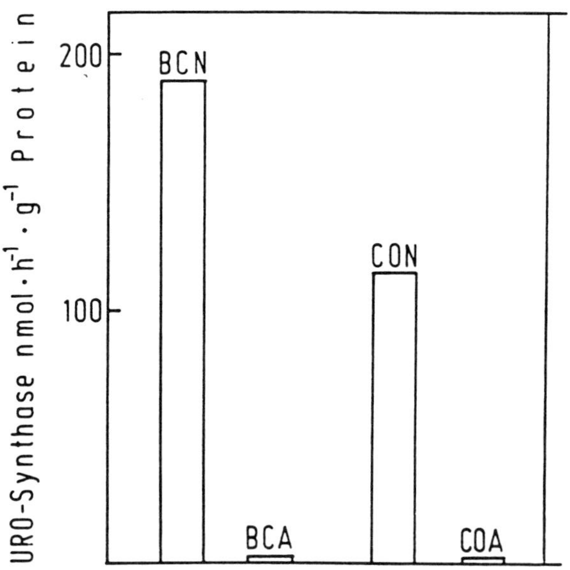

Fig. 6.3. Specific activity of uroporphyrinogen-I (URO) synthase in the flagellates *Blastocrithidia culicis* and *Crithidia oncopelti*. BCN, *Blastocrithidia* with symbionts; BCA, symbiont-free *Blastocrithidia*; CON, *Crithidia* with symbionts; COA, symbiont-free *Crithidia*. (Based on Chang *et al.*, 1975.)

cytoplasm are surrounded by a host membrane (Görtz, 1986). Within 30 min after release of the infectious forms from the micronucleus into the cytoplasm, the bacteria are enveloped into one or even two membranes (Wiemann and Görtz, 1989).

The freshwater amoeba *Pelomyxa palustris* contains at least two types of bacterial endosymbionts. One of these endosymbionts belongs to the methanogenic bacteria and is a strictly anaerobic prokaryote. About 2×10^4 methanogenic bacteria (archaebacteria) are found in one host cell. A close contact of the bacteria and microbodies (hydrogenosomes) has been observed (Lersten, 1975; Stumm and Zwart, 1987). Whether this metabolic type of symbiont has anything to do with the fact that *P. palustris* lacks mitochondria, is unknown.

The metabolic relationship between endosymbiotic bacteria and trypanosome flagellates from insects is better researched. These parasitic protozoans, known as haemoflagellates, generally show defects in their haem biosynthesis so that haemin or haemoglobin are essential for growth. Some haemoflagellates, e.g. *Blastocrithidia culicis* and *Crithidia oncopelti* are able to synthesize haem with the help of endosymbiotic bacteria (Chang *et al.*, 1975). After removal of the prokaryotic endosymbiont with antibiotic treatment the activity of uroporphyrinogen synthase I is reduced to 3% of the value found in flagellates containing symbionts (Fig. 6.3).

6.3 LEAF SYMBIOSES OF THE RUBIACEAE, MYRSINACEAE AND DIOSCOREACEAE

Leaf nodules are formed on leaves of more than 400 species of the genera *Psychotria*, *Pavetta*, *Neorosea* (Rubiaceae) and *Ardisia* (Myrsinaceae) (Miller *et al.*, 1983a,b, 1984). *Pavetta* sp. seems to have the evolutionary oldest form of these leaf symbioses (Lersten, 1975). The identity of the bacterial endophyte is still controversial. After a number of other previous designations (*Bacillus follicola*, *Xanthomonas hortoricola*, *Chromobacterium lividum*, *Flavobacterium* sp.) Knösel (1962) proposed the genus *Phyllobacterium* isolated from mature leaf nodules. *Phyllobacterium* belongs to the family Rhizobiaceae and is a Gram-negative, aerobe rod with polar flagella and chemo-organotrophic metabolism. The G+C content is 60–61 mol %, a narrow range within that of the genus *Rhizobium* (59–64 mol %). It has also been argued (Miller, 1990) that these isolates might be contaminants, since the pleomorphic endophytes might be not viable from this plant stage. However, from *Rhizobium* and *Bradyrhizobium* it is known that under appropriate conditions, up to 90% of the bacteroids can redifferentiate to colony-forming cells. Nevertheless, it may be that the true identity of the endosymbionts still has to be confirmed, considering that no free-living form may exist, due to a continuous life cycle with the plant development

Fig. 6.4. (a) Leaves of *Psychotria bacteriophila* with leaf nodules. Magnification ×0.9. (b) Leaf of *Pavetta indica* with leaf nodules. Magnification ×1.4. (Original photographs, A. Wolff, Marburg.)

transferred through the flowering stage and the seeds (Miller, 1990). Nitrogen fixation does not occur in either pure culture or in leaf nodules. Early reports of this were due to contamination with other N_2-fixing bacteria on the leaf surface.

Leaf nodules are darker than the surrounding leaf tissue and, in *Psychotria* (Fig. 6.4a) and *Pavetta* (Fig. 6.4b), relatively evenly spread over the

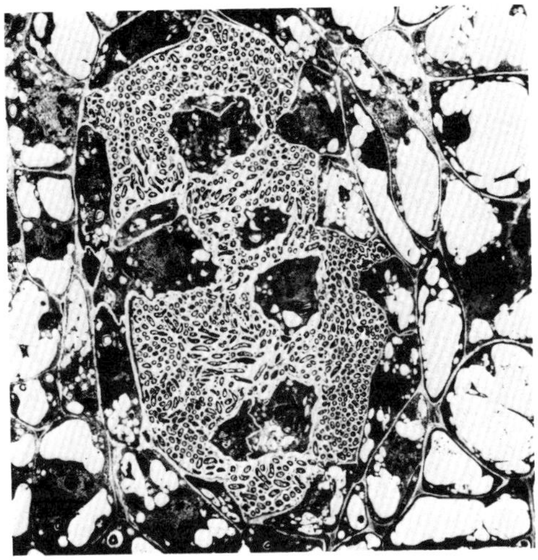

Fig. 6.5. Cross-section through a young leaf nodule of *Psychotria kirkii*. Magnification ×2700. (Original photograph, I. Miller, Glasgow.)

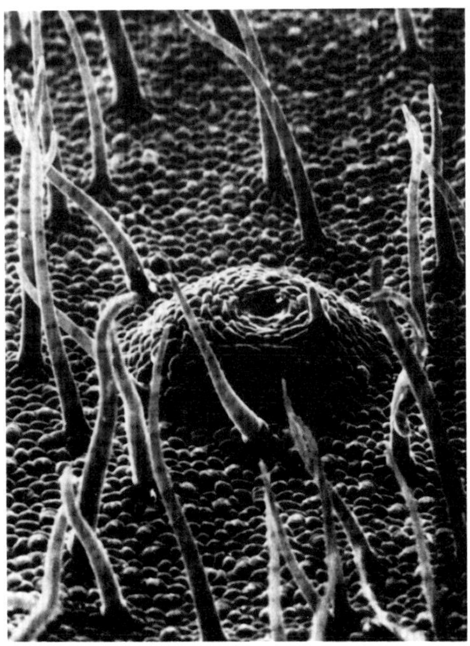

Fig. 6.6. SEM photograph of the surface of a leaf nodule from *Psychotria kirkii*, surrounded by multicellular leaf hairs. Magnification ×110. (Original photograph, I. Miller, Glasgow.)

leaf surface. Before the development of leaves, the phyllobacteria are located in the slime of the hair glands in the shoot tip. From here stomata of single young leaves are infected, resulting in a leaf nodule. The cells surrounding the stomata enter periclinal division. The bacteria divide and fill the growing leaf nodule (Fig. 6.5). Bacteria are surrounded by small, partitioning cells which are rich in cytoplasm. Some of these boundary cells

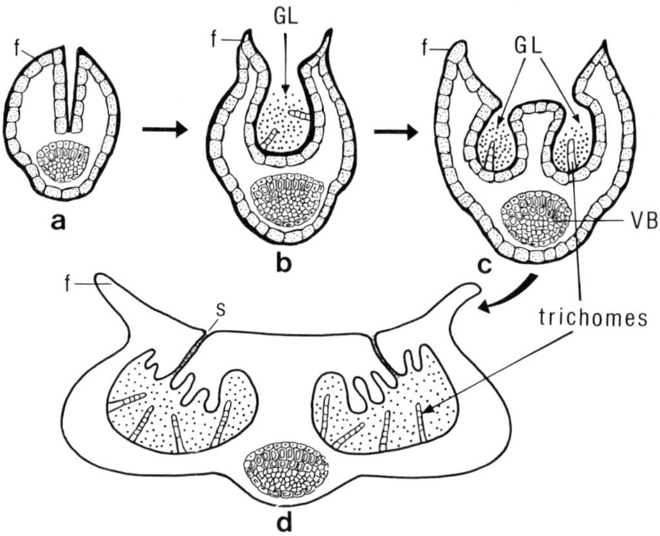

Fig. 6.7. Diagrammatic representation of stages in the development of an acuminate leaf gland of *Dioscorea sansibarensis*. The region marked VB in each diagram represents the main vascular bundle. (a) Transverse section through a very young developing leaf acumen. An invagination or split occurs from the upper epidemis leaving two flanges (f) of tissue on either side. (b) With further development the invagination deepens and widens to become a flask-shaped cavity (GL). Secretory trichomes begin to develop from the inner surface of the gland while the flanges (f) of tissue begin to elongate. (c) The central portion of the floor of the flask-shaped cavity elevates to the level of the upper epidermis forming a ridge of tissue which bisects the original cavity and as such gives rise to two young glands (GL). The inner epidermes of the tissue flanges (f) begin to bulge and grow inward towards the raised ridge of tissue. The trichomes continue to develop and secrete mucilage. (d) At maturity the epidermis of the raised ridge of tissue and the epidermis of the inwardly growing flange (f) become closely appressed as an occluded slit (s) sealing the lumen of the gland (GL) off from the external environment. The epidermis of the flange and raised ridge become highly convolute giving the roof of the gland a very irregular appearance. The trichomes, which are secretory and vermiform in nature, fill the gland with mucilage in which the symbiotic bacteria multiply. Infection of the gland with bacteria presumably takes place when the gland is open to the external environment but yet when there are trichome secretions to attract and maintain them: as such infection must occur at the stages in development represented in (b) and (c). (From Miller, 1990.)

become loose and fall into the bacterial mass. At full growth the nodule measures 750–1500 μm in diameter and stands out from the leaf surface, partially covered by multicellular hairs (Fig. 6.6). Cells of the stomata remain morphologically intact, but are no longer able to function. The rod-shaped bacteria become pleomorphic during nodule ageing. They are embedded in polysaccharide, derived from the Golgi of the surrounding host cells.

Bacteria-free seedlings of *Pavetta* and *Psychotria* develop normally only until the fourth pair of leaves. Further growth and development stops and the plants die. Bacteria-free seedlings of *Ardisisia* develop only to the second leaf stage. It can be concluded from this that these are obligate symbioses. A hypothesis has been developed that the host plants are cytokinin-deficient species, which are dependent, for normal development, on the supply of these phytohormones by the endophytic bacteria (Miller, 1990).

The special development of the leaf glands of *Dioscorea* (Dioscoreaceae) and infection with endophytic bacteria is described in Fig. 6.7. The endophytic bacteria may be different from those in the obligate symbiosis within the Rubiaceae and Myrsinaceae.

6.4 INSECTS, WORMS AND MUSSELS AS HOSTS

Endobiotic bacteria are widespread in insects (Dasch *et al.*, 1984). This appears mostly in those insects which suck blood or plant sap or feed on cellulose. These bacteria are poorly researched, this is partly due to their resistance to isolation and cultivation. Morphologically they are mostly rods. Cocci, spirals and pleomorphs are also observed. G+C content is between 25 and 70 mol%, thus it is apparent that the endobionts are taxonomically widely scattered. The symbionts are mostly found in special cells, the mycetocytes, or in special organs, the mycetomes. Some insects host up to six different symbionts at the same time. The passing on of these symbionts from generation to generation must be a very complex and specialized process. Methylotrophic bacteria have also been identified as intracellular symbionts in deep sea mussels. They are located in specialized gill epithelial cells, surrounded by a peribacterial membrane (Cavanaugh *et al.*, 1987).

6.4.1 Cockroaches and termites

The cockroach *Blatta orientalis* hosts a symbiotic bacterium *Blattabacterium cuenotii* in its fat bodies (Table 6.2). The microsymbiont synthesizes a peptide growth factor. Lack of this leads to dramatically reduced development, growth and fertility rates in the host. Pigmentation with melanin is also symbiont-dependent. This is probably due to the stimulation of

Table 6.2 Symbiotic bacteria in insects. (Based on Dasch *et al.*, 1984 and Potrikus and Breznak, 1977)

Insect	*Symbiont/Host tissue*	*Functional relationship*
Blatta orientalis (Cockroach)	*Blattabacterium cuenotii* in mycetocytes of the fat body	Synthesis of growth factors (peptides) by the symbionts
Reticulitermes flavipes (Termite)	*Enterobacter agglomerans* in the hind-gut	N_2-fixation, metabolism of reduced N by cellulolytic bacteria and by the host itself
Sitophilus (*Calandra*) *oryzae* (Rice beetle)	Rod-shaped bacteria, G + C content 55 mol%, in mycetocytes of the foregut	Generally enhanced nutrient efficiency for the insects in symbiosis
Glossina morsitans (Tsetse fly)	Rod-shaped bacteria of 3–9 μm length, in epithelial cells of the gut	Synthesis of B vitamins by the symbionts
Pediculus humanus (Human headlouse)	Rod-shaped bacteria, in certain stages extremely long (up to 30 μm), in gut mycetomes	Synthesis of pyridoxal phosphate and pantothenic acid by the symbionts
Acyrthosiphon pisum (Pea aphid)	Gram-negative bacteria, often pleomorphic, in mycetocytes in the abdominal haemocoel	Synthesis of lipids, especially cholesterol

tyrosine production. The termite *Reticulitermis flavipes* contains a symbiotic *Enterobacter agglomerans*, which can be cultivated in pure culture. It fixes atmospheric N_2 in the gut of the termite (Potrikus and Breznak, 1977). This fixation is much higher in the larval stage than in the adult, and is correlated with a high protein synthesis and N-demand. The curing of termites from gut bacteria using antibiotics leads to the death of gut protozoans also. Trichomes in the intestine of termites are often covered with bacteria (Fig. 6.8). Apart from N_2-fixing and cellulose-degrading bacteria, termite intestines often contain methanobacteria. These produce CH_4 from H_2 which is set free by the fermentation of cellulose in a manner similar to that described in ruminants (Section 6.6). Also H_2-oxidizing and CO_2-reducing acetogenic bacteria, such as *Sporomusa termitida* sp. nov., have been characterized from gut contents of *Nasutitermes nigriceps* (Breznak *et al.*, 1988).

6.4.2 Beetles

The use of nutrients in the rice-eating beetle, *Sitophilus* (*Calandra*) *oryzae* is improved by symbiosis with rod-shaped bacteria in the mycetocytes.

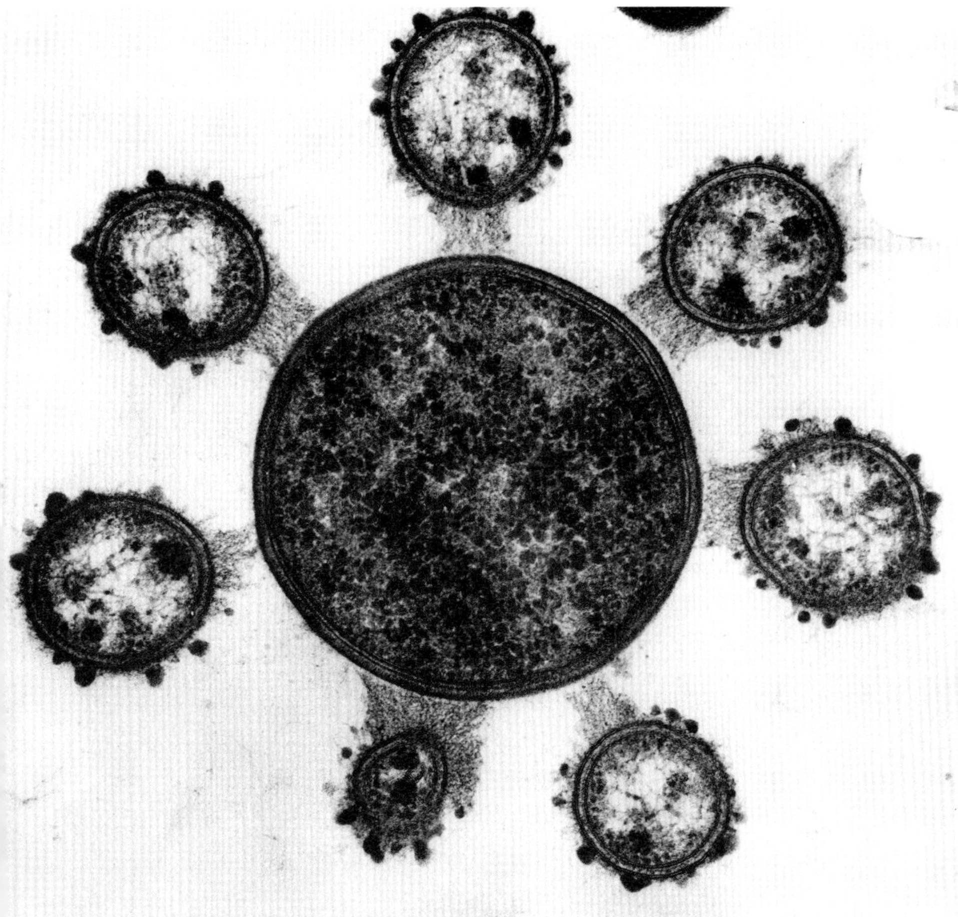

Fig. 6.8. Cross-section through a trichome in the gut of *Reticulitermis flavipes* with regular arrangement of bacteria. (Original photograph, J.A. Breznak and H.S. Pankratz, Michigan.)

Growth of the 4 mm long beetles at 33°C results in cured individuals. These are smaller, less fertile and less pigmented than host beetles. They are restricted in their use of natural sterols as food. The symbiotic bacteria in *Sitophilus oryzae*, localized free in the host cytoplasm, also supply their host with vitamins such as panthotenic acid, biotin and riboflavin (Heddi *et al.*, 1991). Another ectosymbiosis is the association between beetles like *Xyloterinus politus* (Haanstadt and Morris, 1985), *Monarthrum mali* or *Syleborus affinis* with the fungi *Ascoidea asiatica*, *Ambrosiella hartigii*, *Raffaelea ambrosiae* or *Cephaloascus fragrans*. The beetles are wood-boring and the female animal carries the fungus in the mycangia of the prothorax.

These animals inoculate wood fibres in their galleries, and the larvae are fed from the resulting mycelium and associated microorganisms. Obligate associated microflora isolates are the yeasts *Candida* and *Pichia* as well as bacteria from the genera *Flavobacterium*, *Alcaligenes* and *Gluconobacter*. These can be isolated from both beetles and galleries.

6.4.3 Tsetse flies, lice and aphids

Rod-shaped symbionts in the epithelium of tsetse fly (*Glossina morsitans*) middle gut produce excesses of B vitamins. The elimination of symbionts with antibiotics or antiserum specific to the symbiont leads to a degeneration of the ovaries in the fly. This can be avoided by a B vitamin-enriched diet. A similar function is served by symbionts in the human head louse *Pediculus humanus* (Table 6.2) which produce pyridoxalphosphate and pantothenic acid for the host. The pea aphid *Acyrthosiphon pisum* needs lipids, especially cholesterol, for growth, but it feeds on plant juices which contain only very low concentrations of lipids. The aphid overcomes this problem by playing host to pleomorphic symbionts which are able to synthesize such compounds. Adult aposymbiotic aphids excrete more amino acids (glutamine and asparagine) than symbiotic aphids (Sasaki *et al.*, 1990).

In the endosymbionts from the pea aphid *Acyrthosiphon pisum* a predominant protein, symbionin (mol. wt. 63 000 Da), has been found to be homologous to an *Escherichia coli* heat shock protein (Hara *et al.*, 1990).

6.4.4 Worms

An obligate symbiosis between the deep sea worm *Riftia pachyptila* and sulphur oxidizing bacteria has been described (Cavanaugh *et al.*, 1981). Despite having no digestive tract and living on deep sea volcanic outcrops, these worms grow to a length of 150 cm and are 4 cm wide. They grow rapidly (Roux *et al.*, 1989) and achieve a population density of up to 10 kg biomass m^{-2}, take up sulphide from volcanic springs and transfer this to a special symbiotic tissue, the trophosome. The trophosome comprises about 15% of the volume of the tubeworm with a bacterial density of 3.7 – 10 $\times 10^9 \times 10^9$ bacteria g^{-1} tissue (Powell and Somero, 1986). Their symbiotic chemolithotrophic bacteria oxidize sulphide, reduce CO_2 and deliver organic compounds to the animals' circulatory system. In trophosome preparations CO_2 fixation was only stimulated by sulphide and not by thiosulphate, hydrogen or ammonia (Fisher *et al.*, 1989).

Annelids in warm, sulphated water also exhibit similar symbioses. The worms have neither gut tissue nor excretion organs. They are subcutaneously infected with sulphur-bacteria.

A nitrogen-fixing and cellulolytic bacterium has been isolated from specialized glands of the shipworm (*Psiloteredo healdi*) which has strong adhesive properties (Imam *et al.*, 1990).

6.5 THE SYMBIOTIC LIGHT ORGANS OF FISH AND SQUID

Fish of various genera such as *Photoblepharon*, *Leiognathus*, *Monocentris*, *Gazza* or *Secutor* produce light by using symbiotic bacteria (Nealson and Hastings, 1979). These bacteria are concentrated in special light organs which may be in various locations, e.g. under the eye as in *Photoblepharon palpebratus* or on the tip of the lower jaw as in *Monocentris japonicus*. The symbionts are *Photobacterium leiognathi* (in fish from shallow tropical waters), *Photobacterium phosphoreum* (in deep sea fish) or *Vibrio fischeri* (previously called *Photobacterium fischeri*). All these bacteria can be isolated from the surface of other marine animals or from sea water. The light organ of *Monocentris japonicus* can be described as a modified exocrine organ (Tebo *et al.*, 1979). It contains an epithelial tube system, in which the symbiotic bacteria are in a fluid (Fig. 6.9). The base of each tube is sealed with a multilayered epithelium, the mouth with a monolayer of tissue. Many tubes join together at a communal exit to the organ. The base of the tissue is packed with blood vessels and the host cells contain many cristae-rich mitochondria. The blue–green light emitted by *Photobacterium* (max

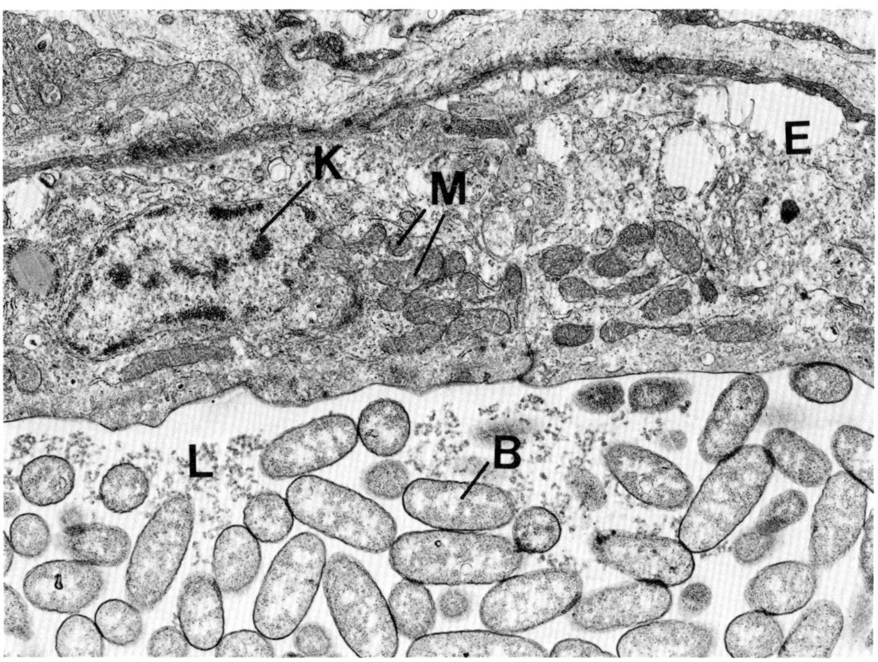

Fig. 6.9. Transverse section of the tubular system in the light organ of *Monocentris japonicus*. B, photobacteria in the tube lumen (L); E, epithelial cell in tube wall; M, mitochondria in epithelial cell. Magnification ×8900. (Original photograph, K.M. Nealson, Milwaukee.)

496 nm) is inside the spectrum which fish see and respond to (Kessel, 1977). Light is produced by the direct oxidation of flavin mononucleotide and a long-chain aldehyde with O_2, without cytochromes taking part, and is catalysed by the enzyme **luciferase**. The long-chain fatty acid formed is reduced again to the aldehyde with NADPH and ATP. Some strains of *Photobacterium* produce luciferase constitutively whereas others induce the enzyme in an O_2-dependent fashion. Luciferase is a well-known enzyme responsible for light production in glow-worms etc. Since the bacteria in a light organ ($5-9 \times 10^9$ bacteria per ml fluid) are remarkably homogeneous (Ruby and Nealson, 1976), it is normally assumed they are cloned from a single organism after initial infection. On the other hand, bacteria can easily be released from the organ into the surrounding water. The mechanisms of infection and the prevention of infection with other bacteria are unknown.

In the darkness of deep seas, the light organ can attract prey or show the host where prey is. Enemies may also be warned off. Light signals may also function as communication, so fishes may find each other for mating etc. Light signals are not simple emittors, but may be modified by the fish. *Photoblepharon* can draw an 'eyelid' over its light organ, cutting off emission. *Anomalops* may turn its light organ over, exposing a darkly pigmented underside. Thus this symbiosis primarily affects the behaviour of the host.

The photobacteria belong to the family Vibrionaceae and exhibit both aerobic and fermentative chemoorganotrophic metabolism. Few compounds besides glucose are metabolised. Glucose and O_2 are supplied to the bacteria in controlled quantities by the blood circulatory system of the host animal.

Vibrio fisheri was identified as the symbiont in the light organ of the sepiolid squid *Euprymna scolopes*. Cultivated outside the light organ association, the maximal luminescence was reduced by a factor of 1000 (Boettcher and Ruby, 1990).

6.6 SYMBIOSES IN THE DIGESTIVE TRACT OF RUMINANTS

The digestive tract of grazing animals accounts for between 8 and 15% of the total body weight. For the 1.2 billion domestic cows in the world this means a rumen volume of 10^{11} litres. With $10^{13}-10^{14}$ bacteria l^{-1} and 10^8-10^9 protozoans l^{-1}, this is one of the most highly populated and economically important biotopes known (Bryant, 1977; Hobson and Wallace, 1982).

6.6.1 Bacteria and phycomycetes

Rumen bacteria are mostly Gram-negative. Their most important biochemical role for the host is the digestion of cellulose to succinate and acetate. A dominating species is *Bacteroides succinogenes*, but *Ruminococcus albus*, *Clostridium lochheadii* and *Butyrivibro fibrisolvens* are also well represented (Table 6.3). Cellulose is also metabolized by Phycomycetes (Chytridiomycetes) such as *Piromonas communis*, *Neocallimastix frontalis* and *Sphaeromonas communis* (Orpin and Letcher, 1979; Orpin and Munn, 1985). The zoospore stage of these species often used to be mistaken for flagellates. The developmental cycle from zoospores to zoosporangia takes about 24–32 h (Lowe *et al.*, 1987). The zoosporangium of the fungi is supported by a branched non-septate rhizomycelium. The fungi have also been isolated from the saliva and faeces (of sheep), suggesting different routes of transfer of anaerobic fungi between animals (Lowe *et al.*, 1987b). Chitinase activity has been used to assess the fungi associated with fibres during forage degradation in ruminants (Akin, 1987). An inhibition of the bacterial population increases the fibre colonization by the fungi (Akin and Rigsby, 1987). The temperature optimum for carboxymethylcellulase and xylanase from these fungi is surprisingly high at 50°C (Lowe *et al.*, 1987a).

The organisms in the rumen show complex metabolic relationships to each other, many using end products or excretions from others. *Bacteroides ruminicola* and *Selenomonas ruminantium* can form up to 20% of the gut bacteria. They ferment succinate and cellobiose to propionate. *B. ruminicola* needs B-vitamins and a peptide growth factor to replicate. This peptide factor is produced by *Clostridium* spp. as a degradation product from protein. N_2-fixation plays no significant role in this nutrient-rich environment. Proteolysis results in the presence of free methionine, an essential growth factor for *Ruminococcus flavefaciens.* Plant secondary products, such as flavones which can be toxic for bacteria, are metabolized by *Butyrivibrio fibrisolvens* (Table 6.3). *Anaerovibrio lipolytica* is specialized in the meta-

Table 6.3 The rumen symbiosis: microorganisms associated with ruminants

Organism	*Substrates*	*Products*	*Growth factors*	*Dependent on*
Bacteroides succinogenes	Cellulose, cellobiose NH_3	Succinate, acetate, CO_2	Isobutyric acid, isovaleric acid biotin, *p*-aminobenzoic acid, valeric acid	*Clostridium* sp., *Bacteroides ruminicola*
Ruminococcus albus	Cellulose, cellobiose, xylose, CO_2	Succinate, acetate, CO_2, H_2	Isobutyric acid, isovaleric acid	*Bacteroides ruminicola*
Ruminococcus flavefaciens	Cellulose, cellobiose, pectins, xylans	Succinate acetate, CO_2, H_2	Methionine	*Bacteroides ruminicola*
Clostridium lochheadii	Cellulose, starch, proteins	Amino acids and their deamination products		

Table 6.3 *Continued*

Organism	*Substrates*	*Products*	*Growth factors*	*Dependent on*
Butyrivibrio fibriosolvens	Cellulose, starch, pectins, lecithin, flavones (rutin), urea	Butyrate	B-vitamins	
Bacteroides ruminicola (*Ruminobacter ruminicola*)	Starch, pectins, sucrose, succinate, isobutyrate, proteins, urea	Propionate, methionine, valine	Haem compounds, B-vitamins, methionine, cysteine, peptides or ammonia	*Bacteroides succinogenes*, *Ruminococcus albus*, *Clostridium* sp.
Selenomonas ruminantium	Cellobiose, succinate nitrate	Propionate		*Bacteroides succinogenes*, *Ruminococcus albus*
Treponema bryantii	Cellobiose, glucose + CO_2	Acetate, formate, succinate		
Lachnospira multiparus	Pectins	Unsaturated galacturonic acid dimers		
Methanobacterium ruminantium	CO_2, H_2, formate, for anabolism: acetate	Methane	2-Mercapto-ethanesulphonic acid	*Ruminococcus*, *Treponema*
Anaerovibrio lipolytica	Phospholipids, triglycerides	Propionate, succinate glycerol		
Vibrio succinogenes	H_2	Succinate fumarate		
Piromonas communis, *Neocallimastix frontalis*, *Sphaeromonas communis* (Anaerobic phycomycetes)	Cellulose, xylans			
Entodinium caudatum (Protozoa)	Starch, bacteria, amino acids	Biomass, amino acids		
Dasytricha (Protozoa)	Bacteria	Biomass		
Isotricha (Protozoa)	Bacteria			
Polyplastron multivesiculatum (Protozoa)	Bacteria and other ciliates			*Epidinium*, *Eudiplodinium*
Anaerobic mycoplasmas	*Butyrivibrio* *Ruminococcus albus*			
Anaerobic Gram-negative rods	Oxalate	CO_2, formate		

bolism of phospholipids, triglycerides and glycerol. Formate produced by *Treponema bryantii* is used by *Methanobacterium ruminantium* with CO_2 and H_2, to produce CH_4. Close attachment of methanogenic bacteria to entodiniomorphid ciliates has been observed (Stumm and Zwart, 1987; Gijzen *et al.*, 1991) (Fig. 6.10).

The digestion of polymer substrates is mostly by exoenzymes. Cultures of *Ruminococcus albus* produce four different exocellulases. The plasmid pattern of this species is correlated with the adaptation to glucose, cellobiose or cellulose as carbon source (Hazlewood and Teather, 1988; Ohmiya *et al.*, 1989). *Lachnospira multiparus* exports extracellular polygalacturonic acid lyase. This results in the production of mostly dimerized unsaturated galacturonic acids. The plant cell wall contains at least eight different polysaccharides (homogalacturonan, rhamnogalacturonan I and II, xyloglucan, arabinogalactan, glucuronarabinoxylan and two other components). Thus cell wall-degrading bacteria must be able to produce many different specific hydrolytic enzymes. Phytopathogenic bacteria produce cell wall-degrading enzymes in the order pectinase, hemicellulase and cellulase. This order is similar to that found in ruminant digestive tracts. Before cellulolytic bacteria attack chewed plant material, tissue with a high pectin content (e.g. white clover leaves) is softened and broken by *Lachnospira multiparus*. Silicates and wax inhibit degradation. Thus, whereas long epidermis cells can be digested from inside, silicated short cells cannot. The colonization of the cell wall structures with bacteria is accomplished within a few minutes, probably by chemotactic responses to released soluble sugars. Attachment is facilitated by bacterial capsular polysaccharide. Experiments with gnotobiotically grown sheep, whose intestines had been infected with the 11 most important gut bacteria, resulted in normal growth for 4 months. After this time most of the animals died. It is thus postulated that further bacteria are essential for ruminants. These bacteria would be present in very small numbers and be continually taken up by the animal.

6.6.2 Ciliates

Rumen protozoans are mostly ciliates belonging to the groups of Holotricha and Entodiniomorpha. The holotrichs utilize primarily soluble carbohydrates, the entodiniomorphid ciliates ingest particulate matter (Williams, 1986). The most important holotrich species are *Isotricha intestinalis*, *Isotricha prostoma* and *Dasytricha ruminantium*. *Isotricha* sp. can utilize starch, but maltose and cellobiose are not effective substrates. *Dasytricha ruminantium* can utilize sucrose, fructose, glucose, raffinose, galactose and inulin. The end product of fermentation by all holotrich species is mainly lactic acid, with much less butyric acid and acetic acid, and only traces of propionic acid. A large percentage of the glucose taken up is stored in the cells as homoglucan with 22 glucose units and a structure, similar to amylopectin. The holotrich ciliates seem to be less sensitive to low oxygen concentrations than other members of the rumen habitat. The protozoans regulate the free sugar concentration, and therefore the rate of bacterial fermentation. Their protein needs are catered for by preying on bacteria, and thus regulating their numbers. The ciliates are not thought to be dependent on specific bacterial species. *Isotricha* sp. can engulf bacteria (e.g. *E. coli*) at a rate of 3000 cells per h for 24 h (Coleman, 1986b). Some ciliates like *Polyplastron multivesiculatum* prey not only on bacteria, but also on other ciliates like *Epidinium* and *Eudiplodinium.* Without protozoans, bacterial titres in the gut increase significantly. Mycoplasmas and bacteriophages also contribute to the control of bacterial numbers. The digestive tract functions well without the presence of ciliates. Feeds containing a high polysaccharide content lead to conditions in the stomachs where ciliate growth is inhibited. Earlier reports that ciliate protein and essential amino acid contents are generally higher than that of gut bacteria has not been reproduced. Growth rates in ciliate-free animals can be better or worse than in control animals, depending on the protein concentration of the feed. In defaunated animals the urea excretion was lower, whereas the formation of allantoin and bacterial nitrogen increased. Apparently a compensation of protozoa by bacteria takes place with a 2- to 3-fold increase of bacteria (Kreuzer, 1986; Kreuzer and Kirchgessner, 1986). Defaunation also leads to a reduced gaseous exchange and a reduction of heat production by about 5% (Kreuzer *et al.*, 1986). Carboxymethylcellulase activity was reduced in defaunated animals. With the cellulolytic species *Epidinium ecaudatum caudatum* and *Eudiplodinium maggii* present, 60–70% of the enzyme activity was associated with the protozoan fraction in sheep (Coleman, 1985, 1986a).

In the X-bacteria, living inside D-amoebae, large amounts of a **67-kDa protein, a groEL analogue and a heat shock protein** has been found and the gene sequenced (Ahn *et al.*, 1991). This was interpreted that the symbiotic bacteria lives under a comparable stress as cells, undergoing a heat shock.

6.6.3 Symbiotic metabolism

The end products of symbiotic metabolism in ruminant guts are acetate, propionate and butyrate. These are absorbed by the animal and used in oxidative metabolism. Methane and CO_2 are excreted as gas. Ethanol and lactate are present in unimportant quantities insomuch as H_2 is incorporated into CH_4 and cannot therefore be used to reduce pyruvate or acetaldehyde. Thus the acetylphosphate can be used to synthesize one extra ATP, resulting in the increased ATP yield of symbiotic fermentation. A total of 4 mol of ATP is then produced per mol of hexose instead of 2 mol as is normal with fermentation to lactate or ethanol. The yield efficiency for

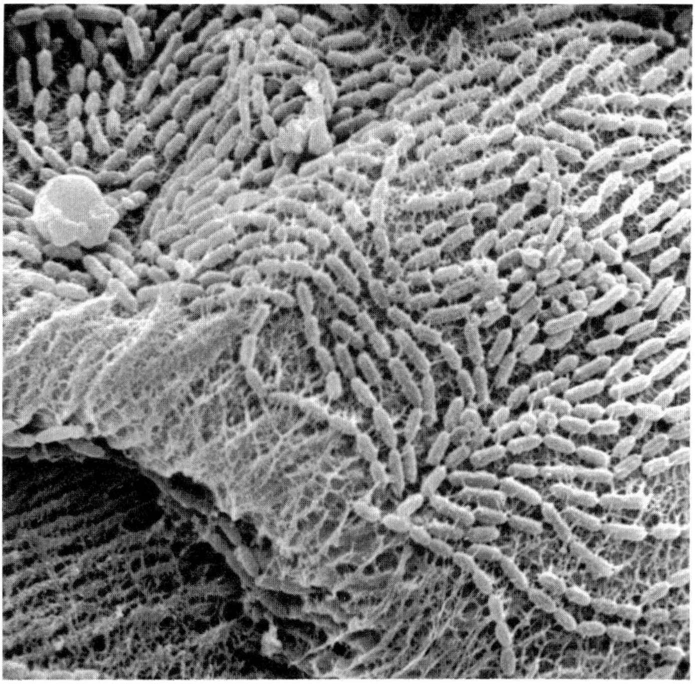

Fig. 6.10. Attachment of methanogenic bacteria to entodiniomorphid ciliates. Modified from Stumm and Zwart, 1987.

typical gut bacteria (50–60 g cells per mol glucose) is higher than for most anaerobic bacteria.

In recent years some new characteristics of gut bacteria have been discovered. For example, phospholipids from *Butyrivibrio lipolytica* contain a unique type of long-chain fatty acid; the so-called **diabol-acids**. Their function is to maintain membrane fluidity in the absence of unsaturated fatty acids. Free linoleic and linolenic acids are rapidly hydrolysed by gut bacteria. This is the explanation for the independence of the composition of animal fat from dietary fat. *Bacteroides ruminicola* uses succinate as intermediate for 2-oxoglutarate, since it does not possess citrate synthase or isocitrate dehydrogenase.

Fermentation can be artificially redirected to favour propionate users by, for example, using the ionophore monensin to inhibit *Ruminococcus* species. This indirectly reduces CH_4 production by stopping H_2-production and therefore increases the amount of carbon compounds for the animal's use.

Host animals provide masticated plant material for the symbionts. Furthermore, digestive juices provide a 0.1–0.2 M-bicarbonate buffered medium, pH 6.5, which neutralizes many metabolic products. Temperature is also held constant at 39°C in an absolutely anaerobic environment, with an O_2 concentration in ruminant stomachs under 10^{-20} M.

6.7 REFERENCES

Ahn, T.I., Leeu, H.K., Kwak, I.H. and Jeon, K.W. (1991) Nucleotide sequence and temperature-dependent expression of XgroEL gene isolated from symbiotic bacteria of *Amoeba proteus. Endocytobiosis and Cell Res.*, **8**, 33–44.

Akin, D.E. (1987) Use of chitinase to assess ruminal fungi associated with plant residues in vitro. *Appl. Environ. Microbiol.*, **53**, 1955–8.

Akin, D.E. and Rigsby, L.L. (1987) Mixed fungal populations and lignocellulosic tissue degradation in the bovine rumen. *Appl. Environ. Microbiol.*, **53**, 1987–95.

Boettcher, K.J. and Ruby, E.G. (1990) Depressed light emission by symbiotic *Vibrio fischeri* of the sepiolid squid *Euprymna scolopes. J. Bacteriol.*, **172**, 3701–6.

Breznak, J.A., Switzer, J.M. and Seitz, H.-J. (1988) *Sporomusa termitida* sp. nov., an H_2/CO_2-utilizing acetogen isolated from termites. *Arch. Microbiol.*, **150**, 282–8.

Brown, M.R.W. and Williams, P. (1985) The influence of environment on envelope properties affecting survival of bacteria in infections. *Ann. Rev. Microbiol.*, **39**, 527–56.

Bryant, M.P. (1977) Microbiology of the rumen, in *Dukes Physiology of the Rumen of Domestic Animals*, 9th edn, (ed. M.J. Swenson), Cornell University Press, Ithaca, London, pp. 287–309.

Cavalier-Smith, T. and Lee, J.J. (1985) Protozoa as hosts for endosymbioses and the conversion of symbionts into organelles. *J. Protozool.*, **32**, 376–9.

Cavanaugh, C.M., Gardiner, S.L., Jones, M.L.S. *et al.*, (1981) Prokaryotic cells in the hydrothermal vent tube worm, *Riftia pachyptila*: possible chemoautotrophic symbionts. *Science*, **213**, 340–2.

Cavanaugh, C.M., Levering, P.R., Maki, J.S. *et al.*, (1987) Symbiosis of methylotrophic bacteria and deep-sea mussels. *Nature*, **325**, 346–8.

Chang, K.P., Chang, C.S. and Sassa, S. (1975) Heme biosynthesis in bacterium–protozoon symbioses: enzymic defects in host hemoflagellates and complemental role of their intracellular symbiotes. *Proc. Natl. Acad. Sci. USA*, **72**, 2979–83.

Coleman, G.S. (1985) The cellulase content of 15 species of entodiniomorphid protozoa, mixed bacteria and plant debris isolated from the bovine rumen. *J. Agric. Sci., Camb.*, **104**, 349–60.

Coleman, G.S. (1986a) The distribution of carboxymethylcellulase between fractions taken from the rumens of sheep containing no protozoa or one of five different protozoal populations. *J. Agric. Sci., Camb.*, **106**, 121–7.

Coleman, G.S. (1986b) The metabolism of rumen ciliate protozoa. *FEMS Microbiol. Rev.*, **39**, 321–44.

Colombo, P.M. (1978) Occurrence of endophytic bacteria in siphonous algae. *Phycologia*, **17**, 148–51.

Dasch, G.A., Weiss, E. and Chang, K.-P. (1984) Endosymbionts of insects, in *Bergeys Manual of Systematic Bacteriology*, Vol. 1, (ed. N.R. Krieg), Williams & Wilkins, Baltimore, pp. 811–33.

Fisher, C.R., Childress, J.J. and Minnich, E. (1989) Autotrophic carbon fixation by the chemoautotrophic symbionts of *Riftia pachyptila. Biol. Bull.*, **177**, 372–85.

Fuhishima, M. and Görtz, H.-D. (1983) Infection of macronuclear 'anlagen' of *Paramecium caudatum* with the macronucleus-specific symbiont *Holospora obtusa. J. Cell Sci.*, **64**, 137–46.

Gijzen, H.J., Broers, C.A.M., Barughare, M. and Stumm, C.K. (1991) Methanogenic bacteria as endosymbionts of the ciliate *Nyctotherus ovalis* in the cockroach hindgut. *Appl. Env. Microbiol.*, **57**, 1630–4.

Görtz, H.-D. (1986) Endonucleobiosis in ciliates. *Int. Rev. Cytol.*, **102**, 169–213.

Görtz, H.-D. and Dieckmann, H. (1980) Life cycle and infectivity of *Holospora elegans* Haffkine, a micronucleus-specific symbiont of *Paramecium caudatum* (Ehrenberg). *Protistologica*, **16**, 591–603.

Görtz, H.-D., Lellig, S., Miosga, O. and Wiemann, M. (1990) Changes in fine structure and polypeptide pattern during development of *Holospora obtusa*, a bacterium infecting the macronucleus of *Paramecium caudatum*. *J. Bacteriol.*, **172**, 5664–9.

Görtz, H.-D. and Wiemann, M. (1989) Route of infection of the bacteria *Holospora elegans* and *Holospora obtusa* into the nuclei of *Paramecium caudatum*. *Eur. J. Protistol.*, **24**, 101–9.

Haanstadt, J.O. and Morris, D.M. (1985) Microbial symbionts of the ambrosia beetle *Xyloterinus politus*. *Microb. Ecol.*, **11**, 267–76.

Hara, E., Fukatsu, T., Kakeda, K. *et al.* (1990) The predominant protein in an aphid endosymbiont is homologous to an *E. coli* heat shock protein. *Symbiosis*, **8**, 271–83.

Hazlewood, G.P. and Teather, R.M. (1988) The genetics of rumen bacteria, in *The Rumen Microbial Ecosystem* (ed. P.N. Hobson) Elsevier Applied Science, London, New York, pp. 323–41.

Heckmann, K. (1983) Endosymbionts of *Euplotes*. *Int. Rev. Cytol.*, Suppl. 14, 111–43.

Heckmann, K., Hagen, R.T. and Görtz, H.-D. (1983) Freshwater *Euplotes* species with a 9 type 1 cirrus pattern depend upon endosymbionts. *J. Protozool.*, **30**, 284–9.

Heckmann, K. and Schmidt, H.J. (1987) *Polynucleobacter necessarius* gen. nov., sp. nov., an obligately endosymbiotic bacterium living in the cytoplasm of *Euplotes aediculatus*. *Int. J. Syst. Bacteriol.*, **37**, 456–7.

Heddi, A., Lefebvre, F. and Nardon, P. (1991) The influence of symbiosis on the respiratory control ratio (RCR) and the ADP/O ratio in the adult weevil *Sitophilus oryzae* (Coleoptera, Curculionidae). *Endocytobiosis and Cell Res.*, **8**, 61–73.

Hobson, P.N. and Wallace, R.J. (1982) Microbial ecology and activities in the rumen. *CRC Crit. Rev. Microbial*, **9**, 165–225; 253–95.

Imam, S.H., Greene, R.V. and Griffin, H.L. (1990) Adhesive properties of a symbiotic bacterium from a wood-boring marine shipworm. *Appl. Environ. Microbiol.*, **56**, 1317–22.

Kessel, M. (1977) The ultrastructure of the relationship between the luminous organ of the teleost fish *Photoblepharon palbebratus* and its symbiotic bacteria. *Cytobiologie*, **15**, 145–58.

Knösel, D. (1962) Prüfung von Bakterien auf Fähigkeit zur Sternbildung. *Zentralb. Bakteriol. Parasitenk. Infektionskr. Hyg.*, 2, **116**, 79–100.

Kreuzer, M. and Kirchgessner, M. (1986) Die Bedeutung der Protozoen im Pansen von Hammeln für Verdaulichkeit, N-Bilanz und N-Fraktionen in Kot und Harn bei unterschiedlichen Gehalten und Arten von Stärke in der Ration. *J. Anim. Physiol. Anim. Nutr.*, **56**, 217–31.

Kreuzer, M. (1986) Methodik und Anwendung der Defaunierung beim wachsenden

Wiederkäuer. *J. Vet. Med.* A, **33**, 721–45.
Kreuzer, M., Kirchgessner, M. and Müller, H.L. (1986) Effect of defaunation on the loss of energy in wethers fed different quantities of cellulose and normal or steamflaked maize starch. *Anim. Feed Sci. Technol.*, **16**, 233–41.
Lee, J.J., Soldo, A.T., Reisser, W. *et al.* (1985) The extent of algal and bacterial endosymbioses in protozoa. *J. Protozool.*, **32**, 391–403.
Lersten, N.R. (1975) Colleter types in Rubiaceae, especially in relation to the bacterial leaf nodule symbiosis. *Bot. J. Linn. Soc.*, **71**, 311–19.
Lowe, S.E., Griffith, G.G., Milne, A. *et al.* (1987) The life cycle and growth kinetics of an anaerobic rumen fungus. *J. Gen. Microbiol.*, **133**, 1815–27.
Lowe, S.E., Theodorou, M.K. and Trinci, A.P.J. (1987a) Cellulases and xylanase of an anaerobic rumen fungus grown on wheat straw, wheat straw holocellulose, cellulose, and xylan. *Appl. Environ. Microbiol.*, **53**, 1216–23.
Lowe, S.E., Theodorou, M.K. and Trinci, A.P.J. (1987b) Isolation of anaerobic fungi from saliva and faeces of sheep. *J. Gen. Microbiol.*, **133**, 1829–34.
Mikami, K. and Görtz, H.-D. (1985) *Holospora elegans*-infected micronuclei of *Paramecium caudatum* still serve a function during vegetative life. *J. Protozool.*, **32**, 736–7.
Miller, I.M. (1990) Bacterial leaf nodule symbiosis. *Adv. Bot. Res.*, **17**, 163–234.
Miller, I.M., Gardner, I.C. and Scott, A. (1983a) The development of marginal leaf nodules in *Ardisia crispa* (Thunb.) A.DC. (Myrsinaceae). *Bot. J. Linn. Soc.*, **86**, 237–52.
Miller, I.M., Scott, A. and Gardner, I.C. (1983b) Leaf nodule development in *Psychotria kirkii* Hiern. (Rubiaceae). *Ann. Bot.*, **52**, 791–802.
Miller, I.M., Scott, A. and Gardner, I.C. (1984) The occurrence of calyx nodules in *Psychotria* spp. (Rubiaceae). *Protoplasma*, **121**, 199–208.
Nealson, K.H. and Hastings, J.W. (1979) Bacterial bioluminescence: its control and ecological significance. *Microbiol. Rev.*, **43**, 496–518.
Ohmiya, K., Hoshino, C. and Shimizu, S. (1989) Cellulose-dependent and penicillin resistant plasmids isolated from *Ruminococcus albus*. Japan. *J. Zootech. Sci.*, **264**, 1461–6.
Orpin, C.G. and Letcher, A.J. (1979) Utilization of cellulose, starch, xylan, and other hemicelluloses for growth by the rumen phycomycete *Neocallimastix frontalis*. *Curr. Microbiol.*, **3**, 121–4.
Orpin, C.G. and Munn, E.A. (1985) *Neocallimastix patriciarum* sp. nov., a new member of the Neocallimasticaceae inhabiting the rumen of sheep. *Trans. Br. Mycol. Soc.*, **86**, 178–81.
Potrikus, C.J. and Breznak, J.A. (1977) Nitrogen-fixing *Enterobacter agglomerans* isolated from guts of wood-eating termites. *Appl. Environ. Microbiol.*, **33**, 392–9.
Powell, M.A. and Somero, G.N. (1986) Adaptations to sulfide by hydrothermal vent animals: sites and mechanisms of detoxification and metabolism. *Biol. Bull.*, **171**, 274–90.
Preer, J.R. Jr, and Preer, L.B. (1984) Endosymbionts of protozoa, in *Bergeys Manual of Systematic Bacteriology*, Vol. 1 (ed. N.R. Krieg), Williams and Wilkins, Baltimore, pp. 795–811.
Reisser, W., Meier, R., Görtz, H.-D. and Jeon, K.W. (1985) Establishment, maintenance, and integration mechanisms of endosymbionts in protozoa. *J. Protozool.*, **32**, 383–90.

Roux, M., Rio, M., Schein, E. *et al.* (1989) Mesures *in situ* de la croissance des bivalves et des vestimentifères et de la corrosion des coquilles au site hydrothermal de 13°N (dorsale du Pacifique oriental). *C.R. Acad. Sci. Paris*, Sér. III, **308**, 121–7.

Ruby, E.G. and Nealson, K.H. (1976) Symbiotic association of *Photobacterium fischeri* with the marine luminous fish *Monocentris japonica*: a model of symbiosis based on bacterial studies. *Biol. Bull.*, **151**, 574–86.

Sasaki, T., Aoki, T., Hayashi, H. and Ishikawa, H. (1990) Amino acid composition of the honeydew of symbiotic and aposymbiotic pea aphids *Acyrthosiphon pisum. J. Insect Physiol.*, **36**, 35–40.

Schnepf, E. and Melkonian, M. (1990) Bacteriophage-like particles in endocytic bacteria of *Cryptomonas* (Chryptophyceae). *Phycologia*, **29**, 338–43.

Soldo, A.T. (1987) The ciliated protozoan and its guests. *BioEssays*, **4**, 86–90.

Soldo, A.T. and Godoy, G.A. (1973) Molecular complexity of *Paramecium* symbiont Lambda deoxyribonucleic acid. Evidence for the presence of a multicopy genome. *J. Mol. Biol.*, **73**, 93–108.

Surek, B. and Melkonian, M. (1983) Intracellular bacteria in the Euglenophyceae: prolonged axenic culture of an algal-bacterial system, in Schenk, H.E.A., W. Schwemmler: *Endocytobiology* II. *Intracellular Space as Oligogenetic Ecosystem* (eds H.E.A. Schenk and W. Schwemmler), W. de Gruyter, Berlin, New York, pp. 475–86.

Stumm, C.K. and Zwart, K.B. (1987) Symbiosis of protozoa with hydrogen-utilizing methanogens. *Microbiol. Sci.*, **3**, 100–5.

Tebo, B.M., Linthicum, D.S. and Nealson, K.H. (1979) Luminous bacteria and light emitting fish: ultrastructure of the symbiosis. *BioSystems*, **11**, 269–80.

Wiemann, M. and Görtz, H.-D. (1989) Release of the endonucleobiotic bacterium *Holospora elegans* from its host cell *Paramecium caudatum*. *Eur. J. Protistol.*, **25**, 100–8.

Williams, A.G. (1986) Rumen holotrich ciliate protozoa. *Microbiol. Rev.*, **50**, 25–49.

7 Cyanobacterial Symbioses (except Lichens)

Nitrogen fixing symbioses are important associations between cyanobacteria (blue–green algae) and higher plants, combining two light harvesting plant types (Böger, 1982). Such symbioses allow plants to be independent of environmental nitrate and ammonium. Connected with this aspect is the noticeably larger number of heterocysts on the algae. Heterocysts are thick-walled differentiated cells which do not possess photosystem II and are the unique site of nitrogenase in these organisms (Stewart *et al.*, 1980; Carr and Whitton, 1982; Bothe *et al.*, 1984; Reisser, 1984; Mollenhauer, 1986).

7.1 ENDOCYANOMS

Endocyanoms are single-celled algae without plastids. The endosymbionts have a blue–green pigmentation and are designated as cyanelles. The endocyanoms comprise only a few members, e.g. *Cyanophora paradoxa*, *Glaucocystis geitleri*, *Gloeochaete wittrockiana*, *Glaucosphaera vacuolata*, *Cryptella cyanophora*, *Paulinella chromatophora* and *Cyanoptyche* (Kies, 1989).

The cyanelles of *Cyanophora paradoxa* are very similar in thylakoid structure and pigment composition to cyanobacteria. The DNA from cyanelles is circular and has a density of $1.692\,g\,ml^{-1}$. The genome size is 12×10^7 Da and around 60 copies are present per cyanelle. In comparison, genomes from higher plant chloroplasts are of $8–9 \times 10^7$ Da with a copy number of 35–100 (Breiteneder *et al.*, 1988). Cyanobacterial DNA from cells like *Synechococcus* is of $2–4 \times 10^9$ Da and has a very low copy number. Thus cyanelles form an inbetween stage, but closer to plastids than cyanobacteria. In two different strains of *Cyanophora paradoxa*, the genome of the cyanelles differs by 9 kb (136 versus 127 kb), which can be explained by numerous small insertions in non-coding regions, whereas gene arrangement and coding sequences are strictly conserved (Löffelhardt *et al.*, 1987). On the cyanelle DNA the genes for the herbicide-binding protein, the subunit of ATP-synthase, the large and the small subunit of RUBISCO and others have been identified (Table 7.1) (Löffelhardt *et al.*, 1987).

Table 7.1 Protein genes identified on cyanelle DNA. (Based on Löffelhardt *et al.*, 1987)

Gene Symbol	*Function/Description*	*mRNA size (bases)*
psbA	Herbicide binding protein	1200
atpA	Alpha subunit ATP synthase	
atpB	Beta subunit ATP synthase	2800
atpE	Epsilon subunit ATP synthase	
atpH	DCCD binding protein, CF_0-III	
psaA	PS I (P700) apoprotein	6000
psaB	PS I (P700) apoprotein	6000
psbB	PS II (55 kd) apoprotein	
psbC	PS II (44 kd) apoprotein	3200
rps19	rbs protein S19	1500
rp12	rbs protein L2	1500
petA	Cytochrome f	1600
petB	Cytochrome b6/f complex, cyt. b6	
petD	Cytochrome b6/f complex, SU4	1400
rbcL	RUBISCO, LSU	2600
rbcS	RUBISCO, SSU	2600
pcyA/B	α, β subunits phycocyanin	
apcA/B	α, β subunits allophycocyanin	

Cyanelles possess oxidative photosynthesis with both photosystems present, in addition to a typical Calvin cycle. They do not fix N_2 but reduce NO_3^- and NO_2^- in a ferredoxin-dependent light-reaction. RUBISCO is basically cyanobacterial, with a molecular weight 525000 in an L_8S_8 subunit structure.

Table 7.2 Photosynthetic rates of endocyanoms, cyanobacteria and unicellular algae. (Based on Kremer *et al.*, 1979 and Philipps and Mitsui, 1982)

Species	*Temperature*	*Light intensity (Klx)*	*CO_2-fixed ($\mu mol\ h^{-1}$ mg chlorophyll^{-1})*
Cyanophora paradoxa	20°C	20	130
Glaucocystis nostochinearum	20°C	20	118
Glaucosphaera vacuolata	20°C	20	137
Gloeochaete wittrockiana	20°C	20	123
Anabaena cylindrica	30°C	10	133
Anabaena flos-aquae	25°C	8.5	180
Nitzschia palea	20°C	18	266
Cyclotella nana	20°C	11	400
Chlorella ellipsoidea	20°C	10	250
Scenedesmus sp.	20°C	7	229

The rate of photosynthetic CO_2-fixation in endocyanoms is similar to that of cyanobacteria with 120–140 mmol CO_2 fixed h^{-1} mg chlorophyll^{-1}. This is a factor of at least 2 lower than that of other single celled algae (diatoms or coccal green algae) under similar conditions (Table 7.2). Comparisons of fine structural and cytological characteristics between *Cyanophora paradoxa*, *Gloeochaete wittrockiana* and *Glaucocystis nostochinearum* suggest that *Glaucocystis* and *Gloeochaete* are relatively similar. However, they are both so different from other algal groups like the Dinophyceae, Cryptophyceae

Table 7.3 Comparison of morphological and fine structural characteristics of *Cyanophora paradoxa*, *Gloeochaete wittrockiana* and *Glaucocystis nostochinearum*. (Based on Kies, 1979)

Characteristic	Cyanophora	Gloeochaete *vegetative cell*	Glaucocystis
Organizational level	Monadoid	Palmelloid	Coccoid
Reproduction	Binary fission	Zoospores	Autospores
Cell wall	–	No cellulose	Cellulose
Apical or subapical pit	+ and longitudinal groove	+	+
Dictyosomes parabasal	+	+	+
Lacunar system	+	+	+
Flagella or gelatinous flagella (number)	2	2	2
Flagellar movement	Heterodynamic	Non-motile	Non-motile
Flagellar shaft (structure)	(9 + 9) + 2	(9 + 9) + 0 or less	(9 + 9) + 0
Basal body at an angle of 120–180°	+	+	+
Flagellar roots (structure)	Cruciform 2 roots, 3 MT 2 roots, ~10 MT	Cruciform 4 roots, 50 MT	Cruciform 4 roots, 50 MT
Multilayered flagellar roots	2 roots with fibrillar ligament	4 MLS	4 MLS
Fragmentation of nuclear membrane during mitosis	+	+	+
Centrioles present during mitosis	–	–	–
Phycoplast	–	–	–
Starch granules extraplastidic	+	+	+
Cyanelles with rudimentary wall	+	+	+

+, present; –, not present; MT, microtubules; MLS, multilayered structure.

and the Chlorophyceae that it has been suggested they should be classified together as a new class, the Glaucophyceae. After treatment of *Gloeochaete* and *Glaucocystis* with penicillin, the cyanelles stop dividing and lose their biliproteins; the symbiophoric vacuole is transformed into a lysosome and finally the cells die (Kies, 1988). Seven penicillin binding proteins (PBP) similar to those from *E. coli* have been characterized in *Cyanophora paradoxa* (Berenger *et al.*, 1987). *Cyanophora* is different from all the organisms mentioned above, due to its lack of cell walls, lack of multilayered flagellar bases and other properties (Table 7.3).

Host cells of *Paulinella chromatophora* are amoeboid and contain two cyanelles each of which is individually enclosed in a vesicle membrane. Functionally cyanelles from *P. chromatophora* are similar to those from other endocyanoms (Figs 7.1 and 7.2).

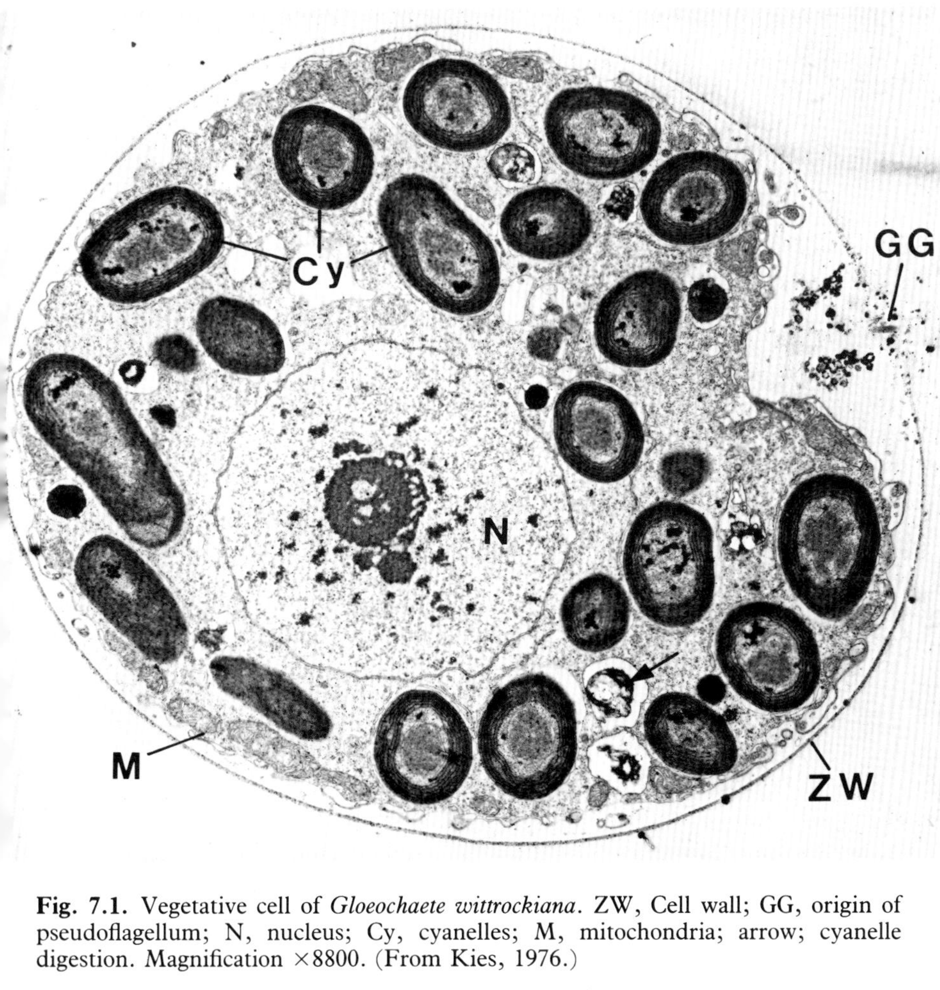

Fig. 7.1. Vegetative cell of *Gloeochaete wittrockiana*. ZW, Cell wall; GG, origin of pseudoflagellum; N, nucleus; Cy, cyanelles; M, mitochondria; arrow; cyanelle digestion. Magnification ×8800. (From Kies, 1976.)

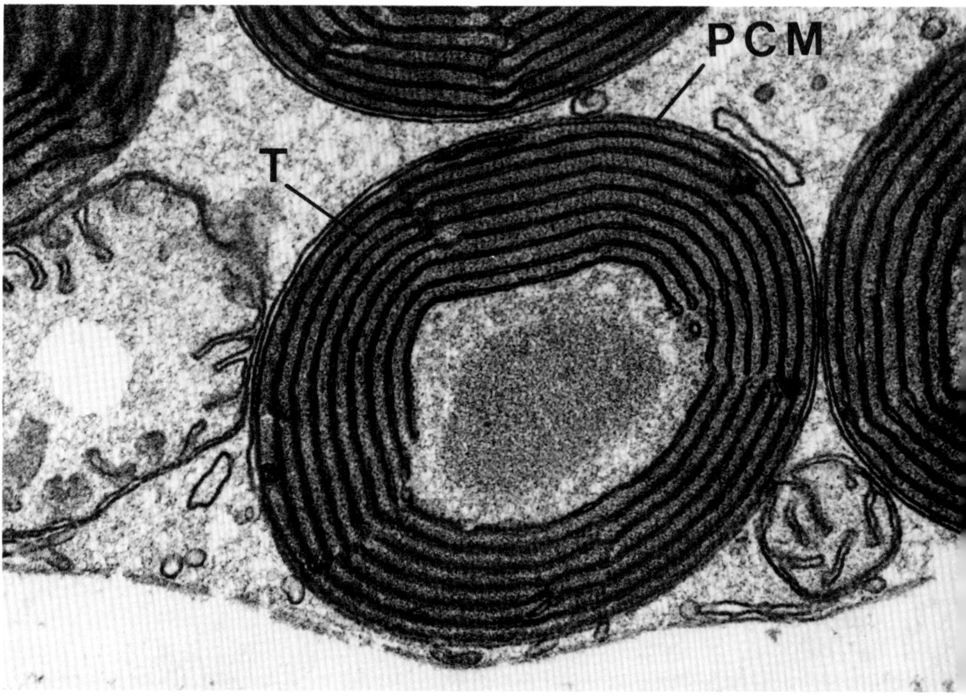

Fig. 7.2. Transverse section through a cyanelle from *Gloeochaete wittrockiana*. PCM, pericyanelle membrane (= membrane of host vesicle); T, thylakoids. Magnification ×44 000. (From Kies, 1976.)

7.2 DIATOM–CYANOBACTERIA SYMBIOSES

Marine planktonic diatoms of the genera *Rhizosolenia* and *Hemiaulus* may contain endosymbiotic cyanobacteria identified as *Richelia intracellularis* (Mague, 1977). The symbiont is in the form of filaments with terminal heterocysts, the ecology of which is hardly known. It is known that they may fix N_2 to the extent of 3 kg N ha^{-1} year^{-1}. Assuming that they occupy one-quarter of the surface of the Pacific Ocean, this means they fix 15×10^6 tonnes of N per year. Given that nitrogen is normally the limiting nutritional factor in water, these diatoms play an important role in oligotrophic oceanic ecology.

Benthic freshwater diatoms such as *Rhopalodia gibba* contain between two and five symbionts similar to cyanobacteria. They fix N_2 at a rate of about 10 nmol C_2H_4 h^{-1} mg protein^{-1} (Floener and Bothe, 1980). This is much lower than the rate for free-living cyanobacteria. It must be remembered, however, that this specific activity also includes the host cell. Oxygen optimum is 4–5% O_2 in the gas phase.

Table 7.4 Specific nitrogenase activities and growth rates of N_2-fixing prokaryotes

Species/Strain	*Maximum nitrogenase activity (nmol C_2H_4 mg protein^{-1} h^{-1})*	*Culture conditions*	*Generation time under diazotrophic conditions*
Anabaena cylindrica	480	5 W m^{-2} (Light intensity)	24–30 h
Nostoc muscorum 7119	300	7–9% heterocysts	
Azotobacter chroococcum NCIB 8003	700	Chemostat culture S-limited, O_2-optimal,	10 h
Enterobacter cloacae	5400	anaerobic	
Alcaligenes latus DSM 1123	800	5% O_2 (Gas phase)	39 h (with NH_4^+ as N-source: 3 h)
Klebsiella oxytoca M 5a1	1800	Chemostat culture sulphate-limited	9 h
Klebsiella planticola	3000	Anaerobic	12–14 h

The intracellular O_2 concentration is not known. It is also not known how nitrogenase is protected from the two oxygen-producing cell zones, the plastids of the diatom and the photosynthetically active membrane of the endosymbiont. Maximal nitrogenase activities of cyanobacteria and chemo-organotrophic bacteria are compared in Table 7.4.

7.3 BRYOPHYTE–*NOSTOC* SYMBIOSIS

Extracellular symbioses between *Nostoc* and some genera of bryophytes are known, but these only take place during the plant's haploid gametophyte generation.

7.3.1 Anthoceros

Anthoceros belongs to the class Anthocerotae (liverworts). Plants are infected through slime pores similar to stomata on the underside of the thalli. Cyanobacteria multiply in the air space and stimulate the surrounding cells to grow into tubes. Some microsymbiotic cells enter neighbouring cells, where they are lysed. At least two *Nostoc* species are symbionts with the genus *Anthoceros*. *Nostoc sphaericum* is found in *A. punctatus*, *A. husnotii* and *A. laevis*. Heterocysts make up to 40% of symbiotic *Nostoc* cells whereas free-living filaments amount to only 3–6% (Rodgers and Stewart, 1977). This explains the high specific N_2-fixing capacity of *Anthoceros* (400 nmol C_2H_4 h^{-1} g fresh weight^{-1}) (Fig. 7.3) (Stewart and Rodgers, 1977).

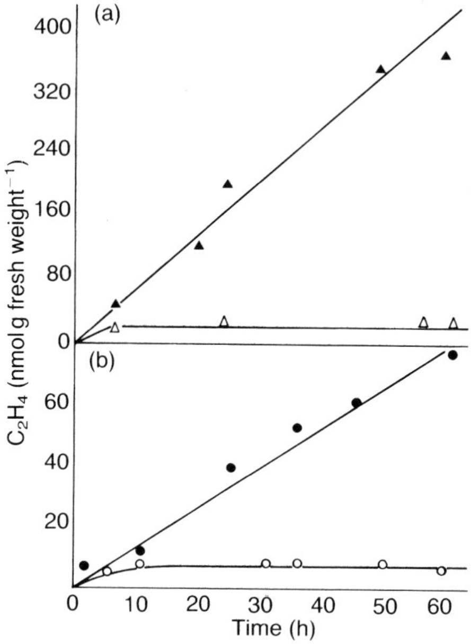

Fig. 7.3. Nitrogenase activity in gametophytes of (a) *Anthoceros punctatus* and (b) *Blasia pusilla* in the light at 10 W m^{-2} (▲, ●) or in the dark (△, ○) at 25°C. (Based on Stewart and Rodgers, 1978.)

7.3.2 Other liverworts: *Blasia* and *Cavicularia*

The infection of *Blasia pusilla* (Fig. 7.4) takes place via the hemispheric auricles which are made from a layer of cells formed into cavities with a narrow entrance, blocked by papillae (Duckett *et al.*, 1977). *Nostoc* filaments grow into these entrances and divide in the cavities which are filled with slime. *Nostoc* from *Anthoceros* can infect *Blasia* gametophytes. As the *Nostoc* cells multiply, tubular branched hyaline cells from the host grow into a position where they have close contact with the symbionts (Stewart and Rodgers, 1978). It is still an open question whether the slime in the cavities is produced by the cyanobacteria or these hyaline cells.

There is a constant relationship between liverwort biomass and infection, with approximately two colonies per mg fresh weight of *Blasia*. *Nostoc* colonies isolated from *Blasia* fix N_2 at the rate of 600 nmol C_2H_4 h^{-1} mg fresh weight^{-1}. This is a very high rate and reflects the high proportion of heterocysts present. Isolated *Nostoc* colonies excrete over 95% of the fixed nitrogen as NH_4^+. This excretion drops to zero at 72 h after isolation. Isolated *Nostoc* colonies show no photosynthetic CO_2 fixation and apparently take their carbon and energy sources from the gametophyte of the

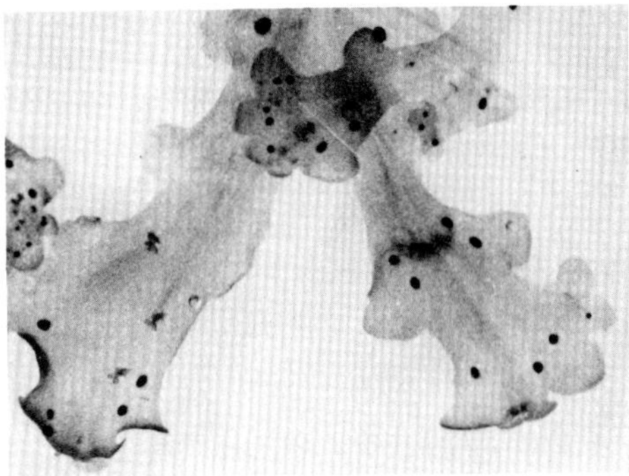

Fig. 7.4. Gametophyte of *Blasia pusilla* with dark *Nostoc* colonies. (From Stewart and Rodgers, 1978.)

liverwort. This is probably in the form of sucrose. It is assumed that this is accomplished through the branched hyaline cells.

7.4 *AZOLLA–NOSTOC* SYMBIOSIS

Azolla is a heterosporous aquatic fern. It belongs to the order Salvinales (Pteridophyta) and is assigned to a monotypic family Azollaceae separate from *Salvinia*. The genus has two subgenera, the New World *Euazolla* (*Azolla filiculoides*, *A. caroliniana*, *A. microphylla* and *A. mexicana*) and the European *Rhizosperma* (*A. pinnata* and *A. nilotica*). The subgenera are divided on the basis of reproductive organs (number of floats in the megaspore apparatus) and the species according to branching and the leaf order (Fig. 7.5) (Peters and Meeks, 1989). The *Azolla* microsymbiont may in fact belong to the genus *Nostoc* (Peters *et al.*, 1986) as originally named by Strasburger (1873).

DNA probes from *Azolla* microsymbionts show a closer relationship to free-living *Nostoc* strains than to free-living *Anabaena* strains (Plazinski *et al.*, 1990).

Because of the symbiotic association with N_2-fixing cyanobacteria of the genus *Nostoc* the fern has a very rapid growth rate. In New Zealand and South Africa it has achieved the status of a persistent weed. In Indochina it has importance and further potential as a green manure, which is of appreciable value on rice paddy fields.

7.4.1 Development of the symbiosis

The general life cycle of *Azolla* and the route taken by the symbiont *Nostoc azollae* are shown in Fig. 7.6. In the sexual cycle equal numbers of mega-

Fig. 7.5. *Azolla caroliniana* (a), *Azolla mexicana* (b), *Azolla filiculoides* (c) and *Azolla pinnata* (d). (From Peters and Calvert, 1983.)

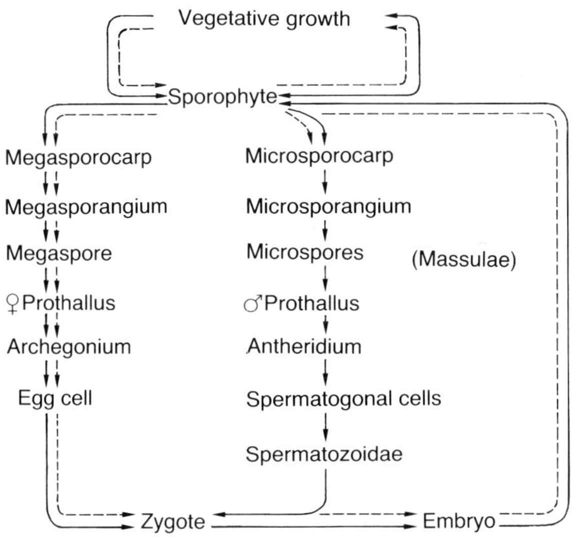

Fig. 7.6. Generation cycle of *Azolla filiculoides* (——) and redistribution of the symbiont *Nostoc* (– – –).

and microsporocarps are formed on the first ventral leaf. Nuclear division in the megasporocarps results in 32 nuclei, all of which die, except one. One megaspore develops from one megasporocarp. In the microsporocarp 8–130 microsporangia are found, from each of which may develop 32–64 microspores. Between 8 and 15 microspores clump together to form a so-called massula which carries hook-shaped glochidia on the surface. These hooks serve as an anchor for a megaspore. The megaspore germinates to form a female prothallus, on which one or more archegonia develop. The microspore germinates to form a male prothallus on which antheridia develop, with spermatogoniae and spermatozoae. Fertilization of egg cells can take place under water or on wet surfaces. The embryo, which develops from a zygote grows into a sporophyte and then increases further by vegetative growth. Vegetative growth can double the biomass in 2 days.

The microsymbiont *Nostoc* is generally assumed to be transmitted via the megasporocarp and archegonia to the zygote as akinetes. Microsporocarps may also contain cyanobacteria, but the following stages do not. At zygote germination the cyanobacterium grows into undifferentiated filaments which colonize a cavity created by epidermal cell growth covering a depression on the basal half of the ventral surface of the lobe primordia. Above the centre of the cavity, epidermal cells meet and form a pore which allows gaseous exchange with the environment. Trapped filaments in the shoot apex colonize the cavity, and the fern and cyanobacteria then develop synchronously. Even when growth of the host is reduced by, for example,

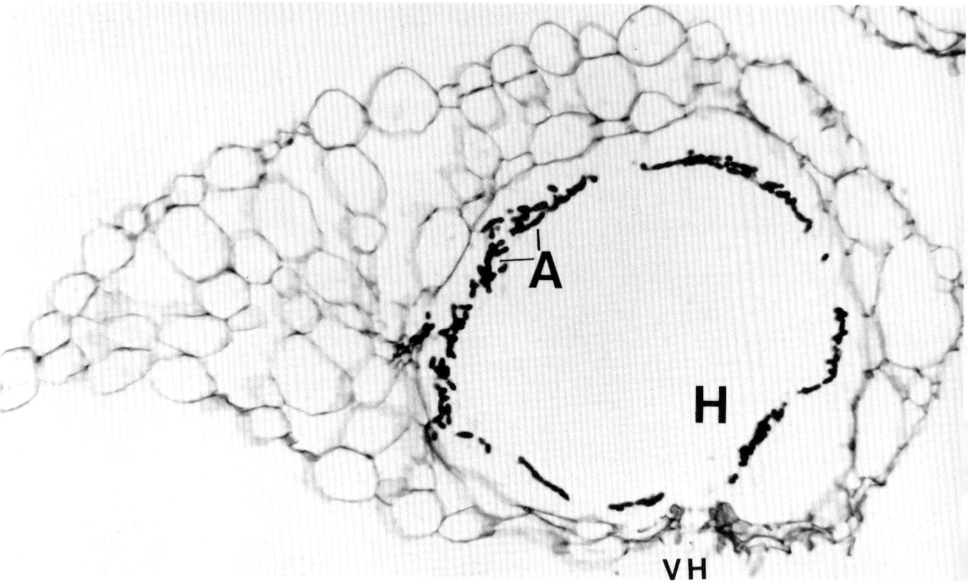

Fig. 7.7. Leaf cavity (H) with *Nostoc* (A) in the dorsal lobe of *Azolla caroliniana* leaves. VH, pore closure of cavity. (From Peters and Calvert, 1983.)

light or temperature limitation, growth of *Nostoc* is correspondingly reduced. Cyanobacterial cells remain on the periphery of the cavity (Fig. 7.7). Every leaf cavity contains one primary hair. The cell walls are formed of a network of fibrils and the cells contain many ribosomes, mitochondria, endoplasmic reticulum (ER) and proplastids. Cells of such hairs are connected by many plasmodesmata and therefore may be transfer cells (Peters *et al.*, 1978; Calvert *et al.*, 1985). Older leaves contain several such structures (Fig. 7.8). It must be noted here that the sexual cycle in *Azolla* is not very important and almost all growth is vegetative. *Azolla* can be cured of *Nostoc* infection by growth for 2–3 weeks in a medium containing nitrate and antibiotics such as chloramphenicol, penicillin and rifampicin. Since a light-induced proton efflux, driven by photosystem I, has been demonstrated in *Nostoc* (Scherer *et al.*, 1988b), this could lead to acidification of the fern cavity. *Azolla* plants both with and without symbiotic cyanobacteria have been shown to possess a lectin. Pure lectin from *A. caroliniana* has a molecular weight of 126 000 Da and consists of six subunits, and has a

Fig. 7.8. Longitudinal section of an unbranched hair cell from the leaf cavity of *Azolla caroliniana*. The hair cell shows structural features of transfer cells (P-J). Vegetive cells (VZ) and heterocysts (H) of *Nostoc* are located in close proximity. M. mitochondria; P, plastid; V, vacuole. (Original photograph, H.E. Calvert, Ohio.)

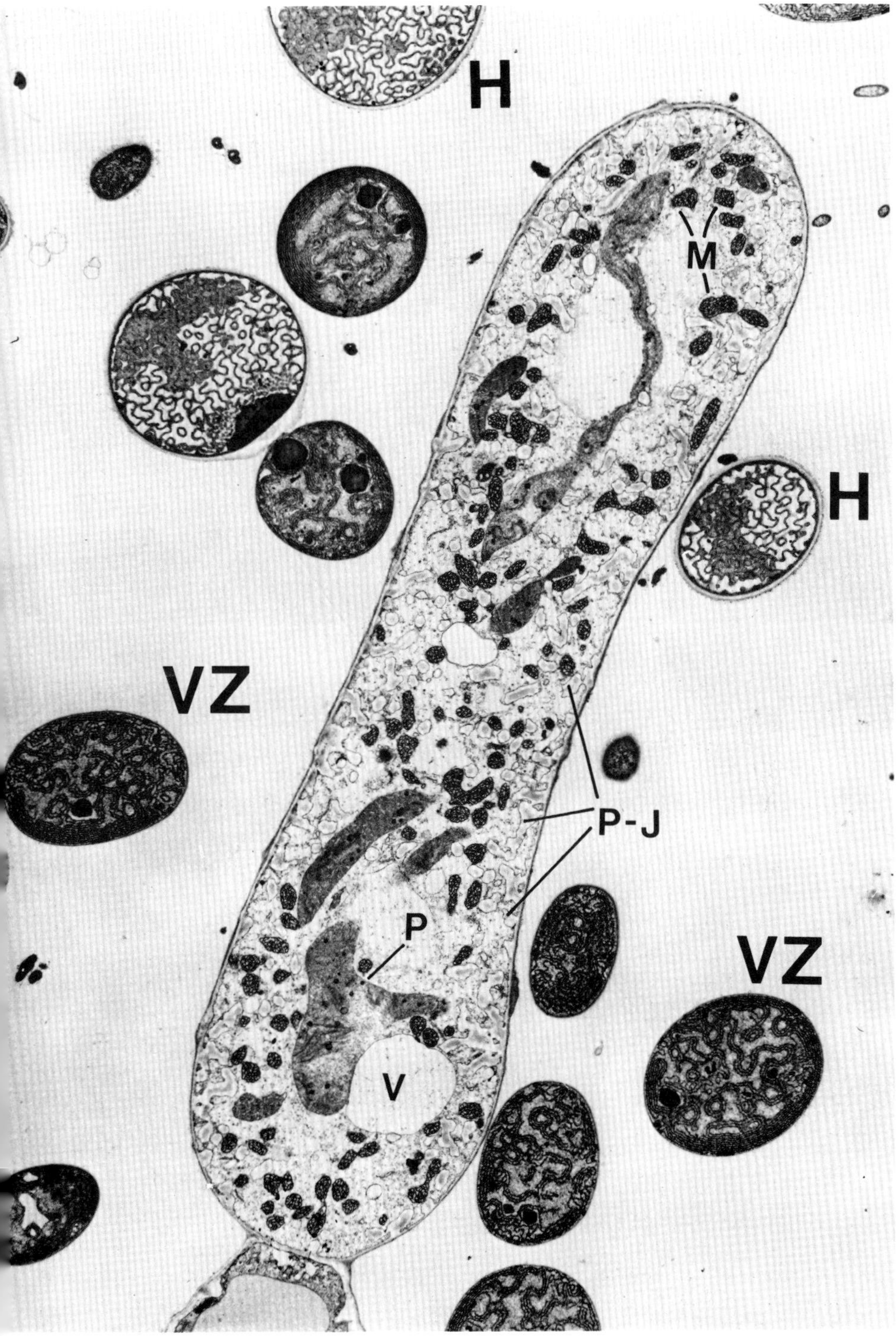
H
M
H
VZ
P-J
P
VZ
V

slightly higher affinity for galactose than for other sugars. The lectin, which is found in the symbiotic cavity around the microsymbiont, binds sugar reversibly to both symbiotic and non-symbiotic *Nostoc* strains (Mellor *et al.*, 1981).

Two other types of bacteria are also present in the leaf cavity of *Azolla*, a helical species and a rod-shaped bacterium (Nierzwicki-Bauer and Aulfinger, 1990) different from *Arthrobacter* sp., which has been previously identified (Petro and Gates, 1987).

7.4.2 Heterocyst formation and function

Symbionts show an increasing differentiation into heterocysts with increasing leaf age. This ranges from 0% in young leaves to 30% in those leaves at greatest distance from the shoot tip (Fig. 7.9) (Hill, 1975). This differentiation also occurs when nitrate is present. Since heterocysts do not divide, division must be less in other cells.

During heterocyst formation a double-layered envelope of polysaccharides and glycolipids is deposited outside the cell wall. Heterocyst formation requires the activation of a number of genes with different promoters which are not transcribed by purified RNA polymerase from vegetative cells (Haselkorn *et al.*, 1990).

Metabolism and nutrient exchange between vegetative cells and heterocysts is shown in Fig. 7.10. Heterocysts have no photosystem II and thus no

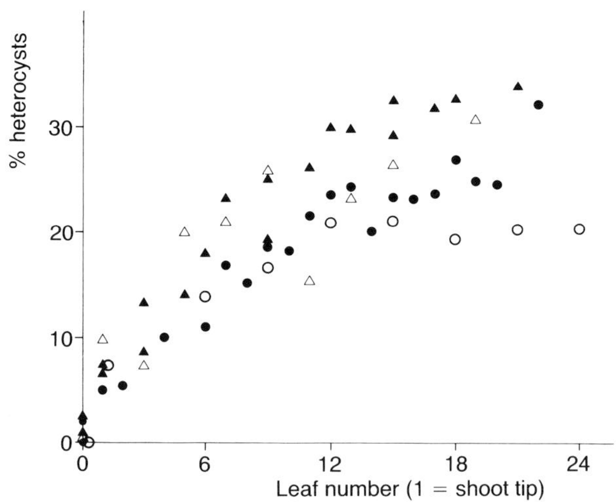

Fig. 7.9. Correlation between percentage heterocystous *Nostoc* cells and the age of *Azolla filiculoides* leaves. Culture conditions: (○) $2\,W\,m^{-2}$ without nitrate; (△) $40\,W\,m^{-2}$ with nitrate; (▲) starting at $40\,W\,m^{-2}$ with nitrate and continued at $4.4\,W\,m^{-2}$ without nitrate; (●) field material. (Based on Hill, 1975.)

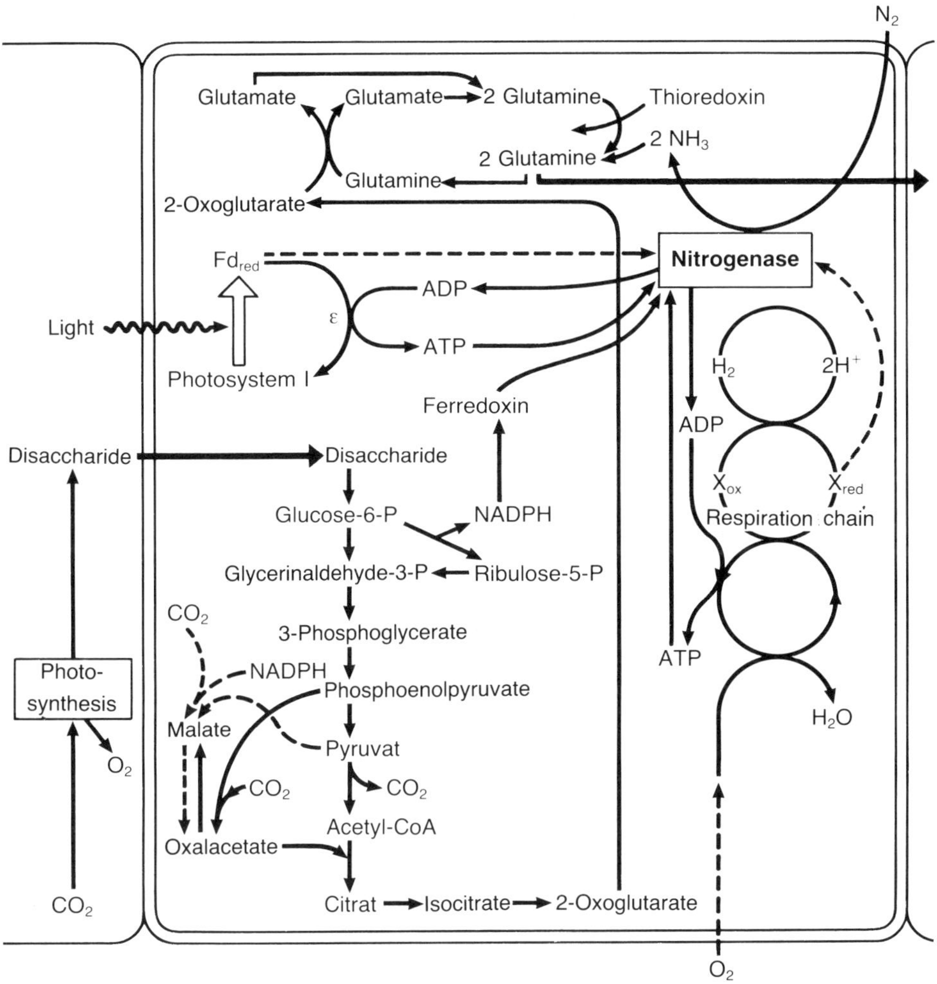

Fig. 7.10. C- and N-metabolism in heterocysts of cyanobacteria, with transport processes to and from neighbouring vegetative cells. (Based on Gallon, 1980 and Bothe *et al.*, 1984.)

photosynthetic O_2-production. Reduced ferredoxin, produced in a light-dependent fashion and ATP, produced by photophosphorylation, can be used directly for the reduction of N_2 to ammonia. A heterocyst-specific form of ferredoxin has been identified (Schrautemeier and Böhme, 1985). This is encoded by the *fdxH* gene, mapping downstream of *nif* K (Haselkorn *et al.*, 1990). Heterocysts also lack a Calvin cycle, the C-source being disaccharides from vegetative cells. These are metabolized to glucose-6-phosphate, then degraded by the oxidative pentose phosphate cycle. The

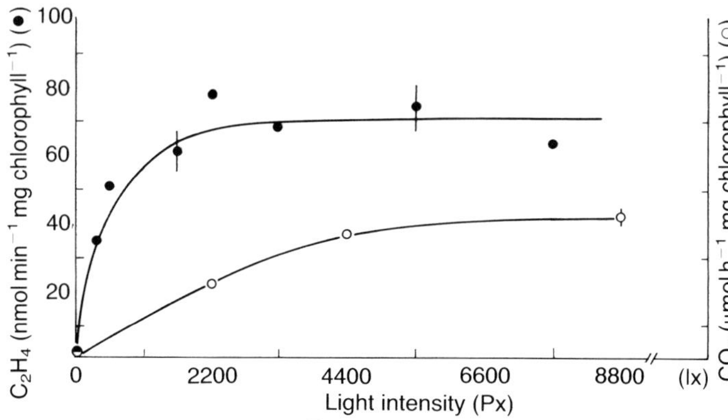

Fig. 7.11. Nitrogenase activity (●) and photosynthetic CO_2 fixation (○) in *Nostoc* isolated from *Azolla*, in relationship to light intensity. (250 lx is equivalent to approx. $1\,W\,m^{-2}$.) (Based on Peters, 1983.)

NADPH thus formed reduces ferredoxin and forms a second source of reduced ferredoxin. A second ATP-source is the respiratory chain in the heterocyst. The cytochrome *c* oxidase from heterocysts has a high cyanide sensitivity and is of the aa_3 type Häfele *et al.*, 1988). The tricarboxylic acid cycle is non-functional in heterocysts and 2-oxoglutarate cannot be further oxidised. It is, however, used in the glutamate synthase reaction. Malate and 2-oxoglutarate are formed by very active malate dehydrogenase and PEP-carboxylase respectively. H_2 is the best tested electron donor for the nitrogenase. This reaction is light-dependent. H_2 can be used in the respiratory chain at low rates because photosynthesis and respiration both use the plastoquinone and cytochrome b_6f complex. An uptake hydrogenase, activated by thioredoxin, can feed in electrons close to the plastoquinone site of thylakoids (Papen *et al.*, 1986). Light-dependent nitrogenase activity is saturated at $8\,W\,m^{-2}$, but photosynthetic CO_2-fixation in the vegetative cells is saturated at $32\,W\,m^{-2}$ (Fig. 7.11).

Photosynthetic electron transport exceeds respiratory oxygen uptake in cyanobacteria 10–20-fold. Cyanide-insensitive oxygen uptake accounts for 6–20% of the total oxygen uptake (Scherer *et al.*, 1988a), and is not coupled to ATP formation. Non-heterocystous N_2-fixing cyanobacteria, growing under a light–dark regime, use a temporal separation between photosynthetic O_2 evolution in the light and N_2-fixation in the dark (Gallon and Chaplin, 1988). This requires an efficient ATP and ferredoxin supply by respiration. Synthesis of nitrogenase mRNA is rhythmic and is found only during the dark period (Huang and Chow, 1990).

It is still unclear whether heterocysts export the majority of nitrogen as ammonia or whether this is fixed by glutamine synthetase first and then

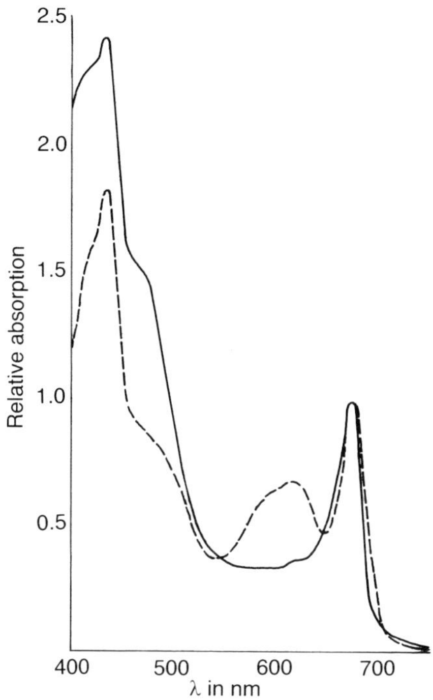

Fig. 7.12. Absorption spectra of cell-free extracts from *Azolla* without endophytes (——) and from *Nostoc* (–––). Spectra are normalized at 673 nm. (Based on Peters and Calvert, 1983.)

exported. Heterocysts of endophytic *Nostoc* (*Anabaena*) contain only 5–10% of the glutamine synthetase concentration found in vegetative cells. In free-living cyanobacterial glutamine synthetase, labelled with specific inimunogold, was much more pronounced in heterocysts than in vegetative cells (Bergman *et al.*, 1985). GDH does not contribute to ammonia assimilation (Meeks *et al.*, 1985).

Genetically the *Nostoc* symbiosis is very different from other prokaryote–plant symbioses. Although the symbiosis is never interrupted, differentiation of the microsymbiont involves a genetic rearrangement. In vegetative cells *nif D* and *nif H* lie together and are separated from *nif K* by an intervening sequence of 11 kbp. In N_2-fixing cells this intervening sequence is removed so that the three genes are in one transcriptional unit (Stewart, 1985). In non-heterocystous N_2-fixing cyanobacteria the *nif H*, *D* and *K* are generally clustered (Gallon and Chaplin, 1988). In *Anabaena*, an alternative (vanadium) nitrogenase has also been detected (Kentemich *et al.*, 1988).

Symbiotic *Nostoc* isolated from *Azolla* plants has a negative CO_2 balance in the light. It must therefore be assumed that the host delivers photosynthates to the symbiont. Sucrose is found in the cyanobacterium, but cannot be synthesized there. It is thus assumed that this is the transported form of carbon.

The symbiosis of *Azolla–Nostoc* can use light better than the separated organisms (Fig. 7.12). In addition to the absorption maximum of the chlorophylls there is a maximum at 620 nm caused by **phycobiliproteins**, especially the phycocyanins. Phycobiliproteins in the symbiont consist of 70% phycocyanin, 17% phycoerythrin and 13% allophycocyanin. The phycobiliproteins are also present in heterocysts (Tyagi *et al.*, 1981). *Azolla*, in addition to photosynthetic pigments, may possess significant concentrations of anthocyanin. *Nostoc* chlorophyll accounts for less than 20% of the total chlorophyll in the symbiosis (Ray *et al.*, 1978). The contribution to the total CO_2-fixation of the association has been estimated to be only 5% (Kaplan and Peters, 1988).

7.4.3 Ecological and agricultural aspects

In China, Vietnam, Indonesia and other South East Asian lands, approx. 2 million hectares of land is cultivated with *Azolla*. This is normally done either with or after the growth of rice in paddy fields. It is the most effective after the rice harvest, when between 60 and 100 kg of biomass per hectare can grow. This biomass fixes 3 kg N ha^{-1} day^{-1} (Padhy, 1985).

7.5 CYCAD–*NOSTOC/ANABAENA* SYMBIOSIS

The genera *Cycas*, *Macrozamia*, *Zamia* and *Encephalartos* of the class Cycadateae exhibit coral root formations. These negatively geotropic coral roots (Fig. 7.13) may function in respiration, or they may become colonized with cyanobacteria. In the latter case they are not so strongly negatively geotropic (Fig. 7.14). Coral roots have often been called nodules. This terminology is false, since they develop independently of infection. They develop from the vascular pericycle of primary roots or hypocotyls. The tip is rounded and the meristem is terminal. Instead of a root cap a secondary layer of rind cells forms over the protoderm, which in turn develops into a single-celled layer in which intercellular infection with cyanobacteria of the genera *Nostoc* (or *Anabaena*) can take place. Infection succeeds when the middle lamellae dissolve or sometimes through dead epidermal cells. In *Macrozamia communis* rind cells are also infected. The symbionts colonize all spaces between host cells (Figs 7.15 and 7.16) which have enlarged nuclei, many mitochondria and a highly developed ER (Neumann, 1977). The heterocysts of the symbiont fix N_2, which is then transported from the coral roots into other parts of the plant (Table 7.5). In a 48 h period after labelling ^{15}N was present in coral roots in a concentration 10 times higher

Fig. 7.13. Coral roots from *Macrozamia communis* free of cyanobacteria. Magnification ×0.55. (From Wittman and Bergersen, 1965.)

Fig. 7.14. Coral roots from *Macrozamia communis* with cyanobacteria and abundant lenticels. Magnification ×0.63. (From Wittman and Bergerson, 1965.)

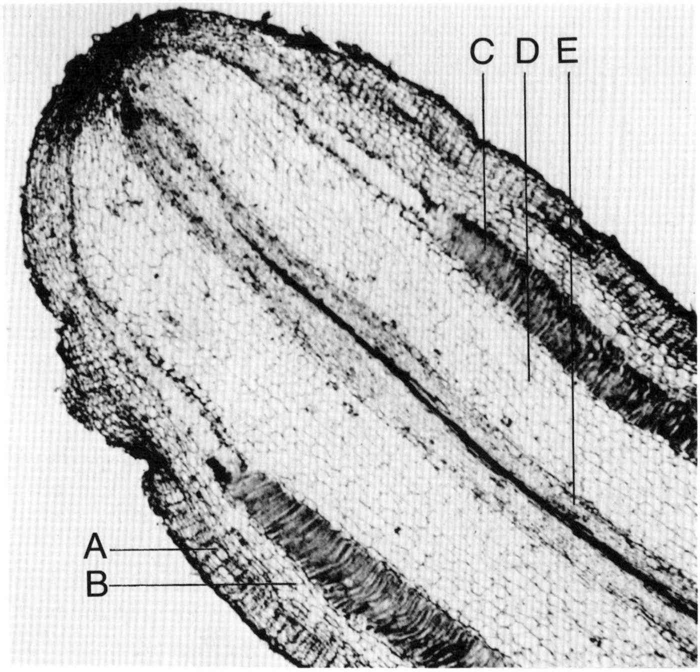

Fig. 7.15. Longitudinal section of a coral root from *Macrozamia communis* with periderm (A), outer cortex (B), zone with cyanobacteria (C), inner cortex (D) and central vascular system (E). (From Wittman and Bergersen, 1965.)

than that in the primary root or trunk. The nitrogen was subsequently transported to basal leaves, trunk, hypocotyl and primary root (Table 7.5 and Fig. 7.17). The morphology of cycad side roots is affected by blue light, independently of symbiotic infection. In *Zamia floridana* this takes the form of nodule-like thickenings (Webb, 1981).

The development of coral root symbioses in *Cycas revoluta* is in principle similar to that in *Macrozamia*. Cells of the algal zone are derived from the rhizoderm and the surrounding cells form a hard persistent root cap. In *Cycas* the host cells are formed from two cell types. The first are small cells rich in cytoplasm. These cells tend to lyse and form large intracellular spaces in which *Anabaena* cells multiply. The second type is elongated and traverses the algal zone. They contain many amyloplasts and mitochondria.

The *Anabaena* symbiont from *Cycas revoluta* coral roots exhibits a remarkable annual cycle in the enrichment of **cyanophycin**. Cyanophycin serves as N-storage and is formed from arginine and asparagine in the molar ratio 1 : 1. In winter the N storage capacity is considerable, in summer low. It is assumed that in summer the N-demand of the plant forbids the

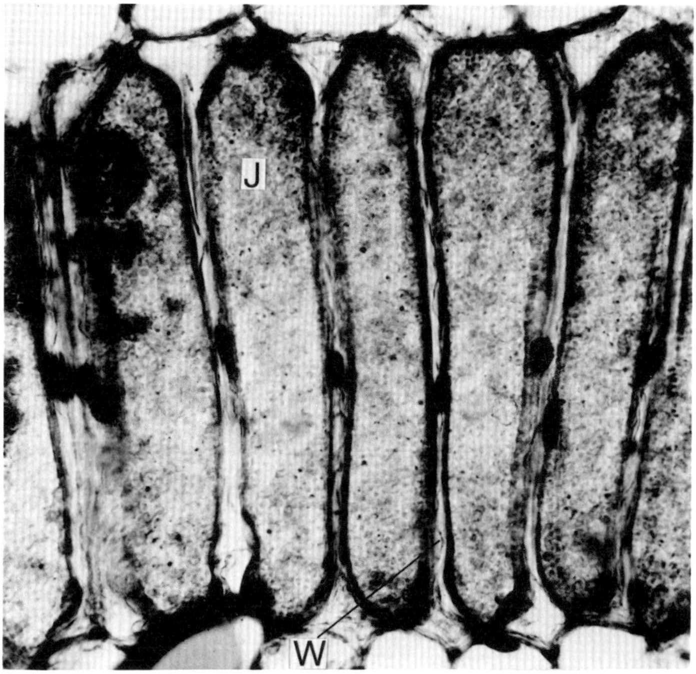

Fig. 7.16. Cyanobacteria in large intercellular spaces (J) between long and narrow cells (W) of the host plant *Macrozamia communis*. (From Wittman and Bergersen, 1965.)

Table 7.5 Allocation of ^{15}N between different organs of *Macrozamia communis*, after 48 h exposure in an atmosphere of 12% N_2 (with 44 atom % $^{15}N_2$), 25% O_2 and 63% Ar. (Based on Bergersen *et al.*, 1965)

Plant organ	*Dry weight (g)*	*Total N (mg)*	^{15}N *(atom % excess)*	^{15}N *(μg excess)*
Leaves	3.8	60.2	0	0
Petioles	2.1	15.2	0.065	9.9
Leaf bases and stem	10.2	102	0.057	66.2
Coral roots	0.8	15.3	0.72	110
Hypocotyl and swollen main root	1.7	21.7	0.064	13.9
Fine roots	1.1	13.4	0.11	15.1

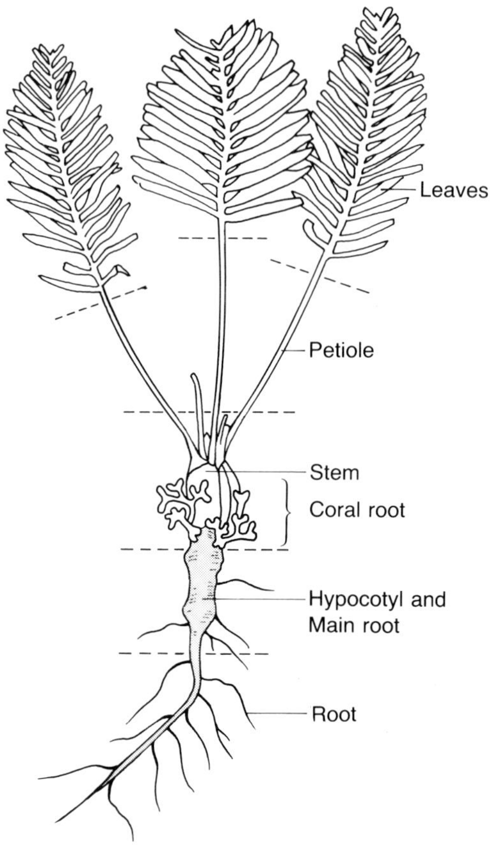

Fig. 7.17. The organs of *Macrozamia communis* as analysed for ^{15}N-partition in Table 7.5.

formation of cyanophycin. Despite careful searches only *Nostoc* and *Anabaena* are found in coral roots of healthy *Cycas* and *Encephalartos* plants. Mechanisms specifically excluding other microorganisms are unknown. One theory, so far unproven, is the participation of phenols in surrounding tissue.

7.6 *GUNNERA–NOSTOC* SYMBIOSIS

The genus *Gunnera*, now in a separate family, the Gunneraceae with perhaps some relationship to the Saxifragales (Doyle and Scogin 1988) enters into an intracellular symbiosis with *Nostoc punctiforme* (Harder, 1917). The symbiotic tissue forms specific, gland-like organs on the bushy axis.

The mode of infection has been studied in detail with endosymbiont-free plants regenerated from tissue culture of *Gunnera macrophylla* (Schmidt, 1990). Independently from infection by the *Nostoc* cells, in the host plant rhizome cells partially digest the middle lamellae and the cell walls and form infection sacs. The apical cells of the glands are also lysing, forming larger pores filled with slime. Only a few microsymbionts successfully invade through the pores of the glands, whereas large numbers are plasmolysed and die. This is analogous to the large number of abortive infections in the root hairs of legumes invaded by rhizobia.

Experiments with *Gunnera arenaria*, *Gunnera strigosa* and *Gunnera albocarpa* have shown that infection is achieved through specific papillae which form infected tissues before the root development stage. The symbionts force their way into the thin-walled meristematic cells at the base of the papillae where they are enclosed by a membrane initially in continuum with the host plasma membrane. This **pericyanobacteroid** membrane may be similar to the symbiosome membrane in the *Rhizobium*–legume symbiosis. The cyanobacteria divide inside these membranes and differentiate into heterocysts. The N_2 fixed by the microsymbiont is transferred to the host plant (Silvester and Smith, 1969) (Fig. 7.18). *Gunnera* can be infected with *Nostoc* strains from *Cycas revoluta*, *Peltigera polydactyla* or *Anthoceros* (Table 7.6). However, strains from *Anthoceros* and *Peltigera* show lowered infectivity against that exhibited by the real symbiont. *Nostoc* strains from

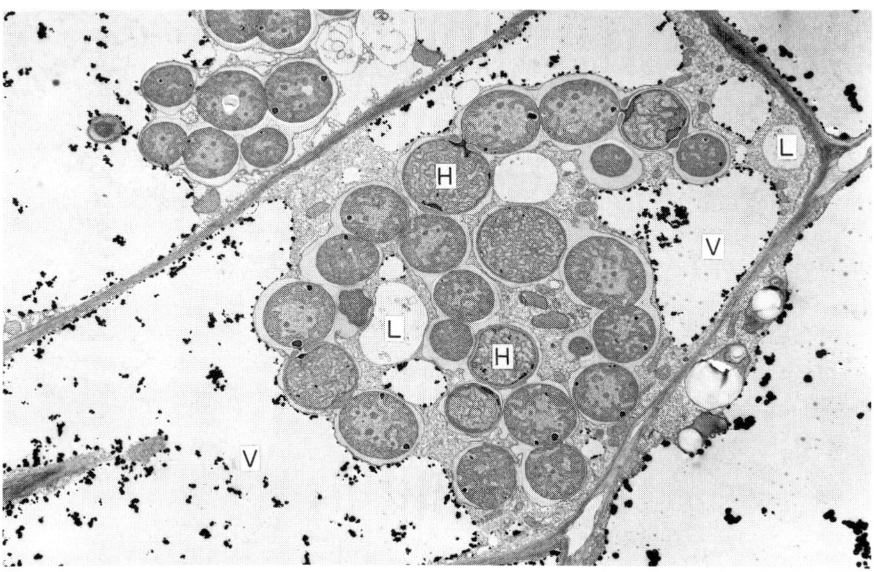

Fig. 7.18. Rhizome cell of *Gunnera macrophylla* with *Nostoc* as endocyanobionts. A large percentage of the *Nostoc* cells are present as heterocysts (H). Some lysosomes (L) and large vacuoles (V) are also visible. Marker: 3 µm. (Original EM micrograph from Schmidt, 1990.)

Table 7.6 Infection of *Gunnera manicata* by various cyanobacterial isolates. (Based on Bonnett and Silvester, 1981)

Cyanobacterium	*Origin*	*Infectivity*
Nostoc sp.	*Gunnera arenaria*	+
Nostoc sp.	*Cycas revoluta*	+
Nostoc sp.	*Anthoceros* sp.	+
Nostoc sp.	*Peltigera polydactyla*	+
Nostoc muscorum	*Encephalartos* sp.	−
Nostoc sp.	*Macrozamia lucida*	−
Anabaena azollae	*Azolla caroliniana*	−

Encephalartos and *Macrozamia* are not infectious on *Gunnera*. *Nostoc azollae* can also not colonize *Gunnera*. An older term for the symbiotic tissue from *Gunnera* is phycorhiza. In view of the similarities with actinorhiza or mycorrhiza this should rather be a **cyanorhiza**.

7.7 ANIMAL SYMBIOSES WITH CYANOBACTERIA AND *PROCHLORON*

The importance of these symbioses for animals rests not so much in N_2-fixation, but in the capacity of the microsymbiont to assimilate CO_2 by photosynthesis.

7.7.1 Sponges

Many marine sponges contain cyanobacteria in the mesophyll in close association to their archaeocytes. As well as contributing to C and N metabolism of the host, cyanobacteria also produce toxic secondary products which protect sponges against predators (Sara, 1971).

Endosymbiotic *Oscillatoria spongeliae* occurs in the horn-sponge *Dysidea herbacea*. Filaments grow up to 20 cells long. The thylakoids are very regular and at right angles to the longer cell walls. Also characteristic are the many star-shaped invaginations in the cytoplasm. N_2-fixing cyanobacteria occur in large, vacuolated cells of the sponge *Siphonochalina tarbernacula*. They account for less than 1% of the sponge's total volume. Cyanobacteria in the ectosome of the sponge *Theonella swinhoei* account for up to 15% of the total ectosome volume. In this case the cyanobacteria are freely scattered in the intercellular matrix. Measurement of N_2-fixation gives figures of around 2–4 nmol C_2H_4 h^{-1} g fresh weight^{-1} (Wilkinson and Fay, 1979). This is very low compared to rates in symbiotic plants.

7.7.2 Tunicates

Tropical tunicates like *Didemnum* spp. or *Lissoclinum patella* contain *Prochloron* in their cloacal cavity (Cox and Dwarte, 1981; Lewin, 1981a,b). Until now *Prochloron* has not been successfully cultivated. Thus all experiments must be performed on freshly isolated material. This genus is special among the prokaryotes in containing the pigments chlorophyll *a* and *b*. For this reason it occupies a monotypic genus alone, separated from the cyanobacteria. The nucleotide sequence of the 5S and 16S rRNA, the immunological properties of its RUBISCO, the biosynthesis of monogalactolipids and the presence of muraminic acid in the cell wall, all indicate a relationship to the cyanobacteria. The pigmentation and the presence of unbranched glucans indicates the presence of eukaryotic, green algal, genes. Taxonomically, *Prochloron* is placed at the very edge of the cyanobacteria due to the presence of these extra genes. The gene for phycobiliprotein has also been lost in these organisms. The symbiont provides photosynthetically produced oxygen for the ascidian host (Lesser and Stockaj, 1990). The symbiont in turn benefits from the high CO_2 concentrations in the cloacal cavity. *Prochloron* can be found either free in the cavity or squeezed between cloacal tissue. *Prochloron* has been shown to fix CO_2 at a rate of 120 μmol h^{-1} mg chlorophyll^{-1} (Kremer *et al.*, 1982).

For photoprotection against higher irradiance, *Prochloron* reacts with increasing activities of superoxide dismutase and ascorbate peroxidase (Lesser and Stokaj, 1990). Another species of the Prochlorophyta, *Prochlorothrix hollandica*, has been detected as a free-living organism which can be cultivated easily and will allow more biochemical studies on this group (Burger-Wiersma *et al.*, 1986, 1989).

7.8 REFERENCES

Berenger, J., Rojo, F. de Pedro, M.A. *et al.* (1987) Penicillin-binding proteins in the cyanelles of *Cyanophora paradoxa*, a eukaryotic photoautotroph sensitive to β-lactam antibiotics. *FEBS Lett.*, **224**, 401–5.

Bergersen, F.J., Kennedy, G.S. and Wittmann, W. (1965) Nitrogen fixation in the coralloid roots of *Macrozamia communis* L. Johnson. *Aust. J. Biol. Sci.*, **18**, 1135–42.

Bergman, B. Lindblad, P., Pettersson, A. *et al.* (1985) Immuno-gold localization of glutamine synthetase in a nitrogen-fixing cyanobacterium (*Anabaena cylindrica*). *Planta*, **166**, 329–34.

Böger, P. (1982) Photosynthese und Nutzung der Sonenenergie, in *Photosynthese* (ed. U. Heber) DFG, Bonn, pp. 191–8.

Bohnert, H.J., Crouse, E.J., Pouyet, J. *et al.* (1982) The subcellular localization of DNA components from *Cyanophora paradoxa*, a flagellate containing endosymbiotic cyanelles. *Eur. J. Biochem.*, **126**, 381–8.

Bonnett, H.T. and Silvester, W.B. (1981) Specificity in the *Gunnera–Nostoc* endosymbiosis. *New Phytol.*, **89**, 121–8.

Bothe, H., Nelles, H., Häger, H.P. *et al.* (1984) Physiology and biochemistry of N_2-fixation by cyanobacteria, in *Advances in Nitrogen Fixation Research* (C. Veeger and W. Newton) Martinus Nijhoff/Dr Junk, The Hague, pp. 199–210.

Breiteneder, H., Seiser, C., Löffelhardt, W. *et al.* (1988) Physical map and protein gene map of cyanelle DNA from the second known isolate of *Cyanophora paradoxa* (Kies-Strain). *Curr. Genet.*, **13**, 199–206.

Burger-Wiersma, T., Stal, L.J. and Mur, L.R. (1989) *Prochlorothrix hollandica* gen. nov., sp., sp. nov., a filamentous oxygenic photoautotrophic procaryote containing chlorophylls *a* and *b*: Assignment to Prochlorotrichaceae fam. nov. and order Prochlorales Florenzano, Balloni, and Materassi 1986, with emendation of the ordinal description. *Int. J. Syst. Bacteriol.*, **39**, 250–7.

Burger-Wiersma, T., Veenhuis, M., Korthals, H.J. *et al.* (1986) A new prokaryote containing chlorophylls *a* and *b*. *Nature*, **320**, 262–4.

Calvert, H.E., Pence, M.K. and Peters, G.A. (1985) Ultrastructural ontogeny of leaf cavity trichomes in *Azolla* implies a functional role in metabolite exchange. *Protoplasma*, **129**, 10–27.

Carr, N.G. and Whitton, B.A. (eds) (1982) *The Biology of Cyanobacteria.* Bot. Monogr., Vol. 19, Blackwell Scientific, Oxford.

Cox, G. and Dwarte, D.M. (1981) Freeze-etch ultrastructure of a *Prochloron* species – the symbiont of *Didemnum molle. New Phytol.*, **88**, 427–38.

Doyle, M.F. and Scogin, R. (1988) A comparative phytochemical profile of the Gunneraceae. *N. Z. J. Bot.*, **26**, 493–6.

Duckett, J.G., Prasad, A.K.S.K., Davies, D.A. and Walker, S. (1977) A cytological analysis of the *Nostoc*-bryophyte relationship. *New Phytol.*, **79**, 349–62.

Floener, L. and Bothe, H. (1980) Nitrogen fixation in *Rhopalodia gibba*, a diatom containing blue-greenish inclusions symbiotically, in *Endocytobiology, Endosymbiosis and Cell Biology*, Vol. 1, (eds W. Schwemmler and H.E.A. Schenk) W. de Gruyter, Berlin, New York, pp. 542–52.

Gallon, J.R. (1980) Nitrogen fixation by photoautotrophs, in *Nitrogen Fixation*, (eds W.D.P. Stewart and J.R. Gallon) Academic Press, London, New York, pp. 197–238.

Gallon, J.R. and Chaplin, A.E. (1988) Recent studies on N_2-fixation by non-heterocystous cyanobacteria, in *Nitrogen Fixation: Hundred Years After* (eds H. Bothe, F.J. de Bruijn and W.E. Newton) G. Fischer Verlag, Stuttgart, New York, pp. 183–94.

Häfele, U., Scherer, S. and Böger, P. (1988) Cytochrome aa_3 from heterocysts of the cyanobacterium *Anabaena variabilis*: isolation and spectral characterization. *Biochim. Biophys. Acta*, **934**, 186–90.

Harder, R. (1917) Ernährungsphysiologische Untersuchungen an Cyanophyceen, hauptsächlich dem endophytischen *Nostoc punctiforme. Z. Bot.*, **9**, 145.

Haselkorn, R., Basche, M., Böhme, H. *et al.* (1990) Nitrogen fixation in filamentous cyanobacteria, in *Nitrogen Fixation: Achievements and Objectives* (eds P.M. Gresshoff, L.E. Roth, G. Stacey and W.E. Newton) Chapman and Hall, New York, London, pp. 497–504.

Hill, D.J. (1975) The pattern of development of *Anabaena* in the *Azolla–Anabaena* symbiosis. *Planta*, **122**, 179–84.

Huang, T.-C. and Chow, T.-J. (1990) Characterization of the rhythmic nitrogen-fixing activity of *Synechococcus* sp. RF-1 at the transcription level. *Curr. Microbiol.*, **20**, 23–6.

Kaplan, D. and Peters, G.A. (1988) Interaction of carbon metabolism in the *Azolla–Anabaena* symbiosis. *Symbiosis*, **6**, 53–68.

Kentemich, T., Danneberg, G., Hundeshagen, B. and Bothe, H. (1988) The occurrence of the alternative nitrogenase in the cyanobacterium *Anabaena variabilis*, in: *Nitrogen Fixation: Hundred Years After* (eds H. Bothe, F.J. de Bruijn and W.E. Newton) Gustav Fischer Verlag, Stuttgart, New York, p. 89.

Kies, L. (1976) Untersuchungen zur Feinstruktur und taxonomischen Einordnung von *Gloeochaete wittrockiana*, einer apoplastidalen capsalen Alge mit blaugrünen Endosymbionten (Cyanellen). *Protoplasma*, **87**, 419–46.

Kies, L. (1979) Zur systematischen Einordnung von *Cyanophora Paradoxa*, *Gloeochaete wittrockiana* und *Glaucocystis nostochinearum*. *Ber. Deutsch. Bot. Ges.*, **92**, 445–54.

Kies, L. (1988) The effect of penicillin on the morphology and ultrastructure of *Cyanophora*, *Gloeochaete* and *Glaucocystis* (Glaucocystophyceae) and their cyanelles. *Endocytobiosis Cell Res.*, **5**, 361–72.

Kies, L. (1989) Ultrastructure of *Cyanoptyche gloeocystis* f. *dispersa* (Glaucocystophyceae). *Pl. Syst. Evol.*, **164**, 65–73.

Kremer, B.P., Kies, L. and Rostami-Rabet, A. (1979) Photosynthetic performance of cyanelles in the Endocyanomes *Cyanophora*, *Glaucosphaera*, *Gloeochaete* and *Glaucocystis*. *Z. Pflanzenphysiol.*, **92**, 303–17.

Kremer, B.P., Pardy, R. and Lewin R.A. (1982) Carbon fixation and photosynthates of *Prochloron*, a green alga symbiotic with an ascidian, *Lissoclinum patella*. *Phycologia*, **21**, 258–63.

Lesser, M.P. and Stokaj, W.R. (1990) Photoadaptation and protection against active forms of oxygen in the symbiotic procaryote *Prochloron* sp. and its ascidian host. *Appl. Environ. Microbiol.*, **56**, 1530–5.

Lewin, R.A. (1981a) The prochlorophytes, in *The Prokaryotes. A Handbook on Habitats, Isolation, and Identification of Bacteria* (eds M.P. Starr, H. Stolp, H.G. Trüper *et al.*) Springer-Verlag, Berlin Heidelberg, pp. 258–66.

Lewin, R.A. (1981b) *Prochloron* and the theory of symbiogenesis. *Ann. N. Y. Acad. Sci.*, **361**, 325–9.

Löffelhardt, W., Mucke, H., Breiteneder, H. *et al.* (1987) The cyanelle genome of *Cyanophora paradoxa*. Chloroplast and cyanobacterial features. *Ann. N. Y. Acad. Sci.*, **503**, 550–2.

Mague, T.H. (1977) Ecological aspects of dinitrogen fixation in blue–green algae, in *A Treatise on Dinitrogen Fixation*, Section IV, (eds R.W.F. Hardy and A.H. Gibson) Wiley, New York, pp. 85–140.

Meeks, J.C., Steinberg, N.A., Joseph, C.M. *et al.* (1985) Assimilation of exogenous and dinitrogen-derived $^{15}NH_4^+$ by *Anabaena azollae* separated from *Azolla caroliniana* Wild. *Arch. Microbiol.*, **142**, 229–33.

Mellor, R.B., Gadd, G.M., Rowell, P. and Stewart W.D.P. (1981) A phytohaemagglutinin from the *Azolla–Anabaena* symbiosis. *Biochem. Biophys. Res. Commun.*, **99**, 1348–53.

Mollenhauer, D. (1986) Blaualgen der Gattung *Nostoc* – ihre Rolle in Forschung und Wissenschaftsgeschichte IV. *Natur Museum*, **116**, 104–20.

Neumann, D. (1977) Ultrastrukturelle Untersuchungen zur Symbiose von Cyanophyceen mit Cycadeen (*Cycas circinnalis* L., *Zamia furfuracea* L.). *Biochem. Physiol. Pflanzen*, **171**, 313–22.

Nierzwicki-Bauer, S.A. and Aulfinger, H. (1990) Ultrastructural characterization of eubacteria residing within leaf cavities of symbiotic and cyanobiont-free *Azolla mexicana. Curr. Microbiol.*, **21**, 123–9.

Padhy, R.N. (1985) Cyanobacteria employed as fertilizers and waste disposers. *Nature*, **317**, 475–6.

Papen, H., Kentemich, T., Schmülling, T. and Bothe, H. (1986) Hydrogenase activities in cyanobacteria. *Biochimie*, **68**, 121–32.

Peters, G.A. and Calvert, H.E. (1983) The *Azolla–Anabaena azollae* symbiosis, in *Algal Symbiosis* (ed. L.J. Goff) Cambridge University Press, Cambridge, pp. 109–45.

Peters, G.A. and Meeks, J.C. (1989) The *Azolla–Anabaena* symbiosis. Basic biology. *Ann. Rev. Plant Physiol. Plant Mol. Biol.*, **40**, 193–210.

Peters, G.A., Toia, R.E., Jr, Calvert, H.E. and Marsh, B.H. (1986) Lichens to *Gunnera* – with emphasis on *Azolla. Plant Soil*, **90**, 17–34.

Peters, G.A. Toia, R.E. Jr, Raveed, D. and Levine, N.J. (1978) The *Azolla–Anabaena azollae* relationship. VI. Morphological aspects of the association. *New Phytol.*, **80**, 583–93.

Petro, M.J. and Gates, J.E. (1987) Distribution of *Arthrobacter* sp. in the leaf cavities of four species of the N-fixing *Azolla* fern. *Symbiosis*, **3**, 41–8.

Philipps, E.J. and Mitsui, A. (1982) Light intensity preference and tolerance of aquatic photosynthetic microorganisms, in Zaborsky, O.R.: *CRC Handbook of Biosolar Resources*, Vol. 1, Part 2, Basic Principles 257–308, CRC Press, Boca Raton, pp. 257–308.

Plazinski, J., Zheng, Q., Taylor, R. *et al.* (1990) DNA probes show genetic variation in cyanobacterial symbionts of the *Azolla* fern and a closer relationship to free-living *Nostoc* strains than to free-living *Anabaena* strains. *Appl. Environ. Microbiol.*, **56**, 1263–70.

Ray, T.B., Peters, G.A., Toia, R.E. and Mayne, B.C. (1978) *Azolla–Anabaena* relationship. *Plant Physiol.*, **62**, 463–7.

Reisser, W. (1984) Endosymbiotic cyanobacteria and cyanellae, in *Cellular Interactions* (eds H.F. Linsken and J. Heslopp-Harrison) Vol. 17, *Encyclopedia of Plant Physiology*, New Series, Springer, Berlin.

Rodgers, G.A. and Stewart, W.D.P. (1977) The cyanophyte-hepatic symbiosis. I. Morphology and physiology. *New Phytol.*, **78**, 441–58.

Sara, M. (1971) Ultrastructural aspects of the symbiosis between two species of the genus *Aphanocapsa* (Cyanophyceae) and *Ircinia variabilis* (Demospongia). *Mar. Biol.*, **11**, 214–21.

Scherer, S., Häfele, U. Krüger, G.H.J. and Böger, P. (1988a) Respiration, cyanide-insensitive oxygen uptake and oxidative phosphorylation in cyanobacteria. *Physiol. Plant.*, **72**, 379–84.

Scherer, S., Riege, H. and Böger, P. (1988b) Light-induced proton efflux of the cyanobacterium *Anabaena variabilis*, in *Biochemistry of the Algae and Cyanobacteria* (eds L.J. Rogers and J.R. Gallon) Clarendon Press, Oxford, pp. 121–9.

Schmidt, H. (1990) Licht- und elektronenmikroskopische Untersuchungen zum Infektionsablauf der *Gunnera–Nostoc*-Symbiose. Dissertation, Hamburg.

Schrautemeier, B. and Böhme, H. (1985) A distinct ferredoxin for nitrogen fixation isolated from heterocysts of the cyanobacterium *Anabaena variabilis*. *FEBS Lett.*, **184**, 304–8.

Silvester, W.B. and Smith, D.R. (1969) Nitrogen fixation by *Gunnera–Nostoc* symbiosis. *Nature*, **224**, 1231.

Stewart, W.D.P. (1985) Differentiating cyanobacteria rearrange their *nif* genes. *Nature*, **314**, 404–23.

Stewart, W.D.P. and Rodgers, G.A. (1977) The cyanophyte-hepatic symbiosis. II. Nitrogen fixation and the interchange of nitrogen and carbon. *New Phytol.*, **78**, 459–71.

Stewart, W.D.P. and Rodgers, G.A. (1978) Studies on the symbiotic blue–green algae of *Anthoceros*, *Blasia* and *Peltigera*. *Ecol. Bull.* (Stockholm), **26**, 247–259.

Stewart, W.D.P., Rowell, P. and Rai, A.N. (1980) Symbiotic nitrogen-fixing cyanobacteria, in *Nitrogen Fixation* (eds W.D.P. Stewart and J.R. Gallon) Academic Press, London, pp. 239–77.

Strasburger, E. (1873) Über *Azolla*. Hermann Dabis Verlag, Jena, 86pp.

Tyagi, V.V.S., Ray, T.B., Mayne, B.C. and Peters, G.A. (1981) The *Azolla–Anabaena azollae* relationship. XI. Phycobiliproteins in the action spectrum for nitrogenase-catalyzed acetylene reduction. *Plant Physiol.*, **68**, 1479–84.

Webb, D. (1981) Effects of light quality on root elongation and nodulation of *Zamia floridana* DC. seedlings in sterile culture. *Z. Pflanzenphysiol.*, **104**, 253–8.

Wilkinson, C.R. and Fay, P. (1979) Nitrogen fixation in coral reef sponges with symbiotic cyanobacteria. *Nature*, **279**, 527–9.

Wittman, W. and Bergersen, F.J. and Kennedy, G.S. (1965) The coralloid roots of *Macrozamia communis* L. Johnson. *Aust. J. Biol. Sci.*, **18**, 1129.

8
Phycosymbioses

8.1 THE PLASTIDS OF CRYPTOPHYTES AND MASTIC AMOEBA AS REDUCED SYMBIOTIC CELLS

Cryptophyceae such as *Cryptomonas*, *Chilomonas* or *Rhodomonas* possess variously coloured plastids. These plastids contain chlorophyll *a* and *c*, carotenoids, phycoerythrin and phycocyanin. Such plastids are not surrounded by two membranes, as is usual, but by four (Fig. 8.1). Outside the chloroplast membrane, but inside the third membrane and the fourth, which is *in continuum* with the nuclear membrane, is a vestigal nucleus, the so called **nucleomorph** (Gibbs, 1983). This special cytology has led to the suggestion that plastids of the Cryptophyceae may be the remains of single-celled Rhodophytes (Gibbs, 1981). The division of nucleomorphs normally occurs before the division of plastids. Microtubules are not involved in this nuclear division which is not inhibited by colchicine. The nucleomorph is bounded by a body similar to the nucleolus and a membrane with pores.

The nucleomorph contains DNA (Hansmann *et al.*, 1985) which can be separated from the DNA of the host nucleus, the mitochondria and the plastids (Eschbach *et al.*, 1991a). RNA is present in the 'fibrogranular body', the nucleolus of the nucleomorph (Hansmann, 1988).

By pulse field gel electrophoresis it can be demonstrated that the nucleomorph in the cryptomonad *Pyrenomonas salina* contains three chromosomes of 195 kb, 225 kb and 240 kb. The combined haploid genome of 660 kb is the smallest eukaryotic genome known (Eschbach *et al.*, 1991a). On the basis of small subunit rRNA gene sequences, the host cells of *P. salina* shows a close relationship to chlorophytes (Eschbach *et al.*, 1991b).

In the mastic amoeba *Chlorachnion reptans* green chloroplasts have probably evolved from a symbiotic green alga, reduced to the plastid, a nucleomorph and 80S ribosomes (Ludwig and Gibbs, 1987).

8.2 DINOPHYTES AS SYMBIONTS

The host spectrum of symbiotic dinoflagellates is extraordinarily wide. It encompasses some protozoans, the Cnidaria, the Plathelminths and the Mollusca (Vernberg, 1974; Goff, 1983).

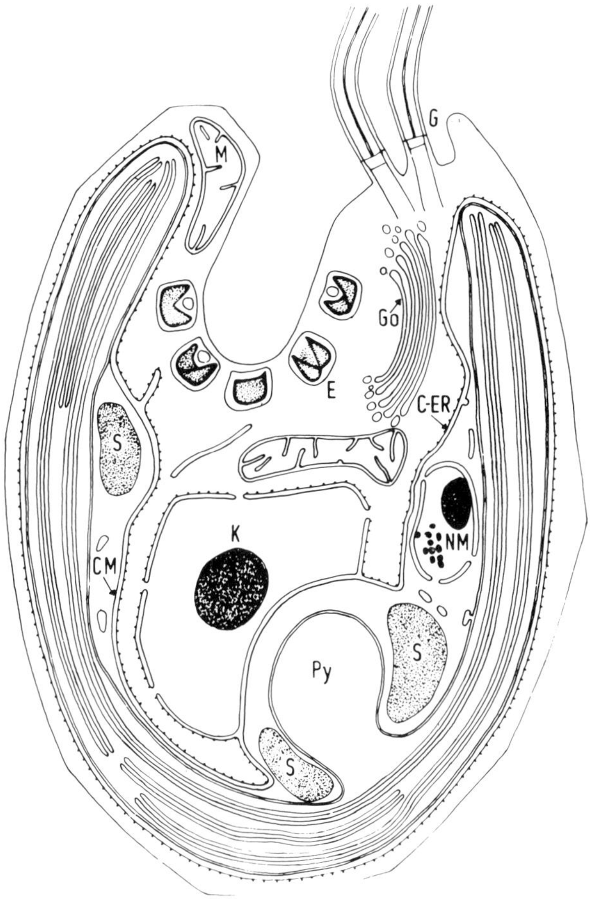

Fig. 8.1. Diagram of longitudinal section of *Cryptomonas*. K, nucleus; NM, nucleomorph; E, ejectosome; Py, pyrenoid matrix; CM, chloroplast membrane; C-ER, chloroplast ER; S, starch; M mitochondrion; Go, Golgi apparatus; G, flagella. (Based on Gibbs, 1983.)

From the number of host species involved and their ecology, most important are Cnidaria, especially the Actiniaria, Madreporaria (scleractinia, ossified corals) and Octocorallia groups.

The symbionts are generally named zooxanthella. They are divided into two groups: (1) *Symbiodinium microadriaticum* strains and (2) *Amphidinium* strains. Isolates from various hosts can be classified according to the number of isoenzymes of esterases, glucose phosphate isomerase and malate dehydrogenase which they exhibit in pure culture (Table 8.1). Thus symbionts from 20 different hosts fall into 12 isoenzyme classes, indicating 12

Table 8.1 Isoenzyme patterns of esterase (EST), glucose phosphate isomerase (GPI) and malate dehydrogenase (MDH) in different strains of the *Symbiodinium microadriaticum* group from different host animals. (Based on Schoenberg and Trench, 1980)

Symbiodinium *strain*	*Isolated from*	*Isoenzymes*		
		EST	*GPI*	*MDH*
A	*Aiptasia tagetes* (Actiniaria)	1	1	1
A	*Gorgonia ventalina* (Octocorallia)	1	1	1
A	*Pseudopterogorgia bipinnata* (Octocorallia)	1	1	1
B	*Bartholomea annulata* (Actiniaria)	6	4	6
C	*Cassiopeia xamachana* (Scyphozoa)	2	3	3
C	*Protopalythoa grandis* (Zoanthidea)	2	3	3
C	*Pseudopterogorgia acerosa* (Octocorallia)	2	3	3
D	*Condylactis gigantea* (Actiniaria)	10	2	7
H	*Bartholomea annulata* (Actiniaria)	4	4	4
H	*Heteractis lucida* (Actiniaria)	4	4	4
H	*Rhodactis sancti-thomae* (Corallimorpharia)	4	4	4
M	*Meandrina meandrites* (Scleractinia)	9	2	1
N	*Mussa angulosa* (Scleractinia)	1	2	2
N	*Pseudopterogorgia americana* (Octocorallia)	1	2	2
O	*Oculina diffusa* (Scleractinia)	7	6	7
P	*Palythoa mammilosa* (Zoanthidea)	5	5	5
Q	*Palythoa mammilosa* (Zoanthidea)	3	4	4
Q	*Protopalythoa grandis* (Zoanthidea)	3	4	4
T	*Tridacna gigas* (Mollusca)	11	2	3
Z	*Zoanthus sociatus* (Zoanthidea)	8	7	8

strains of *Symbiodinium*. There are also significant differences to be found in chromosome number, chromosome volume and in the nuclear volume from symbionts from various host animals (Table 8.2). Taxonomically the *Symbiodinium microadriaticum* group is a miscellany. Similar conclusions can be reached after cross inoculation tests (Table 8.3). Non-infected *Aiptasia tagetes* (Actinia) is reinfected by its natural symbiont to the extent of 2×10^4 algae per mm^3 tentacle within 50 days. The same animal is infected by a strain from *Bartholomea angulata* (2.5×10^3 per 10 mm^3 tentacle) and by a strain from *Heteractis lucida* (250 algae per mm^3 tentacle). A strain from the anemone *Zoanthus sociatus* was not infective (Trench, 1979). Thus infection groups and isoenzyme groups do not correlate with each other. The conclusion seems to be that the *Symbiodinium microadriaticum* group consists of various species subdivided into infection groups with different specificities.

Besides *S. microadriaticum* other new species have been described: *Symbiodinium goreauii*, in symbiosis with the Caribbean sea anemone *Racactis lucida*; *Symbiodinium kawagutii*, a symbiont of the stony coral

Table 8.2 Morphometric data for chromosomes and nuclei (G_2-phase of cell cycle) from strains of the *Symbiodinium microadriaticum* group, isolated from different host species. (Based on Blank and Trench, 1985)

Symbiont isolated from	*Chromosomes*		*Nucleus volume* (μm^3)
	Number	*Volume* (μm^3)	
Cassiopeia xamachana (Jellyfish)	97 ± 2	1.6 ± 0.1	11.3 ± 2.0
Heteractis lucida (Sea anemone)	74	7.8	27.2
Anthopleura elegantissima (Sea anemone)	50 ± 1	3.7 ± 0.3	13.9 ± 1.3
Montipora verrucosa (Stony coral)	26	3.2	7.6

Table 8.3 Infection and recolonization of symbiont-free *Aiptasia tagetes* by strains of the *Symbiodinium microadriaticum* group, isolated from different hosts. (Based on Trench, 1979)

Algal strain	*Isolated from*	*Order of Cnidaria*	*Recolonized algae*★ *(number mm*$^{-3}$ *tentacle)*
A	*Aiptasia tagetes*	Actiniaria	20 000 ± 1600
M	*Mussa angulosa*	Scleractinia	7 000 ± 400
B	*Bartholomea angulata*	Actiniaria	2 500 ± 370
H	*Heteractis lucida*	Actiniaria	250 ± 80
O	*Oculina diffusa*	Scleractinia	200 ± 90
Z	*Zoanthus sociatus*	Zoanthidea	0

★ Determined 50 days after infection.

Montipora verrucosa and *Symbiodinium pilosum*, isolated from the zoanthid *Zoanthus sociatus* (Trench and Blank, 1987).

During their development, several host groups show symbiont-free stages. In the laboratory, cultivated swarms of *Symbiodinium* swim into the mouth of polyps and are endocytosed into host cells. How the fusion of lysosomes with endocytosed symbionts, and the ensuing digestion is prevented, is not known. In other hosts such as some corals and zoanthids, algae are taken into the egg cell shortly after fertilization. Larval stages of such animals do not need to be newly infected and resulting adults show a very high symbiotic specificity. Species living in a continual state of symbiosis are physiologically very stable. This leads to the ecological situation where symbioses are more common in asexually reproducing organisms than those undergoing sexual replication.

Metabolic exchange from *Symbiodinium* cells to the host takes the form of glycerol (Trench, 1979). Small amounts of other products of photosynthesis, namely glucose and alanine are also transferred. Between 20 and 60%

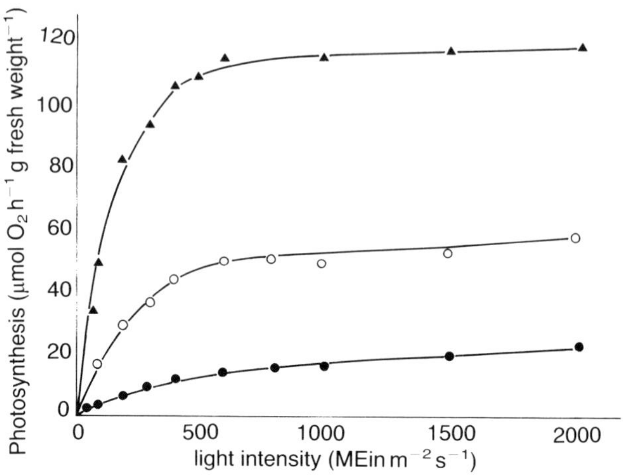

Fig. 8.2. Photosynthetic O_2 production in *Tridacna gigas* with zooxanthellae, in relationship to light intensity. Animal size (wet weight): (●) 1700 g; (○) 1.3 g; (■) six clams with an average weight of 0.014 g. (Based on Fisher *et al.*, 1985.)

of photosynthetically fixed carbon lands in the host organism in most cases. In sea anemones up to 100% of the energy requirements of the host can be supplied by the symbiotic algae (Muscatine *et al.*, 1984). The export of carbohydrate from the symbiont is increased by tissue factors, probably proteins, from the host (Hinde, 1987). This explains discrepancies in export with age. Cultivated symbionts lose the ability to export carbohydrate with age after isolation. Due to commercial importance the giant clam *Tridacna gigas* and its symbionts have been intensively studied. Adult clams have never been found without zooxanthella. The symbiont is mostly restricted to the haemal sinus tissue. This is also called the coat tissue. Young mussels contain about 10 zooxanthellae per μg tissue, larger mussels only 2–3 symbionts. The specific photosynthetic ability of smaller organisms reaches 100 μmol $O_2 h^{-1} g$ fresh weight^{-1} (Fig. 8.2). Due to lack of shadowing, light saturation is reached at 500 μEinstein $m^{-2} s^{-1}$ (with sunlight at sea level 4.6 μEinstein $m^{-2} s^{-1}$ is equivalent to 1 W m^{-2}). In larger animals this may be extended to 1500–2000 μEinstein m^{-2} g fresh weight^{-1}. In comparison, sun-adapted leaves produce 500–700 μmol $O_2 h^{-1}$ g fresh weight^{-1}, leaves in shadows, 60–100 μmol (at 20°C).

Freshly isolated symbionts cannot take up nitrate and do not possess nitrate reductase, whereas cultivated symbionts can (Wilkerson and Trench, 1985). Both forms take up NH_4^+ at equal rates. Zooxanthellae take up organic compounds from the host. Over 40% of radioactivity was found in the symbionts after feeding [^{35}S]methionine to the anemone *Anthopleura eleqantissima*. This exchange influences the behaviour of the host. In a 14

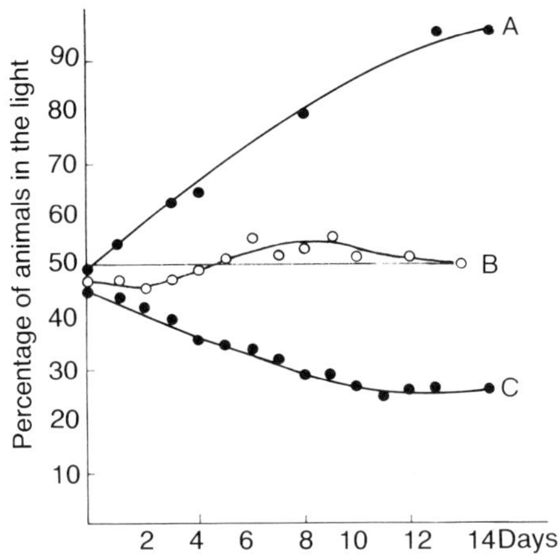

Fig. 8.3. Phototaxis in sea anemones (*Anthopleura elegantissima*). The percentage of animals in the illuminated half of a seawater bath after random distribution at the start of the experiment. A, Sea anemones plus zooxanthellae, originally from sunny sea locations; B, specimens deprived of zooxanthellae; C, specimens plus zooxanthellae, originally from shaded locations. (Based on Buchsbaum-Pearse, 1974.)

day experiment, (Fig. 8.3) anemones from a sunny environment with symbionts showed positive phototaxis. Non-symbiotic animals gave no tactic response and symbiotic animals from an overshadowed environment showed negative phototaxis.

8.2.1 Coral reefs

Coral reefs extend over a 600 000 km^2 area in shallow tropical seas where the temperature is always over 20°C (335 000 km^2 in the Pacific Ocean, 185 000 km^2 in the Indian Ocean and 97 000 km^2 in the Atlantic Ocean) (Smith, 1978). With few exceptions all reef-building corals are associated with *Symbiodinium microadriaticum*. A symbiotic density of $1–5 \times 10^6$ zooxanthellae per cm^2 coral surface and a photosynthetic C-fixing rate of 20–80 $\mu mol\ CO_2\ h^{-1}\ mg\ chlorophyll^{-1}$ results in 2.5 g C fixed per m^2 coral surface. Due to the highly branched nature of some corals, calculations result in a figure of 15 g C fixed per day per m^2 sea surface. This gives a value of 3×10^9 tonnes carbon fixed per year in the coral seas. This is almost 10% of total marine primary production. Annual calcification rate is about 8×10^8 tonnes.

The coral host can improve the photosynthetic efficiency by a mechanism which transforms wavelength from a shorter wavelength to those that match the absorption maxima of the algal symbiont (Schlichter and Fricke, 1990).

By this mechanism, the coral *Leptoseris fragilis* can grow at a depth of 95–135 m below sea level. The IK, the light intensity producing the maximal photosynthetic response, was determined at $10\,\mu Ein\, m^{-2} s^{-1}$. This is only 2% of the light intensity for maximal photosynthesis of the zooxanthella in *Tridacna* (Fig. 8.2).

In the scleractinian coral *Plesiastrea versipora* a host factor, probably a protein, stimulates the release of up to 40% of the ^{14}C fixed from the microsymbiont. The host extracts have no effect on the photosynthetic rate (Sutton and Hoegh-Guldberg, 1990).

Zooxanthellae intercept the ammonia before it is excreted into the environment from the corals. Up to 95% of the total carbon fixed can be transferred to the animals (Falkowski, 1991).

Apart from the excretion of glycerol, glucose and alanine, symbiotic lipid excretion is also important here. Corals use such lipids directly as reserve fat, which can form up to 40% of the total body dry weight. In situations where reserve lipids are metabolized, the symbiont can use the resulting acetate and CO_2 for light-dependent syntheses. Most important in this symbiosis is the interaction of symbiotic metabolism with that of the calcareous skeleton formation, and with that, the building of the coral reef itself. Reef growth is 3–10 times faster in the light than in the dark. Dissolved CO_2 in the sea equilibrates with H_2CO_3, HCO_3^- and CO_3^{2-}

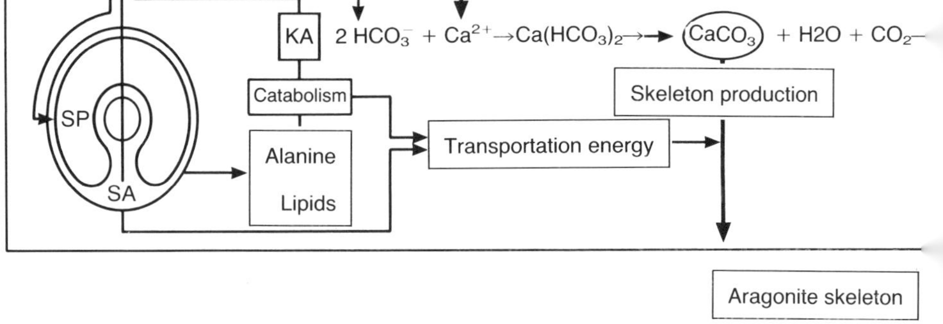

Fig. 8.4. Calcification and skeleton production in reef-building corals with *Symbiodinium microadriaticum* as symbiont. SP, Photosynthesis by symbionts (CO_2 fixation); SA, respiration of symbionts; KA, respiration of corals. (Based on Kremer, 1985.)

(Fig. 8.4). When CO_2 is used photosynthetically, the equilibrium product, HCO_3^-, becomes readily available for structural incorporation. Ca^{2+} is not limiting due to its high concentration in sea water (10 mM) and an efficient uptake system in the animal. Thus atolls containing 500 km^3 solid substance can be produced by this symbiosis.

8.3 CHLOROPHYTES AS SYMBIONTS

The most highly researched symbioses are those of *Hydra*, *Paramecium* and *Convoluta roscoffensis* with chlorophytes (Muscatine, 1989). Symbiotic green algae have, however, also been reported from many other host animals such as the fresh water mussel *Anodonta*, the sea anemone *Anthopleura xanthocamica*, *Radiolaria* of the genus *Thalassolampe* and heliozoans as *Acanthocystis turfacea* (Matzke *et al.*, 1990).

8.3.1 The *Hydra–Chlorella* symbiosis

Digestive cells of the endoderm of *Hydra viridis* take up symbiotic *Chlorella* strains by phagocytosis from the digestive cavity. They are then confined alone in single vacuoles and transported from the apex to the base of the host cell (McAuley and Smith, 1982). Not only symbiotic *Chlorella*, but also other organisms can be taken up. These are enclosed in somewhat larger vacuoles, which remain at the distal end of the cell. Such cells are normally discharged after 24 h. A subpopulation of symbiotic *Chlorella* is also lost in the first 1–3 days. After this time the number of symbionts per absorption cell increases to 12 per cell (Fig. 8.5). Cropping the upper two-thirds of the

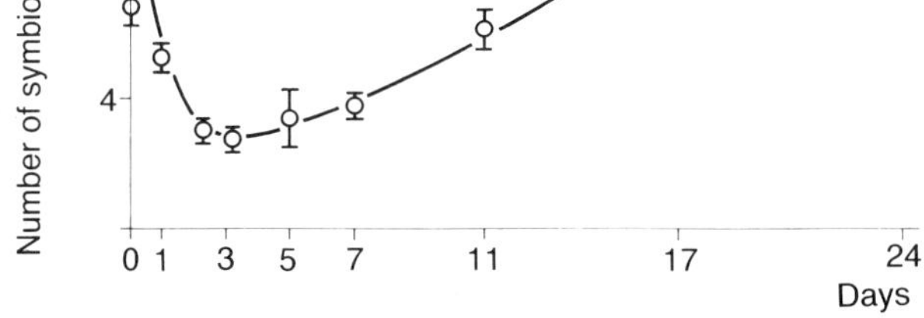

Fig. 8.5. Uptake of freshly isolated symbiotic *Chlorella* by symbiont-free *Hydra*. (Based on Jolley and Smith, 1980.)

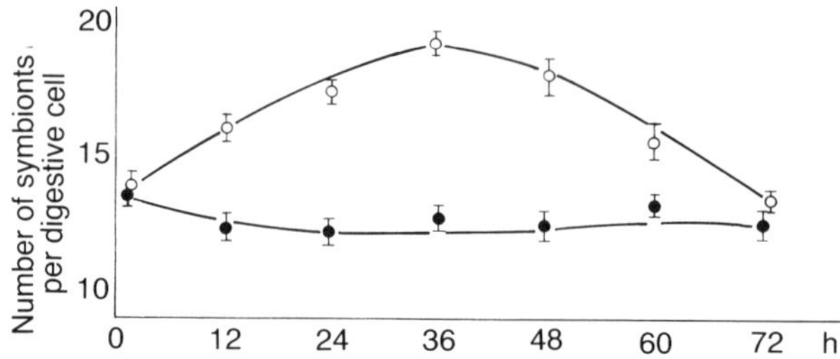

Fig. 8.6. Average number of algae per digestive cell from the peduncle region of *Hydra*, at various times after removal of the upper two-thirds of the animal. (○) Regenerating peduncle; (●) peduncles with grafted heads. (Based on Pardy and Heacox, 1976.)

Hydra results in a further increase 36 h later, then a return to normal levels (Fig. 8.6). Amputated animals, whose trunks are grafted on again, react with a slight decrease in the symbiont number (Fig. 8.6). The symbiont number develops over several weeks to an average of 12–15 per cell. This value is regulated from non-infected parts of the *Hydra* by physiological mechanisms which are as yet unknown. The possibilities being discussed include (McAuley, 1985):

1. Controlling the pH value in the perialgal vacuole. It is known that the export of maltose from *Chlorella* is pH-dependent, with an optimum at pH 4.5. Thus the animal, by controlling the pH of the *Chlorella*-environment may remove more or less carbon from the symbiont.
2. Regulation by a 'cell-division factor' from the host. Biochemically nothing definite is known about this.
3. Regulation by inorganic nutrients. It has been shown experimentally, that symbiotic number in cells increase when nitrate, ammonium, sulphate and phosphate concentrations in the culture medium are raised. This can lead to the death of the host. Asymbiotic organisms are not affected. It is not known if these nutrients directly affect *Chlorella* growth or if they are metabolically processed by the host first.

In the symbiotic state the algae seem to be nitrogen limited (McAuley, 1987).

Hydra viridis must be fed continually and cannot live fully autotrophically. Large numbers of bacteria are found in the symbiotic ectoderm cells which disappear when symbiotic photosynthesis is inhibited. Bacteria outnumber symbiotic algae 10 to one (Margulis *et al.*, 1978). The bacteria are Gram-negative rods with polar flagella.

The infection and recognition of symbionts by the host can be divided into five phases (Pardy and Muscatine, 1973):

1. Contact phase
2. Engulfment phase
3. Recognition phase
4. Intracellular migration phase
5. Multiplication of symbionts

Recognition of symbiotic algae as against non-symbionts and food particles is probably achieved by the microvilli of entoderm cells. Those microvilli leading to endocytosis of symbionts are different from those involved in phagocytosis of food particles, inasmuch as they not appear to possess coated vesicles. Such microvilli make a network over the surface of the alga before internalization takes place. When dead algae or latex beads are used, no such network is seen, but rather a tubular structure is formed from the cell membrane. Thus specificity on behalf of the host is given by the different types of microvilli and their cell-surface properties. Also cell wall structures of the algae play an important role in the formation of the host–microsymbiont interface (Reisser, 1990).

There is no clear border between the group of symbiotic (compatible) and the group of other, non-symbiotic (incompatible) *Chlorella*. Some strains are eliminated quickly from the host cell. Other non-symbionts persevere in the host. *Chlorella* strains containing sporopollenin and capable of mixotrophic growth are generally more likely to be accepted (Jolley and Smith, 1980). Maltose excretion is not a decisive signal in recognition as symbionts which are incapable of this have been isolated.

From 15 taxa of *Chlorella* studied, only *Chlorella saccharophila*, *Chlorella fusca* var. *vacuolata*, *Chlorella kessleri*, *Chlorella luteoviridis* and *Chlorella*

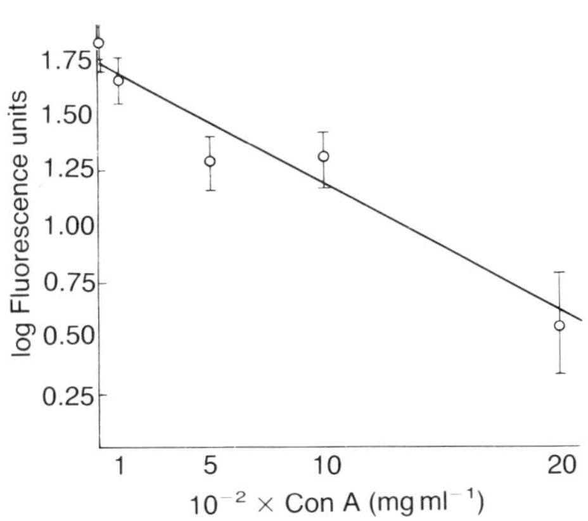

Fig. 8.7. Effect of concanavalin A on the uptake of *Chlorella* by *Hydra viridis*. Determination of chlorophyll fluorescence after re-isolation of symbionts. (Based on Meints and Pardy, 1980.)

protothecoides were able to form a stable symbiosis with *Hydra viridis* (Kessler *et al.*, 1988).

Acid tolerance (growth at pH 3.5–4.0) seems to be one prerequisite for the formation of a stable symbiosis. However, some strains of *Chlorella* growing at this pH, do not form a stable symbiosis. Thus, growth at low pH is not the only essential character for symbiotic strains (Rahat *et al.*, 1989).

The lectin concanavalin A inhibits the uptake of *Chlorella* into *Hydra viridis* in a concentration-related manner (Fig. 8.7). Other lectins, such as wheat germ agglutinin or *Lens culinaris* lectin also show some inhibitory effect. Thus, by similar reasoning to that involved in the *Rhizobium*–legume symbiosis, it is thought that **glycoconjugates** are involved in the recognition process.

8.3.2 The ciliate–*Chlorella* symbiosis

Host specificity in ciliate–*Chlorella* symbiosis, similar to that in the *Hydra*–*Chlorella* system, is very low. For example, *Paramecium bursaria* can be infected by its own symbionts, isolated from infected cells, and also to the same extent by many other free-living species. However, the stability of the symbiosis with those species is somewhat lower. Within the genus *Paramecium* however, only *P. bursaria* can be infected. Transplantation experiments using infected *Stentor polymorphus* and the non-host *Stentor coeruleus* show that *Chlorella* can only survive where an excess of cytoplasm from host cells is present.

Ciliates from various orders may be symbiotic: *Paramecium* from the

Table 8.4 Excretion of photosynthetic products by isolated symbiotic and by free-living *Chlorella* strains. (Based on Reisser, 1984)

Origin of Chlorella *strain*	*Carbon excreted (% of photosynthetically fixed C)*	*Major components of exudate*	*Photosynthetic O_2-production (nmol O_2 min^{-1} mg^{-1} chlorophyll* a + b*)**
Paramecium bursaria	45	Maltose	745
Hydra viridis	37	Maltose	1567
Euplotes daidaleos	6	Fructose, xylose	270
Climacostomum virens	8	Glucose, fructose	419
Free-living *Chlorella vulgaris* strain 211-11b	2	none	246
Free-living *Chlorella fusca* strain 211-8b	0.7	none	824

* Light intensity: 6 W m^{-2} (ca. 1550 lx).

holotricha, *Stentor* from the spirotricha and *Vorticella* from the peritricha. *Chlorella* isolated from symbioses export much more carbohydrate than from free-living strains (Table 8.4) (Weis, 1979). This carbohydrate takes the form of maltose, fructose, glucose or xylose. Photosynthetic oxygen evolution was between 270 and 750 nmol O_2 min^{-1} mg $chlorophyll^{-1}$. This is the same order of magnitude as for free-living strains under the same conditions. As expected, the photosynthetic compensation point in symbiosis (90 μEin m^{-2} s^{-1}) is higher than in isolated algae (6 μEin m^{-2} s^{-1}).

Due to the higher CO_2-concentration from the respiration of the host cell, the rate of photosynthetic oxygen production within the host cells is higher than the rate by algae in the isolated state at normal CO_2-concentrations (Reisser, 1980).

In the light the symbiotic algae accumulate starch (Fig. 8.8) whereas in the dark starch reserves are rapidly used (Fig. 8.9). The maltose exported covers a large part of the carbon and energy demand of the host animal. Algae-free *Paramecium* take up much more carbohydrate than symbiotic strains.

The symbiotic specificity of the heterotrichous ciliate *Climacostomum virens* (Fig. 8.10) for *Chlorella* strains is higher than that exhibited by *Paramecium bursaria*. Only *Chlorella* isolated from the symbiosis may reinfect to form a stable association, whereas other *Chlorella* strains are digested in food vacuoles (Fig. 8.11). The symbiotic ciliates can grow completely phototrophically in an axenic medium containing only simple mineral salts. Dark conditions and feeding with appropriate food organisms leads to the loss of the *Chlorella* (Reisser and Wiessner, 1984). Non-symbiotic *Chlorella* may enter the cell but are digested in phagocytic vacuoles. Another interest-

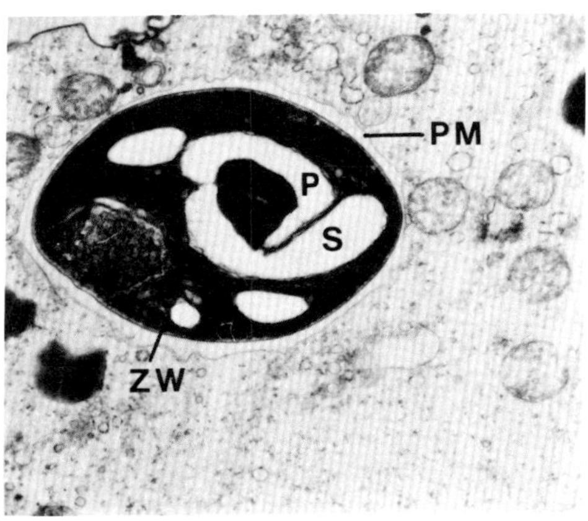

Fig. 8.8. Ultrastructure of *Chlorella* in symbiosis with *Paramecium bursaria*, from illuminated culture. P, Pyrenoid; S, starch; PM, perialgal membrane; ZW, cell wall. Magnification ×16 500. (From Reisser, 1976.)

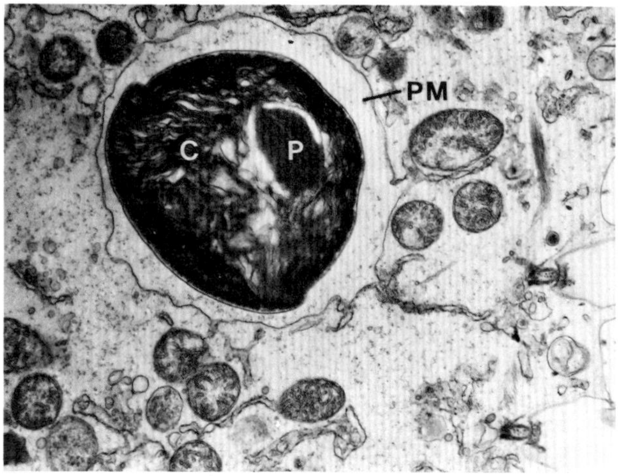

Fig. 8.9. Ultrastructure of *Chlorella* in symbiosis with *Paramecium bursaria*, from dark culture. C, Chloroplast; P, pyrenoid; PM, perialgal membrane. Magnification ×16 500. (From Reisser, 1976.)

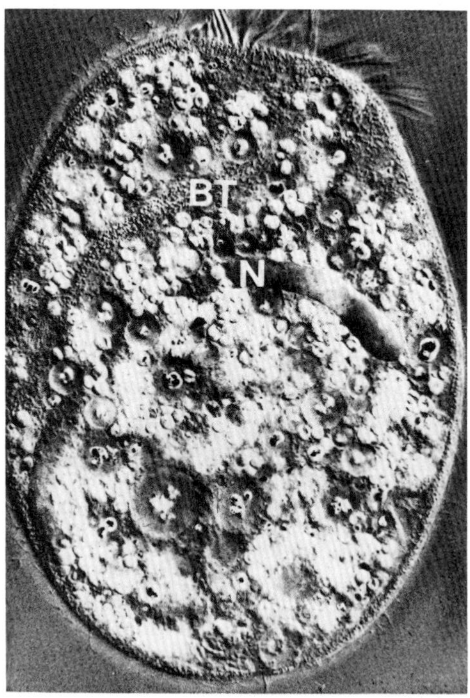

Fig. 8.10. *Climacostomum virens* (Protozoa; Ciliata) with symbiotic *Chlorella*. N, Nucleus of *Paramecium*; BT, buccal apparatus. (Original photograph, W. Reisser, Marburg.)

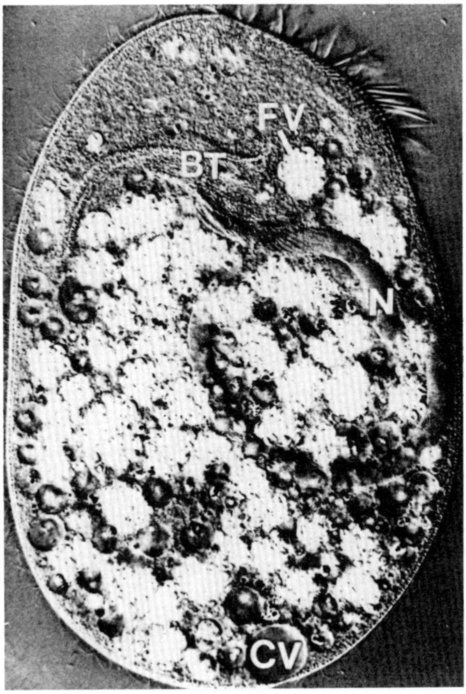

Fig. 8.11. Digestion of *Chlorella* in food vacuoles of *Climacostomum virens*. CV, contractile vacuole; FV, food vacuole; BT, buccal apparatus. (From Reisser and Kurmeier, 1984.)

ing method to discriminate symbiotic from non-symbiotic *Chlorella* strains is the use of phycoviruses (Reisser *et al.*, 1986). A polygonal virus with a prominent capsid, isolated from a freshwater pond, attacked only endosymbiotic *Chlorella*.

8.3.3 The *Convoluta–Tetraselmis* symbiosis

In the egg stage, *Convoluta roscoffensis* (Turbellaria) does not contain any symbionts, as with *Hydra* and *Chlorella*, symbiotic and non-symbiotic algae are taken up during development. Non-symbiotic strains, however, divide slowly and are out-competed by symbiotic strains. An adult *Convoluta* contains 20 000–25 000 cells of *Tetraselmis convolutae*, mostly localized between the epidermal and subepidermal tissues (Höll, 1983). The mouth opening is closed during development so the organism is totally dependent on algal products. Between 10 and 50% of carbohydrate fixed by *T. convolutae* is exported to the host animal, mostly in the form of glutamine (Fig. 8.12). The major reserve in the microsymbiont is mannitol (Muscatine *et al.*, 1974), which remains in the algal cell and is not exported. Algal cells

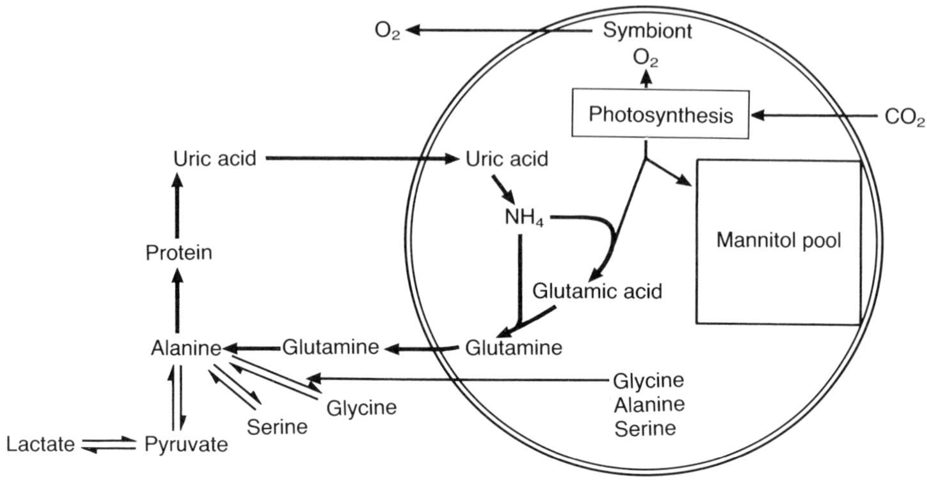

Fig. 8.12. Exchange of metabolites between *Convoluta roscoffensis* and *Tetraselmis convolutea*. (Based on Boyle and Smith, 1975.)

take up urea from the host. When symbiotic photosynthesis is inhibited with DCMU, urea accumulates in the host. Urea degradation releases ammonium, which is used for glutamine synthesis. Phototrophic and heterotrophic cells exchange CO_2 and O_2.

Endosymbiotic *Tetraselmis convolutae* lose their (four) flagella, the eyespot and have a reduced Golgi activity compared to free-living cells (Höll, 1983).

8.3.4 The radiolarian–*Pedinomonas* symbiosis

The green alga *Pedinomonas symbiotica* (a Prasinophyceae) has only been described associated with the jelly surrounding the capsule of the radiolarian *Thalassiolampe margarodes* (Cachon and Caram, 1979). The most similar free-living species from this genus seems to be *Pedinomonas minor* (Eztl and Manton, 1964). More similar may even be *Pedinomonas noctilucae*, which is symbiotic with *Noctiluca miliaris* (Sweeney, 1976).

8.4 CHRYSOPHYTES AS SYMBIONTS

A diatom of the genus *Licmophora* has been identified as living symbiotically in the peripheral parenchyma tissue of *Convoluta convoluta* (Turbellaria). The diatom synthesizes the typical silica shell only after isolation and culture.

Ecologically more important is the symbiosis between Foraminifera (Profozoa, Rhizopoda) and diatoms. All endosymbionts found are relatively

Table 8.5 Symbioses between foraminifers and diatoms

Foraminifer	*Symbiotic diatom*
Amphistegina lobifera	*Fragilaria shiloi*
Amphistegina lessonii	*Fragilaria shiloi* + *Nitzschia laevis*
Amphistegina lessonii	*Fragilaria laevis*
Heterostegina depressa	*Nitzschia panduriformis*
Heterostegina depressa	*Nitzschia valdestriata*

small pennate forms (Lee and McEnery, 1979). Most Foraminifera contain only one species of diatom as symbionts. However, a few can contain two or three different species. The diatom *Fragilaria shiloi* can infect various Foraminifera species, such as *Amphistegina lobifera* and *A. lessonii* (Table 8.5). In Foraminifera, as in *Convoluta*, the diatoms do not have a silica shell. After isolation and cultivation in medium with the essential vitamins B12 (cyanocobalamin), thiamine and biotin, they regenerate their shell and can be identified. Symbiosis is obligatory for *Amphistegina lessonii*, even with extra nutrition (algae, bacteria, protozoans and yeast) because the animal partner cannot grow in the dark. Optimal light conditions are low, 2–3 $W\,m^{-2}$. Other Foraminifera are mostly dependent on particulate prey and the photosynthetic CO_2-fixation accounts for only about 10% of their carbohydrate supply.

8.5 RHODOPHYTES AS SYMBIONTS

Single-celled rhodophytes (Rhodophyceae), probably from the genus *Porphyridium*, are found as microsymbionts in the Foraminifera *Peneroplis planatus* and *Spirolina*. In this case the product of photosynthesis found in the animal host is floridosite (2-*O*-D-glycerin-D-galactopyranoside).

In the hydroids, *Selaginopsis pinnata* and *Tubularia crocea*, the red alga *Rhodochorton membranaceum* was identified. The species develops endozoic basal filaments which ramify between the layers of chitin in the perisarc of the hydroids. Tetrasporangia are formed on upright filaments (West, 1979).

8.6 CHLOROPLASTS OF ALGAE AS ORGANELLES IN GASTROPODS

About 80% of all researched species of aquatic snails of the order Saccoglossa (gastropods), contain algal chloroplasts (Muscatine and Greene, 1973). These chloroplasts remain active for up to three months (Trench and Ohlhorst, 1976). Snails obtain their food by grazing the cells of plants and

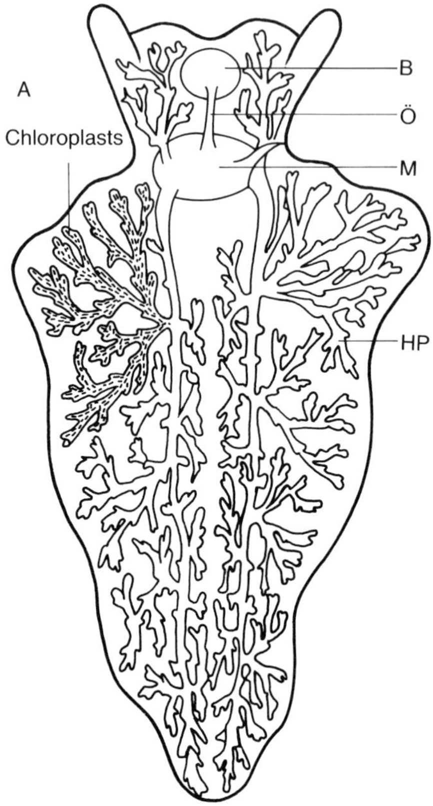

Fig. 8.13. Dorsal view of *Elysia viridis* (Gastropoda). B, buccal cavity; Ö, oesophagus; M, stomach; HP, hepatopancreas. Chloroplasts (from *Codium fragile*) are shown in one part of the hepatopancreas. (Based on Muscatine and Greene, 1973.)

sucking out the contents. Especially favoured as prey are algae of the genera *Codium* and *Caulerpa*. Chloroplasts and other cellular debris are taken into the digestive organ (hepatopancreas) (Fig. 8.13). This is connected to the stomach and forms a highly branched system of tubes through the whole back of the animal (that of *Elysia viridis* is extremely highly branched). Although most snails contain chloroplasts from green algae, some species, like *Limapontia depressa*, contain chloroplasts from *Vaucheria* (Xanthophyceae). In *Hermaea bifida* the chloroplasts are derived from *Griffithsia flosculosa* (Rhodophyceae) (Table 8.6). Chloroplasts from *Elysia viridis* have been well researched. Photosynthetic efficiency is 8–10 $\mu mol\, CO_2\, h^{-1}\, mg$ chlorophyll^{-1}. This is about the same as in the donor alga, *Codium fragile* (Table 8.7). Hepatopancreas tissue synthesizes a factor which stimulates the export of photosynthate from chloroplasts (Gallop, 1974). This export takes the form of glycollate, glucose and alanine.

Table 8.6 Algal chloroplasts as functional organelles in gastropods. (Based on Muscatine and Greene, 1973, and Hinde, 1980)

Gastropod species	*Chloroplast origin*
Elysia atroviridis	*Codium fragile* (Caulerpales = Siphonales)
Elysia viridis	*Codium fragile*
Tridachia crispata	*Caulerpa* sp. (Caulerpales)
Tridachiella diomedea	*Caulerpa* sp.
Placobranchus ocellatus	*Udotea* sp. (Caulerpales)
Limapontia depressa	*Vaucheria* (Xanthophyta)
Hermaea bifida	*Griffithsia flosculosa* (Rhodophyta)

Table 8.7 Photosynthetic CO_2-fixation by the alga *Codium fragile* (Caulerpales) and by *Elysia viridis* (Gastropoda) containing algal chloroplasts, in comparison with CO_2-fixation by isolated chloroplasts. (Based on Smith, 1984; Trench, 1980)

Species	*Photosynthetic CO_2-fixation ($\mu mol\ CO_2\ h^{-1}\ mg\ chlorophyll^{-1}$)**
Codium fragile	11–12
Chloroplasts isolated from *C. fragile*	18–23
Elysia viridis	8–10
Chloroplasts isolated from *E. viridis*	19–21

* Light intensity: 21 500 lx.

It is not known how chloroplasts in the hepatopancreas are protected from digestion. Most chloroplasts are free in the host cell cytoplasm, only a few remain in the phagocytic vacuole. However, there is a low rate of continual digestion which is probably due to lysosomes fusing with phagocytic vacuoles, and lysing the few chloroplasts therein. The earlier suggestion that chloroplasts from *Codium* and *Caulerpa* are particularly robust has not been confirmed. Of course the chloroplasts cannot multiply in the hepatopancreas. Thus this association is not a true symbiosis. The animals need to have their chloroplasts continually replenished. Apart from supplying the animal with supplementary carbohydrate, the chloroplasts give colour to the back of the animal, and thus function as camouflage.

8.7 REFERENCES

Blank, R.J. and Trench, R.K. (1985) Speciation and symbiotic dinoflagellates. *Science*, **229**, 656–8.

Boyle, E. and Smith, D.C. (1975) Biochemical interactions between the symbionts of *Convoluta roscoffensis*, *Proc. R. Soc. Lond. B*, **189**, 121–35.

Buchsbaum-Pearse, V. (1974) Modification of sea anemone behavior by symbiotic zooxanthellae: Phototaxis. *Biol. Bull.*, **147**, 630–40.

Cachon, M. and Caram, B. (1979) A symbiotic green alga, *Pedinomonas symbiotica* sp. nov. (Prasinophyceae), in the radiolarian *Thalassolampe margarodes*. *Phycologia*, **18**, 177–84.

Ettl, H. and Manton, I. (1964) Die feinere Struktur von *Pedinomonas minor* Korchikoff. *Nova Hedwigia*, **8**, 421–51.

Eschbach, S., Hofmann, C.J.B., Maier, U.-G. *et al.* (1991a) A eukaryotic genome of 660 kb: electrophoretic karyotype of nucleomorph and cell nucleus of the cryptomonad *Pyrenomonas salina*. *Nucl. Acid Res.*, **19**, 1779–81.

Eschbach, S., Wolters, J. and Sitte, P. (1991b) Primary and secondary structure of the nuclear small subunit ribosomal RNA of the cryptomonad *Pyrenomonas salina* as inferred from the gene sequence; evolutionary implications. *J. Mol. Evol.*, **32**, 247–52.

Falkowski, P.G. (1992) The paradox of nutrient limitation in zooxanthellate corals. *Symbiosis*, **12**, Lectures from the International Symbiosis Congress, Jerusalem, Nov. 1991.

Fisher, C.R., Fitt, W.K. and Trench, R.K. (1985) Photosynthesis and respiration in *Tridacna gigas* as a function of irradiance and size. *Biol. Bull.*, **169**, 230–45.

Gallop, A. (1974) Evidence for the presence of a 'factor' in *Elysia viridis* which stimulates photosynthate release from its symbiotic chloroplasts. *New Phytol.*, **73**, 1111–17.

Gibbs, S.P. (1981) The chloroplasts of some algal groups may have evolved from eukaryotic algae. *Ann. N.Y. Acad. Sci.*, **361**, 193–208.

Gibbs, S.P. (1983) The cryptomonad nucleomorph: is it the vestigial nucleus of a eukaryotic endosymbiont? in *Endocytobiology* II. *Intracellular Space as Oligogenetic Ecosystem*, (eds H.E.A. Schenk and W. Schwemmler), W. de Gruyter, Berlin, New York, pp. 987–92.

Goff, L.J. (1983) *Algal Symbiosis. A Continuum of Interaction Strategies*. Cambridge University Press, Cambridge.

Hansmann, P. (1988) Ultrastructural localization of RNA in Cryptomonads. *Protoplasma*, **146**, 81–8.

Hansmann, P., Falk, H. and Sitte, P. (1985) DNA in the nucleomorph of *Cryptomonas* demonstrated by DAPI fluorescence. *Z. Naturforsch.*, **40c**, 933–5.

Hinde, R. (1980) Chloroplast 'symbiosis' in saccoglossan molluscs, in *Endocytobiology* (eds W. Schwemmler and H.E.A. Schenk), Vol. I, W. de Gruyter, Berlin, pp. 729–36.

Hinde, R. (1987) Animals with photosynthetic symbionts. *Ann N.Y. Acad. Sci.*, **503**, 355–8.

Höll, W. (1983) Interaction between plants and animal in marine systems, in *Physiological Plant Ecology* III, Vol. 12C. *Encyclopedia of Plant Physiology*, New

Series, (eds O.L.Lange, P.S. Nobel, C.B. Osmond and H. Ziegler) Springer, Berlin, pp. 469–88.

Jolley, E. and Smith D.C. (1980) The green hydra symbiosis. II. The biology of the establishment of the association. *Proc. R. Soc. Lond. B*, **207**, 311–33.

Kessler, E., Huss, V.A.R. and Rahat, M. (1988) Species-specific ability of *Chlorella* strains (Chlorophyceae) to form stable symbioses with *Hydra viridis. Pl. Syst. Evol.*, **160**, 241–6.

Kremer, B.P. (1985) Symbiose von Algen und Korallen. *Naturwiss. Rundschau*, **38**, 508–16.

Lee, J.J. and McEnery, M.E. (1979) Isolation and cultivation of diatom symbionts from larger Foraminifera (Protozoa). *Nature*, **280**, 57–8.

Ludwig, M. and Gibbs, S.P. (1987) Are the nucleomorphs of cryptomonads and *Chlororachnion* the vestigial nuclei of eukaryotic endosymbionts? *Ann. N.Y. Acad. Sci.*, **503**, 198–211.

Margulis, L., Thorington, G., Berger, B. and Stolz, J. (1978) Endosymbiotic bacteria associated with the intracellular green algae of *Hydra viridis. Curr. Microbiol.*, **1**, 227.

Matzke, B., Schwarzmeier, E. and Loos, E. (1990) Maltose excretion by the symbiotic *Chlorella* of the heliozoan *Acanthocystis turfacea. Planta*, **181**, 593–8.

McAuley, P.J. (1985) Regulation of numbers of symbiotic *Chlorella* in digestive cells of green hydra. *Endocyt. C. Res.*, **2**, 179–90.

McAuley, P.J. (1987) Nitrogen limitation and amino-acid metabolism of *Chlorella* symbiotic with green hydra. *Planta*, **171**, 532–8.

McAuley, P.J. and Smith, D.C. (1982) The green hydra symbiosis. V. Stages in the intracellular recognition of algal symbionts by digestive cells. *Proc. R. Soc. Lond. B*, **216**, 7–23.

Meints, R.H. and Pardy, R.L. (1980) Quantitative demonstration of cell surface involvement in a plant–animal symbiosis: lectin inhibition of reassociation. *J. Cell. Sci.*, **43**, 239–51.

Muscatine, L. (1989) Adventures in symbiosis. *Am. Zool.*, **29**, 1203–8.

Muscatine, L., Boyle, J.E. and Smith, D.C. (1974) Symbiosis of the acoel flatworm *Convoluta roscoffensis* with the alga *Platymonas convolutae. Proc. R. Soc. Lond. B*, **187**, 221–34.

Muscatine, L., Falkowski, P.G., Porter, J.W. and Dubinsky, Z. (1984) Fate of photosynthetic fixed carbon in light- and shade-adapted colonies of the symbiotic coral *Stylophora pistillata. Proc. R. Soc. Lond. B*, **222**, 181–202.

Muscatine, L. and Greene, R.W. (1973) Chloroplasts and algae as symbionts in molluscs. *Int. Rev. Cytol.*, **36**, 137–69.

Pardy, R.L. and Heacox, A.E. (1976) Growth of algal symbionts in regenerating hydra. *Nature*, **260**, 809–10.

Pardy, R.L. and Muscatine, L. (1973) Recognition of symbiotic algae by *Hydra viridis*. A quantitative study of the uptake of living algae by aposymbiotic *H. viridis. Biol. Bull. Mar. Biol. Lab.*, **145**, 565–79.

Rahat, M., Kessler, E. and Huss V.A.R. (1989) Is acid tolerance of symbiotic *Chlorella in vitro* an indicator of pH in intracellular perialgal vacuoles of *Hydra viridis*? *Symbiosis*, **7**, 117–26.

Reisser, W. (1976) Die stoffwechselphysiologischen Beziehungen zwischen

Paramecium bursaria Ehrbg. und *Chlorella* spec. in der *Paramecium bursaria*-Symbiose. II. Symbiose-spezifische Merkmale der Stoffwechselphysiologie und der Cytologie des Symbioseverbandes und ihre Regulation. *Arch. Microbiol.*, **111**, 161–70.

Reisser, W. (1984) The taxonomy of green algae endosymbiotic in ciliates and a sponge. *British Phycol. J.*, **19**, 309–18.

Reisser, W. (1990) Participation of algal cell wall surface structures in the formation of the host–symbiont-interface of endocytobiotic systems, in *Cell Walls and Surfaces, Reproduction and Photosynthesis* (eds W. Wiessner, D.G. Robinson and R.C. Starr) Springer Verlag, Stuttgart, New York, pp. 55–68.

Reisser, W. (1980) The metabolic interactions between *Paramecium bursaria* Ehrbg. and *Chlorella* spec. in the *Paramecium bursaria*-symbiosis. III. The influence of different CO_2-concentrations and of glucose on the photosynthetic and respiratory capacity of the symbiotic unit. *Arch. Microbiol.*, **125**, 291–3.

Reisser, W., Becker, B. and Klein T. (1986) Studies on ultrastructure and host range of a *Chlorella* attacking virus. *Protoplasma*, **135**, 162–5.

Reisser, W. and Kurmeier, B. (1984) The endosymbiotic unit of *Climacostomum virens* and *Chlorella* sp. Symbiotic features of the association and host-symbiont regulatory mechanisms. *Protistologica*, **20**, 265–70.

Reisser, W. and Wiessner, W. (1984) Autotrophic eukaryotic freshwater symbionts, in *Cellular Interactions*, Vol. 17, *Encyclopedia of Plant Physiology*, New Series (eds H.F. Linskens and J. Heslopp-Harrison) Springer, Berlin, pp 59–74.

Schlichter, D. and Fricke, H.W. (1990) Coral host improves photosynthesis of endosymbiotic algae. *Naturwissenschaften*, **77**, 447–50.

Schoenberg, D.A. and Trench R.K. (1980) Genetic variation in *Symbiodinium* (=*Gymnodinium*) *microadriaticum* Freudenthal, and specificity in its symbiosis with marine invertebrates. I. Isoenzyme and soluble protein patterns of axenic cultures of *Symbiodinium microadriaticum*. *Proc. R. Soc. Lond. B*, **207**, 405–27.

Smith, D.C. (1974) Transport from symbiotic algae and symbiotic chloroplasts to host cells. *Symp. Soc. Exp. Biol.*, **28**, 485–520.

Smith, S.V. (1978) Coral-reef area and the contributions of reefs to processes and resources of the world's oceans. *Nature*, **273**, 225–6.

Sutton, D.C. and Hoegh-Guldberg, O. (1990) Host–zooxanthella interactions in four temperate marine invertebrate symbioses: assessment of effect of host extracts on symbionts. *Biol. Bull.*, **178**, 175–86.

Sweeney, B.M. (1976) *Pedinomonas noctilucae* (Prasinophyceae), the flagellate symbiotic in *Noctiluca*, in Southeast Asia. *J. Phycol.*, **12**, 460–4.

Trench, R.K. (1979) The cell biology of plant–animal symbiosis. *Ann. Rev. Plant Physiol.*, **30**, 485–531.

Trench, R.K. (1980) Uptake, retention and function of chloroplasts in animal cells, in *Endocytobiology* (eds W. Schwemmler, H.E.A. Schenk), Vol. I, W. de Gruyter, New York, pp. 703–27.

Trench, R.K. and Blank, R.J. (1987) *Symbiodinium microadriaticum* Freudenthal, *S. goreauii* sp. Nov., *S. kawagutii* sp. nov. and *S. pilosum* sp. nov.: gymnodinioid dinoflagellate symbionts of marine invertebrates. *J. Phycol.*, **23**, 469–81.

Trench, R.K. and Ohlhorst, S. (1976) The stability of chloroplasts from siphonaceous algae in symbiosis with saccoglossan molluscs. *New Phytol.* **76**, 99–109.

Vernberg, W.B. (ed.) (1974) *Symbiosis in the Sea*. University of South Carolina, Columbia.

Weis, D.S. (1979) Correlation of sugar release and concanavalin A agglutinability with infectivity of symbiotic algae from *Paramecium bursaria* for aposymbiotic *P. bursaria*. *J. Protozool.*, **26**, 117–19.

West, J.A. (1979) The life history of *Rhodochorton membranaceum*, an endozoic red alga. *Bot. Mar.*, **22**, 111–15.

Wilkerson, F.P. and Trench, R.K. (1985) Nitrate assimilation by zooxanthellae maintained in laboratory culture. *Mar. Chem.*, **16**, 385–93.

9
Lichens and *Geosiphon*

Lichens are symbiotic organisms, forming a large number of new structures from the two bionts: the **photobionts** and the **mycobionts**. Due to their capability to colonize ecosystems with extreme climatic conditions, the ecophysiology of these lichens is especially interesting. Comprehensive monographs and reviews have been published (Henssen and Jahns, 1973; Hawksworth and Hill, 1984; Galun, 1988; Honegger, 1991).

The application of **molecular genetical methods** to lichens is still in the stage of proposals (Ahmadjian, 1991). So far, there is no experimental evidence that a gene transfer between the two lichen bionts exists or should be proposed. But by DNA-subtraction hybridization (see Chapter 3) it might be very interesting to identify the genes that are different in lichen forming species and not lichenized species.

9.1 THE PHOTOBIONTS

About 35 of the genera of green algae (Chlorophyta) and cyanobacteria have been found as photobionts in lichens (Table 9.1). Several species from these genera such as *Nostoc* (Chapter 7) and *Chlorella* (Chapter 8) are also microsymbionts in other symbioses. Other genera seem to be very lichen-specific such as *Trebouxia* or *Myrmecia*. In addition, one genus from the Xanthophyta and one from the Phaeophyta have been reported as algal symbionts. More than 60% of all lichen species have *Trebouxia* as a photobiont. *Trentepohlia* and *Nostoc* are other widespread genera (Table 9.1). The number of species is difficult to estimate, about 100 have been listed (Tschermak-Woess, 1988). The actual number of photobiont species may be much larger in the more than 13 500 lichen species. Taxonomy and systematics of the phototrophic partners in lichens are complicated, for example, by repression of sexual reproduction in the symbiotic stage or by morphological changes. The symbiotic algae form sporangia abundantly after isolation in culture and also in the free-living stage in the field, but not *in situ* in the lichen (Nakano, 1988). Important cytological characters to discriminate species have only been studied in some genera. For *Trebouxia*,

Table 9.1 Genera of photobionts in lichens*

Chlorophyta		*Xanthophyta*	*Phaeophyta*	*Cyanobacteria*
Asterochloris	*Hyalococcus*	*Heterococcus*		*Anabaena*
Cephaleuros	*Leptosira*	*Petroderma*		*Anacystis*
Chlorosarcina	*Myrmecia*			*Aphanocapsa*
Chlorella	*Nanochloris*			*Calothrix*
Chlorellopsis	*Phycopeltis*			*Chroococcus*
Chlorococcum?	*Protococcus?*			*Dichothrix*
Coccobotrys	*Pseudochlorella*			*Fischerella*
Coccomyxa	*Pseudopleurococcus*			*Gloeocapsa*
Dictyochloropsis	*Stichococcus*			*Hyella*
Dilabifilium	*Trebouxia*			*Hyphomorpha*
Elliptochloris	*Trentepohlia*			*Scytonema*
Gloeocystis	*Trochiscia*			*Stigonema*
Heterococcus				

* Modified from Hawksworth (1984) and Tschermak-Woess (1988).

which includes now ‘Pseudotrebouxia’ the following parameters have been used (Gärtner, 1985; Friedl, 1989):

1. the structure of the pyrenoids with different arrangement and forms of pyrenoid tubuli
2. the morphology of the plastids, often with reduced structures in the symbiotic stage
3. cell wall structures
4. asexual reproduction with either autospore formation (daughter cells with cell walls) or zoospore formation (flagellate cells without cell walls)

On the basis of these characteristics, only eight species have been established in the genus *Trebouxia* (Friedl, 1989). In some cases, in addition to a green photobiont additional cyanobacteria can be included in the thallus or are located in special parts of the thallus, called cephalodia. The decisive biochemical characters which are different in lichen photosymbionts and non-symbiotic strains from the same genus or species are unknown.

9.2 THE MYCOBIONTS

Lichens are also called lichenized fungi, indicating the dominant role of the fungi in the development and the morphogenesis of the symbiosis. In the context of this book the fungi are regarded as hosts. Exceptions are some gelatinous lichens in which the fungal hyphae only invade colonies of the photobionts. The ascomycetes are by far the largest fungal group with hosts in 98% all of lichen species. Every second ascomycete described is a lichenized species. Five orders belong exclusively to this group (Hawksworth and Hill, 1984): Graphidales; Gyalectales; Peltigerales; Pertusariales; Teloschistales.

In several orders of the ascomycetes different species can be symbionts in lichens or in the ectomycorrhiza (Chapter 11), pathogens in a wide range of host plants or widespread saprobes. This is especially pronounced in the orders Heliotales, Sphaeriales and Dothideales (Table 9.2). Even within one genus different species can be either a lichen-forming, a parasitic or a saprobic species. Examples are found in the genera *Omphalina* or *Chaenothecopsis* (Table 9.3). In the small group of basidiolichens seven different

Table 9.2 Symbiotic, parasitic and saprobic characters of orders of the ascomycetes (Based on Gäumann, 1964; Bresinsky, 1983; Schlösser, 1983; Siegel *et al.*, 1987)

Order	*Characteristic species and genera*		
	as symbiosis	*as pathogens*	*as saprobes*
Endomycetales			yeasts
Protomycetales		on flowering plants	
Ascosphaerales		on bees	
Taphrinomycetales		on trees	
Laboulbeniomycetales		on insects	
Eurotiales	Ectomycorrhiza	on vertebrates peanuts	*Aspergillus* *Penicillium*
Microascales		on elm trees	
Caliciales	Lichens (many)		
Coronophorales	Lichens		on wood
Meliolales	Lichens		
Erysiphales		higher plants	
Pezizales	Ectomycorrhiza		*Pyronema*
Heliotales	Lichens (few)	on wine berries	*Chlorosplenium*
Phacidiales	Ectomycorrhiza	on roses on leaves	
Lecanorales	Lichens (many)		
Sphaeriales	Lichens	canker on fruit trees on wheat, bananas	*Neurospora*
Diaporthales		on chestnuts	
Xylariales			on wood
Clavicipitales	Grassendophyta	on grasses (cereals)	
Ostropales	Lichens (few)		on plant material
Myriangiales		on plants and insects	
Pseudosphaeriales		on apple fruits	
Dothideales	Lichens (few)	on plant organs	on plant material
Arthoniales	Lichens (many)		
Pyrenulales	Lichens (many)		
Verrucariales	Lichens (many)		

Table 9.3 Examples of genera exhibiting different nutritional types

	Arthonia	*Buellia*	*Chaenothecopsis*	*Dactylospora*	*Lecidea*	*Nectriella*	*Omphalina*	*Pezizella*
Lichen-forming	*tumidela*	*disciformis*	*lignicola*	–	*fuscoatra*	–	*luteovitellina*	*parasitica*
Lichenicolous lichen	–	*pulverulenta*	–	–	*insularis*	–	–	–
Parasymbiont	*intexta*	–	*epithallina*	*parasitica*	*supersparsa*	*robergei*	–	–
Parasite	*fuscopurpurea*	–	*parasitaster*	*saxatilis*	*vitellinaria*	*tincta*	*cupulatoides*	*epithallina*
Fungicolous	–	–	*caespitosa*	–	–	–	–	–
Algicolous	–	–	–	–	–	*laminariae*	–	–
Bryophilous	–	–	–	*scapanaria*	–	–	*sphagnicola*	*polytrichi*
Plant mycorrhizae	–	–	–	–	–	–	–	*ericae*
Saprobe	–	–	*debilis*	*stygia*	–	*fuckelii*	*chrysophylla*	*eburnea*

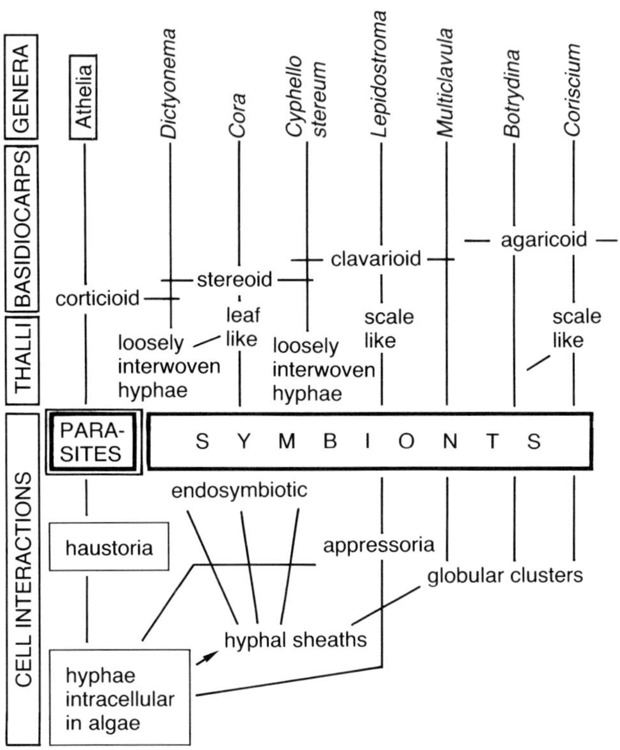

Fig. 9.1. Cell–cell interactions and thallus characters of basidiolichens and *Athelia*. (Modified from Oberwinkler, 1984.)

genera of basidiomycetes have been described as hosts. They can be discriminated by cell interactions, basidiocarps and the thalli (Fig. 9.1).

9.3 THE SYMBIOSIS

9.3.1 The genus and species concept in lichenology

The first lichen genus '*Lichen*' was described by Linnée (Linnaeus, 1753). Since then more than 1600 genera of lichens have been named, but only about 650 genera are considered to be acceptable by recent standards (Hale, 1984). The number of genera introduced in the nineteenth century and still valid varies significantly between authors. Acharius (1803) named 39 genera and about half of them are still used and very well known, such as *Parmelia*, *Ramalina*, *Lecanora* and *Lecidea*. Of the 140 genera described by Massalongo (1856) only about 30 are still accepted. The very active Finnish lichenologist Nylander (1866) proposed about 80 genera; almost half of them are still valid, for example *Peltula*, *Cladia* and *Dactylina*.

Today the main diagnostic characters for the establishment of new genera and species are:

1. ascus structure
2. apothecial ontogeny
3. perithecial ontogeny
4. paraphysis morphology
5. conidiomata location
6. spore development and ultrastructure
7. secondary metabolites (at the level of genera and families)
8. thallus structure and organs for aeration, dispersal and attachment

These terms are defined as follows:

1. Asci are the tube like reproductive cells of the ascomycetes, usually forming eight ascospores. However, between one up to several hundred have been found, e.g. in Acarosporaceae (Henssen and Jahns, 1973);
2. Perithecia are closed fruiting bodies (ascocarps), opening with a porus;
3. Ascocarp is a general term for fruiting bodies of ascomycetes;
4. Apothecia are open ascocarps;
5. Paraphyses are sterile hyphae between the asci;
6. Conidiomata are almost exclusively pycnidia.

Comprehensive work using structural as well as biochemical methods can also lead to the reduction in the number of species. Since no correlation was found between the chemistry and morphology of different chemotypes of *Cladonia arbuscula* and *Cladonia mitis*, an aggregate species was proposed (Ruoss, 1987). The reverse taxonomic development is also observed. The genus *Parmelia* has been split into more than 30 genera such as *Hyptrachnya*, *Pseudoparmelia* and *Xanthoparmelia* (Krog and Swinscow, 1987; Adler, 1990). Thoroughly studied new genera are still proposed nowadays, especially from the tropics and subtropics or from very small forms, which apparently escaped the notice of lichenologists before, such as *Hertella* or *Edwardiella* (Henssen, 1985, 1986).

In lichen taxonomy cladistic dendrograms have been used, especially for lower taxa, e.g. in the subgenus *Cetraria* (Kärnefelt and Mattson, 1987). However, it is doubtful if such phenograms based on total soluble protein banding after isoelectric focusing can avoid large differences in single bands, which may only be due to different physiological stages of the large number of cell types in one species. RFLP (restriction fragment length polymorphism) of DNA (Winnacker, 1987) from isolated discrete cell types of the lichen thallus is a much more promising method. It has been used for several other unicellular species and also for higher plants. Other general problems in lichen taxonomy result from the morphogenetic effect of the photobiont (James and Henssen, 1976; Jahns, 1989; Jahns and Ott, 1990).

9.3.2 Morphology and ultrastructure

Lichen morphology allows the discrimination of four major growth forms.

1. The '**leprose type**' is a primitive crustose type of lichen with granular structure. The protection against desiccation is minimal and the lichens have a powdery appearance. Typical genera are *Lepraria* and *Leproplaca*.
2. The '**crustose areolate type**' is a widespread growth form in almost all orders of lichens. The fungal partner and the photobionts are unstratified (homoiomerous forms). In general very flat and tightly fixed to the substrate, several subtypes have developed:
   (a) the cracked thallus types: irregular cracks cover the surface of the thallus
   (b) the areolate thallus, with the surface divided into polygonal or round areas in a dry state
   (c) the placodioid thallus: the surface is often lobed at the margins and leads to the foliose types
3. The '**foliose type**' is represented in the larger orders of the Lacanorales, the Verrucariales and the Arthoniales. The lichens consist of four distinct layers: the upper cortex, the photobiont layer, a loose fungal hyphae layer (the medulla) and normally a lower cortex. The thallus is therefore stratified (heteromerous). This type is attached to the substrate by organs called rhizines.
4. The '**fruticose type**' is found in species of the same orders as mentioned for the foliose type. The layers may have a radial symmetry with a supporting mycelial tissue in the centre of the axis or in the periphery. A typical species of this type is the pendant *Usnea longissima*, the largest lichen with a length of several metres.

The three last growth types can be present within one family (e.g. Physciaceae, Teloschistaceae).

The intensive use of SEM- and TEM-methods including freeze fracturing during the last years has revealed a number of new ultrastructural characters in lichens, of which a few are mentioned here. The complexity of lichen asci is the consequence of different wall stratifications and many internal differences such as formation of rings and axial bodies (Bellemère and Letrouit-Galinou, 1987). In haustoria at the mycobiont–photobiont interface of cyanolichens, the fungal cell wall is thinner than in other areas. The plasmalemma of haustoria and other hyphae have the same number of invaginations (Büdel and Rhiel, 1987) and also the particle frequency in the plasmamembrane is about the same with 1700 particles nm^{-2} for the EF (exoplasmic fracture) face and 3900 particles nm^{-2} for the PF (protoplasmic fracture) face (Budel and Rhiel, 1987).

Ornamentation of ascospores is relatively rare in lichenized ascomycetes, but the spore type itself is a useful character with a large variation in size and septation (Hawksworth and Hill, 1984).

9.3.3 Development of the symbiosis

Germination of spores of the lichen fungus by germ tubes sets the start for development of a lichen from the separate partners. Spore germination can be stimulated by micromolar concentrations of kinetin, as demonstrated for *Endocarpon pusillum* (Stocker-Wörgötter and Türk, 1989). Since several chemo-organotrophic soil bacteria can produce this type of phytohormone (Chapter 2), it is very likely that third partners besides the fungi and the photobionts are involved *in situ* in the start of the symbiosis. The growing hyphae branch and can form anastomoses (secondary fusions), to form a three-dimensional network of the mycelium. Cell differentiations contribute to the final thallus. Local cell wall growth leads to a number of round, ellipsoidal or elongated cell types. Further stabilization of the plectenchyma formed is obtained by gelatinous substances which are excreted by the mycelia (Jahns, 1988). More advanced tissues with conglutinated cell walls in some lichen species resemble a true parenchyma of higher plants and are called pseudoparenchyma. Structures with parallel cylindrical cells resemble the leaf palisade parenchyma and are called palisade plectenchyma. Before the fungus meets a suitable colony of photosymbionts, the energetics and carbon metabolism have to be maintained by saprophytic nutrition. The recognition and uptake of the phototrophic partner is described below in three examples.

In the lichen *Endocarpon pusillum* (Stocker-Wörgotter and Türk, 1989) two connected ascospores are ejaculated by the asci, together with a colony of up to 40 hymenial algal cells of *Stichococcus*. The algal cells remain attached to the spores by a slime layer. The germ tubes of the ascospores reveal a tigmotrophic behaviour. However, the algae have only a very limited effect as chemoattractant. The hyphae develop as appressoria in contact with the algae, and only in a few cases are true haustoria formed. The algae continue to divide, but remain attached to each other due to the surrounding fungal mycelia. Branching in these hyphae is increased and also the diameter is enlarged. In cultivation experiments the lichen fungus also developed three-dimensional structures without the algal partner, but did not form a true hymenium. Further differentiation to an adult thallus requires a dry period of several weeks, especially to form a cortex. Light intensity also affects the size of the cortex. In cultivation experiments these thallus initials developed within 6–9 weeks and were called 'thalles secondaires' by Werner (1931). The initiation of the next step, the formation of

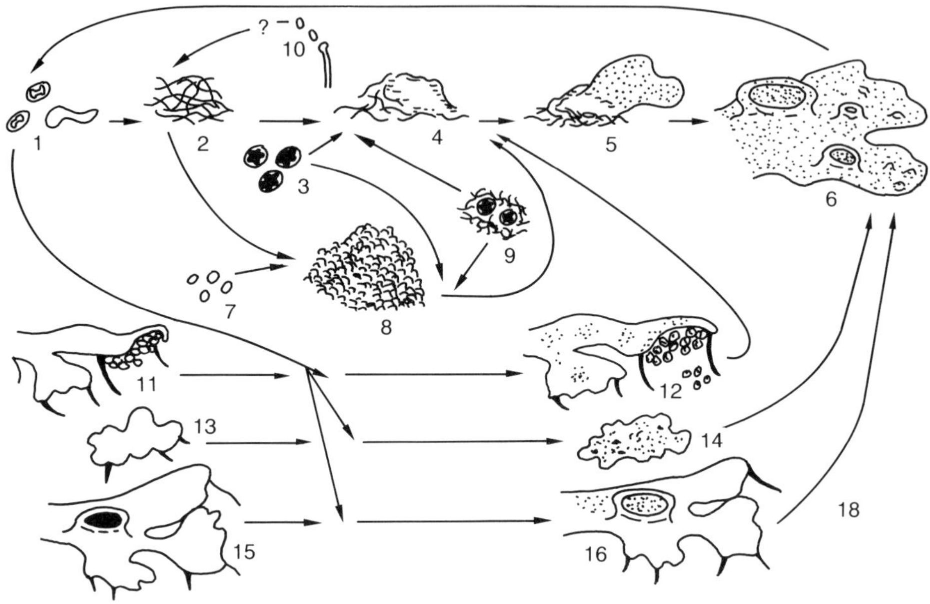

Fig. 9.2. Life cycle of *Xanthoria parietina*. (1) Germinating spores; (2) hyphal network; (3) *Pseudotrebouxia*; (4) secretion of gelatinous substances; (5) young thallus scale; (6) grown thallus; (7) foreign coccal green algae; (8) areolated crust; (9) foreign soredia; (10) conidia; (11)–(16) attack of *Physcia* by *Xanthoria*. The yellow colour of *Xanthoria* is indicated by dots. (Modified from Ott, 1987b.)

the hymenium within perithecia requires a dry period of several weeks. Continuously moistened cultures on soil never developed perithecia. In *Endocarpon pusillum* the hymenial algae are spread throughout the perithecial gelatine. After 9–12 months mature ascospores and colonies of *Stichococcus* in the hymenium are ready for a new cycle.

A much more complicated life cycle is found in *Xanthoria parietina* (order Lecanorales) (Fig. 9.2) (Ott, 1987b). The germinating ascospores develop into a small mycelium, preferentially in the winter months. Their chance to meet a compatible colony of the symbiotic algal partner *Trebouxia* is rather low. Therefore, it looks like a beneficial strategy to take pot luck first with other algae around, such a *Pleurococcus*. The hyphae and the foreign algae form an areolated crust covering larger areas of the substrate such as the bark of trees and preventing other lichens in colonization. These larger crusts now have a much better chance to meet the symbiotic compatible *Trebouxia*. The mechanism which recognizes this specific partner is still unknown as in other algal symbioses (Chapters 7 and 8). The involvement of mycobiont lectins and photobiont carbohydrate structures is considered, but experiments explaining the high degree of specificity in a large number of species compared to the limited number of lectin–carbohydrate pairs leaves many questions open. A fungal glycoprotein has been isolated from cultivated *Xanthoria parietina*, which binds to cultivated phycobionts of this lichen (Bubrick *et al.*, 1985). Therefore, it was called an ABP (algal binding protein). In other lichens this method was not successful (Hersoug, 1983). It can be assumed that lectin–carbohydrate binding is involved in the process, but other components may be necessary in recognition and specificity. One of these other mechanisms and theories emphasizes the low specificity of the photobionts in the association. A large number of phototrophic algae come into contact with the fungi, unappropriate species are killed and a final symbiotic species is distinguished by escaping predation. Non-growing zones of lichen thalli often contain a high percentage (more than 50% in *Parmelia sulcata*) of oversized photobionts, which exceed the cell size of sporulation or cell division (Hill, 1985; Fiechter, 1990). The involvement of the mycobiont in this inhibition of cell division has been postulated (Honegger, 1987).

During the further development of *Xanthoria parietina* the crusts with the foreign algae excrete larger amounts of slime when they come into contact with a *Trebouxia* colony (Fig. 9.2). They develop to mature lichen in several steps (Ott, 1987b), involving upward growth of the tissue in different lobes and further differentiation of the thallus layers as described in section 9.3.2.

An additional strategy to acquire the true symbiotic algae is to take them from soredia of other lichen species, such as *Physcia tenella*. The exploitation of *Physcia* can even be much stronger, resembling parasitic behaviour of *Xanthoria*. A small thallus of *Physcia* is invaded by *Xanthoria* hyphae and the shape and colour of the thallus change rapidly to that of the yellow *Xanthoria parietina*. When a larger specimen of *Physcia* is attacked, the invading hyphae do not change the morphology of the host lichen but only take over the hymenium of the apothecia, whereas the margins of the fruiting bodies remain grey, indicating their *Physcia* origin.

The symbiosis with two photobionts in one lichen species poses additional problems for understanding specificity and development of these interactions. Such thalli are called phycosymbiodemes (Renner and Galloway, 1982) or photosymbiodemes (Ott, 1988) and can often be found in the genera *Lobaria*, *Peltigera* or *Sticta*. The development in *Peltigera venosa* has been studied in some detail (Ott, 1988). Hyphae from the fungus growing on moist loamy soil also covered with cyanobacteria (e.g. *Nostoc*) and green algae (e.g. *Coccomyxa*) develop into the already described crusts, which are also called prothallus. From these crusts develop finger-like black cushions, called squamules, which contain only cyanobacteria of the genus *Nostoc*. Both surfaces of the squamules are formed by an unilayered cortex. Hyphae with a large lumen, called trabeculae, connect the two cortical layers. Between the cortical layers is a homogenous tissue with cyanobacteria, which swell under wet conditions by the formation of gelatinous material.

In addition to these cyanobacterial squamules, green squamules develop from crusts with green algae. These have a dorsiventral structure with an upper multilayered cortex, a layer of the photobionts with *Coccomyxa* and a medulla. In some cases, the squamules with cyanobacteria also contain a few green algae. Similarly, in the upper surface of green thalli, some *Nostoc* cells can be present. These photosymbiodemes demonstrate the morphogenetic effect of the phototrophic partner on the lichen fungus most convincingly.

9.3.4 Synthetic lichens

By experimental synthesis from photobionts and mycobionts, a number of results have been obtained, which are not available by observation of established lichen thalli (Nakano, 1988):

1. The formation of structures in culture, which are not clearly developed under natural conditions;
2. Comparison of developmental stages in lichens, such as sexual reproduction, with those of cultivated ascomycetes or basidiomycetes;
3. Determination of photobiont–mycobiont selectivity and specificity;
4. Establishment of genetically identical minimum units of lichens, developed from single spores and single cells of a mycobiont, i.e. the start of a genetics of lichens;
5. Correlation of synthesis of secondary metabolites with single steps of symbiotic development.

Cultivation of ascospores together with the ejaculated hymenial algae to form fertile perithecia has already been achieved by Stahl (1877).

There is still a very limited number of synthetic lichens with a complete life cycle or at least a significant part of it: *Endocarpon pusillum* (Ahmadjian and Heikkila, 1970), *Xanthoria parietina* (Ott, 1987c; Honegger, 1990) and *Peltigera venosa* (Ott, 1988) have already been described (section 9.3.3). In *Cladonia cristatella* (Ahmadjian, 1966) a very prominent hypothallus was only achieved in synthesis cultures, which is not obvious in natural lichens. On the other hand, in this species, as with *Usnea strigosa*, mature apothecia have not been produced in culture, whereas in natural lichens they are very frequent (Ahmadjian, 1990). A number of synthesis experiments were unsuccessful due to excessive growth of contaminating bacteria, as described for *Peltigera polydactyla*.

Cultivation of lichens in the natural environment or in the laboratory from small, but already symbiotic vegetative propagules, allows the study of the later stages of lichen development (Schuster *et al.*, 1985). **Soredia**, the most widespread form of such propagules, consist of a cluster of photobionts enveloped by a layer of mycobiont hyphae. Often such soredia are formed in discrete patches of the thalli. **Isidia** are another form of vegetative

Table 9.4 Respiration rates of lichens, fungi, algae and cyanobacteria (Based on Gaffron, 1939; Webster and Frenkel, 1953; Kandler, 1954; Reazin, 1954 and Altmann and Dittmer, 1972–74)

Species	*Respiration rate* (mm^3 O_2 h^{-1} *g dry weight*$^{-1}$)
Lichens	
Parmelia acetabulum	21
Cetraria islandica	48
Lobaria linita	72
Cladonia stellaris	85
Fungi	
Aspergillus niger	3 000
Puccinia graminis	3 000
Fusarium graminearum	13 600
Candida albicans	29 000
Algae and cyanobacteria	
Scenedesmus sp. (strain D3)	4 000
Anabaena variabilis	8 400
Chlorella pyrenoidosa	17 000
Ochromonas malhamensis	28 000

propagule, which develop as an outgrowth from the thallus and easily break away.

9.3.5 Physiology, biochemistry and metabolic exchange

Metabolic activity of lichens as indicated by the respiration rate is lower than in fungal and algal species by an order of magnitude of 100 (Table 9.4). The reported data for *Parmelia acetabulum*, *Cetraria islandica*, *Lobaria linita* and *Cladonia stellaris* are between 30 and 85 mm^3 O_2 h^{-1} g dry weight^{-1}. The data for different species of fungi like *Aspergillus* and *Candida* are between 3000 and 29 000 mm^3 O_2 h^{-1} g dry weight^{-1}. The respiratory activity of algae such as *Chlorella* and *Ochromonas* is also between 4000 and 28 000 mm^3 and for *Anabaena variabilis* a value of 8400 mm^3 has been recorded.

Photosynthetic activity in lichens is in the order of 1–2 mg CO_2 h^{-1} g dry weight^{-1} (Table 9.5). The rate of photosynthesis in free-living cyanobacteria and unicellular green algae is between 50 and 360 mg CO_2 g^{-1} h^{-1}. However, in lichens the fungal partner is included in the parameter dry weight and the photobionts make up only 10–15% of the thalline biomass in heteromerous lichens such as *Parmelia sulcata* (Fiechter, 1990). This means that the difference in photosynthetic activity between the photobionts in the lichens and the free-living cells is much smaller but still

Table 9.5 Maximum photosynthetic rate of lichens, algae and cyanobacteria (Based on Brown, 1954; Chapman, 1951; Bliss and Hadley, 1964; Shear and Walsby, 1975; Kappen *et al.*, 1980; Lange and Redon, 1983; Türk, 1983; Thielmann, 1990)

Species	*Photosynthetic rate* *(mg CO_2 g^{-1} h^{-1})*
Lichens	
Litharia vulpina	0.4
Cetraria nivalis	1.0
Cladonia rangiferina	1.2
Heterodermia spinulosa	1.2
Cetraria cucullata	2.0
Cyanobacteria and green algae	
Anabaena flos aquae	50
Chlorella pyrenoidosa	130
Nostoc muscorum	260
Scenedesmus obliquus	350

significant (Chapman, 1951; Brown, 1954; Shear and Walsby, 1975; Thielmann, 1990). The cortical layer can be a barrier for CO_2 uptake, especially at water saturation, although aeration pores (pseudocyphellae) are present in species with a thick cortex.

The photosynthetic metabolite transferred from the photobiont to the fungi is glucose or a glucan in the case of cyanobacteria and ribitol (or erythritol or sorbitol) with green algae (Fig. 9.3) (Galun, 1988). Mechanisms similar to those discussed for symbioses with algae and cyanobacteria can be assumed for the excretion of these carbohydrates. Redevelopment of the photobionts to a free-living state results in a reduction of the release of carbohydrates (Smith, 1974). A high percentage of the exchanged carbohydrate is not immediately used for respiration and energy supply, but is transformed to mannitol and arabitol (Fig. 9.3). The amount of carbohydrate transferred depends on the hydration level of the thalli. At a low water content a high percentage remains in the photobiont, and increasing amounts are transferred to the mycobiont as the level of hydration rises.

Experiments with ^{13}C-NMR studies (Lines *et al.*, 1989) indicate that a larger percentage of carbohydrates is transferred from the photobiont to the mycobiont than previously found with inhibition experiments (Richardson and Smith, 1968). Up to 10% of the thallus dry weight can be polyols (Gorin *et al.*, 1988) but a significant part of these may be washed out after rewetting of dry thalli, indicating an important function in desiccation tolerance.

Light affects lichens in several ways. Lichens with green algae as

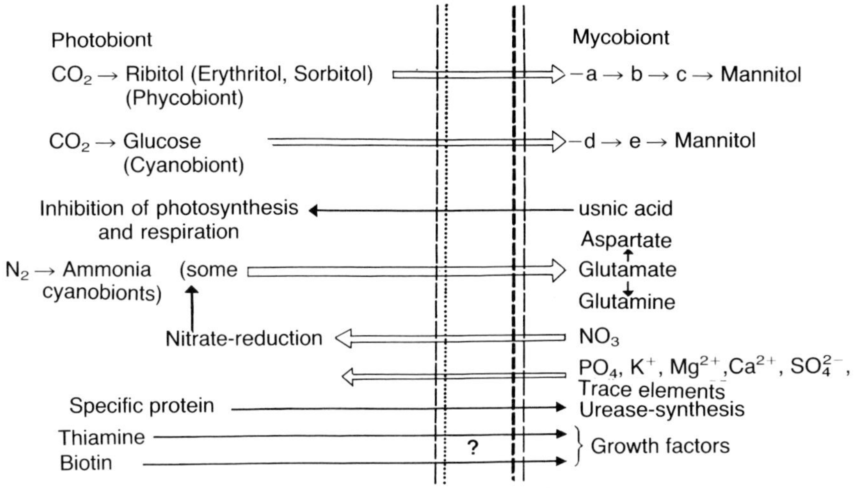

Fig. 9.3. Metabolic exchange between photobionts and mycobionts in lichens. a–b–c, d–e: intermediates.

photobionts can tolerate higher light intensities than species with cyanobacteria. This was explained by a lack of zeaxanthin in these cyanolichens, which prevents them from dissipating radiationless energy in the photochemical apparatus (Demming-Adams *et al.*, 1990). Although the total concentration of carotenoids in lichens with cyanobacteria and in those with green algae is about the same, the xanthophyll cycle components (zeaxanthin, antheraxanthin and violaxanthin) are almost completely absent in the former. In free-living cyanobacteria as well as in *Nostoc* from the lichen *Nephroma laevigatum*, phycobiliproteins are the major light-harvesting pigment for photosystem II (PS II) whereas chlorophyll has this function for PS I (Canaani, 1988). Via phytochrome in the P_{fr} form (far red absorbing form) light regulates the algal nitrate reductase, the fungal orsellinate depside hydrolase and also the D-usnic acid: NAD oxidoreductase (Vicente and Avalos, 1989).

Transfer of nitrogen compounds between the symbiotic partners depends on the type of photobiont (Fig. 9.3). In cyanolichens, N_2 fixed by nitrogenase (EC 1.18.6.1) produces ammonia, which is transferred to the mycobiont and assimilated by glutamate dehydrogenase to glutamate. Alanine is produced via aspartate or directly by the activity of glutamate–pyruvate transaminase (GPT). Alanine seems to be a major transport form from the cephalodia to other parts of the thallus (Rai, 1988). In green algal lichens uptake of nitrate is much more effective than in cyanolichens whereas ammonia uptake is about equal in both types (Rai, 1988). Assimilation of other mineral nutrients such as phosphate, sulphate, potassium, magnesium,

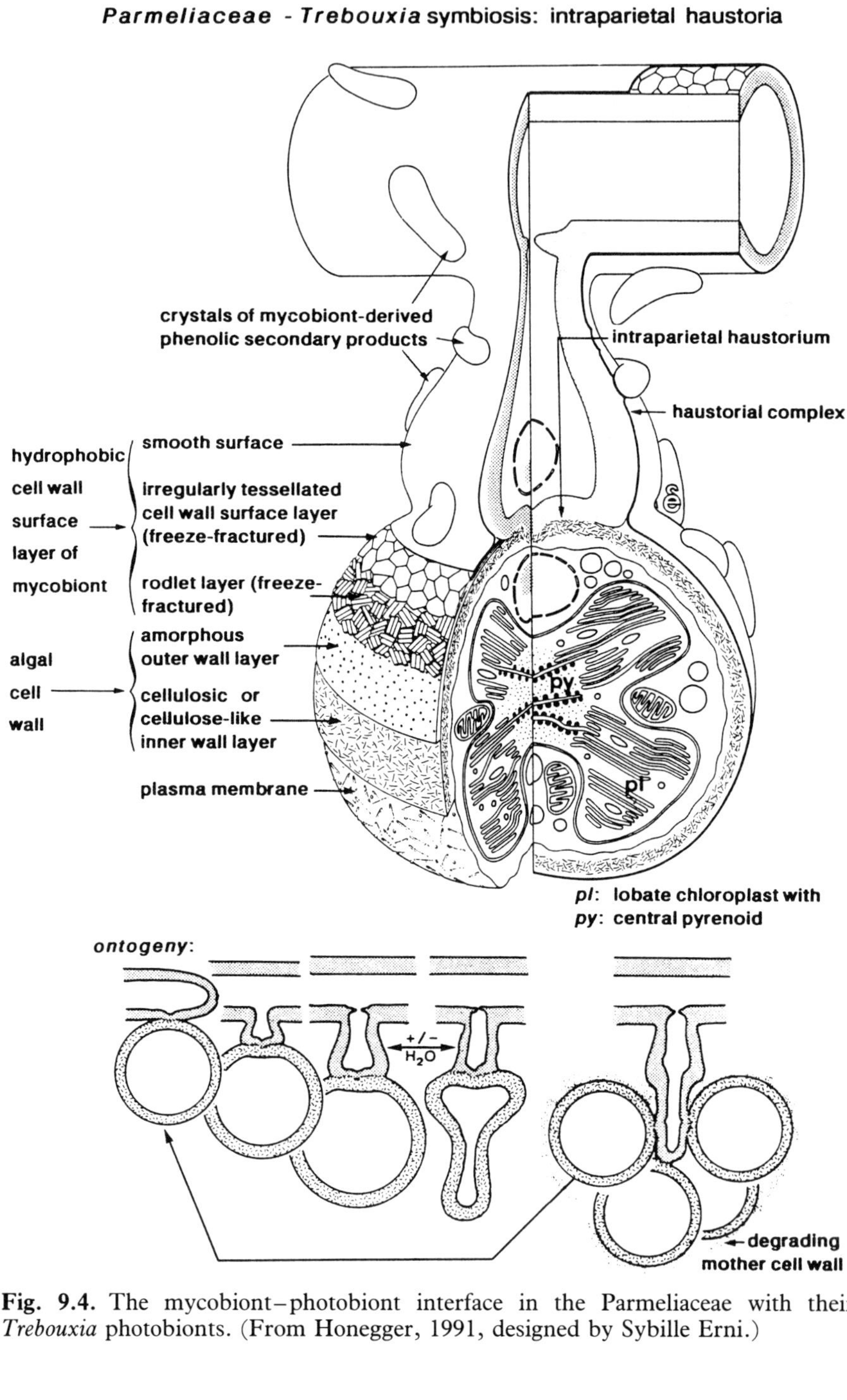

Fig. 9.4. The mycobiont–photobiont interface in the Parmeliaceae with their *Trebouxia* photobionts. (From Honegger, 1991, designed by Sybille Erni.)

calcium and trace elements from the soil or the rock substrate is accomplished by the mycosymbiont (Jones, 1988). The regulatory potential of the fungus towards the photosymbiont by controlling the flow of these nutrients has hardly been studied. Also the control in the other direction by supply of vitamins (Fig. 9.3) from the photobionts to the fungi has not been studied in much detail in lichens (Srere and Mosbach, 1974). Proteins may also be involved in the regulation of the metabolic activity of the symbionts (Vincente and Legaz, 1988). In *Evernia prunastri* the photobiont produces a glycoprotein which can inhibit the synthesis of the fungal urease (Perez-Urria *et al.*, 1989). Different urease activities in cultivated and lichenized fungi have been attributed to the action of this protein.

The mycobiont–photobiont interface in an advanced type of lichen develops from aerial hyphae, forming a haustorial complex (Fig. 9.4). Several layers of the cell walls of the two bionts are in close contact to each other. The structures have been studied by freeze fracturing, revealing below a smooth surface layer a tesselated surface layer and a rodlet layer of the mycobiont, sealing off large parts of the photobiont cell wall with hydrophobic material (Honegger, 1991). This close contact of the fungal haustoria with the photobionts has in early days of lichenology led to the wrong conclusion that the algal cells were spores of the fungal hyphae called 'gonidia'. Little is known about the chemical modification of cell walls and plasmamembrane in the interface area. The surfaces of the two bionts in the haustorial zone remain dry even at water saturation of the lichens (Brown *et al.*, 1987; Honegger, 1988).

9.3.6 Lichen secondary metabolites

The chemical structures of about 350 different secondary metabolites have been characterized in lichens (Huneck, 1984). This is a small number compared to more than 3000 different alkaloids and more than 4000 different flavonoids from higher plants, with a 20 times larger number of species.

The direct analysis of lichens by **tandem mass spectrometry** is a promising method which needs only very small amounts of material. But the interpretation of the results needs special care (Tabacchi *et al.*, 1991).

The major groups of these lichen metabolites are (Fig. 9.5)

1. Acetate–malonate derivatives condensed to
   (a) polyketides such as depsides, depsidones, dibenzofurans, anthraquinones, chromones, xanthones, naphthoquinones and diphenylether
   (b) long-chain fatty acid derivatives: lactone carbonic acids
   (c) macrocyclic lactones
   (d) alkanes

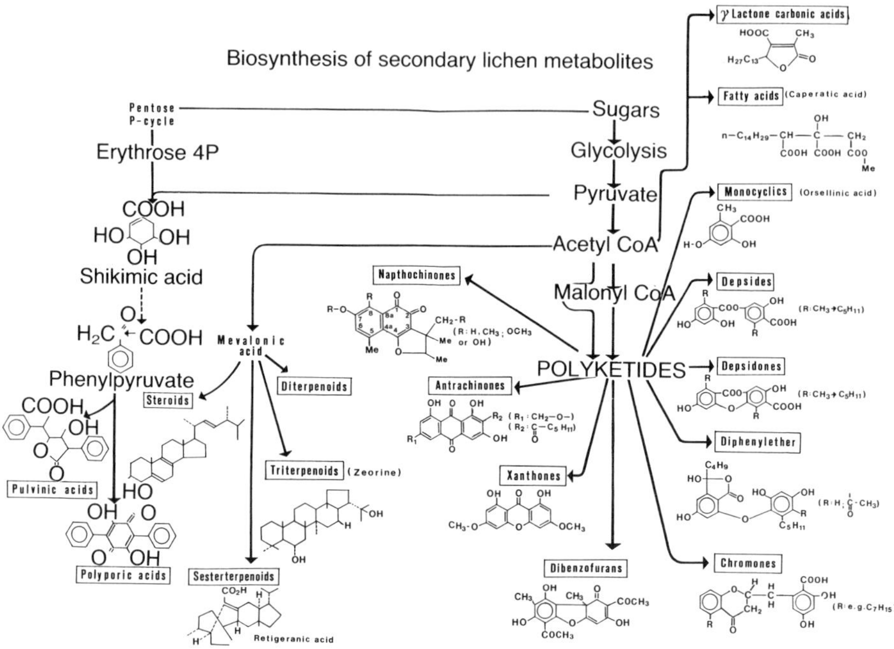

Fig. 9.5. Biosynthesis of secondary lichen metabolites.

2. Mevalonic acid derivatives such as diterpenes, triterpenes, sesquiterpenes, sesterterpenes and steroids
3. Shikimic acid derivatives, pulvinic acid derivatives and polyporic acid derivatives

Only a few biosynthetic pathways have been studied to determine the enzymes involved. Orsellinate synthase, which catalyses the formation of orsellinic acid from polyacetate chains with eight carbon atoms has been purified (Gaucher and Shepherd, 1968). The decarboxylation of orsellinate to orcinol is catalysed by orsellinate decarboxylase (EC 4.1.1.58) (Petterson, 1965). Orsell depside hydrolase (EC 3.1.1.40) is involved in the synthesis of depsides from monocyclic components. The K_m of this very stable protein is 56 μM for lecanoric acid as a substrate (Schultz and Mosbach, 1971). In the biosynthesis of the depside evernic acid (studied by Mosbach and Ehrensvärd, 1966) orsellinic acid is methylated to everninic acid, which condenses with another molecule of orsellinic acid to form evernic acid.

Since there are no confirmed reports of lichen secondary metabolite production by the photobionts, it must be assumed that the mycobionts are encoding the genes for biosynthesis (Renner and Gerstner, 1980; Vicente and Avalos, 1989). However, since only a small number of lichen secondary

Table 9.6 The occurrence and amounts of aromatic compounds in the genus *Cladonia* (modified from Huovinen *et al.*, 1990)

Compound	*Occurrence in species* (n = *51*)	*Contents* (%)	*Range*
Fumarprotocetraric acid	46	2.3	up to 7.8
Protocetraric acid	43	0.1	up to 0.3
Cph-2	41	+	up to 0.2
Atranorin	11	1.4	0.6–3.6
All compounds		2.9	0.6–8.1

metabolites could be detected in cultivated mycobionts or free-living fungi, specific signalling substances from the phototrophic partner for the fungus in the symbiosis may be necessary for induction. In a distinct genus such as *Cladonia*, four different aromatic lichen substances accumulate in significant quantities (Table 9.6) with fumarprotocetraric acid (a depsidone) (2.3% of the dry weight) and atranorin (a depside) (1.4%) as the major compounds. In *Cladonia rappii* up to 8% of the dry weight was found to be fumarprotocetraric acid (Huovinen *et al.*, 1990).

It has been proposed that secondary metabolites such as **triterpernoids**, very abundant in some species such as *Nephroma*, may act as antiherbivore compounds. But their absence in other abundant species such as *Sticta* sp. does not support this idea (Galloway, 1991).

9.3.7 Ecophysiology and growth rates

(a) Water relations

Lichens are poikilohydrous organisms, which can tolerate almost complete desiccation (Matthes and Feige, 1983). Water uptake and release is not under physiological control (Larson, 1979) perhaps with the exception of ice nucleation (see below). However, some species like the cyanolichen *Peltigera horizontalis* are relatively sensitive to desiccation, as indicated by a loss of intracellular potassium due to membrane damage (Brown and Brown, 1990). During the rewetting period a loss of carbon is possible (Farrar and Smith, 1976). Nevertheless, alternations of drying and wetting seem to be essential in the cultivation of lichens, for establishment and for preservation of the symbiosis (Armstrong, 1976). Compared to tissues of higher plants, the water content of lichens *in situ* is much lower at 8–14% of dry weight at 55% relative humidity and 65–80% of dry weight at 97% relative humidity (Jahns, 1984). At all moisture levels studied, *Lecanora muralis* had only about half the water content of *Peltigera canina*.

Such differences were explained by a much higher water-storing capacity of multilayered thalli compared to single-layered species such as *Parmelia saxatilis*. Therefore, Jahns (1984) considered a multilayered thallus to be a good adaptation to dry environments. Thalli of the same species with soredia-like structures are better adapted to humid areas.

The mechanisms that regulate the complete stop of all metabolic activities during desiccation are poorly understood. With cyanolichens it has been established that desiccation induces a functional detachment of phycobilisomes from photosystem II and energy transfer is restored only when liquid water is available for rehydration (Bilger *et al.*, 1989). *Ramalina cruciformis* with green algae as photobionts can also restore this capacity by uptake of water vapour as indicated by chlorophyll fluorescence yield (Lange *et al.*, 1989).

(b) Temperature effects

The optimum temperature for net photosynthesis for lichens from desert areas is around 20°C, and for antarctic species around 10°C (Lange, 1969; Lange and Kappen, 1972). Most lichen species are very cold tolerant, e.g. *Ramalina menziesii* and *Niebla ceruchis* (Kershaw, 1978). Both these species have *Trebouxia* as photobionts, whereas species with *Trentepohlia* as algal partner are less cold tolerant (Nash *et al.*, 1987). Even lichens from tropical areas are to a surprising degree frost tolerant; a positive net photosynthesis is still present at −12°C in *Parmelia africana* and −7°C in *Usnea livida* (Lange, 1965). *In situ* measurement in antarctic lichens (*Usnea sphacelata*) gave a positive net photosynthesis down to −10°C (Kappen, 1989).

Under extreme cold and dry conditions lichens survive as crypto-endolithic forms (Kappen and Friedman, 1983). *Lecidia* sp. and *Buellia* sp. colonize sandstone down to a depth of 10 mm. At −6°C to −16°C these species had a positive net photo-synthesis, although the maximal rate was very low with 0.1 mg CO_2 mg chlorophyll^{-1} h^{-1} equivalent to about 0.2 mg CO_2 g dry weight^{-1} h^{-1} (Table 9.5). Lichens also display ice nucleation activity (INA) (Kieft and Ahmadjian, 1989). Of 14 mycobionts tested, five had ice nucleation activity at −5°C whereas no photobiont showed this characteristic. *Rhizoplaca chrysoleuca* and *Xanthoria elegans* had the highest INA of all species tested. The physiological function of INA may be an additional moisture uptake mechanism and also, in some cases, a better frost tolerance by early formation of small extracellular ice crystals.

(c) Light effects

The light compensation points of lichens have been determined at between 2.6 and 4 W m^{-2} (Table 9.7). This is somewhat lower than the figure of 6 W m^{-2} reported for free-living algae such as *Scenedesmus*, grown under

Table 9.7 Photosynthetic efficiency (at 15°C, 0.035 vol % CO_2 and 60 W m^{-2}), chlorophyll content and light compensation points of sun and shade forms of lichen species (modified from Büttner, 1971)

Species (1) = *sun form* (2) = *shade form*	*Gross photosynthesis* (*mg* CO_2 g^{-1} h^{-1})	*Chlorophyll content* (*mg* (a + b) g^{-1})	*Light compensation point* ($W m^{-2}$)
Cladonia uncialis (1)	2.2	–	2.6
Cladonia uncialis (2)	1.3	–	4.0
Cladonia gracilis (1)	2.25	1.83	3.0
Cladonia gracilis (2)	1.75	1.47	3.0
Cornicularia muricata (1)	2.6	1.67	4.0
Cornicularia muricata (2)	2.1	1.47	4.0

light intensities of 28 W (Humbeck and Senger, 1984). However, under light intensities of only 5 W m^{-2} the compensation point is only 0.5 W m^{-2} in *Scenedesmus*. We can assume that in lichens, the compensation point of species grown under low light intensities is also significantly lower. Saturation of net photosynthesis in species from shaded habitats, e.g. *Cetraria islandica*, was achieved at lower light intensities (6 W m^{-2}) than in species from open habitats, e.g. *Cetraria cucullata* (10–16 W m^{-2}) (Türk, 1981). In the field, photosynthetic activity can be restricted to a part of the light period as found for *Cetraria islandica* from Alaska, where activity under continuous light in the summer months was restricted to 2–10 a.m. Additional rain extended the activity period during the day (Moser and Nash, 1978).

(d) Pollutants, effects and monitoring

During the last 15 years more than a hundred new contributions on air pollution and lichens have been published annually and regularly summarized in *The Lichenologist* (Henderson, 1989). One of the first comprehensive studies on biomonitoring with lichens was done by Sernander in 1926 in which city centres with no lichens on the trunks of trees were defined as 'lichen deserts'. Although nowadays sulphur dioxide (SO_2) is not the most dangerous pollutant anymore, due to very effective SO_2-eliminating installations in power plants, this pollutant is the best studied. Different lichen species can tolerate SO_2 concentrations of between 150 mg$SO_2$$m^{-3}$, e.g. *Lecanora conizaeoides*, and <30 mg, e.g. *Lobaria pulmonaria* and *Usnea florida* (Table 9.8). Even more sensitive (zone 10) are *Usnea articulata* or *Usnea filipendula*. Also germination of ascospores of different lichens species is differentially affected by SO_2 and H_2SO_3 concentrations. Spores of *Peltigera canina* are completely inhibited by 500 μM

Table 9.8 Presence of lichen flora and SO_2 concentration in the air (modified from Huber and Huber, 1990)

Zone	*Lichens*	SO_2 ($\mu g\ m^{-3}$)
1	*Pleurococcus viridis* (basis of stems)	>170
2	*Pleurococcus viridis Lecanora conizaeoides* (basis of stems)	150
3	*Lecanora conizaeoides Lepraria incana* (basis of stems)	125
4	*Hypogymnia physodes, Parmelia saxatilis* (basis of stems) *Lecanora expallens, Chaenotheca ferruginea*	70
5	e.g. *Hypogymnia physodes, Parmelia saxatilis, P. glabrata, Lepraria candelaris, Ramalina farinacea*	60
6	e.g. *Pseudovernia furfuracea, Alectoria fuscescens*	50
7	e.g. *Parmelia caperata, P. revoluta, Usnea subfloridana*	40
8	e.g. *Usnea ceratina, Parmelia perlata, P. reticulata, Normandina pulchella, Rinodina roboris*	35
9	e.g. *Lobaria pulmonaria, L. amplissima, Dimeralla lutea, Usnea florida*	<30
10	e.g. *Lobaria scrobiculata, Sticta limbata, Usnea articulata, Usnea filipendula, Teloschistes flavicans*	'pure'

SO_2 concentrations, whereas spores of *Lecanora conizaeoides* were almost insensitive to this concentration (Belandria *et al.*, 1989). The same concentration of sodium fluoride had a much larger effect on *Peltigera canina* than on *Lecanora conizaeoides*. Heavy metals accumulate in lichens as they do in fungi. They trap these elements from the air (Olmez *et al.*, 1985) and can also be used to monitor dissolved metals in natural waters. By using *Parmelia praesorediosa*, collected near the Gulf of Mexico, a maximum of 3 ppm Hg and 3.2 ppm Cu was detected (Beck and Ramelow, 1990). Much higher concentrations have been found in lichens from areas with heavy air pollution: 1500 ppm zinc (Pilegaard, 1978); 3000 ppm lead (Seaward, 1974); 330 ppm cadmium (Nash, 1971).

(e) Biotic interactions and growth rates

Lichens are considered not to be very competitive at places where other plants can grow (Topham, 1977). Within the lichens, the fruticose type is most competitive, the foliose type form intermediate and the crustose form least competitive (e.g. on rocks in Northern Europe). The most important qualities to assess the competitiveness of a certain species include

1. the growth rate
2. the growth form
3. adaptation to light, temperature and moisture limitations
4. the capacity to scavenge very low concentrations of mineral nutrients
5. the excretions of inhibiting substances (secondary metabolites)

Since the other factors have already briefly been discussed, only some examples for very different growth rates of lichens will be given here. The problems for quantitative assessment of growth in circular lichen thalli have been described in detail by Topham (1977) in a model. During the exponential growth phase (1–10 years), the radial growth increases from 0.14 to 1.15 mm $year^{-1}$. From 15–100 years a constant radial growth of 1.4 mm is identical to a reduction of the relative growth rate from 58% to less than 2%. A maximum radial growth of 27 mm $year^{-1}$ was recorded for the foliose *Peltigera rufescens* (Frey, 1959), contrasting to a very slow growth of the crustose *Rhizocarpon geographicum* with not more than 0.3 mm $year^{-1}$. In cultivation experiments with four different lichen species at different locations, it was found that development was most rapid at the natural location (Ott, 1987a).

Besides the general longevity of lichens, there are also a number of short-living species which are present for only a number of months (Poelt and Vezda, 1990). The very small *Absconditella sphagorum* develops its thalli and spores within a few months on *Sphagnum* sp. and disappears in the spring of the following year. Another example of a lichen species developing on a rapidly changing surface is *Bacidia ephemera*, growing on leaves of *Scirpus sylvaticus*, which develops thalli with sporulating apothecia between autumn and spring.

For a better understanding of the ecophysiology of lichens, the ecology of fungi (Christensen, 1989) and the coevolution of fungi with plants in general (Hawksworth, 1988) need to be studied in more detail.

9.4 *GEOSIPHON*

Geosiphon pyriforme, a phycomycete, forms pear-shaped vesicles which include cyanobacteria from the genus *Nostoc* (Fig. 9.6). The species was detected by von Wettstein in 1915. It has been found continuously at Biebergmünd (Spessart, Germany) where it is cultivated in the Forschungsinstitut Senckenberg (Mollenhauer and Mollenhauer, 1988).

The development of the symbiosis includes the following steps:

1. *Nostoc* hormogonia creep along *Geosiphon* hyphae
2. the cyanobacteria assemble at the hyphal apex
3. the hyphal cell wall below the apex lyses
4. a droplet of the cytoplasm of the fungus emerges
5. phagocytosis of *Nostoc* primorida
6. reestablishment of new fungal cell walls
7. multiplication and heterocyst differentiation of the *Nostoc* symbiont

To establish this symbiosis in cultivation experiments, the fungus needs a supply of growth promoting substances for active growth of the hyphae. It

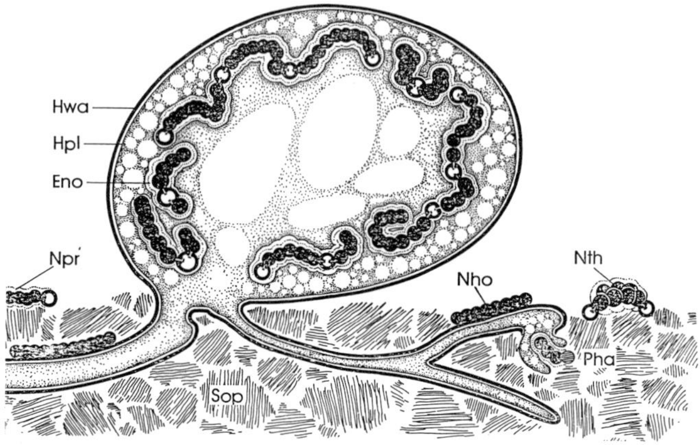

Fig. 9.6. *Geosiphon pyriforme* (cytological details not drawn). Hwa, host cell wall; Hpl, host plasmalemma; Nho, hormogonium of free-living *Nostoc*; Npr, primordium of the vegetative stage of free-living *Nostoc*, i.e. initial phase of very young thalli with 1 or 2 terminal cells just transforming into heterocysts and with the outer gelatinous envelope of the vegetative cells just appearing; Nth, young *Nostoc* thallus with fully developed terminal heterocysts and gelatinous sheath surrounding the vegetative cells; Eno, endosymbiotic *Nostoc* with intercalary heterocysts and gelatinous sheath covering the vegetative cells, the whole enclosed by host plasmalemma (phagosome); Pha, phagocytosis by cell wall lysis just below the hyphal apex and incorporation of a very young *Nostoc* primordium; Sop, soil particle. Original drawing, Mollenhauer and Mollenhauer (1988).

is suggested that on the soil areas where *Geosiphon* is found, these substances are supplied by the moss *Dicranella staphylina*, which is always present where *Geosiphon* is found. Addition of sterile cultivation filtrate of the moss stimulates hyphal growth even from old vesicles (Mollenhauer, 1991). Apparently, in the development of *Geosiphon* three different organisms are involved: a phycomycete, a cyanobacterium and a moss. Other mosses, like *Anthoceros punctatus* and *Blasia pusilla*, which also harbour symbiotic *Nostoc* strains, may also be a continuous source of microsymbionts for *Geosiphon*.

Photosynthesis of *Geosiphon* is very active (Kluge *et al.*, 1991) and N_2-fixation has been shown with the acetylene reduction assay (R. Kape, personal communication).

9.5 REFERENCES

Acharius, E. (1803) *Methodus qua omnes detectos lichenes*. Stockholm.

Adler, M.T. (1990) An artificial key to the genera of the Parmeliaceae (Lichenes, Ascomycotina). *Mycotaxon*, **38**, 331–47.

Ahmadjian, V. (1966) Lichens, in *Symbiosis*, Vol. I, (ed. S. Henry) Academic Press, New York, pp. 35–97.

Ahmadjian, V. (1990) What have synthetic lichens told us about real lichens? *Bibl. Lichenol.*, **38**, 1–10, J.

Ahmadjian, V. (1991) Molecular biology of lichens: A look to the future. *Symbiosis*, **11**, 249–54.

Ahmadjian, W. and Heikkilä, H. (1970) The culture and synthesis of *Endocarpon pusillum* and *Staurothela clopima*. *Lichenologist*, **4**, 259–67.

Altmann, Ph. L. and Dittmer, D.S. (1972–74) *Biology Data Book*, Vols I–II. Bethesda.

Armstrong, R.A. (1976) The influence of the frequency of wetting and drying on the radial growth of three saxicolous lichens in the field. *New Phytol.*, **77**, 719–24.

Beck, J.N. and Ramelow, G.J. (1990) Use of lichen biomass to monitor dissolved metals in natural waters. *Bull. Environ. Contam. Toxicol.*, **44**, 302–8.

Belandria, G., Asta, J. and Nurit, F. (1989) Effects of sulphur dioxide and fluoride on ascospore germination of several lichens. *Lichenologist*, **21**, 79–86.

Bellemère, A. and Letrouit-Galinou, M.-A. (1987) Differentiation of lichen asci including dehiscence and sporogenesis: an ultrastructural survey, in *Bibl. Lichenol.*, **25**, 137–61.

Bilger, W., Rimke, S., Schreiber, U. and Lange, O.L. (1989) Inhibition of energy-transfer to photosystem II in lichens by dehydration: different properties of reversibility with green and blue–green phycobionts. *J. Plant Physiol.*, **134**, 261–8.

Bliss, L.C. and Hadley, E.B. (1964) Photosynthesis and respiration of alpine lichens. *Am. J. Bot.*, **51**, 870.

Bresinsky, A. (1983) Niedere Pflanzen, in *Lehrbuch der Botanik* (ed. E. Strasburger) 32nd edn, Gustav Fischer Verlag, Stuttgart.

Brown, D.H. and Brown, R.M. (1990) Factors influencing desiccation-induced membrane damage in *Peltigera horizontalis*. *Bibl. Lichenol.*, **38**, 33–41.

Brown, D.H., Rapsch, S., Beckett, A. and Ascaso, C. (1987) The effect of desiccation on cell shape in the lichen *Parmelia sulcata* Taylor. *New Phytol.*, **105**, 295–9.

Brown, T.E. (1954) PhD Thesis, Ohio State University, Columbus.

Bubrick, P., Frensdorff, A. and Galun, M. (1985) Selectivity in the lichen symbiosis, in *Lichen Physiology and Cell Biology*, (ed. D.H. Brown) Plenum Press, New York, pp. 319–34.

Büdel, B. and Rhiel, E. (1987) Studies on the ultrastructure of some cyanolichen haustoria. *Protoplasma*, **139**, 145–52.

Büttner, R. (1971) Untersuchungen zur Ökologie und Physiologie des Gasstoffwechsels bei einigen Strauchflechten. *Flora*, **160**, 72–99.

Canaani, O. (1988) Light regulation of electron transport in N_2-fixing cyanobacteria in lichens. *Symbiosis*, **6**, 1–16.

Chapman, H.W. (1951) Absorption of CO_2 by potato leaves. *Am. Potato J.*, **28**, 602–15.

Christensen, M. (1989) A view of fungal ecology. *Mycologia*, **81**, 1–19.

Demming-Adams, B., Adams, W.W. III, Czygan, F.-C. *et al.* (1990) Differences in the capacity for radiationless energy dissipation in the photochemical apparatus of green and blue–green algal lichens associated with differences in carotenoid composition. *Planta*, **180**, 582–9.

Farrar, J.F. and Smith, D.C. (1976) Ecological physiology of the lichen *Hypogymnia physodes*. III. The importance of the rewetting phase. *New Phytol.*, **77**, 115–25.
Fiechter, E. (1990) Thallusdifferenzierung und intrathalline Sekundärstoffverteilung bei Parmeliaceae (Lecanorales, lichenisierte Ascomyceten). Inauguraldissertation, Universität Zürich.
Frey, E. (1959) Die Flechtenflora und Vegetation des Nationalparks im Unterengadin. II. Die Entwicklung der Flechtenvegetation auf photogrammetrisch kontrollierten Dauerflächen. *Ergebn. Wiss. Unters. Schweiz. Natn-Parks*, **6**, 237–319.
Friedl, T. (1989) Systematik und Biologie von *Trebouxia* (Microthamniales, Chlorophyta) als Phycobiont der Parmeliaceae (lichenisierte Ascomyceten). Dissertation, München.
Gaffron, H. (1939) Über Anomalien des Atmungsquotienten von Algen aus Zuckerkulturen. *Biol. Zentralbl.*, **59**, 288–302.
Galloway, D.J. (1991) Chemical evolution in the order peltigerales: triterpenoids. *Symbiosis*, **11**, 327–44.
Galun, M. (1988a) Carbon metabolism, in *Handbook of Lichenlogy* (ed. M. Galun) Vol. I, CRC Press, Boca Raton, pp. 197–200.
Galun, M. (ed.) (1988b) *Handbook of Lichenology*. Vols I, II, III, CRC Press, Boca Raton.
Gärtner, G. (1985) Taxonomische Probleme bei den Flechtengattungen *Trebouxia* und *Pseudotrebouxia* (Chlorophyceae, Chloredales). *Phyton*, **25**, 101–111.
Gaucher, G.M. and Shepherd, M.G. (1968) Isolation of orsellinic acid synthetase. *Biochem. Biophys. Res. Commun.*, **32**, 664.
Gäumann, E. (1964) *Die Pilze*. Birkhäuser Verlag, Basel.
Gorin, P.J.A., Baron, M. and Iacomini, M. (1988) Storage products in lichens, in *Handbook of Lichenology*, (ed. M. Galun) CRC Press, Boca Raton, pp. 9–23.
Hale, M.E. (1984) An historical review of the genus concept in lichenology. *Beiheft Nova Hedwigia*, **79**, 11–23.
Hawksworth, D.L. (1988) Coevolution of fungi with algae and cyanobacteria in lichen, in *Coevolution of Fungi with Plants and Animals* (eds K.A. Pirozynski and D.L. Hawksworth) Academic Press, London, San Diego, pp. 125–48.
Hawksworth, D.L. and Hill, D.J. (1984) *The Lichen-Forming Fungi*, Blackie, Glasgow, London.
Henderson, A. (1989) Literature on air pollution and lichens XXX. *Lichenologist*, **21**, 369–78.
Henssen, A. (1985) *Hertella*, a new lichen genus in the Peltigerales from the Southern Hemisphere. *Mycotaxon*, **22**, 381–97.
Henssen, A. (1986) *Edwardiella mirabilis*, a holocarpous lichen from Marion Island. *Lichenologist*, **18**, 51–6.
Henssen, A. and Jahns, H.M. (1973) Lichenes, eine Einführung in die Flechtenkunde, mit einem Beitrag von Johan Santesson. Thieme Verlag, Stuttgart, 465 pp.
Hersoug, L.G. (1983) Lichen protein affinity toward walls of cultured and freshly isolated phycobionts and its relationship to cell wall cytochemistry. *FEMS Microbiol. Lett.*, **20**, 417–20.
Hill, D.J. (1985) Changes in photobiont dimensions and numbers during co-development of lichen symbionts, in *Lichen Physiology and Cell* (ed. D.H. Brown) Plenum Press, New York, pp. 303–17.

Honegger, R. (1987) Questions about pattern formation in the algal layer of lichens with stratified (heteromerous) thalli. *Bibl. Lichenol.*, **25**, 59–71.

Honegger, R. (1988) The functional morphology of cell-to-cell interactions in lichens, in *Cell to Cell Signals in Plant, Animal and Microbial Symbiosis* (eds S. Scannerini, D.C. Smith, P. Bonfante and V. Gianinazzi-Pearson) NATO ASI Series H, Vol. 17, Springer Verlag, Berlin, pp. 39–53.

Honegger, R. (1990) Mycobiont–photobiont interactions in adult thalli and in axenically resynthesized pre-thallus stages of *Xanthoria parietina* (Teloschistales, lichenized Ascomycetes). *Bibl. Lichenol.*, **38**, 191–208.

Honegger, R. (1991) Functional aspects of the lichen symbiosis. *Ann. Res. Plant Physiol. Mol. Biol.*, **42**, 553–78.

Huber, A. and Huber, W. (1990) Pflanzen als Schadstoffindikatoren. Verhütung von Schäden, in *Schadwirkungen auf Pflanzen. Lehrbuch der Pflanzentoxikologie*, 2nd edn (eds B. Hock and E.F. Elstner) B.I. Wissenschaftsverlag, Mannheim, Wien, Zürich, pp. 47–66.

Humbeck, K. and Senger, H. (1984) The blue light factor in sun and shade plant adaptation, in *Blue Light Effects in Biological Systems* (ed. H. Senger) Springer-Verlag, Berlin, Heidelberg, pp. 343–51.

Huneck, S. (1984) Fortschritte der Chemie von Flechtenstoffen. *Beiheft Nova Hedwigia*, **79**, 793–838.

Huovinen, K., Ahti, T. and Stenroos, S. (1990) The composition and contents of aromatic lichen substances in *Cladonia* section *Cladonia* and group Furcatae. *Bibl. Lichenol.*, **38**, 197–229.

Jahns, H.M. (1984) Morphology, reproduction and water relations – a system of morphogenetic interactions in *Parmelia saxatilis. Beiheft Nova Hedwigia*, **79**, 715–37.

Jahns, H.M. (1988) The lichen thallus, in *Handbook of Lichenology* (ed. M. Galun) Vol. 1, CRC Press, Boca Raton, pp. 95–143.

Jahns, H.M. (1989) Problems in lichen taxonomy. *Abh. naturwiss. Ver. Hamburg*, **28**, 247–52.

Jahns, H.M. and Ott, S. (1990) Regulation of regenerative processes in lichens. *Bibl. Lichenol.*, **38**, 230–9.

James, P.W. and Henssen, A. (1976) The morphological and taxonomic significance of Cephalodia, in *Lichenology: Progress and Problems* (eds D.H. Brown, D.L. Hawksworth and R.H. Bailey) Systematics Association Special Volume no. 8, Academic Press, London, New York, pp. 27–77.

Jones, D. (1988) In *Handbook of Lichenology*, Vol. III, (ed. M. Galun) CRC Press, Boca Raton, pp. 109–26.

Kandler, O. (1954) Gesteigerter Glucoseeinbau im Licht als Indikator einer lichtabhängigen Phosphorylierung. *Z. Naturforsch.*, **9b**, 625–44.

Kappen, L. (1989) Field measurements of carbon dioxide exchange of the Antarctic lichen *Usnea sphacelata* in the frozen state. *Antarctic Sci.*, **1**, 31–4.

Kappen, L. and Friedmann, I. (1983) Kryptoendolithische Flechten als Beispiel einer Anpassung an extrem trocken-kalte Klimabedingungen. *Verh. Ges. Ökol.*, **10**, 517–19.

Kappen, L., Lange, O.L., Schulze, E.D. *et al.* (1980) Ecophysiological investigation on lichens of the Negev Desert. VII. The influence of the habitat exposure on dew imbibition and photosynthetic productivity. *Flora*, **169**, 216.

Kärnefelt, I. and Mattson, J.-E. (1987) Morphological characteristics and affinities of the genus *Cornicularia*. *Bibl. Lichenol.*, **25**, 171–84.

Kershaw, K.A. (1978) The role of lichens in boreal tundra transition areas. *Bryologist*, **81**, 294–306.

Kieft, T.L. and Ahmadjian, V. (1989) Biological ice nucleation activity in lichen mycobionts and photobionts. *Lichenologist*, **21**, 355–62.

Kluge, M., Mollenhauer, D. and Mollenhauer R. (1991) Photosynthetic carbon assimilation in *Geosiphon pyriforme* (Kützing) F. v. Wettstein, an endosymbiotic association of fungus and cyanobacterium. *Planta*, **185**, 311–15.

Krog, H. and Swinscow T.D.V. (1987) New species and new combinations in some parmelioid lichen genera, with special emphasis on East African taxa. *Lichenologist*, **19**, 419–31.

Lange, O.L. (1965) Der CO_2-Gaswechsel von Flechten bei tiefen Temperaturen. *Planta*, **64**, 1–19.

Lange, O.L. (1969) Experimentell-ökologische Untersuchungen an Flechten der Negev-Wüste. I. CO_2-Gaswechsel von *Ramalina maciformis* (Del.) Bory unter kontrollierten Bedingungen im Laboratorium. *Flora*, **158**, 324–59.

Lange, O.L., Bilger, W., Rimke, S. and Schreiber, U. (1989) Chlorophyll fluorescence of lichens containing green and blue–green algae during hydration by water vapor uptake and by addition of liquid water. *Bot. Acta*, **102**, 306–13.

Lange, O.L. and Kappen L. (1972) Photosynthesis of lichens from Antarctica. *Antarct. Terr. Biol.*, *Antarct. Res. Ser.*, **20**, 83–95.

Lange, O.L. and Redon, J. (1983) Epiphytische Flechten im Bereich einer 'Nebeloase' (Fray Jorge). II. Ökophysiologische Charakterisierung von CO_2-Gaswechsel und Wasserhaushalt. *Flora*, **174**, 245.

Larson, D.W. (1979) Lichen water relations under drying conditions. *New Phytol.*, **82**, 713–31.

Lines, C.E.M., Ratcliffe, R.G., Rees, T.A.V. and Southon, T.E. (1989) A ^{13}C-NMR study of photosynthate transport and metabolism in the lichen *Xanthoria calcicola* Oxner. *New Phytol.*, **111**, 447–56.

Linnaeus, C. (1753) *Species plantarum*. Stockholm.

Massalongo, A.B. (1856) Descrizione di alcuni licheni nuovi. *R. Ist. Veneto di Scienze*, *Lettere* e *Arti Atti*, ser. 3, **2**, 351–84.

Matthes, U. and Feige, G.B. (1983) Ecophysiology of lichen symbiosis, in *Encyclopedia of Plant Physiology* (ed. M. Lange, P.S. Nobel, C.B. Osmond and H. Ziegler) New Series **12c**, 423–67.

Mollenhauer, D. (1991) *Geosiphon pyriforme*, in *Algal Symbioses* (ed. W. Reisser) Biogress, Bristol.

Mollenhauer, D. and Mollenhauer, R. (1988) *Geosiphon* cultures ahead. *Endocyt. Cell Res.*, **5**, 69–73.

Mosbach, K. and Ehrensvärd, U. (1966) Studies on lichen enzymes. I. Preparation and properties of a depside hydrolysing esterase and of orsellinic acid decarboxylase. *Biochem. Biophys. Res. Commun.*, **22**, 145.

Moser, T.J. and Nash, T.H. III (1978) Photosynthetic patterns of *Cetraria cucullata* (Bull.) Ach. at Anaktuvuk Pass. Alaska. *Oecologia*, **34**, 37–43.

Nakano, T. (1988) Phycobionts of some Japanese species of the Graphidaceae. *Lichenologist*, **20**, 353–60.

Nash, T.H., III (1971) Effect of effluents from zinc factory on lichens. PhD Thesis, Rutgers University, New Brunswick, NJ.

Nash, T.H., Kappen, L., Losch, R. *et al.* (1987) Cold resistance of lichens with *Trentepohlia-* or *Trebouxia* photobionts from the North American West Coast. *Flora*, **179**, 241–51.

Nylander, W. (1866) Circa novum in studio lichenum criterium chemicum. *Flora*, **49**, 198.

Oberwinkler, F. (1984) Fungus–alga interactions in basidiolichens. *Beiheft Nova Hedwigia*, **79**, 739–74.

Olmez, I., Gulovali, M.C. and Gordon, G.E. (1985) Trace element concentrations in lichens near a coal-fired power plant. *Atmos. Environ.*, **10**, 1663–9.

Ott, S. (1987a) Differences in the developmental rates of lichens. *Ann. Bot. Fenn.*, **24**, 385–93.

Ott, S. (1987b) Sexual reproduction and developmental adaptations in *Xanthoria parietina. Nord. J. Bot.*, **7**, 219–28.

Ott, S. (1987c) Reproductive strategies in lichens, *Bibl. Lichenol.*, **25**, 81–93.

Ott, S. (1988) Photosymbiodemes and their development in *Peltigera venosa. Lichenologist*, **20**, 361–8.

Perez-Urria, E., Rodriguez, M. and Vicente, C. (1989) Algal partner regulates fungal urease in the lichen *Evernia prunastri* by producing a protein which inhibits urease synthesis. *Plant Mol. Biol.*, **13**, 665–72.

Petterson, G. (1965) An orsellinic acid decarboxylaze isolated from *Gliocladium* roseum. *Acta Chem. Scand.*, **19**, 2013–21.

Pilegaard, K. (1978) Airborne metals and SO_2 monitored by epiphytic lichens in an industrial area. *Environ. Poll.*, **17**, 81–92.

Poelt, J. and Vezda, A. (1990) On shortliving lichens. *Bibl. Lichenol.*, **38**, 359–76.

Rai, A.N. (1988) Nitrogen metabolism, in *Handbook of Lichenology*, Vol. I, (ed. M. Galun) CRC Press, Boca Raton, pp. 201–37.

Reazin, G.H. (1954) On the dark metabolism of a golden-brown alga, *Ochromonas malhamensis. Am. J. Bot.*, **41**, 771–7.

Renner, B. and Galloway, D.J. (1982) Phycosymbiodemes in *Pseudocyphellaria* in New Zealand. *Mycotaxon*, **16**, 197–231.

Renner, B. and Gerstner, E. (1980) Mediumabhängige Anthrachinon-Bildung in Mycobiontenkulturen von *Caloplaca ferruginea. Naturwiss. enschaften*, **67**, 352.

Richardson, D.H.S. and Smith D.C. (1968) Lichen physiology. IX. Carbohydrate movement from the *Trebouxia* symbiont of *Xanthoria aureola* to the fungus. *New Phytol.*, **67**, 61–8.

Ruoss, E. (1987) Species differentiation in a group of reindeer lichens (*Cladonia* subg. *Cladina*). *Bibl. Lichenol.*, **25**, 197–206.

Schlösser, E. (1983) *Allgemeine Phytopathologie*. Thieme Verlag, Stuttgart.

Schultz, J. and Mosbach, K. (1971) Studies on lichen enzymes. Purification and properties of an orsellinate depside hydrolase obtained from *Lasallia pustulata. Eur. J. Biochem.*, **22**, 153–7.

Schuster, G., Ott, S. and Jahns, H.M. (1985) Artificial cultures of lichens in the natural environment. *Lichenologist*, **17**, 247–53.

Seaward, M.R.D. (1974) Some observations on heavy metal toxicity and tolerance in lichens. *Lichenologist*, **6**, 158.

Sernander, R. (1926) Granskor och Fiby urskog. *Acta Phytogeogr. Suec.*, **8**, 1.

Shear, H. and Walsby, A.E. (1975) An investigation into the possible light-shielding role of gas-vacuoles in a planktonic blue–green alga. *Br. Phycol. J.*, **10**, 241–51.

Siegel, M.R., Latch, G.C.M. and Johnson, M.C. (1987) Fungal endophytes of grasses. *Ann. Rev. Phytopathol.*, **25**, 293–315.

Smith, D.C. (1974) Transport from symbiotic algae and symbiotic chloroplasts to host cells. *Symp. Soc. Exp. Biol.*, **28**, 485–520.

Srere, P.A. and Mosbach, K. (1974) Metabolic compartmentation: symbiotic, organellar, multienzymic, and microenvironmental. *Ann. Rev. Microbiol.*, **28**, 61–83.

Stahl, E. (1877) *Beiträge zur Entwicklungsgeschichte der Flechten*. II. Über die Bedeutung der Hymenialgonidien, Verlag Arthur Felix, Leipzig.

Stocker-Wörgötter, E. and Türk, R. (1989) Laborversuche zur Kultivierung von Blatt- und Strauchflechten und deren Komponenten. *Nova Hedwigia*, **48**, 207–28.

Tabacchi, R., Allemand, P. and Tsoupras, G. (1991) Direct analysis of lichens by tandem mass spectrometry. *Symbiosis*, **11**, 193–206.

Thielmann, J.J. (1990) Die Anpassung des Photosyntheseapparates der einzelligen Grünalge *Scenedesmus obliquus* an verschiedene Lichtbedingungen unter besonderer Berücksichtigung der beteiligten Photorezeptoren. PhD Thesis, University of Marburg (Germany).

Topham, P.B. (1977) Colonization, growth, succession and competition, in *Lichen Ecology* (ed. M.R.D. Seaward) Academic Press, London, pp. 31–68.

Tschermak-Woess, E. (1988) The algal partner, in *Handbook of Lichenology*, Vol. I, (ed. M. Galun) CRC Press, Boca Raton, pp. 39–92.

Türk, R. (1981) Laboruntersuchungen über den CO_2-Gaswechsel von Flechten aus den Mittleren Ostalpen. I. *Phyton (Austria)* **21**, 203–34.

Türk, R. (1983) Laboruntersuchungen über den CO_2-Gaswechsel von Flechten aus den Mittleren Ostalpen. II. *Phyton (Austria)*, **23**, 1–18.

Vicente, C. and Avalos, A. (1989) Regulation by light of the intracellular levels of cyclic AMP in the lichen *Evernia prunastri*. *Endocytobiosis Cell Res.*, **6**, 203–11.

Vicente, C. and Legaz, M.E. (1988) Lichen enzymology, in *Handbook of Lichenology*, Vol. I (ed. M. Galun) CRC Press, Boca Raton, pp. 239–81.

Webster, G.C. and Frenkel, A.W. (1953) Some respiratory characteristics of the blue–green alga, *Anabaena*. *Plant Physiol.*, **28**, 63–9.

Werner, R.G. (1931) Histoire de la synthèse lichénique. *Mém. Soc. Sci. Nat. Maroc*, **27**, 7–45.

Wettstein, F. von (1915) *Geosiphon* Fr. Wettst., eine neue, interessante Siphonee. *Öster. Bot. Z.*, **45**, 145–56.

Winnacker, E.L. (1987) *From Genes to Clones*, Verlag Chemie, Weinheim.

10
Vesicular-Arbuscular Mycorrhiza

The symbionts of vesicular-arbuscular mycorrhiza (VAM) were grouped together in the order Endogonales (Zygomycetes). However, the placement of VAM-fungi within these taxa was questioned. Because of the occurrence of long-chain unsaturated fatty acids, as $C_{20:4}$, $C_{20:5}$ and $C_{22:5}$, and the replacement of ergosterol by cholesterol derivatives, VAM-fungi resemble protista more than other fungi (Jabaji-Hare, 1988; Pirozynski and Dalpé, 1989). However, these characters are not sufficient for this taxonomic change.

VAM-fungi are obligate symbionts and have not until now been cultivated axenically. However, one or two species of the Endogonales, which do not form VAM symbioses, such as *Endogone pisiformis*, have been cultivated and brought to sporulation. Many obligate parasites, e.g. the mildew fungus *Erysiphe graminis*, can also not be axenically cultivated. The physiology and genetics of VAM-fungi is therefore hardly known.

Vesicles are swollen hyphae. Arbuscules are finely branched haustoria. These structures are typical morphological characteristics of intracellular hyphae.

There is little host specificity of VAM-fungi. The major physiological benefit for the host plant is a better tolerance of nutrient deficiencies and water stress. The beneficial effects on plant growth are enhanced in the presence of some N_2-fixing and phosphate solubilizing rhizosphere bacteria (Linderman, 1988b). Books and reviews on the biology of VA-mycorrhiza have been published by Strullu (1978) Mosse and Haymann (1980), Schenck (1982), Harley and Smith (1983), Smith and Gianinazzi-Pearson (1988), Safir (1987), Bethlenfalvay *et al.* (1988b), Linderman (1988a), Read (1990) and Reid (1990). Other special aspects have been reported by Sanders *et al.* (1975), Tinker (1978), Graw *et al.* (1979), Gianinazzi-Pearson *et al.* (1981), Trappe (1982), Buchenauer (1983), Hayman (1983), Walker (1983), Schenck *et al.* (1984), Hayman and Tavares (1985), Schenk and Perez (1990) and Sicverding (1991).

Table 10.1 Separation of Endogonales and Glomales (Based on Morton and Benny, 1990)

ORDER Endogonales				
		FAMILY Endogonaceae		
			GENUS *Endogone*	
				Sclerogone
ORDER Glomales				
	SUBORDER Glomineae			
		FAMILY Glomaceae		
			GENUS *Glomus*	
				Sclerocystis
		FAMILY Acaulosporaceae		
			GENUS *Acaulospora*	
				Entrophospora
	SUBORDER Gigasporineae			
		FAMILY Gigasporaceae		
			GENUS *Gigaspora*	
				Scutellospora

10.1 THE MICROSYMBIONT

10.1.1 Taxonomy

The order Endogonales shares various characteristics with the related orders Mucorales and Entomophthorales, such as gametangiogamy and the formation of a hypnozygote with subsequent meiosis. The VAM-forming genera and species are now separated into the order Glomales with three families and six genera (Table 10.1). The genus *Endogone* does not form VA-mycorrhiza but does form ectomycorrhiza. The former genus *Gigaspora* contains at least two groups of VAM-species which are different from each other on the basis of spore germination, cell wall structure and the ornamentation of the auxiliary cells and is now separated into the genera *Scutellospora* and *Gigaspora*. The genera *Acaulospora* and *Entrophosphora* are VAM-fungi. The genus *Sclerocystis* is very similar to the genus *Glomus*, the only observable difference being the spore order in the sporocarp. The largest and best-researched group of VAM-fungi is the genus *Glomus*, represented by over 70 species (Table 10.2). The four genera *Acaulospora*, *Entrophospora*, *Glomus* and *Sclerocystis* are much more related to each other than to *Gigaspora* and *Scutellospora*. It is proposed that the latter two genera be placed in a separate family based on phylogenetic trees and cladistic analysis of 57 species (Morton, 1990). Four different wall types in VAM spores have been characterized. However, they are not correlated with the two families and the six genera (Walker, 1983). The spore wall texture is different in the outer layer (2–4 μm thick) and the inner layer (5–6 μm)

(Bonfante-Fasolo and Vian, 1984). Using SDS–polyacrylamide-gel electrophoresis to separate the soluble proteins from chlamydospores of several *Glomus* and *Gigaspora* species, each one was found to have a distinct protein pattern, not affected by the host plant or environmental factors (Giovannetti and Lioi, 1990). A very specific method for discriminating spores of a certain *Glomus* species is an indirect ELISA with monoclonal antibodies (Wright *et al.*, 1987). Antibodies against *Glomus occultum* were not reactive against 29 other mycorrhizal fungi and five non-mycorrhizal fungi. Dot immunoblot assays have also been used to identify *Glomus occultum* colonizing roots (Wright and Morton, 1989).

A four-letter species code for the 136 VAM species described so far was recently proposed by Perez and Schenk (1990), as shown in Table 10.2. The first letter indicates the genus (A = *Acaulospora*, E = *Entrophospora*, G = *Gigaspora*, L = *Glomus*, S = *Sclerocystis* and C = *Scutellospora*). The second letter of the code is always the initial letter of the species epithet, the last two letters are consonants only from the second and third syllable of the species name, whenever possible; *Glomus mosseae* therefore has the code

Table 10.2 Alphabetized genus and species with species code. (Based on Perez and Schenk, 1990)

Species	*Code*	*Species*	*Code*
Acaulospora (27 species)		*trappei*	ATRP
appendicula	AAPD	*tuberculata*	ATBC
bireticulata	ABRT	*undulata*	AUDL
delicata	ADLC	*Entrophospora* (3 species)	
denticulata	ADTC	*colombiana*	ECLB
dilatata	ADLT	*infrequens*	EIFQ
elegans	AELG	*schenckii*	ESHK
foveata	AFVT	*Gigaspora* (7 species)	
gedanensis	AGDN	*albida*	GABD
gerdemannii	AGDM	*calospora*	GCLP
lacunosa	ALCN	*candida*	GCDD
laevis	ALVS	*decipiens*	GDCP
longula	ALGL	*gigantea*	GGGT
mellea	AMLL	*margarita*	GMRG
morrowae	AMRW	*ramisporophora*	GRMS
myriocarpa	AMYC	*rosea*	GRSA
nicolsonii	ANCS	*Glomus* (69 species)	
polonica	APLN	*aggregatum*	LAGR
rehmii	ARHM	*albidum*	LABD
rugosa	ARGS	*ambisporum*	LABS
scrobiculata	ASCB	*arborense*	LABR
spinosa	ASPN	*australe*	LAST
splendida	ASPD	*boreale*	LBRL
sporocarpia	ASPC	*botryoides*	LBTR
thomii	ATHM		

Table 10.2 *Continued*

Species	*Code*	*Species*	*Code*
caledonium	LCLD	*pulvinatum*	LPVN
callosum	LCLL	*pustulatum*	LPST
canadense	LCND	*radiatum*	LRDT
cerebriforme	LCBF	*reticulatum*	LRTC
citricola	LCTC	*scintillans*	LSTL
claroideum	LCRD	*segmentatum*	LSGT
clarum	LCLR	*tenebrosum*	LTNB
constrictum	LCST	*tenerum*	LTNR
convolutum	LCVL	*tenue*	LTNU
delhiense	LDLH	*tortuosum*	LTRT
deserticola	LDST	*tubaeforme*	LTBF
diaphanum	LDPH	*versiforme*	LVSF
dimorphicum	LDMR	*vesiculiferum*	LVCF
dominikii	LDMK	*warcupii*	LWRC
etunicatum	LETC	*Sclerocystis* (10 species)	
fasciculatum	LFSC	*clavispora*	SCVS
fecundisporum	LFCS	*coccogena*	SCCG
flavisporum	LFLS	*coremioides*	SCRM
formosanum	LFMS	*dussii*	SDSS
fragile	LFRG	*indica*	SIDC
fuegianum	LFGN	*microcarpus*	SMRC
fulvum	LFLV	*pachycaulis*	SPCC
geosporum	LGSP	*pakistanica*	SPKS
gerdemannii	LGDM	*rubiformis*	SRBF
globiferum	LGBF	*sinuosa*	SSNS
glomerulatum	LGML	*Scutellospora* (20 species)	
halonatum	LHLN	*alborosea*	CALR
heterosporum	LHTS	*aurigloba*	CARG
hoi	LHOI	*calospora*	CCLS
intraradices	LINR	*coralloidea*	CCRL
invermaium	LIVM	*dipapillosa*	CDPL
lacteum	LLCT	*dipurpurescens*	CDPP
leptotichum	LLPT	*erythropa*	CERT
macrocarpum	LMCC	*fulgida*	CFLG
maculosum	LMCL	*gilmorei*	CGLM
magnicaule	LMGC	*gregaria*	CGRG
manihotis	LMNH	*heterogama*	CHTG
melanosporum	LMLS	*minuta*	CMNT
microaggregatum	LMAG	*nigra*	CNGR
microcarpum	LMRC	*pellucida*	CPLC
monosporum	LMNS	*persica*	CPRS
mosseae	LMSS	*reticulata*	CRTC
multicaule	LMTC	*savannicola*	CSVN
multisubstensum	LMST	*tricalypta*	CTRC
occultum	LOCT	*verrucosa*	CVRC
pallidum	LPLD	*weresubiae*	CWRS
pansihalos	LPSH		
pubescens	LPBS		

LMSS. New species are still being added to this list, e.g. *Glomus nanolumen* (Koske and Gemma, 1989).

As with all other VAM-fungi, the host-specificity of *Glomus* species is very low. Thus *Glycine max* may be infected by *Glomus claroideum*, *Glomus etunicatum* and *Glomus macrocarpum*, although each infects with a different frequency. Several plant species, such as *Glycine max* and *Eupatorium odoratum*, are promiscuous hosts, others select certain *Glomus* spp. as symbionts (Table 10.3). According to some taxonomists *Glomus fasciculatum* shelters several distinct species. These were grouped together before 1974 when single-spore infection tests were not common, and spore mixtures were used. These must be repeated before the taxonomy is confirmed. Generally morphological criteria agree with immunological critera. Indirect ELISA tests show that there are relatively large similarities between *Glomus macrocarpum*, *Glomus caledodinum*, *Glomus clarum* and *Glomus mosseae*, whereas *Glomus tenue* was not so similar. Less similarity has been shown between *Glomus mosseae* and *Gigaspora margarita* and *Acaulospora laevis*.

Table 10.3 Species of the VAM-genus Glomus and their distribution

Species	*Distribution (examples)*
Glomus aggregatum	recultivated soil after 2 years
Glomus albidum	*Capsicum annuum*
Glomus caledonium	*Lactuca sativa, Allium porrum*
Glomus claroideum	*Glycine max, Zea mays*
Glomus clarum	*Allium cepa, Cyamopsis tetragonoloba*
Glomus etunicatum	*Paspalum notatum, Sorghum vulgare, Zea mays, Glycine max*
Glomus margarita	*Paspalum notatum, Sorghum vulgare, Zea mays, Glycine max, Juglans nigra*
Glomus fasciculatum group	*Cyamopsis tetragonoloba, Leucaena leucocephala, Gnaphalium indicum, Sorghum bicolor, Trifolium subterraneum, Citrus limon, Cajanus cajan, Ornithogalum umbellatum, Allium cepa*
Glomus fuegianum (='*Endogone*' *fuegiana*)	*Bidens ternata*
Glomus intraradices	*Citrus limon, Gossypium* sp.
Glomus macrocarpum	*Zea mays, Paspalum notatum, Glycine max, Lycopersicon esculentum*
Glomus melanosporum	*Arachis hypogaea*
Glomus manihotis	*Manihot esculenta*
Glomus monosporum	*Agropyron smithii, Raphanus raphanistrum*
Glomus mosseae group	*Citrus limon, Vigna unguiculata, Liatris aspera, Babtisia leucantha, Capsicum* sp., *Atriplex canescens, Artemisia tridentata*
Glomus occultum	*Trifolium pratense, Ammophila littoralis*
Glomus tenue	*Avena sativa*

Non-VAM fungi, such as *Pezizella ericae*, *Rhizopus oryzae* and *Mucor hiemalis*, were found to be even further separated from *Glomus mosseae*.

The **DNA content** was determined to be between 0.74–0.81 pg per nucleus in *Gigaspora margarita* and 0.25–0.28 pg in *Glomus versiforme*. No large differences were found between several stages of development (spores, germinating hyphae, intraradical hyphae) (Bianciotto *et al.*, 1991). The DNA content of these VAM-fungi is comparable to other obligate biotrophs such as *Erysiphe*.

10.1.2 Physiology and ecology

Some VAM-species seem to be ubiquitously present, e.g. in all tropical soils, whereas others such as *Entrophospora colombiana* and *Gigaspora margarita* are limited to a certain soil pH (Sieverding, 1988).

Spores of VAM fungi (Fig. 10.1) germinate in pure culture on all fungal media to form a mycelium only a few centimetres long. The proximity of a compatible root increases the length somewhat. Proper growth occurs only after the root is infected.

Spore germination in some species is stimulated by diffusible soil components. Substances from roots and tissue culture also show such effects. At any one time only a certain percentage of spores germinate. The ecological importance of this is that there is always a remainder in the soil for further infection.

The flavonoids, naringenin and apigenin, which are *nod* gene-inducing compounds in *Rhizobium*, induced a growth response in spores of *Gigaspora margarita* at micromolar concentrations (Gianinazzi-Pearson *et al.*, 1989).

Glyceollin I, the phytoalexin of soybeans, inhibits VAM spore germination at 50 μM concentrations, but stimulates it at the rather low concentration of 0.5 μM (Morandi *et al.*, 1991).

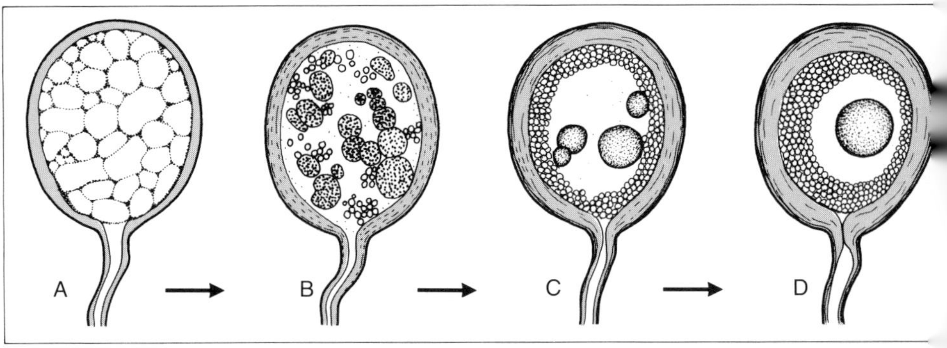

Fig. 10.1. Spore development in *Glomus fasciculatum*. In mature spores (D) the opening to the attached hypha is occluded by deposits of wall material. (Based on Trappe, 1982.)

The cell wall of spores has fibrillar subunits, arranged in regular arcs. The major component is, as expected, chitin (27%), about 25% are protein, 5% are alkali-soluble polysaccharides and 15% are insoluble components after alkali extraction and acid hydrolysis (Bonfante-Fasolo and Grippiolo, 1984).

Spores are relatively easy to isolate. A water/soil mixture is allowed to sediment, whereupon the supernatant is passed through a series of sieves. At a certain net size the spores are retained and the titre can be determined. Further purification can be achieved by sucrose density gradient centrifugation. Banded spores are adsorbed onto glass surfaces and can be plated out after dilution.

The number of spores in soil varies according to crop and plant density. It ranges between 10 and 2×10^4 kg soil^{-1}. Wind is the most important vector for spore distribution. In pasture soil the titre appears to be higher than on wasteland or wheat fields. Tomato roots accumulate between 60 and 150 infections of *Glomus macrocarpum* per g dry stem weight. Thus it appears that the incidence of spores is several orders of magnitude lower in soil than that of prokaryotic symbionts like *Rhizobium* (Chapter 3) or *Frankia* (Chapter 5). Spore counts and infectivity are not correlated to any great extent. This is due to the infectivity of VAM hyphae, and the ability of some fungi, e.g. *Gigaspora gigantea*, to generate several (up to 10) germination hyphae from one spore. In *Gigaspora margarita* only old spores have this ability. VAM fungal activity in soil affects the populations of other microorganisms in the rhizosphere both quantitatively and qualitatively. After addtion of *Glomus fasciculatum*, populations of actinomycetes in tomato rhizoplanes are nearly doubled. The same effect was observed after inoculation with *Azotobacter chroococcum* (Table 10.4). After 60–80 days double inoculation gave an additive effect. This could be explained by better availability of N and P.

The effects of specific VAM-species on the population of rhizosphere bacteria have been reported. The population of fluorescent pseudomonads

Table 10.4 Effect of inoculations with *Glomus fasciculatum* and *Azotobacter chroococcum* on the population of actinomycetes in the rhizosphere of tomato, in unsterilized soil. (Based on Bagyaraj and Menge, 1978)

	Actinomycetes (10^5 g soil dry matter^{-1}) Days after inoculation		
	40	*60*	*80*
Control plants (uninoculated)	2.6	3.7	3.5
With *Azotobacter chroococcum*	9.7	9.9	8.6
With *Glomus fasciculatum*	8.2	8.3	3.4
With *Azotobacter* and *Glomus*	8.6	16.1	12.4

was significantly reduced after infection of young cucumber seedlings with *Glomus intraradices*, but not after infection by *Glomus etunicatum* (Paulitz and Linderman, 1989). With *Glomus fasciculatum* infection of sweet corn (*Zea mays* var. *rugosa*) and clover (*Trifolium subterraneum*), more facultative anaerobes and fewer fluorescent pseudomonads were observed and the population of *Streptomyces* sp. and chitinase-producing actinomycetes was also reduced in the rhizosphere (Meyer and Linderman, 1986b). Full systematic surveys of the effects of VAM on all genera and species of rhizosphere organisms have not yet been reported.

Glomus mosseae and *Gigaspora margarita* have rod-shaped invaginations, similar to bacteria. Their function is unknown but since they are also seen in *Neurospora crassa*, they do not appear to be specific for VAM-fungi.

To achieve a maximal colonization of roots and production of soil-borne spores, the application of nutrient solutions (e.g. Hoagland solution) without phosphate gives the best results (Douds and Schenck, 1990).

10.2 HOST PLANTS (MACROSYMBIONTS) AND 'NON-HOST PLANTS'

With few exceptions species from all angiosperm families can enter into VAM symbioses of one type or another. Whereas over 200 000 angiosperm species are involved, very few gymnosperm species show confirmed infection. These include *Taxus baccata*, *Sequoia gigantea*, *Sequoia sempervirens* and *Ginkgo biloba* (Bonfante-Fasolo and Fontana, 1985). In general, trees at high latitudes form ectomycorrhiza (Chapter 11) whereas in the tropics mostly VAM associations are seen. In Ceylonese rain forests 54 of 63 tree species were associated with VAM fungi. These included *Mangifera zeylanica* (mango tree, Anacardiaceae), *Artocarpus nobilis* (bread fruit tree, Moraceae) and *Elaiocarpus subvillosus* (Elaiocarpaceae). Phylogenetically VAM fungi belong to the oldest symbioses. They infect bryophytes (especially Hepaticae) as well as pteridophytes. In experiments with the horse tail (*Equisetum hyemale*) on sand dunes, some individuals were infected along 90% of their root length, other individuals however, were uninfected.

In a survey of the occurrence of mycorrhizas in the British Isles (Table 10.5) (Moore, 1987), it is evident that in a large number of families between 90 and 100% of the species have mycorrhizal associations, e.g. Gentianaceae, Onagraceae, Plantaginaceae, Liliaceae. The Caryophyllaceae, Polygonaceae and Cyperaceae had less than 40% of species with mycorrhizae. Aquatic plants such as members of the Nymphaceae and the Menyanthaceae have no species with mycorrhizae (Bagyaraj and Menge, 1978).

Due to the wide spread of VAM symbioses among angiosperms interest has been especially focused on the few 'non-host' plants. These are mostly

Table 10.5 Occurrence of mycorrhiza in some major flowering plant families in the British Isles. (Based on Moore, 1987)

	Total species	*Species examined (%)*	*Mycorrhizal species (% of those examined)*
Dicots			
Ranunculaceae	45	56	96
Cruciferae	98	40	46
Caryophyllaceae	83	49	30
Chenopodiaceae	34	44	53
Geraniaceae	20	50	90
Leguminosae	87	47	90
Rosaceae	87	54	96
Onagraceae	21	71	100
Umbelliferae	70	47	79
Euphorbiaceae	18	56	100
Polygonaceae	41	51	38
Salicaceae	34	79	100
Ericaceae	25	84	100
Primulaceae	20	75	100
Gentianaceae	16	50	100
Boraginaceae	34	47	75
Labiatae	63	49	90
Plantaginaceae	8	88	100
Campanulaceae	14	64	78
Rubiaceae	24	50	92
Compositae	160	61	96
Monocots			
Liliaceae	41	71	100
Juncaceae	39	49	63
Iridaceae	15	33	100
Orchidaceae	53	92	100
Cyperaceae	107	54	31
Gramineae	173	60	87

Cyperaceae (Puga, 1985), the Brassicaceae (Yors and Fox, 1979) and Chenopodiaceae. However, some observations have shown that at least a few members of these families can be infected, even when the symbiosis is not complete. *Brassica napus*, after infection with *Glomus fasciculatum*, allows the formation of arbuscules and vesicles. *Brassica junceae* × *Glomus* spp. showed the formation of vesicles and *Raphanus raphanistrum* allowed the formation of arbuscules, vesicles and spores after infection with *Glomus* spp. (Table 10.6). Compared to fully compatible hosts, e.g. barley, maize or potatoes, the symbiotic capacity of the Cruciferae is strongly reduced (Table 10.7). For widely occurring plants the number of infected roots varies between 100% for *Anagallis arvensis* and less than 3% for *Capsella bursa*

Table 10.6 VAM infection in crucifers. (Based on Tommerup, 1985)

Crucifer	*Fungus*	*Arbuscules*	*Vesicles*	*Spores*
Brassica napus	*Glomus fasciculatum*	+	+	–
Brassica juncea	*Glomus* sp.	–	+	–
Raphanus raphanistrum	*Glomus monosporum*	–	+	–
Raphanus raphanistrum	*Glomus tenue*	+	–	–
Raphanus raphanistrum	*Glomus* sp.	+	+	+
Raphanus raphanistrum	*Gigaspora calospora*	+	–	+

Table 10.7 VAM infection of host plants and non-host plants alone or in combination with one another. The soil was sterilized with gamma radiation prior to inoculation with Glomus fasciculatum. Roots were analysed 10 weeks after infection. (Based on Ocampo *et al.*, 1980)

Plant species analysed	*Plant species in intercropping*	*Root infection*		*External mycelium*
		% of root length	*% of cortex*	
Barley	–	79	52	
Barley	+ Rape	94	58	
Barley	+ Onion	82	54	
Maize	–	82	66	
Maize	+ Sugar beet	95	87	
Maize	+ Onion	62	51	
Potato	–	79	59	
Potato	+ Kale	48	39	
Potato	+ Onion	68	44	
Onion	–	82	63	
Onion	+ Swede	90	82	
Onion	+ Barley	83	58	
Cabbage	–		0	1
Cabbage	+ Lettuce		1	12
Rape	–		0	1
Rape	+ Barley		0	9
Sugar beet	–		0	4
Sugar beet	+ Maize		4	23

pastoris (Fig. 10.2). Interestingly the mixing of 'non-host' plants, like rape and sugar beet, between infectable crops (barley, maize, onions) results in a higher infection rate in the compatible species. When barley is planted between rape no increase in root infection in rape could be observed but there was a large increase in the amount of external mycelium on the barley root surface (Table 10.7). Apart from roots, VAM fungi can infect certain leaves. This has been reported in the Zingiberaceae, e.g. *Zingiber officinale* and *Alpinia purpurata*. There they make hyphae, vesicles and spores, but no arbuscules.

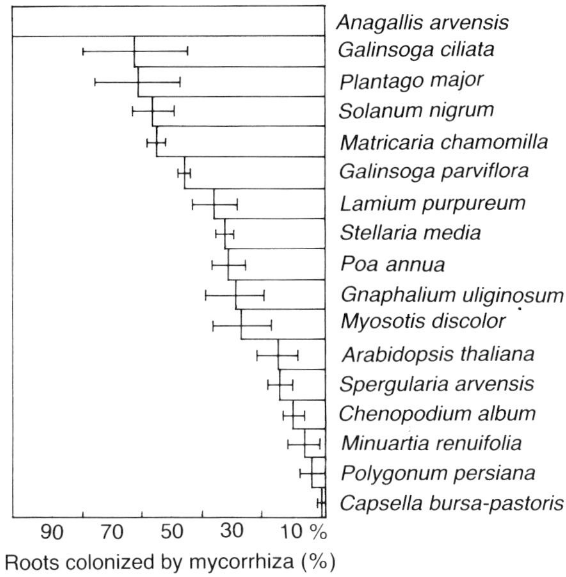

Fig. 10.2. Distribution of VA-mycorrhiza in herbaceous plants. (Based on Kruckelmann, 1975.)

10.3 DEVELOPMENT AND FUNCTION OF THE SYMBIOSIS

Entering into symbiosis with VAM fungi significantly increases the effectivity of roots. Roots profit from increased surface area inasmuch as they can take up more mineral nutrients (phosphate, trace elements). This is coupled with a loss to the fungus of carbon and energy sources. During the vegetative growth period, before seed or fruit setting, most higher plants are not carbon or energy-limited, as shown by the accumulation of starch or lipids. Thus VAM infection results mostly in a profit for the plant, the exchange of its 'excess' (carbon sources) against that of which it has too little (some mineral nutrients).

10.3.1 Infection, differentiation of VA-mycorrhiza structures and of host cells

Young roots are infected mostly 5–20 mm behind the root tip. The infection in this area is higher by a factor of 10 than for whole roots. Germination tubes and the 20–30 μm diameter primary hyphae grow randomly in soil. When these enter the vicinity of an infectible root, the formation of thin (2–7 μm diameter) side hyphae is induced, which grow towards the root. Soil components may inhibit this. The primary hyphae

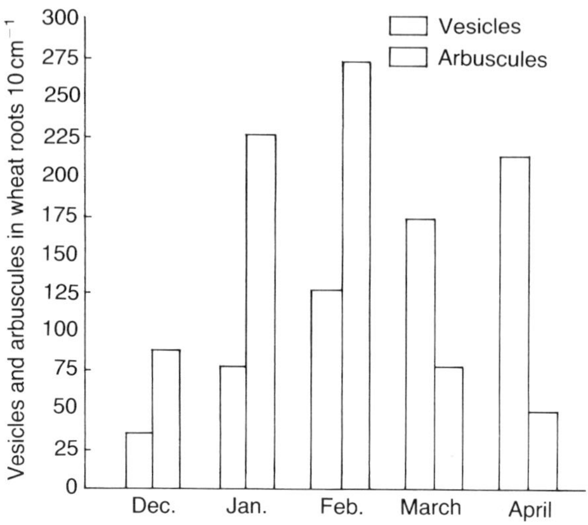

Fig. 10.3. Development of vesicles and arbuscules in wheat roots under field conditions, between December and April. (Based on Kahn, 1975.)

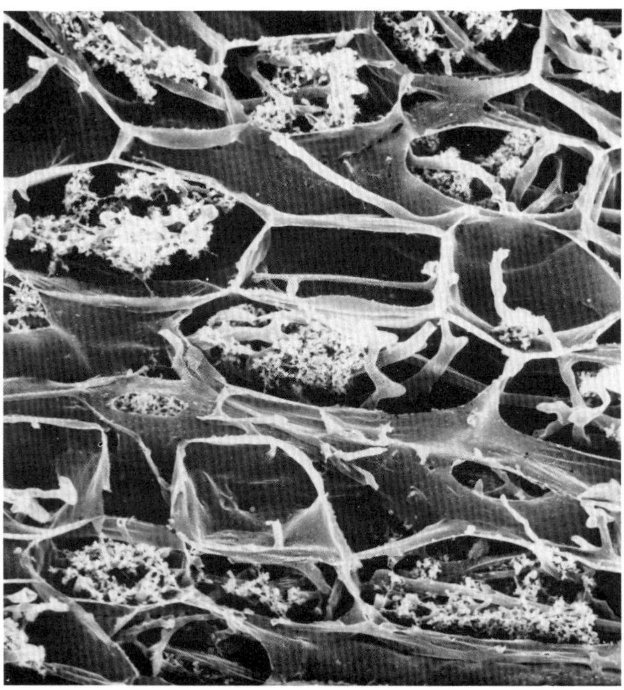

Fig. 10.4. VA-mycorrhiza in *Gingko biloba*, infected with *Glomus caledonium*. Magnification ×600. (Original photograph, P. Bonfante-Fasolo, Turin.)

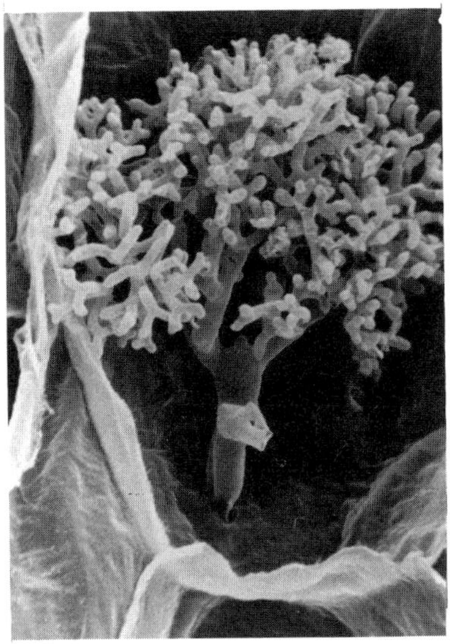

Fig. 10.5. Arbuscule of *Glomus fasciculatum* Gerd. and Tr. in a root of *Zea mays* (1500×). From H. Schüepp and M. Bodmer, Swiss Federal Research Station, Wädenswil and M. Honegger, Institut of Plant Physiology, University of Zürich, see also Schüepp and Bodmer (1991).

can also contact the root. Contact results in the formation of appressoria. The still-growing cell wall of the rhizoderm cell is penetrated, either purely mechanically by the thin penetration hyphae or with the help of glycosidases. After entry hyphae grow and branch into the cortex of, for example *Trifolium subterraneum* at a speed of up to 600 μm per day. Simultaneously an extensive mycelium may develop outside the root. At this time no arbuscules are present. The external mycelium forms a base for secondary infection. The growth of hyphae takes place mostly in the middle and inner root cortex. It is here that the formation of vesicles and arbuscules takes place (Figs 10.3–10.6). The entry of hyphae into host cells does not result in the breaking of the plasmamembrane, but rather a ballooning to form a border membrane. A layer of carbohydrate similar to primary cell wall material is deposited between the host cell boundary membrane and the fungal cell wall. Since many random fibrils are seen in this matrix it is believed that this represents strongly disturbed cell wall synthesis. Plant and fungal cytoplasms are therefore divided by four layers: host plasmamembrane, matrix, fungal cell wall and fungal plasmamembrane. VAM infection does not activate metabolic pathways for the synthesis of new phenolic compounds in the host cell wall (Codignola *et al.*, 1989). In the host plasmalemma and cytoplasm surrounding the arbuscules

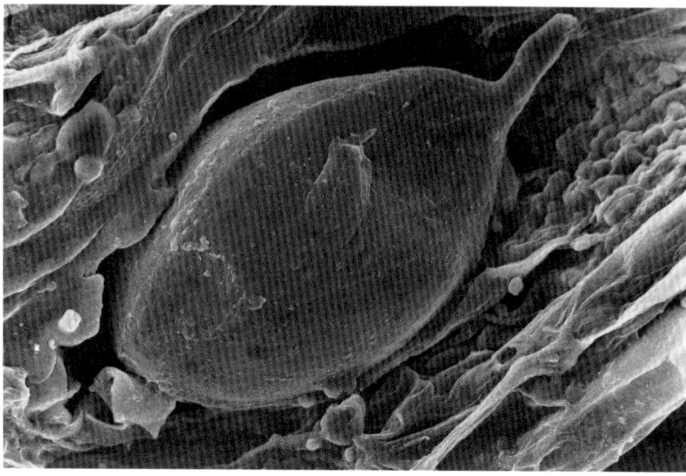

Fig. 10.6. Vesicle of VA-mycorrhiza in *Fragaria* sp. (Original photograph, D.G. Strullo and C. Romand, Angers.)

N-acetyl-D-glucosamine, *N*-acetyl-D-galactosamine, D-mannose and sialic acid have been localized, which are not present in uninfected cells (Jabaji-Hare *et al.*, 1990). The matrix shows a continuum with the host plant cell wall as indicated by cellulose and pectin localization. The cellulose synthase complex in the periarbuscular membrane still seems to be active (Bonfante-Fasolo *et al.*, 1990b). The matrix is often absent at the very thin (0.5–1 μm diam.) tips of the finely branched arbuscules which almost fill the host cell (Fig. 10.5). At this stage the surface area of the plasmamembrane plus boundary membranes is over three times larger than the plasmamembrane in non-infected cells. Infected cells contain 20 times more cytoplasm than non-infected cells (Fig. 10.7).

With a *Glomus* strain it has been observed that inter- and intracellular hyphae, arbuscules and vesicles include multinucleate forms of which the nuclei are extremely small (1–1.5 μm diam.) (Bonfante-Fasolo *et al.*, 1987). Each branch of the arbuscule has a separate nucleus. Chitin is also present in the thin walls of arbuscules (Bonfante-Fasolo *et al.*, 1990a). The arbuscular cell wall is only 50 nm wide, about one-tenth of the dimension of the extraradical hyphal cell walls (Bonfante-Fasolo, 1988).

The life span of arbuscules is between a few days and 2 weeks. Degeneration begins with the degradation of cytoplasm in the tips and subsequent collapse of the fungal cell well. Host cell membrane retains its integrity and the production of matrix material is increased, now with more ordered fibrils. This observation tends to support the hypothesis that active arbuscules are responsible for ultrastructural differences between matrix

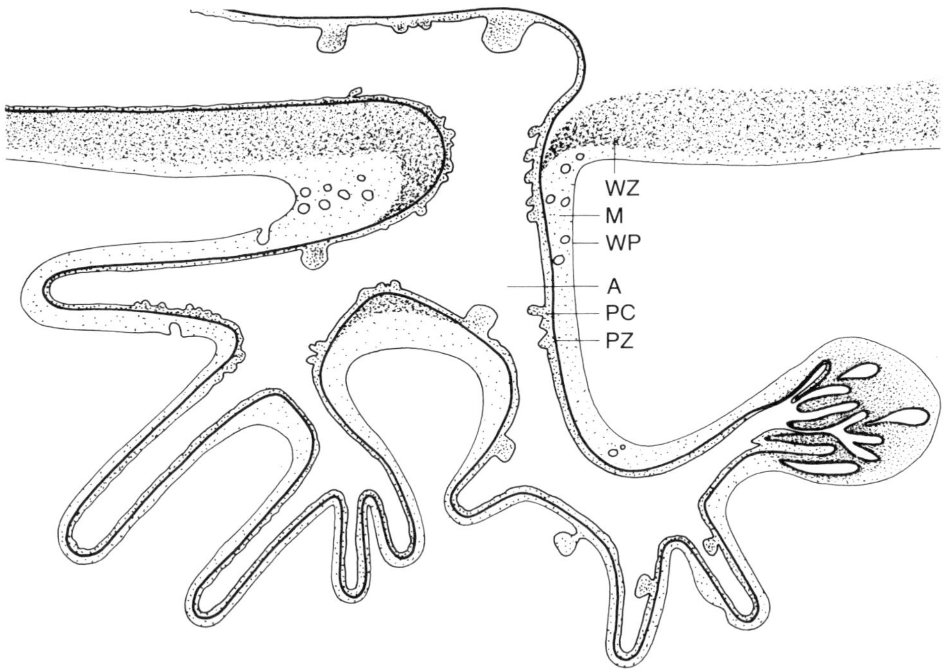

Fig. 10.7. Diagrammatic scheme of an arbuscule. A, Strongly branched arbuscule; PC, fungal cytoplasm; PZ, fungal cell wall; M, matrix with vesicle residues; WP, host plasma membrane; WZ, host cell wall. (Based on Dexheimer, *et al.*, 1979.)

and primary cell wall. The degeneration of arbuscules is seen to be more an ageing process than digestion by the host cell.

Vesicles are present predominantly in the outer and middle cortex. They take the form of terminal swellings of hyphae. They have thickened cell walls, contain large amounts of lipids and are the reserve cells of the microsymbiont. The ratio of vesicles to arbuscules shows an annual cycle. In December to February wheat roots contain twice as many arbuscules as vesicles. In March to April this is reversed by an increase in vesicles and a decline in arbuscules (Fig. 10.3).

In addition to lipids, vesicles can also accumulate large quantities of K, Ca and Mg (Strullu *et al.*, 1983). VAM symbiosis-induced polypeptides, called mycorrhizins, have recently been detected in *Glycine max* after infection with *Glomus mosseae* (Wyss *et al.*, 1990b), using antisera against *Bradyrhizobium*-induced soluble and membrane-bound nodulins (Chapter 3). The soluble polypeptides were not identical to the known *Bradyrhizobium*-induced nodulins. Despite the often observed hypertrophy of host cell nuclei in VAM infected cells, no variation in the DNA content compared to uninfected roots has been found (Berta *et al.*, 1990).

10.3.2 Development and metabolism

The inoculation of *Trifolium subterraneum* with the endogenous VAM fungal population of pasture soil gives rise to successive infections with several VAM fungi, a *Gigaspora* sp., an unknown endophyte and an *Acaulospora* sp. (Fig. 10.8). Calculation of the infection units shows that four infection units per g soil are sufficient to reach maximum possible infection per plant (Fig. 10.9). The infection rate also depends on the rate of root growth. Infection can either be counted by direct cytological methods or by estimating the glucosamine content. Glucosamine content reflects the content of fungal chitin from living and dead cells. To quantify the degree of mycorrhization by a biochemical method, the two sterols **campesterol**

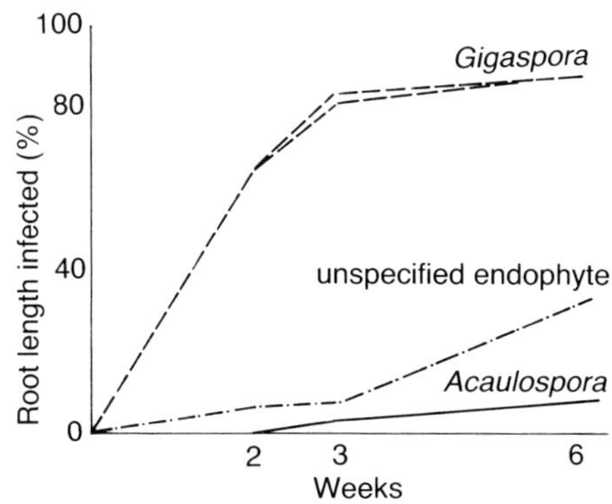

Fig. 10.8. Infection of *Trifolium subterraneum* by endogenous VAM fungi from arable soil, 2–6 weeks after inoculation. (Based on Abbott and Robson, 1982.)

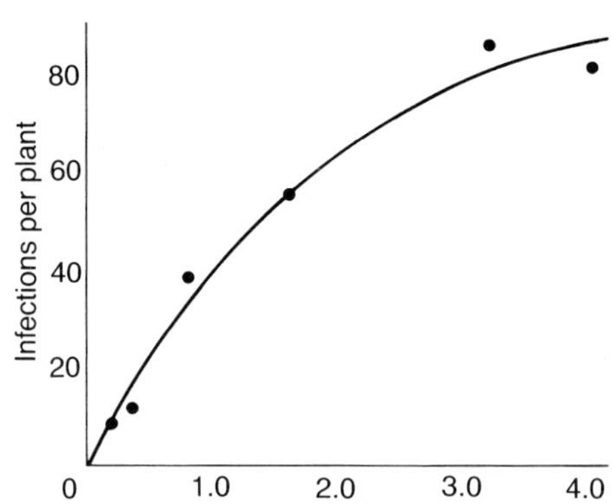

Fig. 10.9. Number of primary VAM infections in *Trifolium subterraneum* in relationship to the titre of infectious units in the soil. (Based on Smith and Walker, 1981.)

and **24-methylenecholesterol** were found to be most closely correlated (Schmitz *et al.*, 1991). These sterols are apparently superior to ergosterol and to the chitin method. Fungi can make up to 20% of total mycorrhizal biomass. VAM fungi are normally able to compete effectively for carbon and energy sources in the root. Reduction in photosynthate delivered to the roots by defoliation reduces the biomass of fungi in soybean roots less than root biomass. On the other hand VAM inoculation in maize can reduce root dry weight (10%), root length (31%) and root hair density (41%) (Kothari *et al.*, 1990). Repeated experiments with nodulated (*Bradyrhizobium japonicum*) plants showed the nodule biomass was even more influenced (Fig. 10.10). VAM fungi can also have a significant positive effect on host photosynthesis, especially in the phase of recovery after dry periods (Fig. 10.11). In *Citrus* the rate of photosynthesis is similar in infected and uninfected plants in dry conditions. In the 4-day long recovery period, however, the infected plants had double the rate of non-infected plants. Other studies, however, have not supported the idea that water relations of *Citrus* are significantly enhanced by VAM (Graham *et al.*, 1987). Six-week-old soya plants (*Glycine max*) doubly infected with *Bradyrhizobium* and VAM exhibited similar sink capacities in both, basic metabolism and respiration for both symbionts (Fig. 10.12). Of the total photosynthate 40%

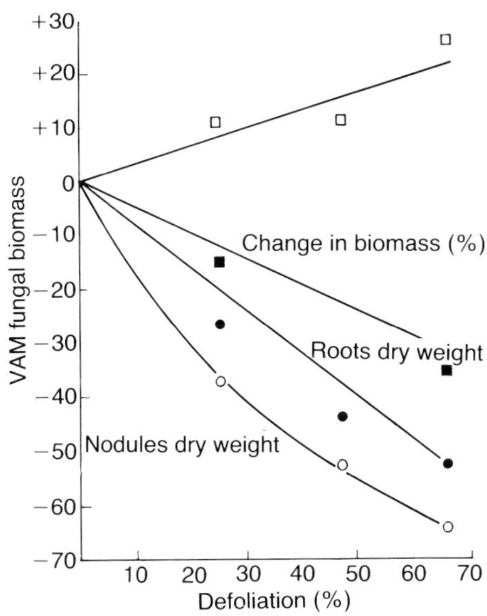

Fig. 10.10. Relative changes in biomass in VAM fungi (■), roots (●), nodules (○) and ratio of VAM fungus/root dry weight (□) in *Glycine max* infected with *Glomus fasciculatum* and *Bradyrhizobium japonicum*, in relation to defoliation of the plants. (Based on Bayne *et al.*, 1984.)

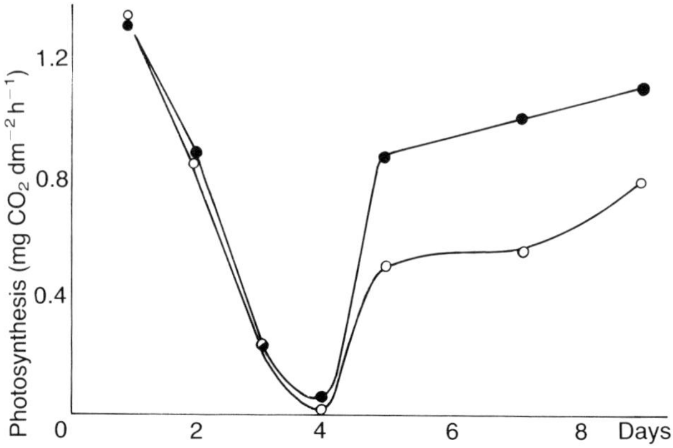

Fig. 10.11. Effect of VAM (●) on the recovery of photosynthetic rate in *Cirtus jambhiri* after water stress, in comparison with uninfected controls (○). (Based on Levy and Krikun, 1980.)

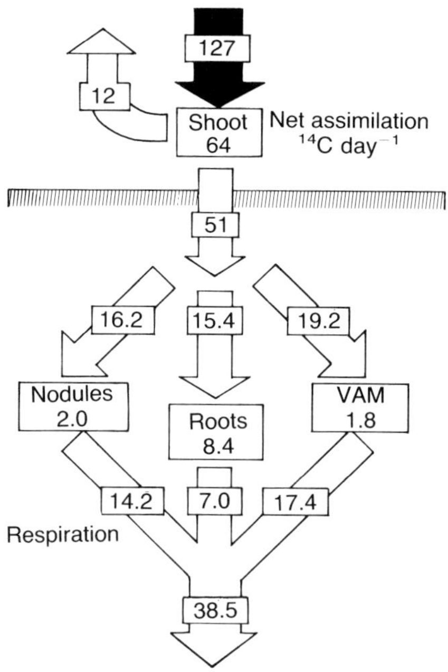

Fig. 10.12. Distribution of assimilation products (127 mg day^{-1} plant^{-1}) in soybean, in symbiosis with VAM and *Bradyrhizobium japonicum*. (Based on Harris *et al.*, 1985, 1987.)

was transported to the roots, a total of 51 mg C day^{-1}. Roots respired 7 mg of this whereas nodules respired 14 mg and VAM 17 mg. Basic metabolism (growth etc.) used 2 mg in nodules and 2 mg in VAM. Both symbioses, therefore, used 30% of the carbohydrate supply of the plant, 80% of which was used for respiration. With cucumber (*Cucumis sativus*) mycorrhizal events were estimated to consume up to 20% of the photoassimilated ^{14}C (Jakobsen and Rosendahl, 1990).

10.3.3 Influence of climatic and soil factors

The temperature optimum for VAM-development is around 30°C. Growth rate increases between infected and non-infected plants also optimally at 30°C (Fig. 10.13). This increase is also oxygen-dependent. The temperature optimum is only found in soils with more than 8% O_2. At 20°C oxygen has hardly any effect. Temperatures of 40°C inhibit the germination of *Acaulospora* and *Glomus* spores. Thus annual cycles with an optimum infection in summer are improbable. Field trials in Utah with *Atriplex gardneri* showed 25% arbuscule development in April, but none in July. Grasses, on the other hand, remain approximately uniformly infected throughout all seasons with a small maximum in summer (Fig. 10.14). Soil depth and pH have little effect on VAM fungi. Infection on *Avena sativa* and *Festuca ovina* was constant between pH 4.5 and pH 7.5. However, the distribution of VAM species involved in infection was influenced, low pH values favoured species with thin hyphae; high concentrations of

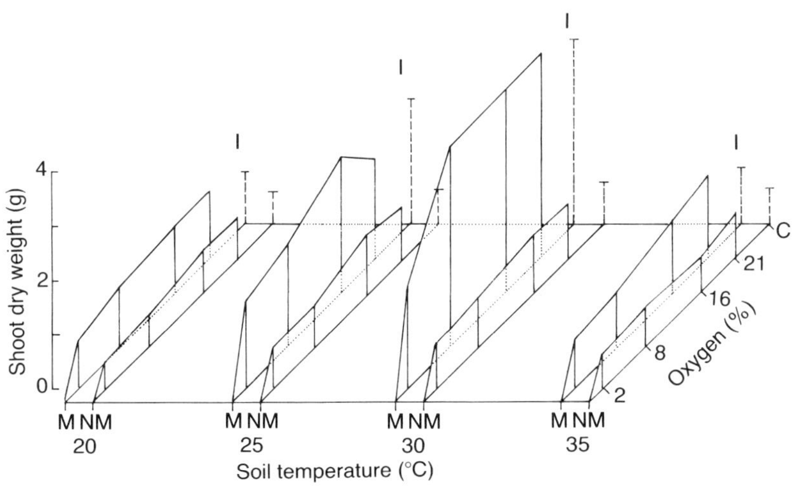

Fig. 10.13. Growth of *Eupatorium odoratum* (shoot dry weight per pot) in relation to soil temperature and oxygen content of soil air, with (M) or without (NM) VAM development. (Based on Saif, 1983.)

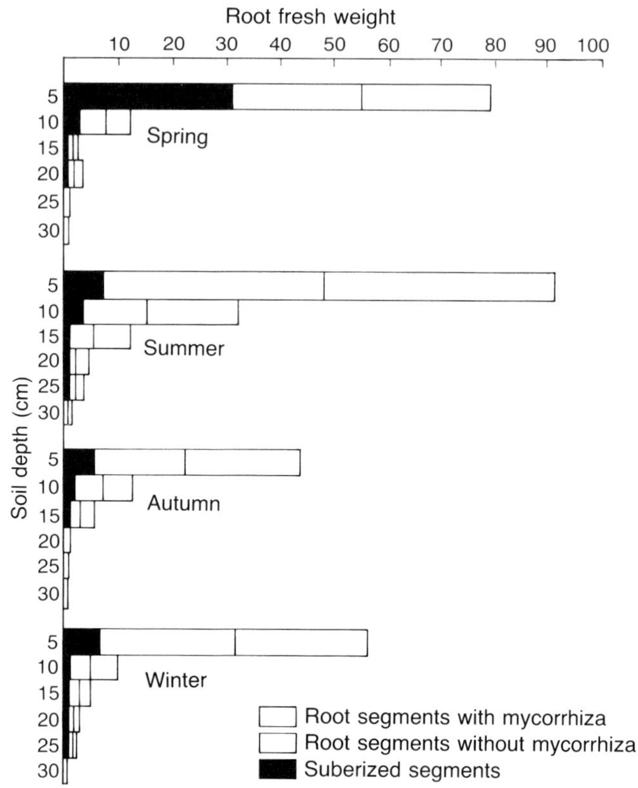

Fig. 10.14. Seasonal changes in root segments with or without VAM and suberized root zones in grasses in various soil depths. (Based on Sparling and Tinker, 1975.)

ammonium fertilizer reduced VAM infection more than similar concentrations of nitrate fertilizer. Inhibition of infection by high phosphate concentrations can be reversed by nitrate. Phosphate, nitrate and ammonium have separate effects on the growth of plant and fungus as well as on the metabolic interactions between the two. It appears, however, that the optimum conditions for infection are not normally present in soil. This is supported by the significant increase in sporocarp formation and plant growth using *Onobrychis viciifolia* (esparsette) and soil supplemented with 0.25% sponge-powder or 0.5% perlite.

Recovery from short drought stress is improved by VAM-infection, as demonstrated by a faster recovery of leaf water potential and leaf turgor (Safir *et al.*, 1972; Hardie and Leyton, 1981). VAM-infected roots can apparently take up water that is not available to the uninfected root (Bethlenfalvay *et al.*, 1988a). Even with very high rooting densities of 20 cm root cm^{-3} soil (0.01 mm root diameter), roots occupy only 2% of the soil pore volume (assumed at 30% of the soil) (Caldwell, 1988). This percentage is greatly increased by a full VAM-infection and explains the better recovery from water stress. By using spores from arid areas a better water use efficiency was achieved than with inoculants from semi-arid areas, although both were cultured for inoculum production on the same host (*Sorghum bicolor*) (Bethlenfalvay *et al.*, 1989) indicating a stable physiological difference.

The water status of the soil affects the distribution of VAM infected plants significantly. In a study from agricultural land, in a dry habitat only 3 out of 43 species were not VAM infected (Kühn, 1991; Kühn *et al.*, 1991), whereas from a wet area, 14 out of 43 species were without VAM structures.

10.3.4 Nutritional metabolic exchange

Apart from the often confirmed improvement in phosphate-supply to the plant, it has been found that VAM-infected plants have less difficulty in taking up other macro- and micronutrients. Important aspects include not only the exchange of carbohydrate for mineral elements between plant and fungus, but also, through the fungus, between plant and plant. Many individual plants in a field or econiche can be connected to each other, by nets of mycelium to make one physiological unit. Although this may be of little quantitative importance, (Smith and Smith, 1990) it can increase the competitiveness of certain species under limitation of a distinct nutrient.

(a) Phosphate

VAM infection in onions grown in normal soil is reduced by 60% when phosphate is added to an end concentration of $0.2\,g\,kg^{-1}$ (Fig. 10.15). Phosphate (0.5–1.0 mM) inhibits 90% of infections of soybean under controlled laboratory conditions. This effect has been explained by a decrease in the concentration of soluble carbohydrates in roots and exudates caused by increasing the phosphate supply (Thomson *et al.*, 1986). *Gigaspora calospora* was more sensitive to this reduced carbohydrate supply than *Glomus fasciculatum*.

VAM inoculation is particularly important to those plants which have inefficient phosphate-uptake systems. In normal soil P content of *Stylosanthes guyanensis* was increased from 24 μg to 360 μg $plant^{-1}$ after VAM infection. When the soil received phosphate fertilizer these figures were 420 μg, 2400 μg, respectively (Table 10.8). Optimal phosphate nutrition is to be seen in the plant nodulation response. VAM-plants or P-fertilized plants formed 6–8 nodules whereas plants with both formed 80. Some plants, e.g. lupin, have a very effective P-uptake system and the effect of VAM-infection is much less noticeable in such species. *Lupinus*

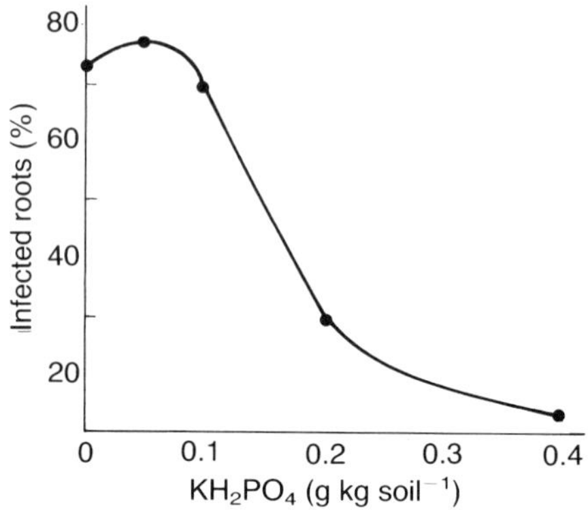

Fig. 10.15. Effect of soluble phosphate on VAM infection in onion. (Based on Sanders and Tinker, 1973.)

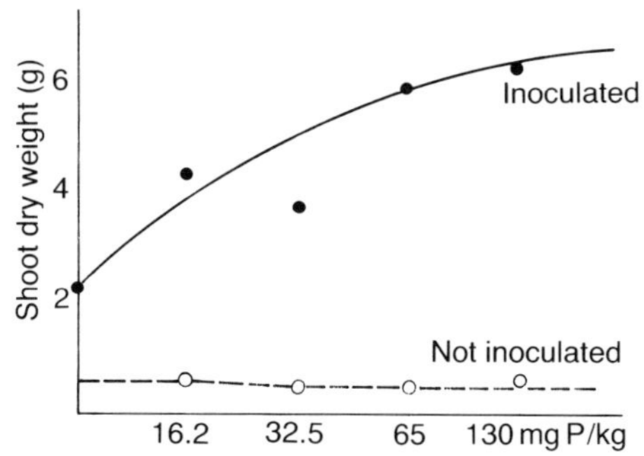

Fig. 10.16. Growth increase in coffee plants after VAM inoculation at various phosphate concentrations in the soil. (Based on Lopes *et al.*, 1985.)

angustifolius and *Lupinus luteus* growing in poor sandy soil had less than 10% of the roots infected with VAM. Arbuscules were also not observed. Coffee (*Coffea arabica*) needs to be infected with a VAM fungus (*Glomus margarita*) for growth, independently of phosphate fertilizer. Coffee plants are then obligate P-mycotrophic (Fig. 10.16). Obligate and facultative P-mycotrophs differ in their use of phosphate (Fig. 10.17). Obligate species continually remove P from roots, allowing continual new infections. The relationship between growth and P-content is almost linear. Facultative mycotrophs on the other hand after VAM-infection or phosphate uptake, tend to store phosphate. They have higher growth rates at specific tissue concentrations of phosphate. In *sensu strictu* there are very few obligate VA-mycotrophic plants.

Since diffusion is the rate-limiting step for movement of phosphate to

Table 10.8 Effects of VAM inoculation and addition of rock phosphate on phosphate content, growth and nodule number of *Stylosanthes guyanensis*. All plants were inoculated with *Rhizobium*. (Based on Mosse *et al.*, 1976)

	Dry matter (*mg plant*$^{-1}$)	*Total P* (μ*g plant*$^{-1}$)	*% VA-infection*	*Nodules per plant*
Control	44	24	0	0
VAM inoculation	450	360	90	6
Rock phosphate	470	420	0	8
VAM inoculation + rock phosphate	660	2400	90	80

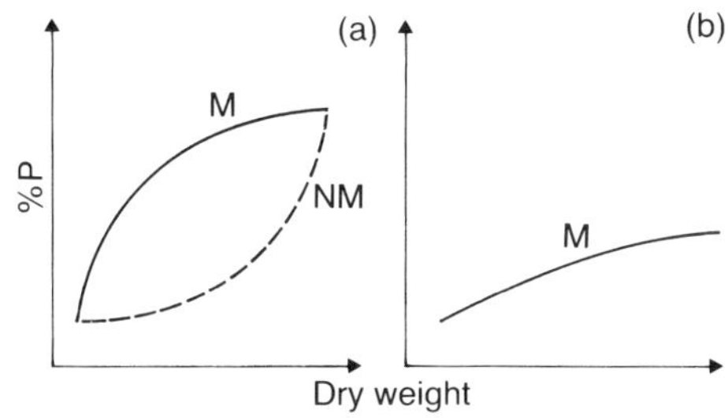

Fig. 10.17. Correlation between phosphate content and growth (dry weight) in facultative (a) or obligate (b) VAM plants. M, VAM plants; NM, controls without mycorrhiza. (Based on Janos, 1985.)

the root, the main effect of VA-mycorrhizae is the large increase of the absorbing surface area. There have been reports of between 0.7 and 14 m hyphae per cm root length (Smith and Gianinazzi-Pearson, 1988). Mass flow contributes only a small percentage of P-uptake in onions and saltwort (*Salsola Kali* L.) (Itoh and Barber, 1983). The exact mechanism by which VAM-infected root can utilize P-sources that are adsorbed and unavailable to uninfected roots is not known (Gianinazzi-Pearson and Gianinazzi, 1988; Smith and Gianinazzi-Pearson, 1988). VAM infection increased the P utilization coefficient 2.7–5.6-fold, independently of the fertilizer type (rock phosphate versus superphosphate) (Blal *et al.*, 1990). The phosphorus concentration in the fungal hyphae is more than 1000 times that in the soil solution, which is often in the range of 1 or a few mM. The movement of phosphate across the arbuscular interface is about 1.3×10^{-14} mol mm^{-2} s^{-1}, a value similar to that reported for free-living fungi (Cox and Tinker, 1976; Beever and Burns, 1980). The storage compound, polyphosphate, is mobilized and broken down by the action of polyphosphatase and polyphosphate kinase (Capaccio and Callow, 1982). Not only is the uptake efficiency of phosphate affected by VAM, but also its efficient use in CO_2-

fixation was increased in soybeans (Brown *et al.*, 1988). The photosynthetic N-use efficiency was also positively affected by VAM-inoculation, but this was dependent on the source of N (symbiotic versus N-fertilization) (Brown and Bethlenfalvay, 1988). With a 'Diagnosis and Recommendation Integrated System (DRIS)', Bethlenfalvay *et al.* (1990) were able to establish a ranking system of different *Glomus mosseae* isolates towards their different efficiency for improved utilization of P, N and K. The suppression of VAM-development by an excess of phosphate does not use the same mechanism as the suppression of nodulation by nitrate in legumes, as shown for supernodulating soybeans (Wyss *et al.*, 1990a).

It has been demonstrated with *Allium cepa* and *Glomus etunicatum* that root colonization is not affected by P addition where plants were N-deficient. However, in plants well supplied with N addition of P had this effect (Sylvia and Heal, 1990).

(b) Trace elements and other macroelements

The leaves, roots and pods of *Glycine max* contain significantly more zinc and copper when the plant roots are VAM infected (Table 10.9). These

Table 10.9 Trace element content in *Glycine max* with either VAM inoculation or phosphate fertilization. Percentage difference (VAM − P)/P × 100. (Based on Pacovsky *et al.*, 1985)

Plant organ	*Plant age (weeks)*		
	6	9	12
		Zinc	
Leaves	+74*	+65*	+21*
Roots	+27*	+ 9*	+18*
Pods		+ 2	+42
		Copper	
Leaves	+56*	+88*	+34*
Roots	+25*	+51*	+19*
Pods		+14	+51*
		Iron	
Leaves	−28*	−34*	−21*
Roots	−28*	−48*	−46*
Pods		−27*	−12
		Manganese	
Leaves	− 8	−43*	−35*
Roots	−15*	−32*	−33*
Pods		−49*	−44*

* Significant difference ($P < 0.05$).

differences are most clear in 9-week-old plants (blossoming). After 12 weeks the difference is highest in pods. Manganese and iron in the same plants show reduced levels compared to P-fed control plants. Zinc and copper enrichment is also found in other VAM-infected plants (Gildon and Tinker, 1983; Tinker and Gildon, 1983). Besides copper, chloride concentration also increased in VAM-infected citrus plants (leaves and roots) compared to the controls (Graham and Sylvertsen, 1989).

Alleviation of manganese toxicity by VAM has been demonstrated in soybeans at a manganese concentration of $40\,\mu g\,g^{-1}$ (Bethlenfalvay and Franson, 1989). VAM-infected plants of corn and lettuce also had a significantly lower cadmium accumulation compared to control plants (Schüepp *et al.*, 1987).

Sulphate can also be transported into the host by VAM, and cause a specific enrichment (Cooper and Tinker, 1978; Rhodes and Gerdemann, 1978). In some VAM plants ^{15}N from $^{15}NH_4^+$ has been found to be transported across VAM-hyphae into the plant. (Ames *et al.*, 1983).

(c) Carbohydrate and lipids

Transfer of carbon and energy to the fungus is achieved mostly through the arbuscules. It is not known if the carbon is in the form of free hexoses or as a hydrolysate of the matrix material. In the latter case the matrix should not be thought of as a physiological barrier, but as a polymerized nutritional 'offer' from the plant (Dexheimer *et al.*, 1979; Harley and Smith, 1983; Wyss *et al.*, 1990a).

For ectotrophic mycorrhizal and VA-mycorrhizal fungi a gradient of C-sources is assumed to exist from the host to the microsymbiont. In VAM-fungi, however, the end-products of fungal synthesis are not trehalose and mannitol but lipids (Mosse, 1973; Cooper and Lösel, 1978; Cooper and Tinker, 1978). This carbohydrate taken up by the fungus is continually removed and the gradient remains constant, favouring the fungus as long as it remains physiologically active.

The rate of transfer in sucrose units has been determined to be about $24\,pg$ sucrose $mm^{-2}\,s^{-1}$ (which is about 1000 times higher than the efflux from other plant cells) (Humphreys, 1973; Harris and Paul, 1987). The fungal arbuscule plasmalemma and the host cell plasmalemma close to the arbuscule show high ATPase activity (Marx *et al.*, 1982).

The total lipid content of VAM-infected roots is 20–60% higher than in non-infected roots. This is also due to the high lipid content of the expanded membrane around the arbuscules. Qualitative differences in fatty acid composition have been found between VAM-infected and non-infected *Citrus* roots. In symbiotic plants three new fatty acids making up 30–40% of the total lipid pool were found.

The fungal biomass important for the C-flux from host to microsymbiont in VAM-infected roots is apparently more dependent on the host plant and the environment than on the fungal species. It varies between 3% in *Sorghum* to 14–16% in *Centrosema* and *Glycine max* (Hepper, 1977; Bethlenfalvay *et al.*, 1982; Harris *et al.*, 1984). These figures have been produced by the chitin method which does not discriminate between active and dead fungal cells. Since the turnover rate of arbuscules is very rapid (2–14 days), the estimation of the active VAM population by the sterol method may be more appropriate (Macauley and Bethel, 1990).

The mechanisms of nutrient transport at the interface of VAM and other symbioses is still poorly understood, due to our limited knowledge of localized ATPase data, and modifications of the membranes at the interfaces (Smith and Smith, 1990). This is especially so in relation to ion channels (Tester, 1990), perhaps modified in symbiotic membranes.

10.3.5 Ecological and agricultural aspects

In the colonization of intensively grazed lands, especially in semi-arid or tropical zones, there is a succession in as much as primary colonizers are mostly non-VAM plants. These are followed by a wave of facultative VAM plants which are, after a longer time, dominated by obligate VAM plants (Fig. 10.18). The establishment of full VA-mycorrhiza is of no importance to the first colonizers of land well supplied with nutrients. However, as the nutritional value of the soil decreases facultative VAM-formation becomes a significant advantage. The subsequent dominance of obligate VAM plants is probably due to their ability to withstand stress caused by a lack of nutrients.

Many tropical trees in particular are obligate mycotrophs which require very effective VAM-strains. The mechanism of how obligate mycotrophs may enhance infection might involve the maintenance of high root exudation

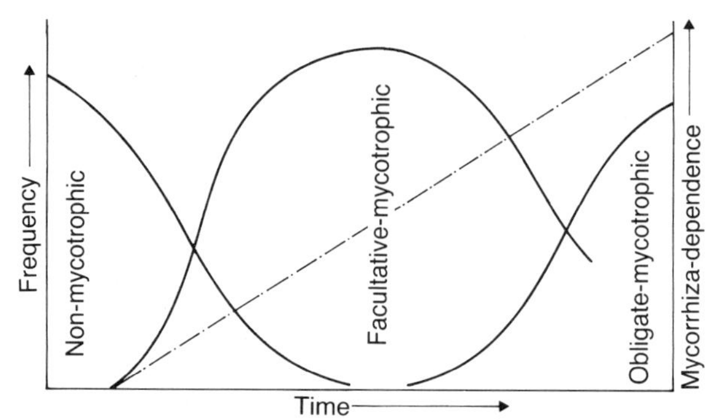

Fig. 10.18. Succession of mycorrhiza-independent plants, facultative and obligate mycorrhiza plants at primary colonization of a biotope. (Based on Reeves, 1985.)

and effective sinks for phosphate in stems or leaves to reduce the phosphate concentration in the root cells (Janos, 1987).

(a) Increased yields in crop plants and trees after VAM-inoculation

Cereals, such as wheat, barley and maize, and legumes, such as soya, lucerne and white clover, all of which have world-wide importance as food crops, show significant increases in growth on poor soils when they are inoculated with VAM-fungi (Table 10.10) (Hayman, 1987). Oranges inoculated with *Glomus fasciculatum* respond with a four times increased growth rate, coffee inoculated with *Gigaspora margarita* has a 10 times increased growth rate. Differences in response depend on the nutritional conditions in the soil and the inoculation method (Table 10.10). In soybean, after infection with *Glomus*, increases in plant growth (shoot and stem) were similar to symbiosis-specific increases in yield (Table 10.11). Both with and without infection with *Bradyrhizobium* the yield increased 11–13 $g\,m^{-2}$. Seed yields, which are N-dependent, were only influenced by the presence of nodules.

In many plants symbiosis-specific growth increase is more evident in the shoot than in the root (Fig. 10.19). In clover, stalk growth is stimulated by a factor of 5 over normal, but plant total weight only by a factor of 3.

Table 10.10 Field trials of inoculation of crop species with VA-mycorrhiza. (Based on Mosse and Hayman, 1980)

Crop plant	*VA-mycorrhiza fungus*	*Inoculation*	*Growth increase factor (control = 1)*
Wheat	*Glomus mosseae*	A, C	3
Barley	*Glomus caledonium*	C	1.3–4
Maize	*Glomus mosseae*	A	2.3
Soybean	*Gigaspora calospora*	A	1–1.5
Alfalfa	*Glomus caledonium*	C, A	4
Clover	*Glomus fasciculatum* and *G. mosseae*	A	1.3
Potato	Endogenous field population	B	1.2
Orange	*Glomus fasciculatum*	D	1.7–4
Peach	*Glomus fasciculatum*	C	1.8
Coffee	*Gigaspora margarita*	A	3–10
Cotton	*Glomus macrocarpum*	A	1.5
Lemon	*Glomus mosseae*	C	2
Castor bean	*Glomus macrocarpum*	A	1.6–2.3
Flax	*Glomus macrocarpum*	A	1.6

Inoculation methods: (A) plant inoculation before transplanting; (B) soil inoculation with soil abundant in VAM; (C) soil inoculation with soil and VAM-roots; (D) seed inoculation.

Table 10.11 Growth and yield of soybeans after inoculation with *Glomus fasciculatum* and *Bradyrhizobium japonicum*, in a soil with 2.4 ppm available phosphate, low endogenous VAM spore titre (200 spores per 50 ml soil) and nearly free of endogenous *Bradyrhizobium japonicum*. (Based on Bagyaraj *et al.*, 1979)

Treatment	*Shoot dry weight* (g)*	*Grain yield (g m^{-2})*	*100-grain weight (g)*
Control (uninoculated)	2.0	38	10.8
Inoculation with *Glomus*	2.6	47	10.9
Inoculation with *Bradyrhizobium*	2.8	71	14.6
Inoculation with *Glomus* and *Bradyrhizobium*	4.7	84	14.6

*60 days after sowing.

Table 10.12 Yield increase of *Leucaena leucocephala* after VAM-inoculation with *Glomus fasciculatum*. All plants were inoculated with *Rhizobium* sp. and analysed after 63 days. (Based on Huang *et al.*, 1985)

	VAM-infected plants	*Controls without mycorrhiza*
Shoot dry weight (g $plant^{-1}$)	14.5	1.9
Root dry weight (g $plant^{-1}$)	6.8	1.9
P-content of shoot (%)	0.17	0.01
Ca-content of shoot (%)	1.6	1.3
Leaf area (cm^2 $plant^{-1}$)	1710	150
Mycorrhiza infection (% of root length)	88	0
Nodule dry weight (mg $plant^{-1}$)	420	0

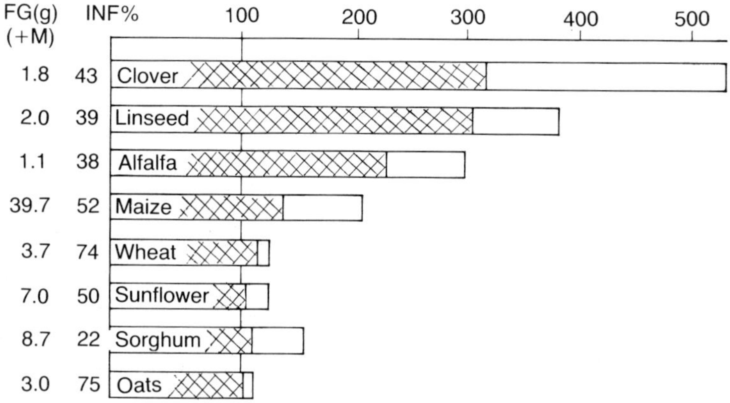

Fig. 10.19. Mycorrhizal dependency of crop plants in soils with low phosphate content. Values are given as fresh weight +VAM/−VAM × 100, based on whole plants (dark area) or shoots (dark + white areas). FG, Fresh weight of mycorrhizal plants; INF, percentage of infected root segments. (Based on Smith, 1985.)

Especially noticeable is symbiosis-specific growth stimulation in subtropical legume trees, such as *Leucaena leucocephala*. Two months after VAM inoculation stem mass was eight times higher than those of control plants, leaf area was eight times increased and the trunk contained 14 times more P (from 0.01% to 0.17% dry weight) than non-VAM plants (Table 10.12). Also noticeable is the increased Ca^{2+} content of VAM-plants. Growth limitation in non-infected plants is made particularly clear in nodule counts when non-VAM legumes may remain unnodulated.

Positive interactions of three different microorganisms have also been reported, with an increase of nodulation by *Rhizobium trifolii* and shoot dry weight in *Trifolium subterranea* after dual inoculation with VAM and *Pseudomonas putida* compared to a single addition by one of these (Meyer and Linderman, 1986a).

Symbiosis-specific growth increases in VAM plants are dependent on the density of planting (Baath and Hayman, 1984). *Allium cepa* planted at a density of two plants per pot show clear growth responses to infection. At 20 plants per pot the effect disappears completely (Fig. 10.20). This is explained by postulating that at very high root densities almost all P is removed from the soil and infection cannot mobilize more. Baylis (1970) published the 'root-hair hypothesis', which states that plants with straight roots and few root hairs, such as onion and *Citrus*, react strongly to VAM-infection. On the other hand plants with highly branched roots and many root hairs show only a small reaction. Many plants have been investigated which support this hypothesis. However, exceptions are found, for exmple,

Fig. 10.20. Effect of plant density on growth stimulation of onion by VAM. Inoculated (INOC.): left 2 pots with 2 plants each, right pot with >20 plants. Controls (uninoculated): right 2 pots with 2 plants each, left with >20 plants. (Original photograph, D.S. Hayman, Rothamsted.)

lupin, which has a low reaction to infection despite unbranched roots, or tropical grasses, like *Paspalum notatum* whose highly branched root system, with many root hairs, is always thickly populated with mycorrhiza. Also in trees such as *Acacia auriculiformis* and *Acacia mangium* a combined inoculation with *Rhizobium* and VAM-fungi as *Gigaspora margarita* increased height (Cruz *et al.*, 1990). A significant positive effect of VAM inoculation on the survival of newly planted seedlings of *Robinia pseudoacacia* and *Medicago arborea* under semi-arid conditions has also been reported (Barea *et al.*, 1990). Endomycorrhiza application is already being used on a wider scale in the citrus industry in California and in blueberry production in New Zealand (Gianinazzi *et al.*, 1990).

Not all growth stimulation is based on better availability of phosphate and other elements being made more available to the plant. In some cases there may also be hormonal effects. Dormancy periods in *Liriodendron tulipifera* may be broken by infection, but not by fertilization. It is especially important that potting compost, which is sterilized to kill pathogens, is subsequently inoculated with VAM-fungi. This is of great commercial use in garden shops, plantations and nurseries. Expanded clay has been used very successfully as a carrier for VAM spores for agricultural plants such as maize (corn) under field conditions (Baltruschat, 1987). Rock wool is another interesting carrier for mass multiplication of VAM fungi (Heinzemann and Weritz, 1990).

(b) Effects of VAM on pathogen resistance

A reduction of phytopathogenic effects by VAM-inoculation has only been observed with soil-borne pathogens, whereas infection of leaves by air-borne fungi and viruses can be increased (Schönbeck and Dehne, 1979; Schenck, 1982).

Some phytoalexin (glyceollin) accumulation has also been reported by VAM-infection of soybean roots (Morandi *et al.*, 1984), but this was much smaller than that reported for soybean nodules with defined mutants of *Bradyrhizobium japonicum* (Chapter 3). Chitinase activity, as another defence reaction, has been found in the early stages of VAM development, whereas, 20 days after infection, the chitinase activity was constantly lower in VAM-infected roots than in the controls (Schlumbaum *et al.*, 1986; Spanu *et al.*, 1988).

Most root pathogenic fungi infect roots much faster than VAM-species can. Simultaneous infection with VAM and *Pythium* or *Phytophthora* or *Fusarium* leads to the VAM-fungus being out-competed. However, when the VAM-infection is already established, pathogenic infection is dramatically reduced. Tomato infected with *Fusarium* 4 weeks after VAM inoculation showed 9% necrotic roots compared to 32% in non-VAM plants. The number of *Fusarium* setting spores was reduced to 5%. These results

Table 10.13 Damage by *Fusarium oxysporum* pv. *lycopersici* to tomato plants with and without mycorrhiza, as measured by wilt index, percentage of leaves turned yellow and chlorophyll content. (Based on Dehne and Schönbeck, 1975, 1979)

Plants infected with		*Wilt index*	*Yellow leaves (%)*	*Chlorophyll content (mg g fresh weight^{-1})*
Fusarium	*Mycorrhiza*			
–	–	0.0	0	28.6
–	+	0.0	0	36.1
+	–	1.84	45	10.1
+	+	1.06	24	19.4

confirm previous experiments in other soils where it was reported that diseases in tomatoes could be prevented by previous VAM infections (Table 10.13).

Root pathogens such as *Fusarium* and *Pythium* are a large problem in the commercial isolation of VAM spores for inoculation. Such isolations are not carried out under sterile conditions and such preparations are often contaminated, leading to pathogenic infections.

VAM infection has positive effects on infections with nematodes. Cotton seed yield is reduced by 45% in plants infected with the nematode

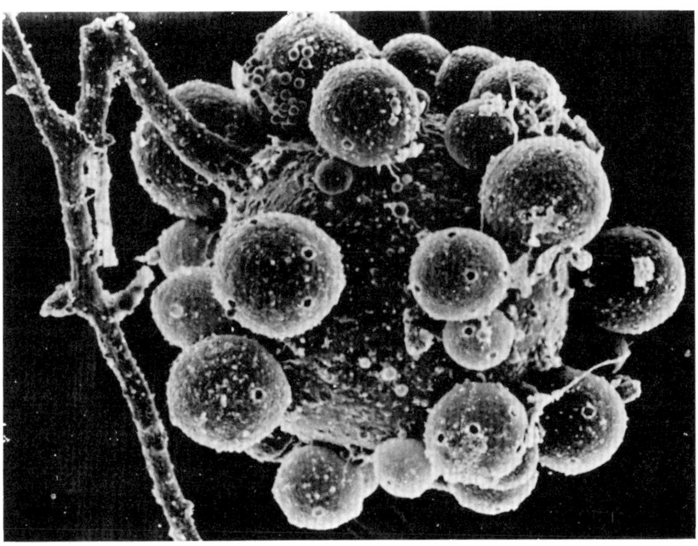

Fig. 10.21. Chlamydospore of *Glomus clarum* infected with numerous zoosporangia of the hyperparasite *Phlyctochytrium plurigibbosum*. (Original photograph, S.S. Tzean, Taipee.)

Meloidogyne incognita. VAM-infected cotton plants with the same disease had the yield lowered by only 25%. The number of nematodes per root was halved. Better root growth can, however, lead to increased nematode populations. Due to better nutrition, some phytopathological viruses replicate better in VAM plants than in non-VAM plants. Virus multiplication in tomatoes, strawberries and other market garden plants is proportional to symbiont-specific increases in the plant phosphate content. VAM spores themselves can fall prey to hyperparasites like *Phlyctochytrium plurigibbosum* (Fig. 10.21).

10.4 REFERENCES

Abbott, L.K. and Robson, A.D. (1982) Infectivity of vesicular arbuscular mycorrhizal fungi in agricultural soils. *Aust. J. Agric. Res.*, **33**, 1049–59.

Ames, R.N., Reid, C.P.P., Porter, L.K. and Cambardella, C. (1983) Hyphal uptake and transport of nitrogen from two ^{15}N-labelled sources by *Glomus mosseae*, a vesicular–arbuscular mycorrhizal fungus. *New Phytol.*, **95**, 381–96.

Baath, E. and Hayman, D.S. (1984) Effect of soil volume and plant density on mycorrhizal infection and growth response. *Plant Soil*, **77**, 373–6.

Bagyaraj, D.J., Manjunath, A. and Patil, R.B. (1979) Interaction between a vesicular–arbuscular mycorrhiza and *Rhizobium* and their effects on soybean in the field. *New Phytol.*, **82**, 141–5.

Bagyaraj, D.J. and Menge, J.A. (1978) Interaction between a VA mycorrhiza and *Azotobacter* and their effects on rhizosphere microflora and plant growth. *New Phytol.*, **80**, 567–73.

Baltruschat, H. (1987) Evaluation of the suitability of expanded clay as carrier material for VA mycorrhiza spores in field inoculation of maize. *Angew. Bot.*, 163–9.

Barea, J.M., Salamanca, C.P. and Herrera, M.A. (1990) The role of VA-mycorrhiza at improving N_2-fixation by woody legumes in arid zones, in *Fast Growing Trees and Nitrogen* Fixing Trees (eds D. Werner and P. Müller) G. Fischer Verlag, Stuttgart, New York, pp. 303–11.

Baylis, G.T.S. (1970) Root hairs and phycomycetous mycorrhizas in phosphorus deficient soil. *Plant and Soil*, **33**, 713–6.

Bayne, H.G., Brown, M.S. and Bethlenfalvay, G.J. (1984) Defoliation effects on mycorrhizal colonization, nitrogen fixation and photosynthesis in the *Glycine–Glomus–Rhizobium* symbiosis. *Physiol. Plant.*, **62**, 576–80.

Beever, R.E. and Burns, D.J.W. (1980) Phosphorus uptake storage and utilization by fungi. *Adv. Bot. Res.*, **8**, 128–219.

Berta, G., Sgorbati, S., Soler, V. *et al.* (1990) Variations in chromatin structure in host nuclei of a vesicular arbuscular mycorrhiza. *New Phytol.*, **114**, 199–205.

Bethlenfalvay, G.J., Brown, M.S., Ames, R.N. and Thomas, R.S. (1988a) Effects of drought on host and endophyte development in mycorrhizal soybeans in relation to water use and phosphate uptake. *Physiol. Plant.*, **72**, 565–71.

Bethlenfalvay, G.J. and Franson, R.L. (1989) Manganese toxicity alleviated by mycorrhizae in soybean. *J. Plant Nutr.*, **12**, 953–70.

Bethlenfalvay, G.J., Franson, R.L., Brown, M.S. and Mihara, K.I. (1989) The *Glycine–Glomus–Bradyrhizobium* symbiosis. IX. Nutritional, morphological and physiological responses of nodulated soybean to geographic isolates of the mycorrhizal fungus *Glomus mosseae*. *Physiol. Plant.*, **76**, 226–32.

Bethlenfalvay, G.J., Franson, R.L. and Brown, M.S. (1990) Nutrition of mycorrhizal soybean evaluated by the diagnosis and recommendation integrated system (DRIS). *Agron. J.*, **82**, 302–4.

Bethlenfalvay, G.J., Pacovsky, R.M., Brown, M.S. and Fuller, G. (1982) Mycotrophic growth and mutualistic development of host plant and fungal endophyte in an endomycorrhizal symbiosis. *Plant Soil*, **68**, 43–54.

Bethlenfalvay, G.J., Thomas, R.S., Dakessian, S. *et al.* (1988b) Mycorrhizae in stressed environments: effects on plant growth, endophyte development, soil stability and soil water, in *Arid Lands. Today and tomorrow* (eds E.E. Whitehead, C.F. Hutchinson, B.N. Timmermann and R.G. Varady) Westview Press, Boulder, Co.

Bianciotto, V., Balestrini, R. and Bonfante, P. (1991) Genome size and nuclear ultrastructural features in *Gigaspora margarita* and *Glomus versiforme*. *Abstracts of the Third European Symposium on Mycorrhizas*, Sheffield, August 1991.

Blal, B., Morel, C., Gianinazzi-Pearson, V. *et al.* (1990) Influence of vesicular–arbuscular mycorrhizae on phosphate fertilizer efficiency in two tropical acid soils planted with micropropagated oil palm (*Elaeis guineensis* jacq.). *Biol. Fertil. Soils*, **9**, 43–8.

Bonfante-Fasolo, P. (1988) The role of the cell wall as a signal in mycorrhizal associations, in *Cell to Cell Signals in Plant, Animal and Microbial Symbiosis*, NATO ASI Series, Vol. H17 (eds S. Scannerini *et al.*,) Springer-Verlag, Berlin, Heidelberg, pp. 219–35.

Bonfante-Fasolo, P., Berta, G. and Fusconi, A. (1987) Distribution of nuclei in a VAM fungus during its symbiotic phase. *Trans. Br. Mycol. Soc.*, **88**, 263–6.

Bonfante-Fasolo, P., Faccio, A., Perotto, S. and Schubert, A. (1990a) Correlation between chitin distribution and cell wall morphology in the mycorrhizal fungus *Glomus versiforme*. *Mycol. Res.*, **94**, 157–65.

Bonfante-Fasolo, P. and Fontana, A. (1985) VAM fungi in *Ginkgo biloba* roots: their interactions at cellular level. *Symbiosis*, **1**, 53–67.

Bonfante-Fasolo, P. and Grippiolo, R. (1984) Cytochemical and biochemical observations on the cell wall of the spore of *Glomus epigaeum*. *Protoplasma*, **123**, 140–51.

Bonfante-Fasolo, P. and Vian, B. (1984) Wall texture in the spore of a vesicular–arbuscular mycorrhizal fungus. *Protoplasma*, **120**, 51–60.

Bonfante-Fasolo, P., Vian, B., Perotto, S. *et al.* (1990b) Cellulose and pectin localization in roots of mycorrhizal *Allium porrum*: labelling continuity between host cell wall and interfacial material. *Planta*, **180**, 537–47.

Brown, M.S. and Bethlenfalvay, G.J. (1988) The *Glycine–Glomus–Rhizobium* symbiosis. VII. Photosynthetic nutrient-use efficiency in nodulated, mycorrhizal soybeans. *Plant Physiol.*, **86**, 1292–7.

Brown, M.S., Thamsurakul, S. and Bethlenfalvay, G.J. (1988) The *Glycine–Glomus–Bradyrhizobium* symbiosis. VIII. Phosphorus-use efficiency of CO_2 and N_2 fixation in mycorrhizal soybean. *Physiol. Plant.*, **74**, 159–63.

Buchenauer, H. (1983) Wirkungsweise moderner Fungizide in Pilzen und

Kulturpflanzen. *Ber. Dtsch. Bot. Ges.*, **96**, 427–57.

Caldwell, M.M. (1988) Plant root systems and competition, in *Proceedings of the XIV International Botanical Congress* (eds W. Greuter and R. Zimmer) Koeltz, Königstein/Taunus, 385–404.

Capaccio, L.C.M. and Callow, J.A. (1982) The enzymes of polyphosphate metabolism in vesicular–arbuscular mycorrhizas. *New Phytol.*, **91**, 81–91.

Codignola, A., Verotta, L., Spanu, P. *et al.* (1989) Cell wall bound phenols in roots of vesicular–arbuscular mycorrhizal plants. *New Phytol.*, **112**, 221–8.

Cooper, K.M. and Lösel, D.M. (1978) Lipid physiology of vesicular arbuscular mycorrhiza. I. Composition of lipids in roots of onion, clover and ryegrass infected with *Glomus mosseae. New Phytol.*, **80**, 143–51.

Cooper, K.M. and Tinker P.B. (1978) Translocation and transfer of nutrients in vesicular–arbuscular mycorrhizas. II. Uptake and translocation of phosphorus, zinc and sulphur. *New Phytol.*, **81**, 43–52.

Cox, G., Sanders, F.E., Tinker, P.B. and Wild, J.A. (1975) Ultrastructural evidence relating to host–endophyte transfer in a vesicular–arbuscular mycorrhiza, in *Endomycorrhizas* (eds F.E. Sanders, B. Mosse and P.B. Tinker) Academic Press, London, pp. 297–312.

Cox, G. and Tinker, P.B. (1976) Translocation and transfer of nutrients in vesicular–arbuscular mycorrhizas. I. The arbuscule and phosphorus transfer: a quantitative ultrastructural study. *New Phytol.*, **77**, 371–8.

Cruz, R.E. de la, Lorilla, E.B. and Aggangan, N.S. (1990) Growth of *Acacia auriculiformis* and *Acacia mangium* in a marginal grassland in response to inoculations with VA mycorrhiza and/or *Rhizobium*, in *Fast Growing Trees and Nitrogen Fixing Trees* (eds D. Werner and P. Müller) G. Fischer Verlag, Stuttgart, New York, p. 321.

Dehne, H.W. and Schönbeck, F. (1975) The influence of the endotrophic mycorrhiza on the fusarial wilt of tomato. *Z. Pflanzenkr. Pflanzensch.*, **82**, 630–3.

Dehne, H.W. and Schönbeck, F. (1979) Untersuchungen zum Einfluß der endotrophen Mycorrhiza auf Pflanzenkrankheiten. III. Phenolstoffwechsel und Lignifizierung. *Phytopathol. Z.*, **95**, 210–16.

Dexheimer, J., Gianinazzi, S. and Gianinazzi-Pearson, V. (1979) Ultrastructural cytochemistry of the host fungus interface in the endomycorrhizal association *Glomus mosseae/Allium cepa. Z. Pflanzenphysiol.*, **92**, 191–206.

Douds, D.D., Jr and Schenck, N.C. (1990) Increased sporulation of vesicular–arbuscular mycorrhizal fungi by manipulation of nutrient regimens. *Appl. Environ. Microbiol.*, **56**, 413–18.

Gianinazzi, S., Trouvelot, A. and Gianinazzi-Pearson, V. (1990) Role and use of mycorrhizas in horticultural crop production. XXIII *International Horticultural Congress*, I.S.H.S., Firenze, pp. 25–30.

Gianinazzi-Pearson, V., Branzanti, B. and Gianinazzi, S. (1989) *In vitro* enhancement of spore germination and early hyphal growth of a vesicular–arbuscular mycorrhizal fungus by host root exudates and plant flavonoids. *Symbiosis*, **7**, 243–55.

Gianinazzi-Pearson, V. and Gianinazzi, S. (1988) Phosphorus metabolism in mycorrhizas, in *Nitrogen, Phosphorus and Sulphur Utilization by Fungi* (eds L. Boddy, R. Marchant and D.J. Read) Cambridge University Press, Cambridge, pp. 227–41.

Gianinazzi-Pearson, V., Morandi, D., Dexheimer, J. and Gianinazzi, S. (1981) Ultrastructural and ultracytochemical features of a *Glomus tenuis* mycorrhiza. *New Phytol.*, **88**, 633–9.

Gildon, A. and Tinker, P.B. (1983) Interactions of vesicular–arbuscular mycorrhizal infections and heavy metals in plants. II. The effects of infection on uptake of copper. *New Phytol.*, **95**, 263–8.

Giovannetti, M. and Lioi, L. (1990) Variation of electrophoretic patterns in proteins of VAM fungi chlamydospores, in *Abstracts of the Fourth International Mycological Congress*, (eds A. Reisinger and A. Bresinsky) University of Regensburg, p. 323.

Graham, J.H. and Sylvertsen, J.P. (1989) Vesicular–arbuscular mycorrhizas increase chloride concentration in citrus seedlings. *New Phytol.*, **113**, 29–36.

Graham, J.H., Sylvertsen, J.P. and Smith, M.L. Jr (1987) Water relations of mycorrhizal and phosphorus-fertilized non-mycorrhizal *Citrus* under drought stress. *New Phytol.*, **105**, 411–19.

Graw, D., Moawad, M. and Rehm, S. (1979) Untersuchungen zur Wirts- und Wirkungsspezifität der VA-Mycorrhiza. *J. Agron. Crop Sci.*, **148**, 85–98.

Hardie, K. and Leyton, L. (1981) The influence of vesicular–arbuscular mycorrhiza on growth and water relations of red clover. *New Phytol.*, **89**, 599–608.

Harley, J.L. and Smith, S.E. (1983) *Mycorrhizal Symbiosis*. Academic Press, London.

Harris, D., Pacovsky, R.S. and Paul, E.A. (1984) Carbon flow and N_2 fixation in *Sorghum*, *Azospirillum*, *Glomus* associations, in *Proceedings 6th N.A.C.O.M. Meeting*, University of Oregon, Corvallis, p. 374.

Harris, D., Pacovsky, R.S. and Paul, E.A. (1985) Carbon economy of soybean, *Rhizobium*, *Glomus* associations. *New Phytol.*, **101**, 427–40.

Harris, D. and Paul, E.A. (1987) Carbon requirements of vesicular–arbuscular mycorrhizae, in *Ecophysiology of VA Mycorrhizal Plants* (ed. G.R. Safir) CRC Press, Boca Raton, pp. 93–105.

Hayman, D.S. (1983) The physiology of vesicular–arbuscular endomycorrhizal symbiosis. *Can. J. Bot.*, **61**, 944–63.

Hayman, D.S. (1987) VA mycorrhizas in field crop systems, in *Ecophysiology of VA Mycorrhizal Plants* (ed. G.R. Safir), CRC Press, Boca Raton, pp. 171–92.

Hayman, D.S. and Tavares, M. (1985) Plant growth responses to vesicular–arbuscular mycorrhiza. XV. Influence of soil pH on the symbiotic efficiency of different endophytes. *New Phytol.*, **100**, 367–77.

Heinzemann, J. and Weritz, J. (1990) Steinwolle: ein neuer Trägerstoff zur Vermehrung von vesikulär-arbuskulären Mycorrhizapilzen. *Angew. Bot.*, **64**, 271–4.

Hepper, C.M. (1977) A colorimetric method for estimating vesicular–arbuscular mycorrhizal infection in roots. *Soil Biol. Biochem.*, **9**, 15–18.

Huang, R.-S., Smith, W.K. and Yost, R.S. (1985) Influence of VA mycorrhiza on growth, nutrient absorption and water relations in *Leucaena leucocephala*, in *Proc. 6th North American Conference on Mycorrhizae* (ed. R. Molina) Forest Research Laboratory, Corvallis, Oregon, p. 411.

Humphreys, T.E. (1973) Sucrose transport at the tonoplast. *Phytochemistry*, **12**, 1211–19.

Itoh, S. and Barber, S.A. (1983) Phosphorus uptake by six plant species as related to root hairs. *Agron. J.*, **75**, 457.

Jabaji-Hare, S.H. (1988) Lipids and fatty acids profiles of some vesicular–arbuscular mycorrhizal fungi: contribution to taxonomy. *Mycologia*, **80**, 622–9.

Jabaji-Hare, S.H., Thérien, J. and Charest, P.M. (1990) High resolution cytochemical study of the vesicular–arbuscular mycorrhizal association, *Glomus clarum* × *Allium porrum*. *New Phytol.*, **114**, 481–96.

Jakobsen, I. and Rosendahl, L. (1990) Carbon flow into soil and external hyphae from roots of mycorrhizal cucumber plants. *New Phytol.*, **115**, 77–83.

Janos, D.P. (1985) Mycorrhizal fungi: agents or symptoms of tropical community composition? in *Proc. 6th North American Conference on Mycorrhizae* (ed. R. Molina) Forest Research Laboratory, Corvallis, Oregon, pp. 98–103.

Janos, D.P. (1987) VA mycorrhizas in humid tropical ecosystems, in *Ecophysiology of VA Mycorrhizal Plants* (ed. G.R. Safir), CRC Press, Boca Raton, pp. 107–34.

Kahn, A.G. (1975) Growth effects of VA mycorrhiza on crops in the field, in *Endomycorrhizas* (eds F.E. Sanders, B. Mosse and P.B. Tinker) Academic Press, London, pp. 419–35.

Koske, R.E. and Gemma, J.N. (1989) *Glomus nanolumen* (Endogonaceae), a new species from Hawaii. *Mycologia*, **81**, 935–8.

Kothari, S.K., Marschner, H. and George, E. (1990) Effect of VA mycorrhizal fungi and rhizosphere microorganisms on root and shoot morphology, growth and water relations in maize. *New Phytol.*, **116**, 303–11.

Kruckelmann, H.W. (1975) Effect of fertilizer, soils, soil tillage and plant species on the frequency of endogone chlamydospores and mycorrhizal infection in arable soils, in *Endomycorrhizas*, (ed. F.E. Sanders, B. Mosse and P.B. Tinker) Academic Press, London, pp. 511–25.

Kühn, K.-D. (1991) Distribution of vesicular-arbuscular mycorrhizal fungi on a fallow agriculture site. II. Wet habitat. *Angew. Botanik*, **65**, 187–203.

Kühn, K.-D., Weber, H.-Chr., Dehne, H.-W. and Gworgwor, N.A. (1991) Distribution of vesicular-arbuscular mycorrhizal fungi on a fallow agriculture site. I. Dry habitat. *Angew. Botanik*, **65**, 169–85.

Levy, Y. and Krikun, J. (1980) Effect of vesicular–arbuscular mycorrhiza on *Citrus jambhiri* water relations. *New Phytol.*, **85**, 25–31.

Linderman, R.G. (1988a) Mycorrhizal interactions with the rhizosphere microflora: the mycorrhizosphere effect. *Phytopathology*, **78**, 366–71.

Linderman, R.G. (1988b) VA (vesicular–arbuscular) mycorrhizal symbiosis. *ISI Atlas of Science*: *Plant and Animals*, **1**, 183–8.

Lopes, E.S., Dias, R., Toledo, S.V. *et al.* (1985) Nutrient uptake by coffee inoculated with *Gigaspora margarita*, in *Proc.* 6th *North American Conference on Mycorrhizae* (ed. R. Molina) Forest Research Laboratory, Corvallis, Oregon, p. 256.

Macauley, B.J. and Bethel M. (1990) Ergosterol as an indicator of biomass of *Agaricus brunnescens* mycelium grown under varying cultural conditions, in *Abstract of the Fourth International Mycological Congress* (eds A. Reisinger and A. Bresinsky) University of Regensburg, p. 331.

Marx, C., Dexheimer, J., Gianinazzi-Pearson, V. and Gianinazzi, S. (1982) Enzymatic studies on the metabolism of vesicular–arbuscular mycorrhizas. IV. Ultracytoenzymological evidence (ATPase) for active transfer processes in the host–arbuscule interface. *New Phytol.*, **90**, 37–43.

Meyer, J.R. and Linderman, R.G. (1986a) Response of subterranean clover to dual inoculation with vesicular–arbuscular mycorrhizal fungi and a plant growth-promoting bacterium, *Pseudomonas putida*. *Soil Biol. Biochem.*, **18**, 185–90.

Meyer, J.R. and Linderman, R.G. (1986b) Selective influence on populations of rhizosphere or rhizoplane bacteria and actinomycetes by mycorrhizas formed by *Glomus fasciculatum*. *Soil Biol. Biochem.*, **18**, 191–6.

Moore, P.D. (1987) Distribution of mycorrhiza throughout the British flora. *Nature*, **327**, 100.

Morandi, D., Bailey, J.A. and Gianinazzi-Pearson, V. (1984) Isoflavonoid accumulation in soybean roots infected with vesicular–arbuscular mycorrhizal fungi. *Physiol. Plant Pathol.*, **24**, 357–64.

Morandi, D., Branzanti, B. and Gianinazzi-Pearson, V. (1991) Effect of glyceollin I, coumestrol and quercitin on *in vitro* germination of *Gigaspora margarita* spores. *Abstracts of the Third European Symposium on Mycorrhizas*, Sheffield, August 1991.

Morton, J.B. (1990) Evolutionary relationships among arbuscular mycorrhizal fungi in the endogonaceae. *Mycologia*, **82**, 192–207.

Morton, J.B. and Benny, G.L. (1990) Revised classification of arbuscular mycorrhizal fungi (Zygomycetes): a new order, Glomales, two new suborders, Glomineae and Gigasporineae, and two new families, Acaulosporaceae and Gigasporaceae, with an emendation of Glomaceae. *Mycotaxon*, **37**, 471–91.

Mosse, B. (1973) Advances in the study of vesicular–arbuscular mycorrhiza. *Ann. Rev. Phytopathol.*, **11**, 171.

Mosse, B. and Hayman, D.S. (1980) Mycorrhiza in agricultural plants, in *Tropical Mycorrhiza Research*, (ed. P. Mikola) Clarendon Press, Oxford, pp. 213–30.

Mosse, B., Li, C., Powell, D.S. and Hayman, D.S. (1976) Plant growth response to vesicular–arbuscular mycorrhiza. IX. Interactions between VA mycorrhiza, rock phosphate and symbiotic nitrogen fixation. *New Phytol.*, **76**, 331–42.

Ocampo, J.A., Martin, J. and Hayman, D.S. (1980) Influence of plant interactions on vesicular–arbuscular mycorrhizal infections. I. Host and non-host plants grown together. *New Phytol.*, **84**, 27–35.

Pacovsky, R.S., Fuller, G. and Paul, E.A. (1985) Developmental effects on micronutrient distribution in mycorrhizal and P-fertilized soybeans, in *Proc. 6th North American Conference on Mycorrhizae* (ed. R. Molina) Forest Research Laboratory, Corvallis, Oregon, p. 373.

Paulitz, T.C. and Linderman, R.G. (1989) Interactions between fluorescent pseudomonads and VA mycorrhizal fungi. *New Phytol.*, **113**, 37–45.

Pérez, Y. and Schenck, N.C. (1990) A unique code for each species of VA mycorrhizal fungi. *Mycologia*, **82**, 256–60.

Pirozynski, K.A. and Dalpé, Y. (1989) Geological history of the Glomaceae with particular reference to mycorrhizal symbiosis. *Symbiosis*, **7**, 1–36.

Puga, C. (1985) Influence of vesicular–arbuscular mycorrhizas on competition between corn and weeds in Panama. MS. Thesis, University of Miami, Coral Gables.

Read, D.J. (1990) Mycorrhizas in ecosystems, in *Abstracts of the Fourth International Mycological Congress* (eds A. Reisinger and A. Bresinsky), University of Regensburg, p. 2.

Reeves, F.B. (1985) Survival of VA mycorrhizal fungi-interactions of secondary succession, mycorrhizal dependency in plants, and resource competition, in *Proc. 6th North American Conference on Mycorrhizae* (ed. R. Molina) Forest Research Laboratory, Corvallis, Oregon, pp. 110–13.

Reid, C.P.P. (1990) Mycorrhizas, in *The Rhizosphere* (ed. J. Lynch) Wiley, New York, London, pp. 281–315.

Rhodes, L.H. and Gerdemann, J.W. (1978) Hyphal translocation and uptake of sulphur by vesicular–arbuscular mycorrhizae of onion. *Soil. Biol. Biochem.*, **10**, 355–60.

Safir, G.R. (ed.) (1987) *Ecophysiology of VA Mycorrhizal Plants*. CRC Press, Boca Raton.

Safir, G.R., Boyer, J.S. and Gerdemann, J.W. (1972) Nutrient status and mycorrhizal enhancement of water transport in soybean. *Plant Physiol.*, **49**, 700–3.

Saif, S.R. (1983) Soil temperature, soil oxygen and growth of mycorrhizal and non-mycorrhizal plants of *Eupatorium odoratum* L. and development of *Glomus macrocarpus*. *Angew. Bot.*, **57**, 143–55.

Sanders, F.E., Mosse, B. and Tinker, P.B. (1975) *Endomycorrhizas*. Academic Press, London.

Sanders, F.E. and Tinker, P.B. (1973) Phosphate flow into mycorrhizal roots. *Pestic. Sci.*, **4**, 385–95.

Schenck, N.C. (ed.) (1982) *Methods and Principles of Mycorrhizal Research*. The American Phytopathological Society, St Paul, MN.

Schenck, N.C. (1989) Interactions between herbaceous legumes and their associated rhizobia, mycorrhizae, pests, and parasites, in *Advances in Legume Biology* (eds C.H. Stirton, and J.L. Zarucchi) Monograph in Systematic Botany, Missouri Botanical Garden, **29**, 595–605.

Schenck, N.C. and Perez, Y. (1990) A culture collection of obligate symbiotic mycorrhizal fungi: current status, in *Abstracts of the Fourth International Mycological Congress* (eds A. Reisinger and A. Bresinsky) University of Regensburg, p. 340.

Schenck, N.C., Spain, J.L., Sieverding, E. and Howeler, R.H. (1984) Several new and unreported vesicular–arbuscular mycorrhizal fungi (Endogonaceae) from Colombia. *Mycologia*, **76**, 685–99.

Schlumbaum, A., Mauch, F., Vögeli, U. and Boller, T. (1986) Plant chitinases are potent inhibitors of fungal growth. *Nature*, **324**, 365–7.

Schmitz, O., Danneberg, G., Hundeshagen *et al.* (1991) Quantification of vesicular–arbuscular mycorrhiza by biochemical parameters. *J. Plant Physiol.*, **139**, 106–14.

Schönbeck, F. and Dehne H.W. (1979) The influence of endotrophic mycorrhiza on plant diseases. 4. Fungal parasites on aerial plant parts, *Olpidium brassicae*, TMV. *Z. Pflanzenkr. Pflanzensch.*, **86**, 103–12.

Schüepp, H., Dehn, B. and Sticher, H. (1987) Interaktionen zwischen VA-Mykorrhizen und Schwermetallbelastungen. *Angew. Bot.*, **61**, 85–96.

Schüepp, H. and Bodmer, M. (1991) Complex response of VA-mycorrhizae to xenobiotic substances. *Toxicological and Environmental Chemistry*, **30**, 193–9.

Sieverding, E. (1988) Ecology of VA mycorrhizal fungi in tropical agroecosystems, in *Abstracts of the 2nd European Symposium on Mycorrhizae*, Institute of Landscape Ecology, Prague, Czechoslovakia, p. 97.

Sieverding, E. (1991) *Vesicular-Arbuscular Mycorrhiza Management*. Deutsche Gesellschaft für Technische Zusammenarbeit (GTZ), Eschborn (Germany).

Smith, S.E. (1985) The concept of effectiveness in symbiotic relationships, in *Proc. 6th North American Conference on Mycorrhizae* (ed. R. Molina) Forest Research Laboratory, Corvallis, Oregon, pp. 146–9.

Smith, S.E. and Gianinazzi-Pearson, V. (1988) Physiological interactions between symbionts in vesicular–arbuscular mycorrhizal plants. *Ann. Rev. Plant Physiol. Plant Mol. Biol.*, **39**, 221–44.

Smith, S.E. and Smith, F.A. (1990) Tansley Review No. 20. Structure and function of the interfaces in biotrophic symbioses as they relate to nutrient transport. *New Phytol.*, **114**, 1–38.

Smith, S.E. and Walker, N.A. (1981) A quantitative study of mycorrhizal infection in *Trifolium*: separate determination of the rates of infection and of mycelial growth. *New Phytol.*, **89**, 225–40.

Spanu, P., Bonfante-Fasolo, P. and Boller, T. (1988) Chitinase activity and VA-mycorrhiza development, in *Abstracts of the 2nd European Symposium on Mycorrhizae*, Institute of Landscape Ecology, Prague, Czechoslovakia, p. 101.

Sparling, G.P. and Tinker, P.B. (1975) Mycorrhizas in pennine grassland, in *Endomycorrhizas* (eds F.E. Sanders, B. Mosse and P.B. Tinker) Academic Press, London, pp. 545–60.

Strullu, D.G. (1978) Histologie et cytologie des endomycorrhizas. *Physiol. Vég.*, **16**, 657–69.

Strullu, D.G., Chamel, A., Eloy, J.F. and Gourret, J.P. (1983) Ultrastructure and analysis, by laser probe mass spectrography, of the mineral composition of the vesicles of *Trifolium pratense* endomycorrhizas. *New Phytol.*, **94**, 81–8.

Sylvia, D.M. and Heal, L.H. (1990) Nitrogen affects the phosphorus response of VA mycorrhiza. *New Phytol.*, **115**, 301–10.

Tester, M. (1990) Tansley Review No. 21. Plant ion channels: whole-cell and single-channel studies. *New Phytol.*, **114**, 305–40.

Thomson, B.D., Robson, A.D. and Abbott, L.K. (1986) Effects of phosphorus on the formation of mycorrhizas by *Gigaspora calospora* and *Glomus fasciculatum* in relation to root carbohydrates. *New Phytol.*, **103**, 751–65.

Thomson, B.D., Robson, A.D. and Abbott, L.K. (1990) Mycorrhizas formed by *Gigaspora calospora* and *Glomus fasciculatum* on subterranean clover in relation to soluble carbohydrate concentrations in roots. *New Phytol.*, **114**, 217–25.

Tinker, P.B. (1978) Effects of vesicular–arbuscular mycorrhizas on plant nutrition and plant growth. *Physiol. Vég.*, **16**, 743–51.

Tinker, P.B. and Gildon, A. (1983) Mycorrhizal fungi and ion uptake, in *Metals and Micronutrients* (eds D.A. Robb and W.S Pierpoint) Academic Press, London, p. 21.

Tommerup, I.C. (1985) Development of VA mycorrhizal infection in Cruciferae and *Trifolium subterraneum*, in *Proc. 6th North American Conference on Mycorrhizae* (ed. R. Molina) Forest Research Laboratory, Corvallis, Oregon, p. 333.

Trappe, J.M. (1982) Synoptic keys to the genera and species of zygomycetous mycorrhizal fungi. *Phytopathology*, **72**, 1102–8.

Walker, C. (1983) Taxonomic concepts in the Endogonaceae: spore wall characteristics in species descriptions. *Mycotaxon*, **18**, 443–55.

Wright, S.F. and Morton, J.B. (1989) Detection of vesicular–arbuscular mycorrhizal fungus colonization of roots by using dot–immunoblot assay. *Appl. Environ. Microbiol.*, **55**, 761–3.

Wright, S.F., Morton, J.B. and Sworobuk, J.E. (1987) Identification of a vesicular–arbuscular mycorrhizal fungus by using monoclonal antibodies in an enzyme-linked immunosorbent assay. *Appl. Environ. Microbiol.*, **53**, 2222–5.

Wyss, P., Mellor, R.B. and Wiemken, A. (1990a) Mutants of soybean (*Glycine max*) unable to suppress nodulation in the presence of nitrate retain the ability to suppress mycorrhization in the presence of phosphate. *J. Plant Physiol.*, **136**, 507–9.

Wyss, P., Mellor, R.B. and Wiemken, A. (1990b) Vesicular–arbuscular mycorrhizas of wild-type soybean and non-nodulating mutants with *Glomus mosseae* contain symbiosis-specific polypeptides (mycorrhizins) immunologically cross-reactive with nodulins. *Planta*, **182**, 22–6.

Yors, R.S. and Fox, R.L. (1979) Contribution of mycorrhizae to P nutrition of crops growing on an Oxisol. *Agron. J.*, **71**, 903.

11
Ectomycorrhiza, Ericoid- and Orchidmycorrhiza

11.1 ECTOMYCORRHIZA

The first to use the term mycorrhiza was the botanist A.B. Frank in 1885. What he described is now known as ectomycorrhiza and is characterized by an outer coat of fungal hyphae which surrounds the host root, by a layer of hyphae between the root rind cells, called the 'Hartig net' and by a stocky and compacted growth of the ectorrhizal root. With few exceptions the fungi remain extracellular.

There has been a considerable number of reviews on this form of fungal–plant interaction. (Frank, 1885; Fellner, 1988; Dorr and Kollmann, 1969; Marx, 1972; Harley, 1978; Mikola, 1980; Trappe *et al.*, 1984; Clarkson, 1985; Meyer, 1986, 1987; Schmitt, 1987; Hoffmann, 1988; Institute of Landscape Ecology, 1988; Berliner and Torrey, 1989; Rousseau and Reid, 1989b; Berliner, 1990).

11.1.1 Microsymbionts and hosts

Most symbiotic fungi of the ectomycorrhiza are Ascomycetes or Basidiomycetes (Table 11.1). There are also several species from the genus *Endogone* (Zygomycetes). Typical ectomycorrhizal fungi include edible mushrooms such as *Cantharellus cibarius*, *Boletus edulis* and *Russula* sp. Species from at least 65 genera have been identified as symbionts by using thorough field experiments or *in vitro* inoculations. Many more types of mycorrhizal mycelium have been found, but, since in many cases fruiting body formation was not observed, they could not be classified.

Host species are found among 140 genera of seed-setting plants (Table 11.2). This form of symbiosis is especially common in the families Dipterocarpaceae, Pinaceae, Fagaceae, Myrtaceae, Salicaceae and Betulaceae. It is absent from monocots (Newman and Reddell, 1987). In total 3% of all seed-setting plant species are known to have ectomycorrhiza (EM) symbioses. This is much less than the number able to form vesicular–arbuscular mycorrhiza (VAM) symbioses. Trees and shrubs dominate the

Table 11.1 Fungal genera with species involved in the formation of ectomycorrhiza. (Based on Harley and Smith 1983 and Singer *et al.*, 1983)

Family	*Genus*
Zygomycetes	
Endogonaceae	*Endogone*
Ascomycetes	
Balsamiaceae	*Balsamia*
Elaphomycetaceae	*Elaphomyces*
Geneaceae	*Genea*
Geoglossaceae	*Cudonia, Spathularia*
Helvellaceae	*Helvella*
Hydnotryaceae	*Bassia, Choiromyces, Hydnotrya*
Otidiaceae	*Otidia*
Pyronemaceae	*Geopora, Lachnea, Sepultaria*
Rhiziniaceae	*Gyromitra*
Terfeziaceae	*Mukomyces, Picoa, Terfezia, Tirmania*
Basidiomycetes	
Agaricaceae	*Lepiota*
Amanitaceae	*Amanita, Amanitopsis*
Boletaceae	*Boletinus, Boletus, Pulveroboletus, Fistulinella, Gyrodon, Gyroporus, Krombholzia, Leccinum, Suillus, Tylopilus, Xerocomus*
Cortinariaceae	*Alnicola, Cortinarius, Hebeloma, Inocybe, Rozites*
Gomphideaceae	*Gomphidius*
Hygrophoraceae	*Hygrophorus*
Paxillaceae	*Paxillus*
Rhodophyllaceae	*Clitopilus, Rhodophyllus*
Russulaceae	*Lactarius, Russula*
Strobilomycetaceae	*Boletellus, Strobilomyces*
Tricholomataceae	*Laccaria, Leucopaxillus, Lyophyllum, Tricholoma*
Cantharellaceae	*Cantharellus, Craterellus*
Hydnaceae	*Hydnum*
Thelephoraceae	*Corticium, Thelephora*
Hymenogastraceae	*Alpova, Rhizopogon*
Geastraceae	*Geastrum, Astraeus*
Lycoperdaceae	*Calvatia, Lycoperdon*
Phallaceae	*Clathrus, Phallus*
Pisolithaceae	*Pisolithus*
Sclerodermataceae	*Scleroderma*
Fungi Imperfecti	
	Cenococcum

Table 11.2 Genera of seed plants with ectomycorrhiza. (Modified from Marks and Kozlowski, 1973 and Harley and Smith, 1983)

Family	*Genus*
Aceraceae	*Acer*
Betulaceae	*Alnus, Betula, Carpinus, Corylus, Ostrya, Ostryopsis*
Bignoniaceae	*Jacaranda, Phyllarthron*
Caesalpiniaceae	*Afzelia, Aldina, Anthonota, Bauhinia, Brachystegia, Cassia, Eperua, Gilbertiondendron, Julbernardia, Monopetalanthus, Paramacrolobium, Swartzia*
Caprifoliaceae	*Sambucus*
Casuarinaceae	*Casuarina*
Cistaceae	*Helianthemum, Cistus*
Combretaceae	*Terminalia*
Compositae	*Lactuca, (Mycelis)*
Cupressaceae	*Cupressus, Juniperus*
Cyperaceae	*Kobresia*
Dipterocarpaceae	*Anisoptera, Balanocarpus, Cotylelobium, Dipterocarpus, Dryobalanops, Hopea, Monotes, Shorea, Valica*
Elaeaganaceae	*Shepherdia*
Ericaceae	*Arbutus, Arctostaphylos, Chimaphila, Gaultheria, Kalmia, Ledum, Leucothoe, Rhododendron, Vaccinium*
Euphorbiaceae	*Poranthera, Uapaca*
Fabaceae	*Bartonia, Brachysema, Chorizema, Daviesia, Dillwynia, Eutaxia, Gompholobium, Hardenbergia, Jacksonia, Kennedya, Mirbelia, Oxylobium, Platylobium, Pultenaea, Robinia, Vicia, Viminaria*
Fagaceae	*Castanea, Castanopsis, Fagus, Lithocarpus, Nothofagus, Pasania, Quercus, Trigonobalanus*
Globulariaceae	*Globularia*
Gnetaceae	*Gnetum*
Goodenaceae	*Brunonia, Goodenia*
Hammamelidaceae	*Parrotia*
Juglandaceae	*Carya, Juglans, Pterocarya*
Lauraceae	*Sassafras*
Mimosaceae	*Acacia*
Myricaceae	*Comptonia, Myrica*
Myrtaceae	*Angophora, Callistemon, Camponesia, Eucalyptus, Leptospermum, Melaleuca, Tristania*
Nyctinaginaceae	*Neea, Pisonia, Torrubia*
Oleaceae	*Fraxinus*
Pinaceae	*Abies, Cathaya, Cedrus, Keteleeria, Larix, Picea, Pinus, Pseudolarix, Pseudotsuga, Tsuga*
Platanaceae	*Platanus*
Polygonaceae	*Coccoloba, Polygonum*
Pyrolaceae	*Pyrola*
Rhamnaceae	*Cryptandra, Pomaderris, Rhamnus, Spyridium, Trygmalium*
Rosaceae	*Chamaebatia, Circocarpus, Crataegus, Dryas, Malus, Prunus, Pyrus, Rosa, Sorbus*
Rubiaceae	*Galium, Opercularia, Rubia, Psychotria*
Salicaceae	*Populus, Salix*

Table 11.2 *Continued*

Family	*Genus*
Sapindaceae	*Allophylus, Nephelium*
Sapotaceae	*Glycoxylon*
Saxifragaceae	*Ribes*
Sterculiaceae	*Lasiopetalum, Thomasia*
Stylidiaceae	*Stylidium*
Thymeliaceae	*Pimelia*
Tiliaceae	*Tilia*
Ulmaceae	*Ulmus, Celtis*
Vitaceae	*Vitis*
Zygophyllaceae	*Peganum*

EM-plants. The specificity of host and symbiont to each other is small. *Picea abies* can form EM-symbioses with over 100 different fungal species (Trappe, 1962). The fly agaric (*Amanita muscaria*) may infect roots of trees as varied as birch, eucalyptus, spruce and Douglas fir (Table 11.3). The butter mushroom (*Boletus grevillei*) can form symbiotic association not only with its normal host, the larch (*Larix*), but also with *Pinus* species and *Pseudotsuga menziesii*.

There are some examples of incompatible associations, as between *Pinus sylvestris* and *Suillus grevillei* or *Boletinus clavipes* (Finlay, 1988). Several trees from the temperate regions have both ectomycorrhizas and VA-mycorrhizas, e.g. poplars, alders and some eucalypts (Lopez-Aguillon and Garbaye, 1988). Within one species, such as *Pisolithus tinctorius*, there may be races adapted to a certain tree, as shown by inoculation of *Eucalyptus urophylla* with strains collected from pine and from eucalypts (Malajczuk *et al.*, 1990).

In spruce stands the total root system reaches a length of over 100 000 km ha^{-1} (Table 11.4). Only 15% of the total root biomass is accounted for by fine roots, but these are responsible for 60% of root growth. The number of root tips per litre of soil is dependent on the soil type; in *Fagus sylvatica* it varies from about 4000 in a silty loam soil to more than 60 000 in a podsol soil (Meyer, 1986). This physiologically active root surface can be magnified many times by mycorrhizal extensions. Moreover trees of the same or of different species can be united in an underground network of mycorrhizal hyphae.

11.1.2 Structure and development of ectomycorrhiza

Ectotrophic mycorrhiza on trees prefer, or are exclusively confined to, secondary or tertiary roots. It is necessary that this root has a low growth

Table 11.3 Symbionts and host spectrum of *Pinus sylvestris* and *Amanita muscaria*. (Based on Harley and Smith, 1983)

Pinus sylvestris
×
Amanita muscaria, A. pantherina
Cenococcum graniforme (geophilum)
Clitopilus prunulus
Cortinarius glaucopus, C. mucosus
Lactarius deliciosus, L. helvus
Lyophyllum immundum (Tricholoma fumosum)
Paxillus involutus
Rhizopogon roseolus, R. luteolus
Rhodophyllus rhodopolius
Russula emetica
Scleroderma aurantium
Suillus bovinus, S. flavidus, S. granulatus, S. grevillei, S. luteus, S. variegatus
Tricholoma flavobrunneum, T. flavovirens, T. imbricatum, T. pessundatum, T. saponaceum, T. vaccinium

Amanita muscaria
×
Betula pendula, B. pubescens
Eucalyptus camaldulensis, E. calophylla, E. dalrympleana, E. diversicola, E. maculata, E. marginata, E. obliqua, E. regnans, E. st. johnii, E. sieberii
Larix decidua, L. occidentalis
Picea abies, P. sichensis
Pinus contorta, P. echinata, P. monticola, P. mugo, P. patula, P. ponderosa, P. radiata, P. strobus, P. sylvestris, P. taeda, P. virginiana
Pseudotsuga menziesii

Table 11.4 Production, biomass and spatial distribution of the root system of *Picea sitchensis* in a 16-year-old stand. (Based on Last *et al.*, 1983)

Root diameter (mm)	*Root biomass*		*Annual production*		*Spatial distribution*
	kg ha^{-1}	*kg tree*$^{-1}$	*kg ha*$^{-1}$	*kg tree*$^{-1}$	
0–2 (Fine roots)	3530	0.93	5240	1.38	ca. 90 000 km ha^{-1}
2–5	1370	0.36	38	0.01	
>5 (Structural roots)	20 000	5.3	3150	0.83	
Total	25 000	6.6	8430	2.22	>100 000 km ha^{-1}

rate. The coating of the root with mycelium reduces the growth lengthways even more, and they are therefore often called short roots. In genera where the main root axis is also infected (e.g. *Fagus* and *Eucalyptus*) root tips grow out of their mycelial coat during the vegetative period and are only later cloaked with fungal hyphae. The morphology of typical mycorrhiza can be

Fig. 11.1. (a) *In vitro* mycorrhizal infection of rooted spruce cuttings with C-deficient *Pisolithus tinctorius*. Macroscopic view through culture vessel 3 months after inoculation. Numerous hyphae and rhizomorphs reach from the fungal sheath into the substrate (vermiculite/peat). (b) Uninfected control abundant in root hairs. (Original photograph, B. Hock, Munchen-Weihenstephan.)

either single or dichotomously branched (Figs 11.1–11.3). The structure of the ectomycorrhiza is determined more by the fungal species than by the host plant (Melville *et al.*, 1987).

The root cap and meristem remain uninfected. They are connected to the mycorrhiza-infection-zone (MIZ) which reaches the degenerating primary bark cells. As the root elongates the MIZ appears to move upwards, that is, it grows acropetally. The fungus in the MIZ grows faster than the root and wraps the root up in a pseudoparenchymal layer of hyphae, called the coat (Fig. 11.3). This coat has a relatively compact surface with few radial mycelia (Fig. 11.1), but can also have a more loosened form (Fig. 11.3). Hyphae in the inner layer of the coat contain more cell organelles and cytoplasm than those in the outer layers. The inner hyphae are often embedded in a carbohydrate matrix. This part of the coat often shows similarities with fruiting bodies. The coat is 20–60 μm thick and accounts for 20–30% of the volume of the mycorrhizal root and 30–40% of the dry weight.

The fungal mycelium which is between the root dermal cells is called

Fig. 11.2. *Fagus sylvatica* mycorrhiza with numerous extending hyphae. (Original photograph E. Zimmermann, Marburg.)

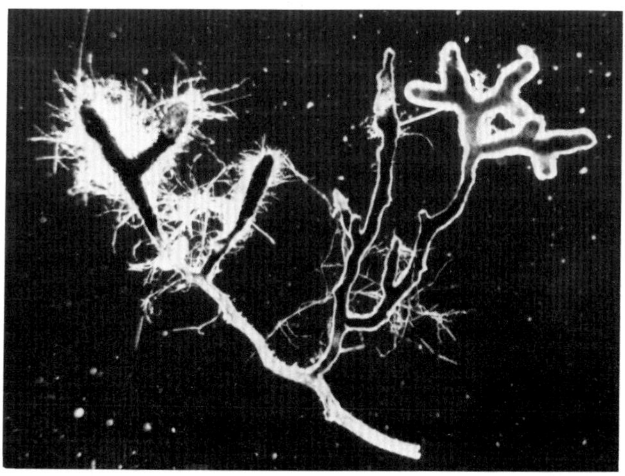

Fig. 11.3. Dark-field photograph of mycorrhiza from *Fagus sylvatica* with bright fungal sheath and hyphae. (Original photograph, E. Zimmermann, Marburg.)

the 'Hartig net', after the forester and botanist T. Hartig (*Vollständige Naturgeschichte der Forstlichen Kulturpflanzen Deutschlands*, Berlin, 1851). It may be restricted to the outermost cell layers of the root or can reach in as far as the endodermis. Hyphae grow between cells, presumably by forcing their way mechanically and by excreting pectinases. Also, plant enzymes may be involved, since the mycelia develop in intracellular spaces. In

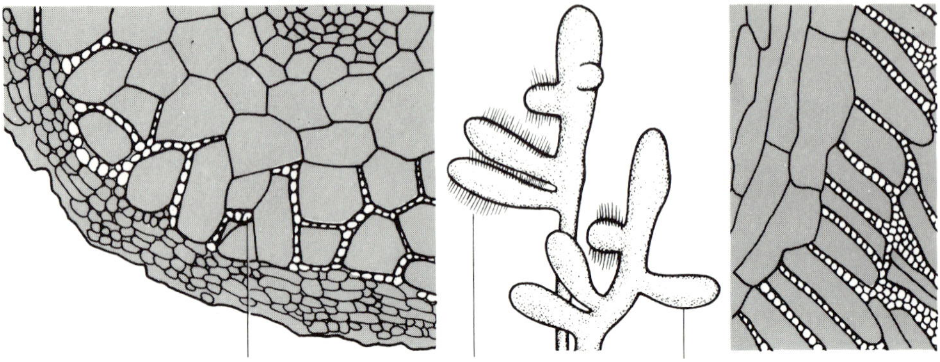

Fig. 11.4. Diagram of mycorrhiza from *Fagus sylvatica*. (a) Root cross-section with intercellular hyphae of the Hartig net. (b) Club-shaped swollen root tips with extending hyphae. (c) Longitudinal section of root with intercellular hyphae.

longitudinal section these hyphae between cells have a ladder-like appearance (Fig. 11.4). The connections between plant cells, the plasmodesmata, remain undisturbed. Fungal hyphae and radial cell walls in the region of the Hartigs net remain unchanged in *Pinus radiata*. However, in other trees, such as *Betula* and *Pseudotsuga*, they may be modified to form a particularly close contact between the symbiotic partners. The number of roots cells and their volume is the same in mycorrhizae as in non-infected roots. Often, however, cells are shortened on their long axis, which results in an increase in the radial diameter. Where secondary and tertiary roots develop from a coated root the new growth is directly infected from the old. The reduced growth rate of the side roots allows the fungal coat growth to keep pace. Primary infection of seedlings is significantly different in that roots are colonized from spores or hyphae. The life of an EM-root is around 9–14 months.

A well-developed sheath is apparently not an absolute prerequisite for the formation of the Hartig net. Hyphal branches of the Hartig net have several characteristics of transfer cells, such as a dense cytoplasm, a well-developed rough endoplasmic reticulum and numerous cristae-rich mitochondria. The occurrence of several nuclei distinguishes the cells as coenocytic transfer cells (Kottke and Oberwinkler, 1986).

Development of an ectomycorrhiza can be very rapid. In *Eucalyptus*, after one day chemical signal exchange is observed, after two days coat formation at the root tips starts and on day 4 all characters of an ectomycorrhiza are present (Horan *et al.*, 1988).

Mycorrhiza-development shows an annual cycle. *Pinus taeda*, infected in April with spores of *Pisolithus tinctorius*, showed mycorrhiza on 90% of roots by August. There were 16 fruiting bodies per tree by autumn (Fig. 11.5).

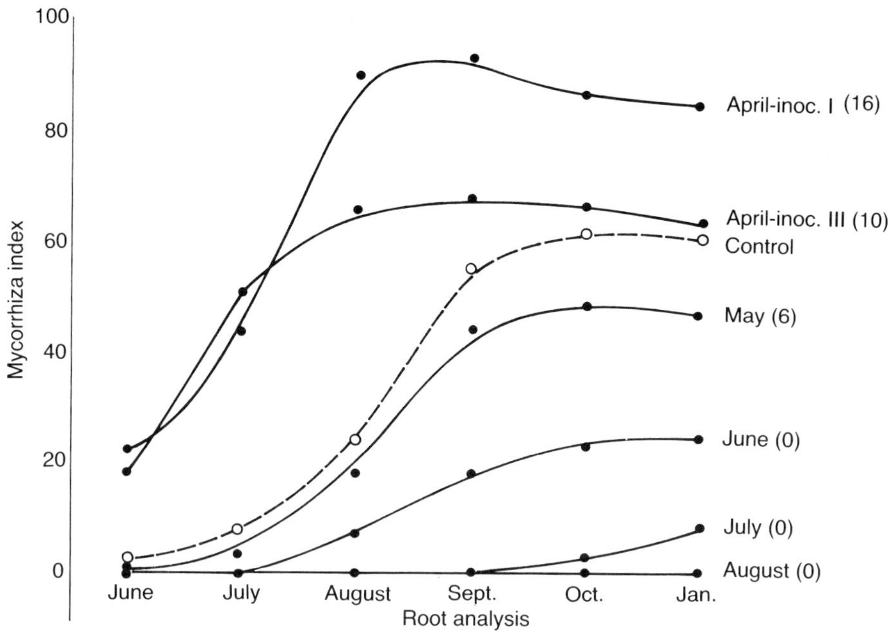

Fig. 11.5. Development of mycorrhiza after inoculation of *Pinus taeda* with spores of *Pisolithus tinctorius*. Inoculations between April and August. Figures in brackets indicate numbers of fruiting bodies. Control: uninfected woodland soil.

Inoculum in June led to 20% infected roots in October, but this was too late for fruiting body formation that year. Infection in August led to no mycorrhizal infection. Possible soil and climatic effects are schematically presented in Fig. 11.6. Due to their complex interactions, quantitation is not really possible. Plants entering symbioses with EM-fungi, as with VAM-fungi, are directly affected by soil temperature, water content, O_2-concentration and mineral nutrition. These factors indirectly affect plant roots, root exudate and root differentiation. In addition, they all affect the fungal partner. Other soil microorganisms can also be present as competitors or parasites and their exudates can also exert effects as antibiotics.

The infiltration of soil by fungal hyphae is much better than that achieved by root hairs (Fig. 11.7). The hyphae branch and enter the smallest soil capillary spaces and thus develop a huge active surface area. The zone where fruiting bodies are formed in mature stands correlates with minima of soil temperature and maxima of soil moisture content between trees (Fig. 11.8). In a 5-year-old *Pinus radiata* stand, *Suillus luteus* fruiting bodies are formed at regular 2–3 m intervals between trees.

Methods to characterize ectomycorrhizas *in situ* have been described in detail by Agerer (1988, 1989).

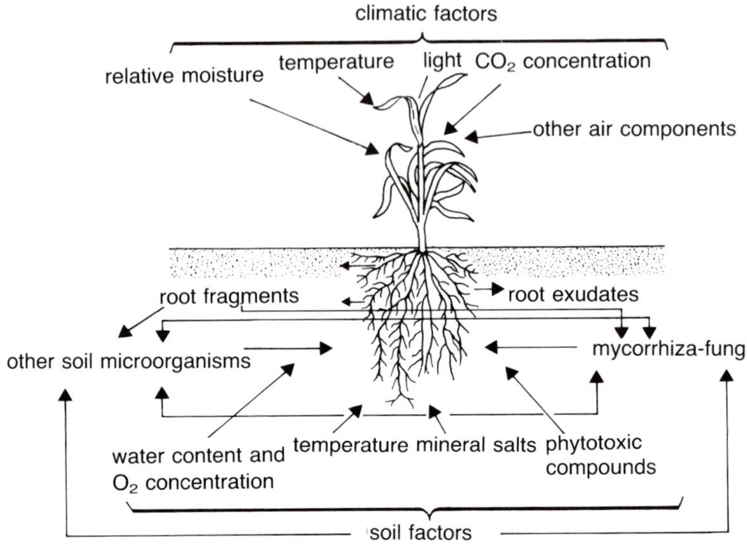

Fig. 11.6. Interactions between mycorrhiza fungi and other soil micro-organisms as affected by root exudates, root debris, soil factors and by climatic factors via the shoot.

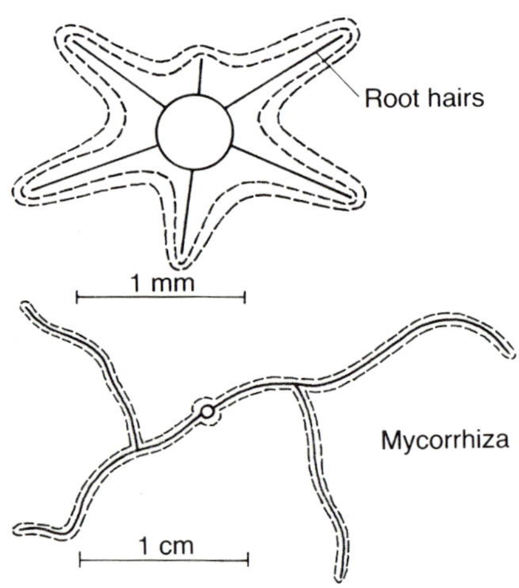

Fig. 11.7. Spatial extension of a root with root hairs and of a mycorrhizal root. The dashed lines indicate zones of nutrient uptake. (Based on Tinker, 1975.)

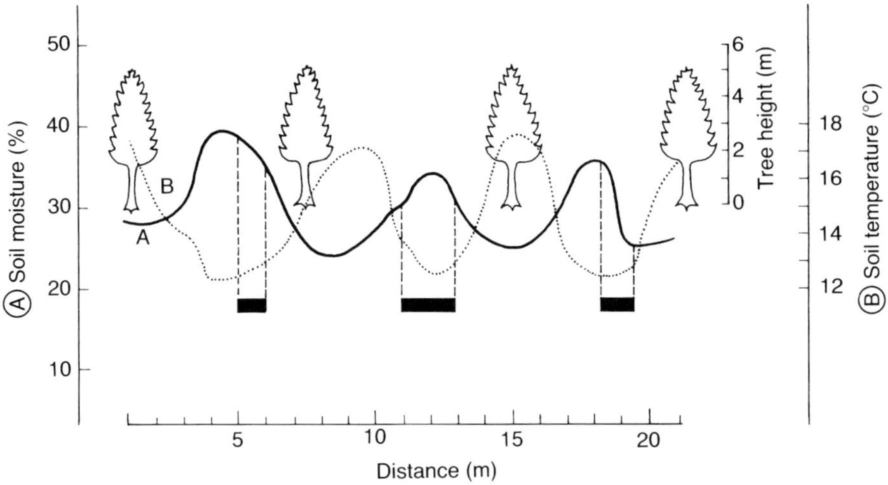

Fig. 11.8. Fructification in *Suillus lateus* (■) in 5-year-old plantations of *Pinus radiata*, in relation to soil moisture (A) and soil temperature (B). (Based on Peredo *et al.*, 1983.)

They include

1. morphological characters such as ramification
2. outer surface of the mantle
3. middle layer of the mantle
4. inner layer of the mantle
5. surface of the tip
6. rhizomorphs
7. emanating hyphae

The most convincing method of identification is the direct tracing of mycorrhizas to fruiting bodies via rhizomorphs (Hacskaylo, 1965). These methods can be complemented by physiological parameters such as iso-enzyme pattern or mycorrhiza specific proteins (Keller, 1990). Ten specific proteins, formed only in an ectomycorrhizal association have been detected in a *Pisolithus tinctorius–Eucalyptus* symbiosis (Hilbert and Martin, 1988).

Mycorrhizas of *Picea abies* with *Laccaria amethystea* had significantly reduced concentrations of **hydroxybenzoic acid glucoside**, **picein** and **catechin** compared to non mycorrhizal short roots (Münzenberger *et al.*, 1991)

11.1.3 The physiology of symbiotic fungi

Spores of EM-forming Basidomycetes germinate with a very low frequency, often under 1%. Other fungi such as *Torulopsis sanguinea* or *Cenococcum grandiforme* excrete components which stimulate spore germination. The presence of hyphae of the same species can also stimulate spore germination so that autocatalytic effects are often observed. In certain species, like *Leccinum scabrum* and *Leccinum aurantiacum*, the stimulation of spore germination is only possible through the near presence of hyphae of the same species. Root exudate from EM-trees such as *Pinus sylvestris* and *Betula pendula* can also stimulate germination of basidiospores, whereas non-EM-tree root exudate shows no such effect. These effective substances seem on the one hand to be widely spread, but on the other, relatively species specific. Spore germination in *Suillus* sp. is enhanced by diterpenoids, which are effective at concentrations as low as 10^{-7} M (Fries *et al.*, 1987). One report that isovalerianic acid is involved has not been confirmed. Germination rates of oidia and chlamydospores is generally somewhat higher than for basidiospores.

A further important part of the fungal life-cycle is the sclerotia. Sclerotia can survive extreme soil conditions and infect roots which spores cannot reach. *Pisolithus tinctorius* forms sclerotia 4 weeks after initial infection on *Pinus strobus* roots. They are about 0.5 mm in diameter and contain reserves in the form of carbohydrate, lipid and storage protein. The surface is formed from especially closely interwoven hyphae. Many of these are highly branched and show swollen ends. Single hyphae can grow out of sclerotia.

The nutritional physiology of EM-fungi, in contrast to that of VAM-fungi, is relatively well known because they can be cultivated axenically. They need simple carbohydrates as carbon sources: growth is equally good on sucrose, maltose, cellobiose, xylose and arabinose. Galacturonic acid is not a growth substrate for typical EM-fungi. Because the degradation of cellulose and lignin plays no nutritional role in the fungi, this does not mean that penetration and formation of the Hartig Net is achieved without the help of pectinases. Contradictory results concerning the stimulatory and/or inhibitory effect of amino acids and other chemical reagents on growth were caused by the use of concentrations far too high for mycorrhizal fungi as inoculum. Purely saprophytic species can, however, tolerate very high substrate concentrations. The phosphate concentration of 50 mM used in many media, is in many cases already sufficient to inhibit growth of EM-fungi. The EM-fungi, as fungi in general, prefer ammonium as the nitrogen source. Some species can also reduce NO_3^-. The degradation of amino acids is very different between species. *Paxillus involutus* grows equally well on glutamate or arginine instead of ammonium sulphate; however, there is no growth with leucine, tyrosine or histidine. *Lactarius rufus*, on the other hand, cannot use glutamate. Whether these differences are caused by the presence/absence of specific carriers or catabolic enzymes, is not known. Thiamine or intermediates leading to thiamine are essential for many EM-fungi.

Especially important for the symbiotic interaction is the production of plant hormones by EM-fungi. Growth on tryptophan as single carbon source results in the production of indole acetic acid (IAA), or indole propionic acid or indole butyric acid in a few species. However, this does not prove that auxins play an essential role *in situ*, since tryptophan is not available at the root surface in the same concentrations as used in cultivation experiments. However, direct measurements of IAA concentration in mycorrhizal roots of *Pinus echinata* (with *Pisolithus tinctorius*) revealed a threefold higher concentration compared to non-mycorrhiza short roots (Mitchell *et al.*, 1986). This has been generally confirmed with Norway spruce (*Picea abies*) inoculated with a number of different mycorrhizal fungi (Hock *et al.*, 1988). When inhibitors of plant IAA oxidase are also present, levels of IAA can be reached where root growth is influenced. This takes the form of inhibition of root hair formation, branching of roots and a morphological restructuring of the outer root cortex cells. *In situ*, these phenomena are not caused by auxin alone, inasmuch as EM-fungi also produce cytokinins, gibberellins and ethylene. With *Pinus mugu* mycorrhiza inoculated with *Laccaria laccata*, significantly higher ethylene concentrations were found than in the control. As aminocyclopropane-carboxylic acid, a direct precursor of ethylene biosynthesis in higher plants, does not stimulate ethylene production in the fungus *Laccaria laccata*, and an inhibitor of plant ethylene production does not inhibit ethylene biosynthesis in the fungus, it has been assumed that ethylene biosynthesis in the two symbiotic partners may follow different pathways (Mudge, 1987). Other experiments have indicated that auxins from the fungal partner could stimulate ethylene production by the host plant. Microorganisms casually associated with EM-fungal hyphae may also excrete phytohormones, vitamins and amino acids, which could have effects on both microsymbiont and host (Fig. 11.6). This is not so in cases where EM-species produce antibiotics. *Amanita muscaria* and *Pisolithus tinctorius* produce no antibiotics, but *Leucopaxillus* (*Clitocybe*) *cerealis* produces diatetryn-nitrile (a polyacetylene) which has both fungistatic and bacteriostatic properties.

11.1.4 Exchange between symbiont and host

The most important process between host plant and microsymbiont is the exchange of carbon- and energy-sources for essential mineral nutrients. This is the case in EM-fungal symbiosis as well as in VAM-fungal symbiosis. In the establishment and structural maintenance of the symbiosis, exchange of nutrients and phytohormones may depend on each other (Nylund, 1988).

(a) Carbohydrate

Whereas sucrose is the main transport form of photosynthate in plants, glucose and fructose appear to be the metabolites transferred to the fungus.

Table 11.5 Distribution (%) of photosynthetically fixed $^{14}CO_2$ in various compartments of *Pinus taeda* seedlings at different ages: comparison of mycorrhizal (MYC) and non-mycorrhizal (NM) plants. (Based on Reid *et al.*, 1983)

Seedling age and status	*Shoot*			*Root*			*Total ^{14}C fixed (10^3 dpm)*
	Needles	*Stem*	*Respiration*	*Woody*	*Short MYC*	*Respiration*	
2 months MYC	28.6	27.3	5.5	32.2	0.1	6.0	77
2 months NM	26.9	53.0	3.8	13.2	–	2.8	270
4 months MYC	36.8	30.1	6.3	14.6	0.1	11.8	856
4 months NM	32.4	48.0	4.8	12.0	–	2.4	587
6 months MYC	15.7	13.9	5.7	33.5	0.1	30.4	1528
6 months NM	23.8	42.5	5.2	20.4	–	7.6	698
10 months MYC	29.2	25.4	6.9	14.3	0.1	23.5	4222
10 months NM	28.7	38.6	6.1	14.3	–	11.6	1701

Glucose taken up by the hyphae is then further processed to trehalose and glycogen, fructose to mannitol. This is supported by the finding that invertase activity is increased in mycorrhizal roots. The cleaving of sucrose occurs either in the host cells or in the extracellular boundary layers. Export from the host cell is passive, but uptake by the fungus is active. The concentration gradient remains stable in favour of the fungus due to immediate phosphorylation of carbon skeletons in the fungal cytoplasm. Mannitol increases export of glucose and fructose due to its inhibitory effect on plant glucokinase and fructokinase. Thus the sugar pool is increased and diffusion outwards increases. However, it must be noted that the effect of mannitol on fungal enzymes is unknown. Apart from sugar transfer, amino acids and organic acids are taken up by hyphae. This is especially important in those parts of the mycorrhiza where intact hyphae are in contact with dying or disintegrating plant cells. The measurement of photosynthetically fixed $^{14}CO_2$ in organs of mycorrhizal and non-mycorrhizal Scots pine showed that in symbiotic roots very little ^{14}C was immobilized (Table 11.5), but was respired at a higher rate in both lignified and non-lignified parts of the root system. Thus the symbiosis represents a higher 'sink', a property noticeable in the whole root system.

During the differentiation from the saprophytic to the symbiotic stage, the **protein content in the cell wall decreases** significantly. This could be also produced in cultivating the fungus (*Cenococcum geophilum*) at a low sugar concentration, indicating a **carbon limitation** of the symbiotic stage (Paris *et al.*, 1991). A surplus of externally supplied carbohydrates increased the growth of the fungi too much and disturbed the balance of the symbiosis (Duddridge *et al.*, 1988). Photosynthetic capacity is nearly doubled in 6–10-month-old *Pinus taeda* plants infected with EM-fungi (Table 11.5). Mycorrhization results in a shift from the glycolytic to the oxidative pentose phosphate pathway (Namysl *et al.*, 1988). Differences in respiration rates caused by mycorrhiza are strongly affected by species and age. Mycorrhization increases the longevity but reduces the growth rate of the roots. Branching in roots also has a large effect on respiration. Mycorrhizal respiration (from *Fagus*) is not inhibited by inhibitors of electron transport to cytochrome oxidase (cyanide, azide). On the contrary, rates may be stimulated by this treatment. This is explained by the fungi possessing a functional by-pass from ubiquinone to oxygen, which is not affected by such inhibitors.

(b) Mineral nutrients and water uptake

Opposite to the situation in VAM-symbiosis, in EM-symbiosis all mineral nutrients must be transported through the hyphal coat before they reach the root. Thus phosphate uptake is only three to five times more efficient in EM-roots as in normal roots. Of the phosphate taken up 80–90% is found

Table 11.6 Phosphate uptake and storage in mycorrhizas of *Fagus sylvatica*. (Based on Harley and McCready, 1952, and Harley and Smith, 1983)

	Phosphate concentration (M)			
	3×10^{-8}	3×10^{-7}	3×10^{-6}	3×10^{-5}
Uptake by 1200 mg mycorrhizal roots (μg phosphate d^{-1})	3.3	32	266	967
Amount stored in fungal sheath (μg)	3.0	29	240	870
Amount absorbed into core (μg)	0.3	3	26	100

first in the hyphal coat where it is stored as polyphosphate (Ashford *et al.*, 1986). Only by non-ecologically high P concentrations can passive P transport to the root be observed. Roots can, however, participate in the mobilization of stored phosphate in mycorrhiza after the free phosphate concentration in hyphae is reduced. Thus the coat functions as a phosphate buffer. At concentrations of 3×10^{-6} M P, mycorrhiza (1.2 g dm^{-3} soil) can take up over 250 μg phosphate in 24 h. At P concentrations of 3×10^{-8} M this value was only 3 μg phosphate (Table 11.6). Polyphosphate granules contain high concentrations of calcium, and developmental accumulation of this element closely follows that of phosphate. Because phosphate in soil is normally organically bound, phosphatase and phytase activities in mycorrhizas are especially important. Phosphatase activity in *Fagus* mycorrhiza is eight times that found in non-infected roots.

A case study with inoculation of *Eucalyptus diversicolor* with *Laccaria laccata* gave a low frequency of infection with no added phosphate, an optimum with 2–4 mg P kg^{-1} of soil and no infection above 36 mg P (Bougher *et al.*, 1990). In loblolly pine, response of photosynthesis to low and medium levels of mycorrhiza infection is attributed to a better phosphate supply. However, at high levels of mycorrhization other mechanisms are postulated. A maximum net photosynthesis rate of 11 mg CO_2 h^{-1} g^{-1} was found in non-mycorrhiza plants with 8–12 μg P g^{-1} added and a similar rate at 0.1 μg P g^{-1} and high mycorrhization (Rousseau and Reid, 1990). A simultaneous high level of phosphate fertilization and high mycorrhization resulted in a 12% decrease of the biomass (with *Pinus taeda* seedlings) (Rousseau and Reid, 1989a).

The rate of nitrate uptake by mycorrhizal and non-mycorrhizal plants of *Pinus pinaster* was the same with 0.20 μmol nitrate absorbed h^{-1} g fresh $weight^{-1}$. Also xylem transport of reduced nitrogen was similar whereas the rate of phloem transport was doubled in the mycorrhizal plants (Scheromm *et al.*, 1990).

Ammonium assimilation is achieved mostly through glutamine synthesis.

This involves high levels of dark-fixation of CO_2, needed for the PEP-carboxylase reaction with produces organic acid intermediates used in ammonium fixation. Despite this, the amino acid compositions of infected and non-infected roots in *Pinus sylvestris* are very similar; 68–73 μmol arginine, 38–41 μmol asparagine, and 23–25 μmol glutamine g dry weight^{-1}. Amounts of all other free amino acids are very small (Krupa and Bränström, 1974).

Ectomycorrhizal mycelia also transport water to the trees. In experimental studies with pine and birch seedlings infected with *Suillus bovinus* and *Paxillus involuta*, the separation of these external mycelia from the seedlings reduced the transpiration rate of the plants by more than 40% (Boyd *et al.*, 1986).

11.1.5 Ecological distribution

Ectomycorrhizas are widespread in deciduous and evergreen forests of the northern hemisphere, the mountainous regions of Africa and South America and the *Eucalyptus* and *Nothofagus* forests of Australia and New Zealand. More than half the afforested surface of the Earth is on poor soil. For these 20 million km^2, this symbiosis is an important factor in colonization and stability of forest ecosystems (Malloch *et al.*, 1980).

Ectotropic mycorrhiza are also abundant in the forests of Mediterranean lands. *Quercus* and *Pinus* species are widespread, e.g. *Quercus ilex*, *Quercus suber* (cork trees), *Quercus coccifera*, *Pinus pinea*, *Pinus pinaster* and *Pinus radiata* (originally imported from California). EM-trees are dominant on the tree-line of subalpine areas, in Europe, *Pinus mugo*, *Pinus cembra* and *Larix decidua*, in the Himalaya *Pinus griffithii*, *Abies webbiana*, *Larix griffithiana* and *Betula utilis*, and in North America *Pinus contorta*, *Pinus flexilis*, *Picea engelmanii* and *Abies lasiocarpa*. On the tree-line of the Asian steppes the EM-trees which dominate the flora include deciduous *Quercus robur* and *Populus tremula* and in Alaska *Populus tremuloides* prevails. Obligate EM-species mainly belong to the genera *Pinus*, *Picea*, *Abies*, *Larix*, *Quercus*, *Fagus* and *Carpinus*, whereas facultative EM-species belong especially to the genera *Acer*, *Alnus*, *Betula*, *Corylus*, *Cupressus*, *Eucalyptus*, *Juniperus*, *Salix* and *Ulmus*.

The distribution of EM-fungi reflects the incidence of their host trees (Dighton *et al.*, 1986). In pure stands of one tree species a succession of fruiting bodies can be observed. Thus in a new birch plantation *Hebeloma* spp. are seen in the first few years. After 6 years *Leccinum* spp. dominate, to be succeeded after 10 years by *Russula* spp. (Table 11.7). During the development of *Pinus contorta* trees a characteristic succession of fungi has been observed by their fruiting bodies: *Paxillus–Laccaria–Cortinarius/Inocybe* to *Russula* spp. VA-mycorrhiza precedes the ectomycorrhiza in those trees where both forms can be present, as in *Eucalyptus* sp. (Chilvers

Table 11.7 Succession of fruiting bodies in a stand of birches. (Based on Last *et al.*, 1983)

Years after planting	*Fungi*
1	–
2	*Hebeloma crustuliniforme*, *Laccaria* sp.
3	*Thelephora terrestris*
4	*Hebeloma fragilipes*, *H. sacchariolens*, *H. mesophaeum*, *Inocybe lanuginella*, *Lactarius pubescens*
6	*Cortinarius* sp., *Hebeloma leucosarx*, *Hymenogaster tener*, *Inocybe petiginosa*, *Leccinium roseofracta*, *L. scabrum*, *L. versipelle*, *Peziza badia*, *Ramaria* sp.
7	Other *Cortinarius* spp., other *Hebeloma* spp., *Lactarius glyciosmus*, *Leccinium subleucophaeum*
10	*Hebeloma vaccinum*, *Russula betularum*, *R. grisea*, *R. versicolor*

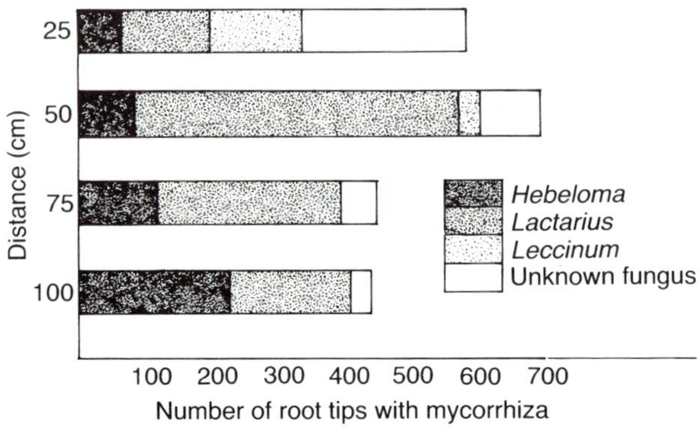

Fig. 11.9. Number of root tips with four different mycorrhizal fungi, at 25–100 cm distance from the stem of *Betula pubescens*. (Based on Deacon and Donaldson, 1983.)

et al., 1987). In mixed forests where young trees continually spring up, it is more difficult to observe such successions and a more mixed spectrum of species is stably produced. Using *Betula pubescens* it has been observed how mycorrhiza-fungi infecting side roots change in accordance to the distance from the tree trunk (Fig. 11.9). At 25 cm from the trunk 10% of roots were infected with *Hebeloma* spp. At a distance of 100 cm the infection rate was increased to over 50%. *Lactarius* spp. reached an optimum at 50 cm distance whereas *Leccinum* mycelium could not be found 75–100 cm away. Also with increasing soil depth the percentage of mycorrhizal root tips was affected differently according to the tree species. Using *Fagus sylvatica* 70% of possible infection sites were occupied even at a depth of 2 m. This figure was only 30% on *Pinus sylvestris* roots (Fig. 11.10).

Hypogeous macrofungi, such as *Mesophelia* sp. and *Castorium* sp., develop their sporocarps in close connection with the host plant root (e.g. *Eucalyptus diversicolor*). A single sporocarp of *Mesophellia trabbalis* can contain up to 5 m of roots with a surface of 45 cm^2 (Dell *et al.*, 1990). To observe these and other forms of ectomycorrhiza *in situ* in undisturbed soil a 'Root Window Method' has been developed and especially designed to contribute to the relationship between forest damage and air pollution (Egli and Kälin, 1988) (section 11.1.6c).

The best form of humus for EM-fungi is a slightly acid leaf mull with a low P and N content but good aeration and reasonable moisture content. Increasing soil moisture above a certain level results in reduced mycorrhiza formation with different widely used fungi (Bougher and Malajczuk, 1990).

11.1.6 Agricultural and Practical Uses

Ectomycotrophic trees are among the most important sources of wood worldwide. More than 3 billion m^3 of wood is felled annually. This has a marked value of over 150 billion US dollars. New plantations may benefit considerably from EM-fungus inoculation, on virgin ground where the EM-fungi are not already present.

Another application of mycorrhiza research is the cultivation of the involved fungi, especially those with a high commercial value. However, despite a ten-year effort, the cultivation of the Burgundy truffle (*Tuber unicinatum*) has not been successful (Chevalier and Frochot, 1988). With the

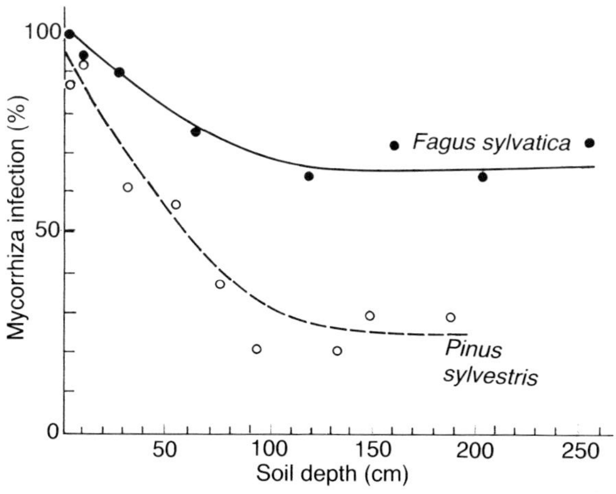

Fig. 11.10. Decrease in mycorrhiza infection in *Fagus sylvatica* and *Pinus sylvestris* with increasing soil depth. (Based on Lyr, 1960.)

desert truffle, *Terfezia leonis*, however, a significant improvement of the association with a host plant (*Helianthemum sessiflorum*) has been achieved in culture by addition of charcoal (Roth-Bejerano *et al.*, 1990).

(a) The Production and Use of Inocula

The simplest and most widespread method used to inoculate new areas is the digging in of soil containing infected roots (Marx and Cordell, 1989). For large-scale operations in distant lands, however, this can lead to high transport costs. The soil must also be protected from overheating and desiccation, since most EM-fungi are rather sensitive to these factors. Other adverse factors are technical inasmuch as quarantine regulations in many countries forbid the import of non-sterilized soil due to the risk of pathogenic infections. In such cases pure cultures of mycorrhizal fungi must be used. This can be achieved using either mycelium or spores. Seedlings are first infected and then planted out between other trees, so that direct infection occurs through the soil. This involves a selection pressure on the fungus which produces mycelium that is best for the soil and climate.

Pisolithus tinctorius has been intensively researched as inoculum because of its wide host spectrum (over 50 species of trees) (Marx *et al.*, 1984, 1989a; Marx and Bell, 1985). In competition with the endogenous soil microflora the fungus can colonize 50–70% of all side roots of *Pinus taeda*, *Pinus ponderosa* or *Quercus rubra* (Table 11.8). This results, even in young plants, in significant increases in plant height, biomass and trunk diameter. Sometimes there is a lag period after successful infection when the loss of assimilate from the plant is not compensated for by better mineral nutrition. *Pisolithus tinctorius* is normally found on slag heaps, rubbish dumps and other poor bush land where hardy pioneer species grow. *Pisolithus tinctorius*

Table 11.8 Growth and ectomycorrhizal development in *Pinus* species and *Quercus rubra* after inoculation with *Pisolithus tinctorius* (Pt) (Based on Marx *et al.*, 1984)

Tree species	*Inoculum*	*Height (cm)*	*Stem diameter (mm)*	*Fresh weight (top + root) (g)*	*Short roots with ectomycorrhiza (%)*	
					Pt	*All fungi*
Pinus taeda	Pt	26.8	5.4	20.9	43	56
Pinus taeda	–	24.0	4.2	11.1	0	39
Pinus ponderosa	Pt	24.2	7.5	37.9	25	30
Pinus ponderosa	–	22.9	7.3	32.8	0	10
Quercus rubra	Pt	40.2	7.7	34.1	17	31
Quercus rubra	–	32.1	6.8	25.2	0	26

can tolerate very low pH values, dry periods, high concentrations of heavy metals and a high soil temperature. In pure culture it can grow at 40°C, the hyphae dying at 45°C. At the optimal temperature of 30°C it grows rapidly on all normal fungal media in solid or liquid states. The yellow–green colour of the hyphae mean that mycorrhiza development on roots can be easily followed by the naked eye. The plantation of trees in extreme areas can be considerably improved by inoculation with this fungus. However, tree nurseries specializing in forest species prefer to inoculate the samplings with *Telephora terrestris* which is well suited to the soil conditions of nurseries. In cases where such trees are to be used to stock poor areas, a second inoculation with fungi, such as *Pisolithus*, is especially important. Inocula are produced by letting *Pisolithus tinctorius* grow at room temperature on a mixture of vermiculite and humus which has been soaked with a medium containing glucose. The inoculum is then dried to 15–20% moisture content at 25–30°C. Mycelium in such preparations remains infectious for several weeks.

There has been less testing of the direct inoculation of seeds with basidiospores of *Pisolithus tinctorius* or other ectomycorrhizal fungi. Seeds are encapsulated in a carrier material containing 1 mg ($=10^6$) basidiospores. This may also contain a fungicide to which the EM strain is resistant. After sowing the encapsulated seeds must be watered daily in order to set the spores free and improve seed germination.

A vegetative inoculum with *Pisolithus tinctorius* produced a better mycorrhization index than the use of basidiospores (Marx and Cordell, 1989b). Only very high spore inoculum (at $1.8\,g\,m^{-2}$) resulted in infection indices above 50% (Marx and Cordell, 1990a).

Preservation of mycorrhizal fungi in liquid nitrogen can be used for almost all fungi (except the very slow growing). They are grown on a suitable medium with the addition of 5% glycerol. Then a small piece of PVC tube (a 'straw') is pressed several times into the fungal colony, until it is filled, placed into a cryovial, frozen at −80°C and transferred to liquid nitrogen. For regrowth, the straw is placed directly on an agar medium for rapid thawing (Hoffmann, 1988). Good results for preservation have been also obtained with mycelia or spores transferred to pure sterile water and kept at room temperature (Heinonen-Tanski, 1988).

(b) Ectomycorrhiza and root diseases

Protection against root pathogens is more effective in EM-roots than in VAM-roots (section 10.3.5(b)). The survival rate of *Pinus taeda* was significantly increased after infection with *Rhizoctonia solani* when the plant was in symbiosis with *Pisolithus tinctorius*. *Mycelium radicis atrovirens* (*Phialocephala dimorphosphora*) caused growth inhibition and chlorosis in *Picea mariana*. This can be totally cured by simultaneous infection with the

EM-fungus *Suillus granulatus*. Infections by pathogens can be reduced by EM-fungi by the following mechanisms:

1. The excretion of antifungal and antibiotic substances;
2. The stimulation of other microorganisms which themselves inhibit pathogens;
3. The production of antibiotics by the host, under the control of the symbiont;
4. The reduction of carbon- and energy-sources on the root surface to the extent that spores of pathogens can no longer germinate;
5. The excretion of siderophores, resulting in an iron deprivation of the pathogen; such limitation is known for bacteria;
6. The structural protection of the root by the thick fungal coat.

EM-fungi may use one or many combinations of these strategies (Zak, 1964; Meyer, 1974). Thus antibiotics are produced by 80% of the *Tricholoma* sp. but none of the 40 *Russula* sp., *Cortinarius*, *Hebeloma* and *Lactarius* have antibacterial activities; *Boletus* and *Clitocybe*, on the other hand, have antimycotic activities. Very few antibiotics from EM-fungi have been chemically identified or their mode of action explained. Development of mycorrhizal roots of birch results in a symbiosis-specific enrichment of non-pathogenic organisms like *Penicillium* in the rhizosphere which leads to a decrease in the number of pathogens, especially *Pythium* and *Fusarium* (Katznelson *et al.*, 1962). The carbon concentration at which balance between uptake and export of carbohydrate is achieved, the C_{min}-value, has hardly been investigated for mycorrhizal fungi. Roots infected with EM-fungi contain up to eight times more volatile terpenes and sesquiterpenes than normal, non-infected roots. Monoterpene inhibits growth of pathogens like *Phytophthora cinnamomi* (Krupa and Nylund, 1972). EM-fungi may enhance root phytoalexin production, but this is still speculative at present.

The number of bacteria in association with mycorrhiza roots is 10–100 times larger than in non-mycorrhizas (Kilbertus *et al.*, 1978). However, the number of bacteria in the inner part of the mantle is low. The microflora associated with the ectomycorrhizosphere can suppress root rot, i.e. infection by *Phytophthora cinnamomi* (Malajczuk, 1979).

The closed mycorrhizal coat around the root forms a very effective protection. *Pinus taeda* and *Pinus echinata* with the coat remain uninfected by *Phytophthora cinnamomi*, whereas the removal of the coat results in rapid pathogenic infection.

(c) Ectomycorrhiza and air pollution

Trees with an obligate ectomycorrhiza, such as *Picea* sp., *Abies* sp., *Quercus* sp. or *Fagus* sp. are more sensitive to pollution effects than trees with an endomycorrhiza or without mycorrhization (Meyer, 1982). Several mycorrhiza-forming fungi and mycorrhizas are very sensitive bioindicators of air pollution (Dörfelt and Brawn, 1980; Meyer, 1984; Schlechte, 1986; Marx and Shafer, 1989b). *Russula mustelina* in spruce climax forests has been found to be a characteristic signalling bioindicator (Fellner, 1988).

A survey of the macrofungi of forests in the Netherlands showed that, over the last decade, almost half the 126 species of ectomycorrhizal fungi were reduced, whereas the saprophytic fungi were not significantly changed (Arnolds, 1988). A reduction in the mycorrhiza fungi precedes the obvious symptoms of tree damage by pollutants on both Ca-rich and Ca-poor soils (Schmitt, 1987).

Some ectomycorrhizas seem to be tolerant of O_3, whereas others are more tolerant to SO_2 (Garrett *et al.*, 1982). Any factor combination which reduces photosynthate allocation to the root will affect this symbiosis (Winner *et al.*, 1987). But also factors changing the levels of heavy metals and pH in the soil can directly affect the ectomycorrhizas (Wasserman *et al.*, 1987). Distinct proteins in *Pisolithus tinctorius* and *Cenococcum geophilum* may be responsible for the heavy metal tolerance of these fungi (Morselt *et al.*, 1986). However, at higher concentrations of Cu, Zn or Pb in the soil, the number of active mycorrhiza root tips decreases (Bell *et al.*, 1988). Since trees, such as *Picea abies*, damaged by pollution can also suffer from a limited supply of K, Ca and Mg (Krivan *et al.*, 1986), a damaged ectomycorrhiza may enhance the symptoms by a changed supply of these elements. Direct fertilization with K, Ca and Mg was found to increase the content in needles of damaged stands of spruce trees significantly and to reduce stress symptoms (Isermann and Henjes, 1986). In general, it can be concluded that a damaged ectomycorrhiza aggravates nutrient deficiencies and water stress, and thereby intensifies the symptoms of pollutant-damaged trees (Meyer, 1989).

11.2 ECT-ENDOMYCORRHIZA IN CONIFERS

This type of mycorrhiza is characterized by some properties described above for ectomycorrhiza. In ectomycorrhiza hyphae only enter dead host cells. In ect-endomycorrhiza, however, there is also an infection inside living host cells. Ect-endomycorrhizae remain viable over the year in infected cortex cells, thus reaching a similar active age as with EM infections. Ect-endomycorrhizas are restricted mostly to *Pinus* (Hofstein, 1969) and *Picea*, and to a less extent *Larix*. The fungi are probably ascomycetes. Fungal hyphae penetrate the cell walls by apressoria. They then grow into the cell lumen. Meristem cells are not infected. In older roots ect-endomycorrhizas may be out-competed by EM-infections. Experiments have been reported in which the same fungal species formed an ect-endomycorrhiza on *Pinus* and an ectomycorrhiza on *Picea*.

11.3 MYCORRHIZA OF THE ERICALES

Mycorrhiza in this plant order range from ecto-types through ect-endomycorrhiza to pure endo-associations. These are highly specialized forms of mycotrophy which enable the colonization of soil which is especially poor in N and P. This explains the wide distribution of Ericaceae in inhospitable environments such as heath land, bogs and evergreen forests all over the Earth.

11.3.1 Ericaceae

Symbiotic development has been best researched in the genera *Calluna*, *Vaccinium* and *Rhododendron*. These members of the Ericaceae possess especially thin roots, often called hair roots, with a diameter between 40 and 100 μm. Mycorrhiza hyphae grow on the root surface then grow between rhizodermis cells and penetrate cell walls. At this stage hyphae are surrounded by a matrix material (a 'collar') and are separated from host plasmamembrane and cytoplasm. This is an infection similar to that

Table 11.9 Fungi isolated from roots of *Calluna vulgaris* and *Vaccinium myrtillus* by direct plating or after maceration. (Based on Pearson and Read, 1973, 1975)

Fungi	*Occurrence of species (cases)*			
	Calluna vulgaris		Vaccinium myrtillus	
	Direct plating	*Maceration*	*Direct plating*	*Maceration*
Trichoderma sp.	41	3	33	4
Mucor ramannianus	30	0	26	3
Penicillium expansum	27	4	29	6
Mortierella sp.	11	0	14	0
Cephalosporium sp.	4	0	2	0
Aspergillus sp.	6	0	17	2
Fusarium sp.	3	1	5	0
Cladosporium sp.	1	0	1	0
Oidiodendron sp.	1	1	0	0
Unidentified light-coloured mycelia	12	0	3	0
Unidentified fast-growing dark mycelia	16	2	4	2
Slow-growing dark mycelia with segmenting hyphae	3	97	1	89
Slow-growing dark mycelia without segmenting hyphae	4	94	4	85

described for VAM (Chapter 10). Hyphae grow and branch in the host cell to form tight coils. At this time host cell cytoplasm also increases and surrounds all parts of the coils. A swollen nucleus and active cell organelles in the rhizodermis cell suggest that an intensive metabolic exchange is taking place. All infected rhizodermis cells have their own hyphal connections to the soil or growth medium. Approx. 200 such hyphae can be observed per 1 mm of a *Calluna* root.

However, such fungi are not dominant on the root surface. Other soil fungi, such as *Trichoderma*, *Mucor ramannianus*, *Penicillium expansum* and *Aspergillus* sp. (Table 11.9) dominate the microflora. In order to identify the dark mycelium, with or without septa, roots have to be macerated. Koch's postulates (absence of other infecting microorganisms) have been satisfied by the reinfection of sterile seedlings and fruiting body formation, which show that one of these mycelia is in fact *Pezizella ericae*. The basidiomycete *Clavaria vermicularis* is also often found in close association with Ericaceae, but it is still unclear if this is also a mycorrhiza fungus. Infection by dark septate hyphae has also been observed in *Carex* species (*Carex curvula* and *Carex firma*) (Haselwandter, 1987). Reinfection of seedling with the isolated fungi led to an increase in shoot phosphorus.

Ericaceae can also grow without mycorrhiza. In poor soil, however, there is a clear increase in growth rate upon infection. Mycorrhizal *Vaccinium macrocarpon* and *Calluna vulgaris* have 50–100% higher N-content than non-infected plants. Differences in P-content are less pronounced (Fig. 11.11). An improved NH_4^+-uptake by mycorrhizal plants is explained not only by the increased active surface area involved, but also by the fact that

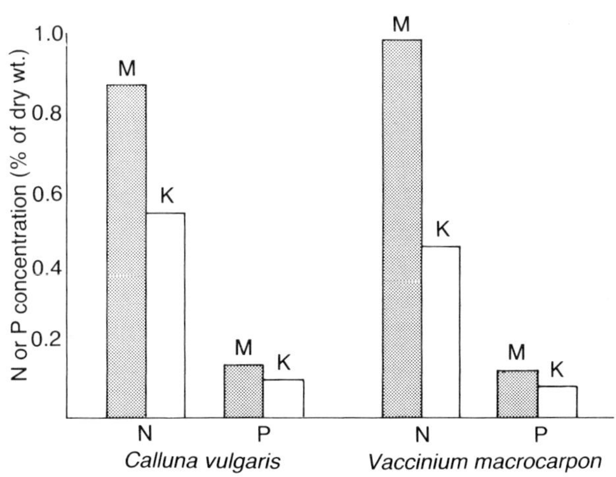

Fig. 11.11. N-content and P-content in 6-month-old plants of *Calluna vulgaris* and *Vaccinium macrocarpon*. M, plants with mycorrhiza; K, controls. (Based on Read and Stribley, 1973.)

fungal NH_4-uptake systems have a higher affinity (lower K_m value) for NH_4^+ than the plant system. This difference has been demonstrated in *Neurospora*, but not yet in these mycorrhizal fungi. The pathway of ^{14}C-labelled photosynthate through the plant leading to trehalose and mannitol in the fungus is similar to that described for the ectomycorrhiza.

Basically the mycorrhiza of *Rhododendron* is comparable with the ericoid mycorrhiza of *Calluna*. The structure of infected roots is essentially the same as that of uninfected roots. Immature tissues around the meristem and some tissue beyond the root tip do not contain fungi. Occasionally infected pericycle cells are observed, but the fungus is usually restricted to non-sterile tissue. High concentrations of phenolic compounds have been found in the epidermis, cortex, and pericycle, only immature cortical cells being free of phenolics. Obviously these phenolic substances do not prevent infection; no correlation could be shown between the accumulation of phenolic compounds and fungal infection. In contrast to findings in *Calluna*, at least two fungi were observed in *Rhododendron* roots, with different types of septa. Penetration of the epidermal cells was found both from the exterior and from one cell to another (Peterson *et al.*, 1981).

In ericoid mycorrhizas the formation of a hyphal weft around the roots is followed by penetration of the epidermal cells. The presence of an electron-lucent zone in advance of the invading fungus suggests that penetration is by enzymatic activity. Fungal isolates were all ascomycetes, one of which could be identified as *Pezizella ericae* Read. Degeneration of host cytoplasm, which usually occurs within 7 weeks, precedes fungal degeneration, which can be identified by vacuolization and the reduction of the cytoplasmic contents. In contrast to orchidaceous and VA mycorrhizas where the host cell remains viable after lysis of the fungus and can be reinfected, in ericoid mycorrhizas the host cell dies after infection. This may be either the result or the cause of the breakdown of the association (Duddridge and Read, 1981).

Of the total nitrogen present in a typical moor humus soil, a habitat of ericaceous plants, 71% is hydrolysable organic N, whereas only 0.4% is present as exchangeable or non-exchangeable NH_4^+. The ericaceous fungi, like those of ectomycorrhizas, are unable to use the complex humic N compounds directly. Hence their significance seems to be the ability of utilizing the simpler organic compounds or their breakdown products. Indeed, little is known about the occurrence or behaviour of these substances in organic soils. In addition to the release of amino acids from humified organic materials, these substances are released from freshly fallen litter. It has been shown that in pure culture the mycorrhizal endophyte can utilize glutamic acid and alanine as effectively as it can use ammonium or nitrate (Pearson and Read, 1975; Stribley and Read, 1980). Ericoid mycorrhizal fungi are more effective in degrading lignin than fungi forming an ectomycorrhiza (Haselwandter *et al.*, 1990). Mycorrhiza infection in

Calluna vulgaris increases the iron content in shoots by 225% (in the presence of sufficient calcium) (Leake *et al.*, 1990).

11.3.2 *Arbutus* and *Monotropa*

The mycorrhiza of *Arbutus* and related genera like *Arctostaphylos* has the typical structure of an ectomycorrhiza: a coat 20–80 μm thick and a fully developed Hartig net, from which hyphae force their way into cells of the rhizoderm and form coils. The symbiosis then combines the properties of ecto- and Erica-mycorrhizas. Symbionts include the basidiomycetes *Cortinarius zakii*, *Cenococcum graniforme*, *Laccaria laccata*, *Telephora terrestris* and *Poria terrestris*. These are typical EM-fungi.

Monotropa (Ericales) represents an end point in mycotrophic development. These chlorophyll-free holosaprophytic plants use their obligate mycorrhiza not only to accumulate minerals, but also to tap into the carbon and energy pools of nearby roots of conifers or Fagaceae.

Side roots exhibit a fungal coat and a Hartig net. From the outer hyphal coat extra haustoria develop, which force their way into cells of the rhizodermis (Fig. 11.12). At first the cell wall surrounds the haustorium, which, in conjunction with an inward growth of the wall, takes on the characteristics of a transfer cell. During blossoming and seed setting the haustoria burst, and empty their contents into the host cell cytoplasm. [^{14}C]Glucose has been transferred experimentally from neighbouring trees to *Monotropa* plants. This transfer did not result in the labelling of nearby symbiotic Ericaceae. ^{32}P injected into *Monotropa* was also recovered in neighbouring trees. Mycelium isolated from *Monotropa* can produce normal ectomycorrhiza on *Pinus* roots. Thus it is proven that both developmentally and by metabolic exchange, the mycorrhiza provides the path of nutritional transport between trees and holoparasites.

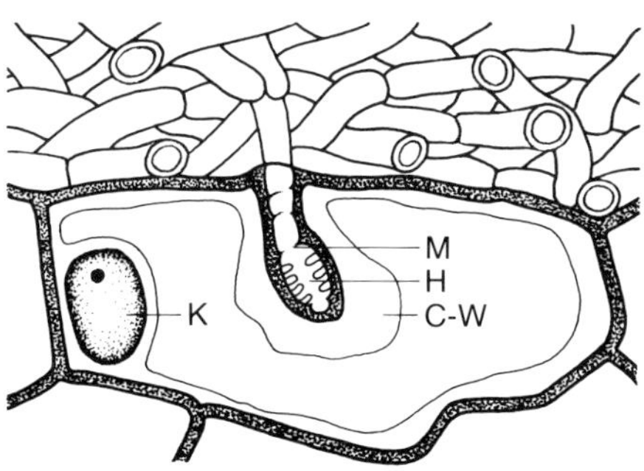

Fig. 11.12. Mycorrhiza in *Monotropa hypopitys*. H, swollen fungal haustorium; C-W, host cytoplasm; M, matrix (transformed cell wall); K, host nucleus. (Based on Francke, 1934.)

The typical features of arbutoid mycorrhizas are consistently present in the six species of *Pyrola* examined (Robertson and Robertson, 1984). No organized sheath was ever observed in this work though there are reports of the presence of a sheath in *Pyrola rotundifolia* and *Pyrola picta*. Obviously this is a variable feature and should not be used to characterize arbutoid mycorrhizae. Penetration frequently occurs through the side walls. The process of intracellular development in *Pyrola* seems to be similar to that described in *Arbutus*, *Calluna*, *Rhododendron* and *Vaccinium*. The presence of an interfacial matrix with a honeycomb appearance in *Pyrola secunda* has provided additional support for the concept that the matrix is of host origin. Apart from *Pyrola secunda* which is associated with an ascomycete, *Pyrola* mycorrhizae are usually formed with basidiomycetes (Robertson and Robertson, 1984).

Arbutus unedo L. mycorrhiza is slightly different from the typical arbutoid type; it is characterized by a few hyphae which extend along the surface of the root without forming a true mantle. The associated fungus is an ascomycete (Fusconi and Bonfante-Fasolo, 1984).

11.4 ORCHID MYCORRHIZA

Due to their very small seeds (0.3–14 μg per seed) many orchids depend for germination in the soil on carbon and energy supplies from fungal hyphae and are obligate mycotrophs during this stage. In later stages the endomycorrhiza in photosynthetic orchids provide the supply of minerals. Heterotrophic species remain obligately mycotrophic, using the fungi as channels to utilize carbohydrates from the soil (Burgeff, 1932).

11.4.1 Host plants and microsymbionts

The more than 20 000 species of orchids are mainly distributed in the tropics and subtropics, growing as epiphytes. In Middle Europe only about 90 species of soil orchids are present, especially from the genera *Orchis*, *Ophrys*, *Epipactis* and *Dactylorhiza* (Danesch and Danesch, 1962). These orchids can grow on very different soil types, to which the mycorrhizas are also adapted: *Epipactis atropurpurea* colonizes calcareous, dry soils, *Ophrys apifera* calcareous, moist habitats, *Liparis loeselii* moorland and *Goodyera repens* coniferous forests with mosses.

The morphological and physiological adaptation of epiphytic orchids are especially diverse and the relationship to the symbiosis is not known. Leaves and stems can be completely reduced and the roots modified to assimilatory organs as in *Campylocentrum* and *Taeniophyllum*. The rhizome can be reduced to a large extent (*Cattleya* and *Dendrobium*) and in many other genera the leaves are modified to organs for water storage. No other plant family has such an abundance of flower forms (Adams, 1959).

The mycorrhizal fungi in orchids are Basidiomycetes. Identified species include *Pellicularia filamentosa, Thanatephorus cucumis* (previously described as *Rhizoctonia solani*, an imperfect form), *Marasmius coniatus, Xerotus javanicus, Sebacina vermifera, Tulasnella calospora, Armillaria mellea* and *Corticium catonii* (Warcup, 1975). *Tulasnella calospora* (*Rhizoctonia repens*) is considered to be a universal orchid symbiont. Some of these species are phytopathogens, such as *Armillaria* in conifer trees. How orchids manage to modify the pathogenic character of these fungi is an open interesting question.

Some orchids have been found to be more receptive than others to a variety of endophytes and some fungi are more active than others both in the degree to which they stimulate symbiotic development and the range of hosts which they infect. *Dactylorhiza purpurella* established a symbiosis with most fungi tested whereas *Coeloglossum viride*, *Goodyera repens* and *Cymbidium canaliculatum* were less able to form symbiotic associations but established compatible infections with many isolates (Burgeff, 1909). Among orchids with photosynthetic protocorms, *Epidendrum radicans* was compatible with several fungi whereas three other species of *Epidendrum* together with *Spathoglottis plicata* and a *Laeliocattleya* cultivar were symbiotic only with *Tulasnella calospora* and certain *Ceratobasidium* isolates. *Ceratobasidium* species also show a broad spectrum of symbiotic activity. Yet, *Ceratobasidium cornigerum* is not symbiotic with most orchids. Isolates of *Thanatephorus cucumis* were mostly symbiotic with *Dactylorhiza purpurella* but less frequently with other orchids tested (Hadley, 1970).

Ammonium and nitrate are nitrogen sources for some orchid fungi. Fungi of tropical holosaprophytes grow better on organic nitrogen sources such as urea, amino acids, peptones, proteins and nucleic acids. Some orchid endophytes can utilize or require an exogenous source of amino acids, e.g. glutamic acid or asparagine. Several orchid endophytes require an exogenous supply of vitamins. Thiamine and *p*-aminobenzoic acid are required for distinct isolates of *Tulasnella calospora* and endophytes of *Arundina graminifolia*. Others are self-sufficient for vitamins. Endophytes of *Cymbidium* and *Dactylorhiza purpurella* have been shown to produce niacine and thiamine. They release these vitamins which increase the growth of the orchid seedlings into the medium. Cytokinin is produced by the mycorrhizal fungi and it has been shown *in vitro* that this hormone influences growth differently, depending on the orchid species and the stage of development. Concerning host specifity it is important to notice that orchids and fungi differ markedly in the range of partners with which they form effective symbiosis (Arditti, 1979).

11.4.2 Development and structures of the mycorrhiza

Seed germination starts with degradation of lipids and the accumulation of starch in the large basal cell in the centre of the embryo (Burgeff, 1959). In many species seeds must be soaked for several weeks, perhaps to extract germination inhibitors from the seed coat. Non-symbiotic fungi can stimulate germination of soil orchids which are difficult to grow such as *Epipactis* and *Cephalantera* sp. A three-month cold treatment of the seeds was also found to be necessary (Weinert, 1990). Fungal hyphae infect the germinating seed through a hair or another epidermal cell close to the suspensor. The plasmalemma and the host cytoplasm enclose the invading hyphae. The host cell remains active as indicated by an obvious protoplasmic streaming. The infected cells have a 2- to 4-fold increased DNA content. The hyphae then grow through the suspensor cells into the neighbouring embryo cells forming dense clusters (Fig. 11.13). The interface between fungus and plant cell has two layers, a compact inner layer and a loose outer layer. The outer layer could be homologous to the matrix in the VAM (Chapter 10). The hyphae remain enclosed by the plant plasmalemma. The host cytoplasm is spread in a very small layer, and the plant tonoplast and plasmalemma are very close to each other. At this stage the hyphae are packed with cytoplasm, ribosomes and mitochondria and have only small vacuoles. The cell wall is only 60 nm wide, the additional inner and outer layer add another 100–200 nm. By light microscopy, the interface between the fungus and the host cell can be hardly separated at this stage.

Infection of protocorms induces DNA synthesis and increases activities of polyphenol oxidase, ascorbic acid oxidase and catalase as well as a higher

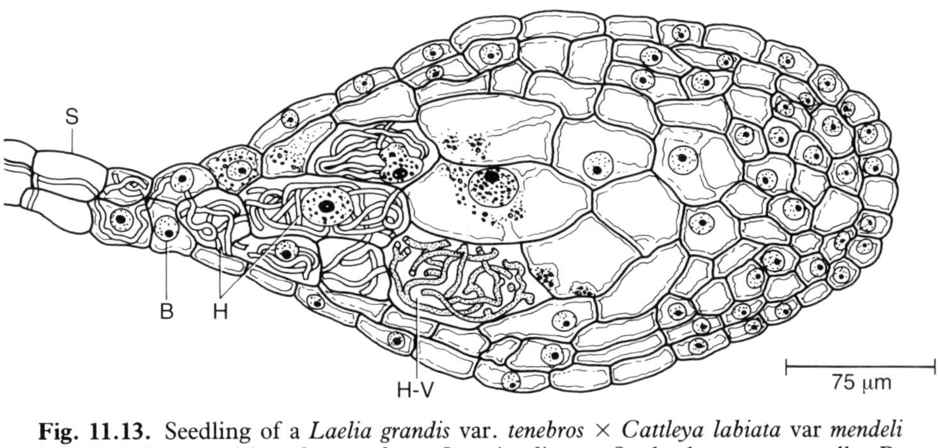

Fig. 11.13. Seedling of a *Laelia grandis* var. *tenebros* × *Cattleya labiata* var *mendeli* hybrid, infected with a fungus from *Serapias lingua*. S, dead suspensor cells; B, basal embryo cells with tetra- and octoploidal nuclei; H, functional hyphal coil; H-V, hyphal coil in digestion. (Based on Burgeff, 1959.)

respiration. Starch disappears from infected regions of seedlings due to enhanced hydrolysis. These changes together indicate that growth and development are initiated by infection. Without infection and in the absence of soluble sugars there is no development since protocorms are not able to hydrolyse all glycosidic bond.

A special character of the orchid mycorrhiza is the early digestion of the fungi by the host plant; this is the case in seedlings as well as in grown plants after new infections. The fungal hyphae develop larger vacuoles and thick cell walls and the cytoplasm degenerates. Finally the cells collapse and clump together. The remaining components with chitin from the cell walls are only very slowly further digested. This characteristic order of cell layers with an active symbiosis and neighbouring digesting cell layers has been called by Burgeff '**tolypophagie**' (Fig. 11.14). In another type, '**ptyophagie**', single hyphae grow into the digestive zone, the tips burst and the cytoplasm is digested. The digestive zone can be only one cell layer wide as in *Gastrodia callosa*.

Coated vesicles which play a role in the biogenesis of cell walls and plasma membrane occur frequently in root cells. They are dispersed in the host cytoplasm, associated with dictyosomes, with irregular arrays of membranes and with the host's sequestration plasmalemma. The contents of the coated vesicles do not stain with the PATAg test for polysaccharides or with ruthenium red. This result seems to indicate that coated vesicles, at least in *Ophrys lutea*, are not involved in polysaccharide transport of the interface. Furthermore, these results do not support any involvement of coated vesicles in the exocytosis of polysaccharide precursors or ruthenium red-reactive pectins. Other works have reported on the role of coated vesicles in the receptor-mediated endocytosis of hormones and bacterial toxins by animal cells. When cultured *in vitro* endophytes of *Ophrys lutea* are able to synthesize indole acetic acid (IAA) and indole ethanol (IEt). These substances may be released into the interfacial matrix and transferred to the host cell. Coated vesicles may be somehow concerned with the endocytosis of these hormones (Barroso and Pais, 1987).

An analysis of the mycelium of the isolated fungus of *Ophrys lutea* and of the culture medium was performed by gas chromatography–mass spectrometry. The results demonstrated that the endophyte of *Ophrys lutea* produces IEt and IAA. Since considerable amounts of IAA could be identified in either the mycelium or the culture medium, it was concluded that this auxin might be transferred from the endophyte to the host. Apparently, germination of *Ophrys lutea* seeds depends at least partly on the production of IAA by the mycorrhizal fungi and its transfer to the host (Barroso *et al.*, 1986).

The analysis of infected and uninfected tubers of *Ophrys lutea* revealed the presence of ergosterol, camposterol and sitosterol; cycloartenol was also identified. Under the conditions used no free sterols could be detected. This

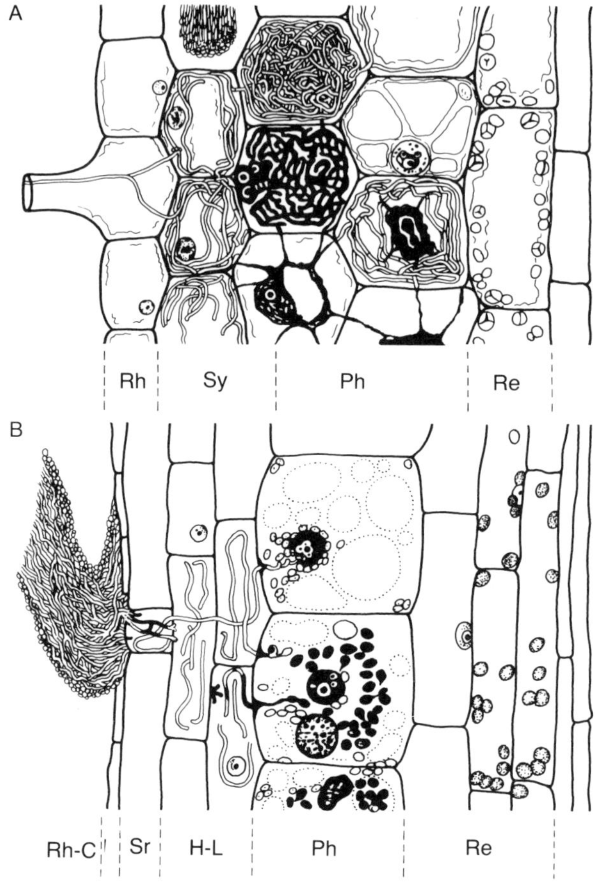

Fig. 11.14. (a) Tolypophagy in substrate root of the orchid *Platanthera chlorantha*. The real number of cortex layers is higher than shown. Rh, Rhizodermis with base root hair cells; Sy, symbiotic layer; Ph, cell layers with phagocytosis of hyphae; Re, reserve layer with storage compounds. (b) Ptyophagy in the root of *Gastrodia callosa*. Rh-C, collapsed rhizodermis; Sr, subrhizodermal layer; HL, hyphal layer; Ph, layer of phagocytosis with large cells where resorption of fungal cytoplasms takes place; Re, reserve layer with storage compounds (based on Burgeff, 1959).

may have been due to the very low concentrations in which they are supposed to occur. Differences in sterol concentration were found: infected tubers were found to have less ergosterol than uninfected tubers. At the same time the ratio of either camposterol or sitosterol/free dimethylsterols increased. These results indicate that infection induces the synthesis of free sterols, though the experiments did not determine which of the two partners produces these compounds. It is suggested that the increase of sterol ratios is a consequence of an increase in membrane production after infection, as sterols are considered to be membrane components. Cholesterol, campesterol and sitosterol are believed to have a stabilizing effect on membranes. The increase in stability probably brings about a decrease in plasmalemma permeability, which in turn may be interpreted as a defence mechanism of the host cells against toxins of the endophyte (Barroso *et al.*, 1987).

11.4.3 Symbiotic nutrition

By using of ^{14}C-labelled glucose it can be demonstrated that the fungi transport sugars into the orchid tissue; ^{14}C-labelled sucrose which is not synthesized by the fungus can be found in the host tissue. After labelling with $^{14}CO_2$, significant label is found in glucose and fructose, only in mycorrhiza-infected roots and shoots due to an increase in invertase activity (Fig. 11.15). Trehalose, a carbohydrate from the fungi, can also be used by the host plant.

Movement of carbon into non-photosynthetic protocorms via the external mycelium was observed while using ^{14}C in the form of an insoluble carbon source to simulate field conditions. Since considerable amounts of ^{14}C were present in plantlets without external mycelium, it seems that plantlets are not entirely dependent on the endophyte for a supply of carbon. In contrast adult plants seem to be fully autotrophic and do not receive carbon via their endophyte even when stressed for carbon by being kept in the dark. The

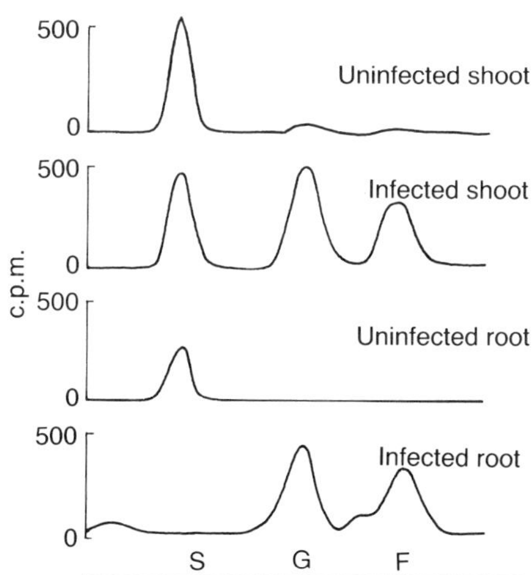

Fig. 11.15. ^{14}C-labelled sugar (sucrose, S; glucose, G; fructose, F) in shoots and roots of *Dactylorhiza purpurella* with or without mycorrhiza infection. (Modified from Purves and Hadley, 1975.)

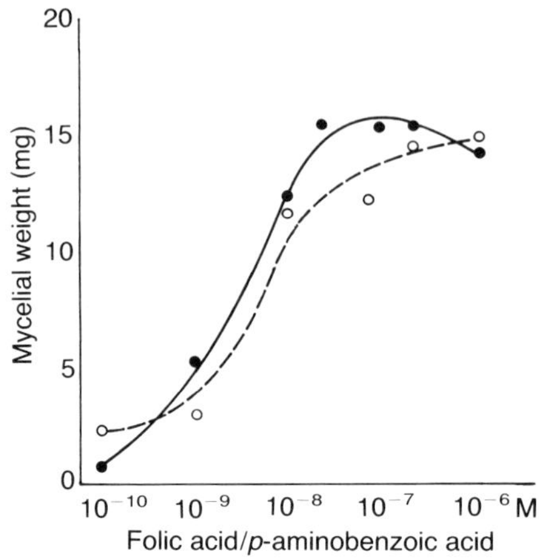

Fig. 11.16. Growth of *Rhizoctonia* sp. (isolated from *Cymbidium*) in relation to the concentration of folic acid (●) and *p*-aminobenzoic acid (○). (Based on Hijner and Arditti, 1973.)

developmental stage at which this movement of carbon from fungus to *Goodyera repens* stops has not as yet been defined, but it appears to be later than the development of the first leaf. No detectable movement of carbon from the autotrophic adult plant to the fungus was observed (Alexander and Hadley, 1985).

Between 1922 and 1930 Knudsen (1924, 1930) had succeeded in germinating and growing orchids to flowering without fungi. He used relatively simple media with glucose or sucrose, ammonium nitrate as N-source and a sufficient buffer capacity. In cultivated protocorms an increase in the nitrogen concentration from $10\,mg\,l^{-1}$ to $100\,mg\,l^{-1}$ induced parasitic

Table 11.10 Interrelationships of orchid mycorrhizae (Arditti, 1979)

Partner	*Effect*	*Condition*
Tree	May be damaged by fungus and/or orchids which become parasitic	Epiphytosis
Fungus	Benefits from tree	Parasite
	Benefits from and contributes to orchids	Mutualistic partner
	Benefits from debris	Saprophyte
Orchid	Benefits indirectly from tree parasitized by fungus	Epiparasite
	Benefits from and contributes to fungus	Mutualistic partner
	Derives indirect nourishment from debris	Parasite on saprophytic fungus
Debris	Broken down and/or utilized by orchid endophyte	Fungus is saprophytic

behaviour in the fungi and the orchids died (Penningsfeld, 1990). Most species do not need added vitamins. Some species use folic acid and *p*-aminobenzoic acids as growth factors (Fig. 11.16). In relation to the normal supply of phosphate and nitrogen sources, orchid mycorrhizal fungi have a similar function as described for the other mycorrhiza types. This is especially important for those species with a compact and little ramified root system. By the continuous digestion of fungal hyphae, the plants also have an additional source of N, P and trace elements. Epiphytic plants may depend directly on this type of mineral nutrition.

The interrelationships of orchids and mycorrhizal fungi are outlined in Table 11.10.

11.5 REFERENCES

Adams, H.H. (1959) Aspects of variation in the Orchidaceae, in *The Orchids. A Scientific Survey* (ed. C.L. Withner) The Ronald Press, New York, pp. 73–154.

Agerer, R. (ed.) (1988) *Colour Atlas of Ectomycorrhizae*, 2nd edn. Einhorn-Verlag, Schwäbisch Gmünd.

Agerer, R. and Weiss, M. (1989) Studies on ectomycorrhizae. XX. Mycorrhizae formed by *Thelephora terrestris* on Norway spruce. *Mycologia*, **81**, 444–53.

Alexander, C. and Hadley, G. (1985) Carbon movement between host and mycorrhizal endophyte during the development of the orchid *Goodyera repens* Br. *New Phytol.*, **101**, 657–65.

Arditti, J. (1979) Aspects of orchid. *Adv. Bot. Res.*, **7**, 422–657.

Arnolds, E. (1988) The changing macromycete flora in the Netherlands. *Trans Br. Mycol. Soc.*, **90**, 391–406.

Ashford, A.E., Peterson, R.L., Dwarte, D. and Chilvers, G.A. (1986) Polyphosphate granules in eucalypt mycorrhizas: determination by energy dispersive x-ray microanalysis. *Can. J. Bot.*, **64**, 677–87.

Barroso, J., Chaves Neves, H. and Pais, M.S. (1986) Production of indole-3-ethanol and indole-3-acetic acid by the mycorrhizal fungus of *Ophrys lutea* (Orchidaceae). *New Phytol.*, **103**, 745–9.

Barroso, J., Chaves Neves, H. and Pais, M.S. (1987) Production of free sterols by infected tubers of *Ophrys lutea* CAV.: Identification by gas chromatography-mass spectrometry. *New Phytol.*, **106**, 147–52.

Barroso, J. and Pais, M.S. (1987) Coated vesicles in the cytoplasm of the host cells in *Ophrys lutea* CAV. mycorrhizas (Orchidaceae). *New Phytol.*, **105**, 67–70.

Bell, R., Evans, C.S. and Roberts, E.R. (1988) Decreased incidence of mycorrhizal root tips associated with soil heavy-metal enrichment. *Plant Soil*, **106**, 143–5.

Berliner, R. (1990) The role of mycorrhizae and soil in plant establishment in natural ecosystems. *Symbiosis*, **8**, 5–8.

Berliner, R. and Torrey, J.G. (1989) Studies on mycorrhizal associations in Harvard Forest, Massachusetts. *Can. J. Bot.*, **67**, 2245–51.

Bougher, N.L., Grove, T.S. and Malajczuk, N. (1990) Growth and phosphorus acquisition of karri (*Eucalyptus diversicolor* F. Muell.) seedlings inoculated with ectomycorrhizal fungi in relation to phosphorus supply. *New Phytol.*, **114**, 77–85.

Bougher, N.L. and Malajczuk, N. (1990) Effects of high soil moisture on formation of ectomycorrhizas and growth of karri (*Eucalyptus diversicolor*) seedlings inoculated with *Descolea maculata*, *Pisolithus tinctorius* and *Laccaria laccata*. *New Phytol.*, **114**, 87–91.

Boyd, R., Furbank, R.T. and Read, D.J. (1986) Ectomycorrhiza and the water relations of trees, in *Mycorrhizae: Physiology and Genetics*, 1st ESM, INRA, Paris, pp. 689–93.

Burgeff, H. (1909) *Die Wurzelpilze der Orchideen*. The Ronald Press, New York.

Burgeff, H. (1932) Saprophytismus und Symbiose. G. Fischer Verlag, Jena.

Burgeff, H. (1959) Mycorrhiza of orchids, in *The Orchids* (ed. C.L. Withner) The Ronald Press, New York, pp. 361–95.

Chevalier, G. and Frochot, H. (1988) Ecology and possibility of culture in Europe of the Burgundy truffle (*Tuber uncinatum chatin*), in *Abstracts of the 2nd European Symposium on Mycorrhizae*, Institute of Landscape Ecology, Czechoslovak Academy of Sciences, Prague, pp. 21–2.

Chilvers, G.A., Lapeyrie, F.F. and Horan, D.P. (1987) Ectomycorrhizal vs endomycorrhizal fungi within the same root system. *New Phytol.*, **107**, 441–8.

Clarkson, D.T. (1985) Factors affecting mineral nutrient acquisition by plants. *Ann. Rev. Plant Physiol.*, **36**, 77–115.

Danesch, O. and Danesch, E. (1962) *Orchideen Europas*. Verlag Hallwag, Bern, Stuttgart.

Deacon, J.W. and Donaldson, S.J. (1983) Sequences and interactions of mycorrhizal fungi on birch. *Plant Soil*, **71**, 257–62.

Dell, B., Malajczuk, N., Grove, T.S. and Thomson, G. (1990) Ectomycorrhiza formation in *Eucalyptus*. IV. Ectomycorrhizas in the sporocarps of the hypogeous fungi *Mesophellia* and *Castorium* in eucalypt forests of Western Australia. *New Phytol.*, **114**, 449–56.

Dighton, J., Poskitt, J.M. and Howard D.M. (1986) Changes in occurrence of Basidiomycete fruit bodies during forest stand development: with specific reference to mycorrhizal species. *Trans. Br. Mycol. Soc.*, **87**, 163–71.

Dörfelt, H. and Braun, U. (1980) Untersuchungen zur Bioindikation durch Pilze in der Dübener Heide, in *Bioindikation auf der Ebene der Population und Biogeozönosen* 1 (Bioindikation 4) (eds R. Schubert and J. Schuh) Wissen. Beitr. Martin-Luther-Universität Halle Wittenberg, **27**, 15–20.

Dörr, I. and Kollmann, R. (1969) Fine structure of mycorrhiza in *Neottia nidus avis*. *Planta*, **89**, 372–5.

Duddridge, J.A., Finlay, R.D., Read, D.J. and Söderström, B. (1988) The structure and function of the vegetative mycelium of ectomycorrhizal plants. III. Ultrastructural and autoradiographic analysis of inter-plant carbon distribution through intact mycelial systems. *New Phytol.*, **108**, 183–8.

Duddridge, J. and Read, D.J. (1981) An ultrastructural analysis of the development of mycorrhizas in *Rhododendron ponticum*. *Can. J. Bot.*, **60**, 2345–56.

Egli, S. and Kälin, I. (1988) The root window – a technique for the *in vivo*-observation of mycorrhizae in the field, in *Abstracts of the 2nd European Symposium on Mycorrhizae*, Institute of Landscape Ecology, Czechoslovak Academy of Sciences, Prague, p. 31.

Fellner, R. (1988) Mycorrhizae-forming fungi as bioindicators of air pollution, in *Abstracts of the 2nd European Symposium on Mycorrhizae*, Institute of Landscape

Ecology, Czechoslovak Academy of Sciences, Prague, p. 34.

Finlay, R.D. (1988) Functional aspects of incompatible ectomycorrhizal associations, in *Abstracts of the 2nd European Symposium on Mycorrhizae*, Institute of Landscape Ecology, Czechoslovak Academy of Sciences, Prague, p. 36.

Francke, H.-L. (1934) Beiträge zur Kenntnis der Mykorrhiza von *Monotropa hypopitys* L. Analyse und Synthese der Symbiose. *Flora*, **129**, 1–52.

Frank, A.B. (1885) Über die auf Wurzelsymbiose beruhende Ernährung gewisser Bäume durch unterirdische Pilze. *Ber. Dtsch. Bot. Ges.*, **3**, 128–45.

Fries, N., Serck-Hanssen, K., Dimberg, L.H. and Theander, O. (1987) Abietic acid, an activator of basidiospore germination in ectomycorrhizal species of the genus *Suillus* (Boletaceae). *Exp. Mycol.*, **11**, 360–3.

Fusconi, A. and Bonfante-Fasolo, P. (1984) Ultrastructural aspects of host–endophyte relationships in *Arbutus unedo* L. mycorrhizas. *New Phytol.*, **96**, 397–410.

Garrett, H.E., Carney, J.L. and Hedrick, H.G. (1982) The effects of ozone and sulfur dioxide on respiration of ectomycorrhizal fungi. *Can. J. For. Res.*, **12**, 141–5.

Hacskaylo, E. (1965) *Thelephora terrestris* and mycorrhizae of Virginia pine. *Forest Sci.*, **11**, 401–4.

Hadley, G. (1970) Non-specifity of symbiotic infection in orchid mycorrhiza. *New Phytol.*, **69**, 1015–23.

Harley, J.L. (1978) Ectomycorrhizas as nutrient-absorbing organs. *Proc. R. Soc. Lond. B*, **203**, 1–21.

Harley, J.L. and McCready, C.C. (1952) The uptake of phosphate by excised mycorrhizal roots of the beech. II. Distribution of phosphorus between host and fungus. *New Phytol.*, **51**, 56–64.

Harley, J.L. and Smith, S.E. (1983) *Mycorrhizal Symbiosis*. Academic Press, London.

Haselwandter, K. (1987) Mycorrhizal infection and its possible ecological significance in climatically and nutritionally stressed alpine plant communities. *Angew. Bot.*, **61**, 107–14.

Haselwandter, K., Bobleter, O. and Read, D.J. (1990) Degradation of ^{14}C-labelled lignin and dehydropolymer of coniferyl alcohol by ericoid and ectomycorrhizal fungi. *Arch. Microbiol.*, **153**, 352–4.

Heinonen-Tanski, H. (1988) Maintenance methods for ectomycorrhizal fungi, in *Abstracts of the 2nd European Symposium on Mycorrhizae*, Institute of Landscape Ecology, Czechoslovak Academy of Sciences, Prague, p. 45.

Hijner, J.A. and Arditti, J. (1973) Orchid mycorrhiza: vitamin production and requirements by the symbionts. *Am. J. Bot.*, **60**, 829–35.

Hilbert, J.L. and Martin, F. (1988) Regulation of gene expression in ectomycorrhizas. I. Protein changes and the presence of ectomycorrhiza-specific polypeptides in the *Pisolithus–Eucalyptus* symbiosis. *New Phytol.*, **110**, 339–46.

Hock, B., Liebmann, S. and Beyrle, H. (1988) Mycorrhizal synthesis and phytohormone status, in *Abstracts of the 2nd European Symposium on Mycorrhizae*, Institute of Landscape Ecology, Czechoslovak Academy of Sciences, Prague, p. 47.

Hoffmann, P. (1988) Preservation of mycorrhizal fungi in liquid nitrogen, in *Abstracts of the 2nd European Symposium on Mycorrhizae*, Institute of Landscape

Ecology, Czechoslovak Academy of Sciences, Prague, p. 48.

Hofstein, A. von (1969) The ultrastructure of mycorrhiza I. Ectotrophic and ectendotrophic mycorrhiza of *Pinus sylvestris*. *Sv. Bot. Tidskr.*, **63**, 455–63.

Horan, D.P., Chilvers, G.A. and Lapeyrie, F.F. (1988) Time sequence of the infection process in eucalypt ectomycorrhizas. *New Phytol.*, **109**, 451–8.

Institute of Landscape Ecology Prague (1988) *Abstracts of the 2nd European Symposium on Mycorrhizae*, Czechoslovak Academy of Sciences, Prague.

Isermann, K. and Henjes, G. (1986) Pflanzen- und Bodenanalyse als Grundlage zur Diagnose und Therapie des Komplexes 'neuartige Waldschäden' aus der Sicht der Waldernährung. Kongreβband 1985, VDLUFA-Verlag, Darmstadt, pp. 413–29.

Katznelson, H., Rouatt, J.W. and Peterson, E.A. (1962) The rhizosphere effect of mycorrhizal and nonmycorrhizal roots on yellow birch seedlings. *Can. J. Bot.*, **40**, 377–82.

Keller, G. (1990) Characterization of ectomycorrhizal fungi (*Suillus* spp.) of *Pinus cembra* L. by isozyme electrophoresis, in *Abstracts of the Fourth International Mycological Congress*. (eds A. Reisinger and A. Bresinsky Regensburg, p. 327/2.

Kilbertus, G., Kiffer, E. and Proth, J. (1978) Influence des racines asymbiotiques et des mycorhizes de *Picea abies* L. Karst sur la microflore tellurique. *Rev. Ecol. Biol. Sol*, **15**, 297–310.

Knudson, L. (1924) Further observations on nonsymbiotic germination of orchid seeds. *Bot. Gaz.*, **77**, 212–19.

Knudson, L. (1930) Flower production by orchid grown non-symbiotically. *Bot. Gaz.*, **89**, 192–9.

Kottke, I. and Oberwinkler F. (1986) Mycorrhiza of forest trees – structure and function. *Trees*, 1–24.

Krivan, V., Lüttge, U. and Schaldach, G. (1986) Profile von Makro- und Mikromineralnährstoffen in gesunden und kranken Fichten (*Picea abies* (L.) Karst.) auf verschiedenen Standorten in Südwestdeutschland. *Angew. Bot.*, **60**, 373–89.

Krupa, S. and Bränström, G. (1974) Studies on the nitrogen metabolism in ectomycorrhizae. II. Free and bound amino acids in the mycorrhizal fungus *Boletus variegatus*, in the root systems of *Pinus sylvestris* and during their association. *Physiol. Plant.*, **31**, 279–83.

Krupa, S. and Nylund, J.-E. (1972) Studies on ectomycorrhizae of pine. III. Growth inhibition of two root pathogenic fungi by volatile organic constituents of ectomycorrhizal root systems of *Pinus sylvestris* L. *Eur. J. For. Pathol.*, **2**, 88–94.

Last, F.T., Mason, P.A. and Wilson, J. (1983) Fine roots and sheating mycorrhizas: their formation, function and dynamics. *Plant Soil*, **71**, 9–21.

Leake, J.R., Shaw, G. and Read, D.J. (1990) The biology of mycorrhiza in the Ericaceae. XVI. Mycorrhiza and iron uptake in *Calluna vulgaris* (L.) Hull in the presence of two calcium salts. *New Phytol.*, **114**, 651–7.

Lopez-Aguillon, R. and Garbaye, J. (1988) Some aspects of a double symbiosis with ectomycorrhizal and VAM fungi, in *Abstracts of the 2nd European Symposium on Mycorrhizae*, Institute of Landscape Ecology, Czechoslovak Academy of Sciences, Prague, p. 66.

Lyr, H. (1960) Über die Abnahme der Mycorrhiza- und Knöllchenfrequenz mit zunehmender Bodentiefe. *Int. Mykorrhiza Symp.*, pp. 303–13.

Malajczuk, N. (1979) The microflora of unsuberized roots of *Eucalyptus calophylla* R. Br. and *Eucalyptus marginata* Donn ex Sm. seedlings grown in soils suppressive and conductive to *Phytophthora cinnamomi* Rands. II. Mycorrhizal roots and associated microflora. *Aust. J. Bot.*, **27**, 255–72.

Malajczuk, N., Lapeyrie, F. and Garbaye, J. (1990) Infectivity of pine and eucalypt isolates of *Pisolithus tinctorius* on roots of *Eucalyptus urophylla in vitro*. 1. Mycorrhiza formation in model systems. *New Phytol.*, **114**, 627–31.

Malloch, D.W., Pirozynski, K.A. and Raven, P.H. (1980) Ecological and evolutionary significance of mycorrhizal symbioses in vascular plants (A Review). *Proc. Natl. Acad. Sci. USA*, **77**, 2113–18.

Marks, G.C. and Kozlowski, T.T. (eds.) (1973) *Ectomycorrhizae*, Academic Press, New York, London.

Marx, D.H. (1972) Ectomycorrhizae as biological deterrents to pathogenic root infections. *Ann. Rev. Phytopathol.*, **10**, 429–34.

Marx, D.H. and Bell, W. (1985) Formation of *Pisolithus* ectomycorrhizae on loblolly pine seedlings with spore pellet inoculum applied at different times. USDA Forest Serv., Southeastern Forest Exp. Sta., Res. Paper SE-249, Asheville.

Marx, D.H. and Cordell, C.E. (1989) The use of specific ectomycorrhizas to improve artificial forestation practices, in *Biotechnology of Fungi for Improving Plant Growth* (eds. J.M. Whipps and R.D. Lumsden) Symposium of the British Mycological Society, Cambridge University Press, Cambridge.

Marx, D.H. and Cordell, C.E. (1990a) *Development of Pisolithus tinctorius ectomycorrhizae on loblolly pine seedlings from spores sprayed at different times and rates*. United States Department of Agriculture, Forest Service, Research Note SE-356, Southeastern Forest Experiment Station.

Marx, D.H. and Cordell, C.E. (1990b) *Inoculation of fall- and spring-sown longleaf pine seedlings with Pisolithus tinctorius*. United States Department of Agriculture, Forest Service, Research Note SE-358, Southeastern Forest Experiment Station.

Marx, D.H., Cordell, C.E., Kenney, D.S. *et al.* (1984) Commercial vegetative inoculum of *Pisolithus tinctorius* and inoculation techniques for development of ectomycorrhizae on bare-root tree seedlings. *Forest Sci.* (Suppl.) **30**, Monograph 25.

Marx, D.H., Cordell, C.E., Maul, S.B. and Ruehle, J.L. (1989a) Ectomycorrhizal development on pine by *Pisolithus tinctorius* in bare-root and container seedling nurseries. I. Efficiency of various vegetative inoculum formulations. *New Forests*, **3**, 45–56.

Marx, D.H. and Shafer, S.R. (1989b) Fungal and bacterial symbioses as potential biological markers of effects of atmospheric deposition on forest health, in *Biological Markers of Air-Pollution Stress and Damage in Forests*, National Academy Press, Washington, pp. 217–32.

Melville, L.H., Massicotte, H.B. and Peterson, R.L. (1987) Morphological variations in developing ectomycorrhizae of *Dryas integrifolia* and five fungal species. *Scanning Microsc.*, **1**, 1455–64.

Meyer, F.H. (1974) Physiology of mycorrhiza. *Ann. Rev. Plant Physiol.*, **25**, 567–86.

Meyer, F.H. (1982) *Bäume in der Stadt*, E. Ulmer, Stuttgart pp. 214–16.

Meyer, F.H. (1984) Mykologische Beobachtungen zum Baumsterben. *Allgemeine Forst Zeitung*, **9/10**, 212–28.

Meyer, F.H. (1986) II. Symbiosen: Mykorrhiza, a) Ektomykorrhiza. *Progr. Bot.*, **48**, 406–14.

Meyer, F.H. (1987) Der Verzweigungsindex, ein Indikator für Schäden am

Feinwurzelsystem. *Forstwiss. Zentralbl.*, **106**, 84–92.
Meyer, F.H. (1989) I. Symbiosen: Mykorrhiza, a) Ektomykorrhiza. *Progr. Bot.*, **51**, 375–90.
Mikola, P. (1980) *Tropical Mycorrhiza Research.* Clarendon Press, Oxford.
Mitchell, R.J., Garrett, H.E. Cox, G.S. and Atalay, A. (1986) Boron and ectomycorrhizal influence on indole-3-acetic acid levels and indole-3-acetic acid oxidase and peroxidase activities of *Pinus enchinata* Mill. roots. *Tree Physiol.*, **1**, 1–8.
Morselt, A.F.W., Smits, W.T.M. and Limonard, T. (1986) Histochemical demonstration of heavy metal tolerance in ectomycorrhizal fungi. *Plant Soil*, **96**, 417–20.
Mudge, K.W. (1987) Hormonal involvement in ectomycorrhizal development, in *Mycorrhizae in the Next Decade. Practical Applications and Research Priorities* (eds D.M. Sylvia, L.L. Hung and J.H. Graham), Institute of Food and Agricultural Sciences, University of Florida, Gainesville, pp. 228–30.
Münzenberger, B., Heilemann, J., Strack, D. *et al.* (1991) Phenolics of mycorrhizas and non-mycorrhizal roots on Norway spruce. *Abstracts of the Third European Symposium on Mycorrhizas*, Sheffield, August 1991.
Namysl, C., Chalot, M., Dell, B. *et al.* (1988) Biochemical pathways of carbohydrate and nitrogen assimilation in spruce ectomycorrhizas. Activities of a few key enzymes in each partner and in the association, in *Abstracts of the 2nd European Symposium on Mycorrhizae*, Institute of Landscape Ecology, Czechoslovak Academy of Sciences, Prague, p. 74.
Newman, E.I. and Reddell, P. (1987) The distribution of mycorrhizas among families of vascular plants. *New Phytol.*, **106**, 745–52.
Nylund, J.-E. (1988) The regulation of mycorrhiza formation – carbohydrate and hormone theories reviewed. *Scand. J. For. Res.*, **3**, 465–79.
Paris, F., Dexheimer, J. and Lapeyrie, F. (1991) Cytochemical evidence of a fungal cell wall modification during infection of *Eucalyptus* roots by the ectomycorrhizal fungus *Cenococcum geophilum. Abstracts of the Third European Symposium on Mycorrhizas*, Sheffield, August 1991.
Pearson, V. and Read, D.J. (1973) The biology of mycorrhiza in the Ericaceae. I. The isolation of the endophyte and synthesis of mycorrhizas in aseptic culture. *New Phytol.*, **72**, 371–9.
Pearson, V. and Read, D.J. (1975) The physiology of the mycorrhizal endophyte of *Calluna vulgaris. Trans. Br. Mycol. Soc.*, **64**, 1–7.
Penningsfeld, F. (1990) Anzucht und Brnährung heimischer Orchideen (Teil II). *Die Orchidee*, **41**, 221–5.
Peredo, H., Oliva, M. and Huber, A. (1983) Environmental factors determining the distribution of *Suillus luteus* fructifications in *Pinus radiata* grazing-forest plantations. *Plant Soil*, **71**, 367–70.
Peterson, T.A., Mueller, W.C. and Englander, L. (1981) Anatomy and ultrastructure of a *Rhododendron* root–fungus association. *Can. J. Bot.*, **58**, 2421–33.
Purves, S. and Hadley, G. (1975) Movement of carbon compounds between the partners in orchid mycorrhiza, in *Endomycorrhizas* (eds F.E. Sanders, B. Mosse and P.B. Tinker), Academic Press, London, pp. 175–94.

Read, D.J. and Stribley, D.P. (1973) Effect of mycorrhizal infection on nitrogen and phosphorus nutrition of Ericaceous plants. *Nature*, **244**, 81–2.

Reid, C.P.P., Kidd, F.A. and Ekwebelam, S.A. (1983) Nitrogen nutrition, photosynthesis and carbon allocation in ectomycorrhizal pine. *Plant Soil*, **71**, 415–32.

Robertson, D.C. and Robertson, J.A. (1984) Ultrastructural aspects of *Pyrola* mycorrhizae. *Can. J. Bot.*, **63**, 1089–98.

Roth-Bejerano, N., Livne, D. and Kagan-Zur, V. (1990) *Helianthemum–Terfezia* relations in different growth media. *New Phytol.*, **114**, 235–8.

Rousseau, J.V.D. and Reid, C.P.P. (1989a) Carbon and phosphorus relations in mycorrhizal and non-mycorrhizal pine seedlings. *Ann. Sci. For.*, **46** Suppl., 715s–17s.

Rousseau, J.V.D. and Reid, C.P.P. (1989b) Measurement of carbon cost in Ectomycorrhizae, in *Applications of Continuous and Steady-State Methods to Root Biology* (eds J.G. Torrey and L.J. Winship), Kluwer Academic, Dordrecht, 183–96.

Rousseau, J.V.D. and Reid, C.P.P. (1990) Effects of phosphorus and ectomycorrhizas on the carbon balance of loblolly pine seedlings. *Forest Sci.*, **36**, 101–12.

Scheromm, P., Plassard, C. and Salsac, L. (1990) Nitrate nutrition of maritime pine (*Pinus pinaster* Soland in Ait.) ectomycorrhizal with *Hebeloma cylindrosporum* Romagn. *New Phytol.*, **114**, 93–8.

Schlechte, G. (1986) Zur Mykorrhizapilzflora in geschädigten Forstbeständen. *Z. Mykol.*, **52**, 225–32.

Schmitt, J.A. (1987) Funktion, Bedeutung und Situation der Pilze in saarländischen Wäldern. – 'Pilzsterben'? Zum Rückgang der Pilzarten und Pilzfruktifikation im Saarland, in *Aus Natur und Landschaft im Saarland. Atlas der Pilze des Saarlandes*, Part 2, *Nachweise, Ökologie, Vorkommen und Beschreibungen* (eds H. Derbsch and J.A. Schmitt), Sonderband, **3**, pp. 23–78.

Singer, R., Araujo, I. and Ivory, M.H. (1983) The ectotrophically mycorrhizal fungi of the neotropical lowlands, especially central Amazonia. *Beih. Nova Hedwigia*, **77**.

Stribley, D.P. and Read, D.J. (1980) The biology of mycorrhiza in the Ericaceae. VII. The relationship between mycorrhizal infection and the capacity to utilize simple and complex organic nitrogen sources. *New Phytol.*, **86**, 365–71.

Tinker, P.B. (1975) Soil chemistry of phosphorus and mycorrhizal effects on plant growth, in *Endomycorrhizas* (eds F.E. Sanders, B. Mosse and P.B. Tinker), Academic Press, London, pp. 352–71.

Trappe, J.M. (1962) Fungus associates of ectotrophic mycorrhizae. *Bot. Rev.*, **28**, 538–60.

Trappe, J.M., Molina, R. and Castellano, M. (1984) Reactions of mycorrhizal fungi and mycorrhiza formation to pesticides. *Ann. Rev. Phytopathol.*, **22**, 331–59.

Warcup, J.H. (1975) Factors affecting symbiotic germination of orchid seed, in *Endomycorrhizas* (eds F.E. Sanders, B. Mosse and P.B. Tinker) Academic Press, London, pp. 37–104.

Wasserman, J.L., Mineo, L., Majumdar, S.K. and Van Tyne, C. (1987) Detection of heavy metals in oak mycorrhizae of northeastern Pennsylvania forests, using

X-ray microanalysis. *Can. J. Bot.*, **65**, 2622–7.

Weinert, M. (1990) Keimungsfördernde Faktoren bei schwerkeimenden europäischen Orchideen: 1. Bodenpilze und Agarbedeckung. *Die Orchidee*, **41**, 127–33.

Winner, W.E., Cotter, I.S., Powers, H.R. Jr and Skelly, J.M. (1987) Screening loblolly pine seedling responses to SO_2 and O_3: analysis of families differing in resistance to fusiform rust disease. *Environ. Pollut.*, **47**, 205–20.

Zak, B. (1964) Role of mycorrhizae in root disease. *Ann. Rev. Phytopathol.*, **2**, 377–92.

Index